T0344807

Metabolism and Bacterial Pathogenesis

Metabolism and Bacterial Pathogenesis

Edited by
Tyrrell Conway | Paul Cohen

ASM
PRESS

Washington, DC

Library of Congress Cataloging-in-Publication Data

Metabolism and Bacterial Pathogenesis/ Executive Editors: Paul S. Cohen, University of Rhode Island, Kingston, RI; Tyrrell Conway, Oklahoma State University, Stillwater, OK.
pages cm
Includes bibliographical references and index.
ISBN 978-1-55581-886-9 alk. paper -- ISBN 978-1-55581-888-3 e-ISBN 1. Microbial metabolism.
2. Pathogenic bacteria. 3. Virulence (Microbiology) I. Cohen, Paul S., 1939- editor. II. Conway, Tyrrell, editor.
  QR88.M48 2015
  572'.429--dc23
2015028186

doi:10.1128/9781555818883

Printed in the United States of America

10 9 8 7 6 5 4 3 2 1

Address editorial correspondence to: ASM Press, 1752 N St., N.W., Washington, DC 20036-2904, USA.

Send orders to: ASM Press, P.O. Box 605, Herndon, VA 20172, USA.

Phone: 800-546-2416; 703-661-1593. Fax: 703-661-1501.

E-mail: books@asmusa.org

Online: http://estore.asm.org

Cover: An artistic representation of the retroviral intasome engaged with target DNA in the nucleus of a host cell. The illustration is based on the crystal structure of the prototype foamy virus strand transfer complex (Protein Databank ID 3OS0; for details see Maertens *et al.*, Nature, 2010, 468, 326-9). Image provided by Dr. Peter Cherepanov, Cancer Research UK, London Research Institute, London, EN6 3LD United Kingdom.

Contents

Contributors

Christopher J. Alteri
Department of Microbiology, University of Michigan, Ann Arbor, MI 48109

Sandra K. Armstrong
Microbiology, University of Minnesota, Minneapolis, MN 55455

Ian Blomfield
Biosciences, University of Kent, Kent, CT2 7HL United Kingdom

Maksym Bobrovskyy
Department of Microbiology, University of Illinois, Urbana, Illinois 61801

E. Fidelma Boyd
Biological Sciences, University of Delaware, Newark, DE 19716

Fabian Commichau
General Microbiology, University of Göttingen, Göttingen, 37077, Germany

Arianna P. Corona
Microbiology and Immunology, New York Medical College, Valhalla, NY 10595

Wolfgang Eisenreich
Biochemistry, Technische Universität München, Garching, Germany

Werner Goebel
Max von Pettenkofer-Institut für Hygiene und Medizinische Mikrobiologie, Ludwig-Maximilians-Universität München, München, Germany

B.L. Haines-Menges
Biological Sciences, University of Delaware, Newark, DE 19716

Jürgen Heesemann
Max von Pettenkofer-Institut für Hygiene und Medizinische Mikrobiologie, Ludwig-Maximilians-Universität München, München, Germany

Nicholas Stephen Jakubovics
School of Dental Sciences, Newcastle University, Newcastle upon Tyne, NE2 4BW United Kingdom

J.B. Lubin
Biological Sciences, University of Delaware, Newark, DE 19716

Harry L.T. Mobley
Department of Microbiology and Immunology, University of Michigan Medical School, Ann Arbor, MI 48109

James P. Nataro
Pediatrics, University of Virginia, Charlottesville, VA 22908

Alline R. Pacheco
University of California Davis, Davis, CA 95616

Gregory C. Palmer
College of Natural Sciences Freshman Research Initiative, University of Texas at Austin, Austin, TX 78712

Gregory R. Richards
University of Wisconsin-Parkside, Kenosha, WI 53144

Anthony R. Richardson
Department of Microbiology and Immunology, University of North Carolina, Chapel Hill, NC 27599

Thomas Rudel
Lehrstuhl für Mikrobiologie, Biozentrum Universität Würzburg, Würzburg, Germany

Ira Schwartz
Microbiology and Immunology, New York Medical College, Valhalla, NY 10595

Greg A. Somerville
School of Veterinary Medicine and Biomedical Sciences, University of Nebraska, Lincoln, NE 68588

Abraham L. Sonenshein
Dep;artment of Molecular Biology and Microbiology, Tufts University, Boston, MA 02111

Vanessa Sperandio
Department of Microbiology, UT Southwestern Medical Center, Dallas, TX 75390

Bärbel Stecher
Max von Pettenkofer Institute, LMU Munich, Munich,

Jörg Stülke
General Microbiology, University of Göttingen, Göttingen, 37077, Germany

Carin K. Vanderpool
Department of Microbiology, University of Illinois, Urbana, Illinois 61801

W.B. Whitaker
Biological Sciences, University of Delaware, Newark, DE 19716

Marvin Whiteley
University of Texas at Austin, Austin, TX 78712

Alan J. Wolfe
Stritch School of Medicine, Loyola University Chicago, Maywood, IL 60153

Executive Editors

Paul Cohen
Cell and Molecular Biology, University of Rhode Island, Kingston, RI 02881

Tyrrell Conway
Department of Microbiology and Molecular Genetics, Oklahoma State University, Stillwater, OK 74078

Preface

In the 1980's, Rolf Freter, a true pioneer in the field of intestinal colonization, concluded that although several factors could theoretically contribute to a microorganism's ability to colonize the intestinal ecosystem, effective competition for nutrients is paramount to success. Freter considered this concept to apply equally to bacterial commensals and pathogens. He considered nutrient acquisition to be as critical for the success of a bacterial pathogen in its host as its ability to produce virulence factors. Despite the general acceptance of Freter's ideas, until recently, metabolism and bacterial pathogenesis were considered to be two distinctly different fields of study. Even the title of this book: ***Metabolism and Bacterial Pathogenesis*** might be interpreted as meaning that these fields are separate entities. Nothing could be further from the truth. There is no doubt that the discovery of pathogen-specific virulence factors such as fimbriae that allow adhesion to mucosal surface receptors, secreted toxins, iron acquisition systems, motility, mechanisms geared to avoid immune responses etc., have been instrumental in understanding bacterial pathogenesis and in some instances in devising ways to interfere with the pathogenic process. Nevertheless, it is becoming increasingly clear that bacterial metabolism, while not a virulence factor per se, is essential for pathogenesis and that interfering with pathogen specific metabolic pathways used during infection might lead to effective treatments. Moreover, recent studies have shown that nutrient acquisition and pathogenesis can be intimately associated, i.e., utilization of specific nutrients for growth can also regulate virulence factor expression.

Although this integrative field of research is in its infancy, it is time to bring together the current thinking of scientists breaking new ground in understanding how bacterial metabolism is foundational to pathogenesis. This book begins with a chapter devoted to an overview of bacterial metabolism and a chapter devoted to an overview of bacterial pathogenesis. Ensuing chapters describe the role of metabolism in the pathogenesis of many different bacterial pathogens in many different host environments. The contents of this book should be beneficial to specialists in bacterial pathogenesis and specialists in metabolism as well as to molecular biologists, physicians, veterinarians, dentists, graduate and undergraduate students and technicians. Finally, we would like to express our sincere gratitude to the authors for their willingness to contribute to this book and their hard work in producing their excellent chapters.

Paul Cohen
Tyrrell Conway

Glycolysis for Microbiome Generation

ALAN J. WOLFE[1]

BENEATH BEHAVIOR LIES METABOLISM

There is no life without metabolism. There is nothing surprising about this statement; it is blatantly obvious and true for both host and bacteria, whether commensal or pathogen. What is surprising is the delayed general recognition that metabolism plays a, or perhaps, *the* central role in pathogenesis, which is simply a manifestation of the need for certain "bad" bacteria to grow and divide on or in a host. Perhaps this delay is natural, as researchers tend to focus on particularities, in this case, cellular processes unique to pathogenesis. Another reason for this delay is likely the aversion of late 20th Century microbiologists, who came to science after the heyday of bacterial metabolic research and who were forced to memorize whole swaths of the metabolic chart, usually out of context and with little understanding of the intricate linkages between metabolic pathways and their connections to other cellular processes.

This certainly had been my experience, at least until the day Pat Conley and I added acetate to *Escherichia coli* cells "gutted" for all but one of the chemotaxis proteins (CheY) and unexpectedly observed flagellar motors intermittently rotate clockwise instead of incessant counterclockwise

[1]Department of Microbiology and Immunology, Stritch School of Medicine, Health Sciences Division, Loyola University Chicago, Maywood, Illinois.

Metabolism and Bacterial Pathogenesis
Edited by Tyrrell Conway and Paul Cohen
© 2015 American Society for Microbiology, Washington, DC
doi:10.1128/microbiolspec.MBP-0014-2014

rotation (1). Although we strongly suspected that this behavior required that the acetate be metabolized, we had no idea how. So, in the days before the Internet, we went looking for the metabolic chart, which we quickly discovered was pristine, still in its plastic wrapper within its cylinder, behind one of the lab doors. Apparently, this biophysics lab (headed by Howard Berg) had had no prior need for metabolism. The lab was studying bacterial behavior—chemotaxis and motility—not metabolism. However, on that day, I began to investigate the metabolism that underlay that bacterial behavior.

From the metabolic chart, we learned that *E. coli* cells convert acetate into acetyl coenzyme A (acCoA) by means of the reversible acetate kinase (AckA)-phosphotransacetylase (Pta) pathway, whose intermediate is acetyl phosphate (acP) or through the irreversible acetyl CoA synthetase (Acs), whose intermediate is acetyladenylate (acAMP) (Fig. 1). From my subsequent reading of the "ancient" literature, typically JBC volumes stored horizontally on the top shelf in Harvard's Biolabs library and covered in years of dust, I discovered Fritz Lipmann, Feodor Lynen, Hans Krebs, and others who had sought the "activated acetate" that we now know to be acCoA (2, 3). Whether derived from glucose via glycolysis or from acetate via Acs or the AckA-Pta pathway, the resultant acCoA replenishes the tricarboxylic acid (TCA) cycle and the glyoxylate shunt to generate energy and provide building blocks for the synthesis of amino acids, nucleotides, and other essential compounds. AcCoA also plays direct roles in the synthesis of fatty acids, amino acids, and most secondary metabolites, including many antibiotics. As such, acCoA could be considered the keystone molecule of central metabolism.

In our 1988 report, we provided evidence that an activated acetate molecule was responsible for our acetate effect on flagellar rotation. We thought that it was acetyladenylate (acAMP), the intermediate of the Acs pathway (1). Subsequently, others determined that multiple mechanisms were at work, that Acs could acetylate the two-component response regulator CheY, presumably using acAMP as the acetyl donor, that acP could donate either its phosphoryl or acetyl group to CheY or that acCoA could donate its acetyl group. Each posttranslational modification (phosphorylation and acetylation) inhibits the other, but both independently increase the probability that CheY will bind the flagellar motor and induce clockwise rotation (1, 4–17). We now know that, under physiologically relevant conditions, acP can donate its phosphoryl group to and activate other response regulators, including NtrC, OmpR, RcsB, CpxR, RssB, SirA/UvrY, Rpr2, DegU, and FlgR from *E. coli, Salmonella enterica, Yersinia pestis, Campylobacter jejuni, Listeria monocytogenes*, and *Borrelia burgdorferi* (18–37).

Following the initial reports that CheY could be acetylated (4, 6), Jorge Escalante-Semerena and his student Vincent Starai reported that a protein acetyltransferase (known as Pat in *S. enterica* and YfiQ, Pka, or PatZ in *E. coli*) catalyzes the Nε-lysine acetylation of Acs using acCoA as the acetyl donor (38). They had earlier linked acetylation to central metabolism by showing that the reversal of Acs acetylation required CobB (39), a member of the NAD^+-dependent sirtuin family of lysine deacetylases (40). It is now known that lysine acetyltransferases and deacetylases are ubiquitous in bacteria (41–44).

More recently, Choudhary's group and ours reported that acP can donate its acetyl group to thousands of lysines on hundreds of proteins, many of which are central to pathogenesis. Amazingly, this process does not require an enzyme (45, 46) but does require that the molecular environment of the lysine residue permit binding of the phosphoryl group and the activation

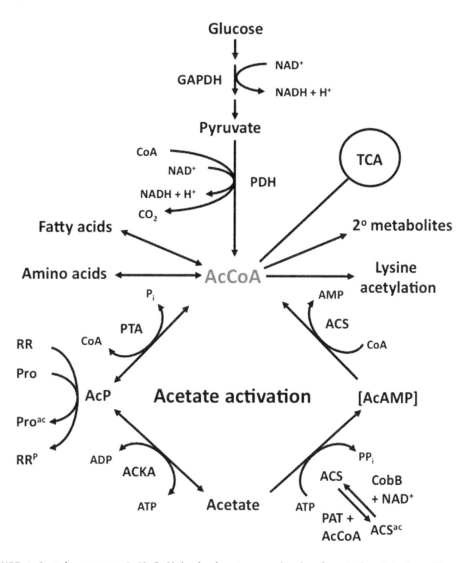

FIGURE 1 Acetyl-coenzyme A (AcCoA) is the keystone molecule of central metabolism. Glucose is metabolized via the EMP pathway to AcCoA in an NAD^+-dependent manner. The AcCoA is interconverted with amino acids and fatty acids. It replenishes the NAD^+-dependent tricarboxylic acid (TCA) cycle. It is the substrate for most secondary metabolites and the acetyl donor for some lysine acetylations, such as the PAT-dependent acetylation of ACS (acCoA synthetase). Acetate dissimilation requires the Pta-AckA pathway. The enzyme PTA (phosphotransacetylase) converts AcCoA and inorganic phosphate (P_i) into coenzyme A (CoA) and the high-energy pathway intermediate AcP. AcP donates its phosphoryl group to certain two-component response regulators (RR). AcP also can donate its acetyl group to hundreds of proteins. The enzyme ACKA (acetate kinase) converts AcP and ADP to acetate and ATP. The acetate freely diffuses across the cell envelope into the environment. Acetate assimilation requires the high-affinity enzyme ACS. In a two-step process that involves an enzyme-bound intermediate (acAMP), Acs converts acetate, ATP, and CoA into AMP, pyrophosphate (PP_i), and acCoA. ACS activity is inhibited by acetylation of a conserved lysine catalyzed by the lysine acetyltransferase (PAT, also known as YfiQ and Pka). Reactivation is catalyzed by the NAD^+-dependent deacetylase CobB. Adapted from Hu et al., 2010. GAPDH, glyceraldehyde-3-phosphate dehydrogenase; PDH, pyruvate dehydrogenase. doi:10.1128/microbiolspec.MBP-0014-2014.f1

(deprotonation) of the lysine (45). The full impact of protein acetylation remains to be investigated, but several studies have hinted that it could affect pathogenesis (20–22, 47, 48).

A GLYCOLYSIS PRIMER

Having made an argument that one small but central part of metabolism likely plays a role in bacterial pathogenesis, I will attempt to make that metabolism a bit more accessible. I will specifically devote the rest of this chapter to glycolysis. In this context, you should notice that NAD$^+$, acCoA, and acP are mentioned often.

Metabolism refers to biochemical pathways that either generate biologically usable energy (catabolism) or consume that energy to permit growth (anabolism). Catabolism converts chemical or electromagnetic energy into the high-energy bonds of ATP. Cells generate ATP via two distinctly different mechanisms: substrate-level phosphorylation and oxidative phosphorylation. Substrate-level phosphorylation is a process that synthesizes ATP by converting an organic molecule from one form to another (Fig. 2). In contrast, oxidative phosphorylation generates ATP via an ATP synthetase that uses a proton motive force established by

driving electrons through a membrane-bound electron transport system associated with respiration, photosynthesis, or some other type of bacterial metabolism. Anabolism uses the energy of ATP to synthesize cellular components. Some pathways are strictly catabolic, while some are strictly anabolic. The central metabolic pathways, however, tend to be amphibolic; they contribute both energy (catabolism) and biosynthetic precursors (anabolism). Glycolysis is amphibolic.

Given the knowledge that the human microbiome consists of thousands of different species (49–51) that are mostly uncharacterized, it is important to remember that different metabolic programs exist. Some bacteria are strict anaerobes, others are strict aerobes, and facultative anaerobes can do both. Some are strict fermenters, others are strict nonfermenters (i.e., they rely on respiration), and some can do both. In this context, note that diverse glycolytic strategies are available. These include (but are not limited to) the Embden-Meyerhof-Parnas, the Pentose Phosphate, and the Entner-Doudoroff pathways, which are commonly used by pathogens of the family Enterobacteriaceae, such as *E. coli* and *S. enterica*. Other strategies include the homolactic acid and heterolactic acid pathways and the Bifidobacterium shunt, strategies used by species of the genera *Lactobacillus*, *Bifidobacterium*, and *Gardnerella*, which have been found in diverse niches of the human body, including the gut (52–54), vagina (55–59), and bladder (60–64).

By convention, glycolytic pathways are depicted with glucose as the substrate, because this simple sugar requires the fewest catalytic steps to enter central metabolism via glycolysis. However, glycolysis can metabolize other carbon sources. For example, many hexoses can enter glycolysis after being isomerized to the activated (phosphorylated) forms of glucose or fructose, while pentoses must be converted to the activated form of xylulose. Whereas

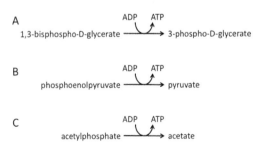

FIGURE 2 Three substrate-level phosphorylations. The first two examples are steps in the lower half of the EMP pathway. The third is a step in acetate fermentation. doi:10.1128/microbiolspec.MBP-0014-2014.f2

acid sugars first must be reduced, sugar alcohols must be oxidized. Similar scenarios hold true for pathway intermediates and their derivatives.

The Embden-Meyerhof-Parnas Pathway

The Embden-Meyerhof-Parnas (EMP) pathway is the most recognizable glycolytic pathway, primarily because it is the pathway taught in biochemistry and cell biology classes. The reason for this choice is simple. Eukaryotes use it to produce the pyruvate that mitochondria convert to the acCoA that replenishes the TCA cycle. Some yeasts and bacteria use the EMP pathway to synthesize ethanol and CO_2, while tumor cells, lymphocytes, neuroblasts, and lactic acid bacteria use it to produce lactic acid. Other bacteria use the EMP pathway to produce diverse gases, fatty acids, and alcohols.

The upper portion of the EMP pathway invests two ATP molecules to activate (phosphorylate) the hexose sugar glucose and rearrange it for cleavage into two triose phosphate molecules (Fig. 3). The key enzyme is 6-phosphofructokinase (PFK). In the lower portion of the EMP pathway, each triose phosphate molecule is phosphorylated and oxidized (via NAD^+) in a reaction that is catalyzed by glyceraldehyde dehydrogenase (GAPDH). Two subsequent substrate-level phosphorylations yield 4 ATP and 2 pyruvate molecules (65, 66).

The aforementioned description is a bit simplistic, however. For example, in many bacteria, including *E. coli* and its relatives, glucose and other hexoses are transported and phosphorylated simultaneously using PEP as the phosphoryl donor instead of ATP. Thus, one of the two PEP molecules generated from glucose by the EMP pathway is used to transport and phosphorylate another glucose molecule (65). Also, the yield per glucose is always smaller because intermediates are extracted from the EMP for entry into anabolic pathways.

For example, dihydroxyacetone phosphate, glyceraldehyde-3-phosphate, and pyruvate are precursors for the biosynthesis of lipids, vitamin B_6, and certain amino acids, respectively. To supply these and other precursors for biosynthesis, flux through the EMP pathway must be maintained. The NADH generated by the EMP pathway also plays an anabolic role, as it can reduce $NADP^+$ to NADPH, which is the primary reducing agent for biosynthesis.

The Pentose Phosphate Pathway

The Pentose Phosphate (PP) Pathway (also known as the phosphogluconate or hexose monophosphate pathway) oxidizes glucose-6-phosphate to pentose phosphates (Fig. 3). It is distinctive for several reasons. First, it uses a different set of reactions than the EMP pathway. Second, it oxidizes sugars with $NADP^+$ rather than NAD^+. As mentioned earlier, the resultant NADPH is a major source of electrons for diverse biosynthetic (anabolic) processes. Third, it produces D-ribose-5-phosphate, sedoheptulose-7-phosphate, and erythrose-4-phosphate, which function as precursors for the biosynthesis of amino acids, nucleic acids, and other macromolecules, including ATP, coenzyme A, NADH, and $FADH_2$. Thus, the PP pathway is an essential central metabolic pathway.

The PP pathway begins with three molecules of one glycolytic intermediate, β-D-glucose 6-phosphate, and ends with formation of three others, two molecules of β-D-fructose 6-phosphate and one of D-glyceraldehyde 3-phosphate. The PP pathway is often divided into its preliminary oxidative and subsequent nonoxidative portions. In the former, β-D-glucose 6-phosphate is oxidized by $NADP^+$ to D-ribulose 5-phosphate with the evolution of CO_2; in the latter, a series of transaldolase and transketolase reactions convert the D-ribulose 5-phosphate to β-D-fructose 6-phosphate and D-glyceraldehyde 3-phosphate. EMP pathway enzymes complete the metabolism

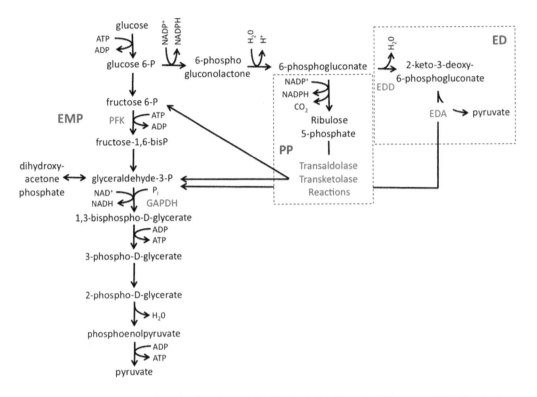

FIGURE 3 Glycolysis in *E. coli* and related bacteria. The Embden-Meyerhoff-Parnas (EMP), the Pentose Phosphate (PP), and Entner-Doudoroff (ED) pathways. The boxes highlight reactions unique to the PP and ED pathways. When glucose is metabolized by the EMP, the lower half of the pathway is repeated twice. In *E. coli* and related bacteria, glucose is transported and phosphorylated using PEP as the phosphoryl donor. Thus, one of the two PEP molecules generated by the EMP pathway is used to transport and phosphorylate another glucose molecule. For the PP pathway to function, two glucose molecules must be metabolized. Pentose sugars and sugar acids can be metabolized via the PP and ED pathways, respectively. PEP, phosphoenol pyruvate; PFK, phosphofructokinase; GAPDH, glyceraldehyde 3-phosphate dehydrogenase; EDD, 6-phosphogluconate dehydratase; EDA, 2-keto 3-deoxy-D-gluconate 6-phosphate aldolase. doi:10.1128/microbiolspec.MBP-0014-2014.f3

of both. Because one-sixth of the carbon is released as CO_2, the net yield per glucose is 1.8 ATP (66).

The Entner-Doudoroff Pathway

Two enzymes form the core of the Entner-Doudoroff (ED) Pathway: 6-phosphogluconate dehydratase (EDD) and 2-keto 3-deoxy-D-gluconate 6-phosphate (KDPG) aldolase (EDA). The former enzyme oxidizes D-gluconate 6-phosphate (from the PP pathway or by phosphorylation of gluconate) to KDPG, which is cleaved by the latter enzyme to

pyruvate and D-glyceraldehyde 3-phosphate (Fig. 3). The D-glyceraldehyde 3-phosphate is oxidized to pyruvate by the enzymes of the EMP pathway, with 1 net ATP produced per glucose (66, 67). As with the EMP pathway, intermediates are extracted for anabolic processes.

The ED pathway is critical to many pseudomonads. These bacteria generally lack PFK and thus cannot use the EMP pathway. Instead, the core of central metabolism is formed by the ED pathway, which operates in a cyclic manner. By means of certain enzymes of gluconeogenesis, the

D-glyceraldehyde 3-phosphate is recycled to D-gluconate 6-phosphate (68). In *E. coli* and other bacteria, the ED pathway plays a more peripheral role. In these organisms, the EMP and PP pathways form the core of central metabolism. However, when gluconate or other acid sugars are present, the ED pathway is induced along with the gluconate transporter and gluconokinase, which activates the gluconate to D-gluconate 6-phosphate (67, 69).

Fermentation

Pyruvate is the end result of the EMP, PP and ED pathways. In the absence of oxygen, this pyruvate (or its derivatives) is further metabolized by fermentation, which uses substrate-level phosphorylation to synthesize energy during the partial oxidation of an organic compound. To perform this partial oxidation, pathway intermediates act as electron donors and electron acceptors. The fundamental fermentation logic is as follows: activate a substrate, use an electron acceptor to partially oxidize that activated substrate, use some of the energy released by oxidization to generate ATP, and recycle the electron acceptor by reducing the oxidized substrate (Fig. 4A).

An advantage of fermentation is that it is fast. Although fermentation generates only 3% to 7% of the 38 ATPs that oxidative phosphorylation can potentially produce, fermentation produces ATP at about 100 times the rate of oxidative phosphorylation. The faster rate is such an advantage that many cells ferment in the presence of glucose instead of respire, even in the presence of oxygen. This behavior, called the Crabtree effect, aerobic fermentation, or overflow metabolism, was first described in tumor cells that performed lactic acid fermentation instead of aerobic respiration. It also powers fast-growing eukaryotic cells such as neuroblasts and lymphocytes (70–73). And it occurs in many bacteria (74–76). For example, when presented with high

concentrations of glucose, *E. coli* ferments even when oxygen is present (77, 78). The opposite behavior is called the Pasteur effect (79). When growth conditions favor a shift from fermentation to aerobic respiration, cells lower their rate of catabolism. This occurs because aerobic respiration is more efficient and generates greater energy per glucose molecule. The mechanisms that regulate the "choice" to ferment or respire remain controversial (36, 80, 81).

The strategy for reducing the pyruvate produced by glycolysis determines the fermentation product and the fermentation pathway name. More importantly, this "choice" of fermentation end product determines the balance between the net ATP and net recycled electron acceptors (i.e., NAD^+ and $NADP^+$). A familiar fermentation strategy produces lactic acid. This strategy, called homolactic

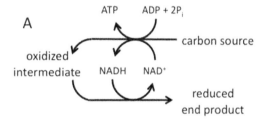

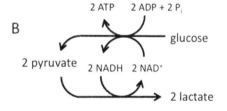

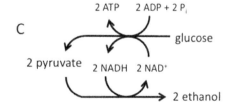

FIGURE 4 Fermentation. A) General strategy, B) Homolactic acid fermentation, and C) Ethanol fermentation. doi:10.1128/microbiolspec.MBP-0014-2014.f4

acid fermentation (82–84), is used by many of the so-called lactic acid bacteria, such as *Lactobacillus, Lactococcus* and many Streptococci. It nets 2 lactate molecules and 2 ATP per glucose (Fig. 4B). It uses the EMP pathway to oxidize one glucose molecule to 2 pyruvates, 2 ATP, and 2 NADH. Both pyruvates are reduced to lactate by NADH, which is oxidized to NAD^+. The lactate is excreted into the surrounding environment. Another familiar pathway is ethanol fermentation by *Saccharomyces* (85). In this strategy (Fig. 4C), each pyruvate molecule is reduced to ethanol and CO_2 via acCoA. Per glucose, the net products are 2 ethanols, 2 CO_2, 2 ATP, and 2 NAD^+.

Members of the family Enterobacteriaceae tend to perform mixed-acid fermentations (77, 78, 86). These fermentation products can include lactate and ethanol but also acetate and succinate, and CO_2 (Fig. 5). The major advantage of mixed-acid fermentation is its ability to generate additional ATP while recycling NAD^+. ATP generation occurs by channeling one pyruvate through acCoA to acP, which is converted by AckA to acetate and ATP. This strategy avoids excess NADH production with the added advantage of generating another ATP (87). NAD^+ recycling happens when pyruvate is converted to products such as lactate, ethanol, or succinate. Like ethanol fermentation, succinate fermentation recycles two NAD^+ per three-carbon intermediate but at the cost of an extra ATP because PEP is not converted to pyruvate.

Another mixed-product fermentation involves acetoin, butanediol, and ethanol. This strategy is common to *Enterobacter, Serratia, Erwinia*, and some *Bacillus* species. Much of the pyruvate is converted to acetoin, which is reduced by NADH to 2,3-butanediol. Ethanol is also produced, as are small amounts of acids. Most fermentation products are organic acids, which acidify the environment, often to the detriment of the fermenting organism. Thus, the advantage of fermenting to the neutral end products acetoin, butanediol, and ethanol is that the organism avoids acidification of its environment (88, 89). Other fermentation strategies exist. For example, *Corynebacteria, Propionibacterium*, and *Bifidobacterium* convert pyruvate

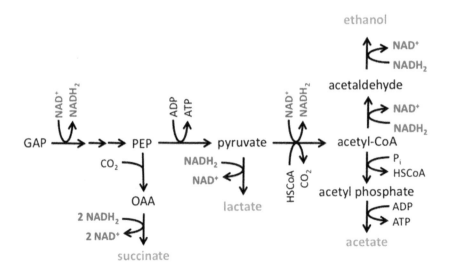

FIGURE 5 A mixed-acid fermentation. Blue steps consume NAD^+, whereas red steps recycle NAD^+. Green compounds are excreted. GAP, glyceraldehyde 3-phosphate; PEP, phosphoenol pyruvate; HSCoA, coenzyme A; P_i, inorganic phosphate. doi:10.1128/microbiolspec.MBP-0014-2014.f5

to propioniate via succinate (90, 91), while *Clostridia* convert it to butyrate or to butanol and acetone (89).

Some lactic acid bacteria (*Lactobacillus* and *Bifidobacterium*) possess the hetero-lactic acid pathway (Fig. 6), which uses a unique enzyme (phosphoketolase, EC 4.1.2.9) to cleave a pentose sugar to a 3-carbon phosphate and a 2-carbon phosphate (84, 92). As in the EMP and PP pathways, the first step activates glucose. Unlike the EMP, 2 NAD^+-mediated oxidations occur before the cleavage. β-D-glucose-6-phosphate is oxidized to D-gluconate 6-phosphate, which is then oxidized and decarboxylated. The resultant xylulose-5-phosphate is cleaved to D-glyceraldehyde 3-phosphate and acP. As in the EMP pathway, the D-glyceraldehyde 3-phosphate is converted to lactate, producing 2 ATP by substrate-level phosphorylation. The acP is converted to ethanol by 2 NADH-mediated reductions that balance the 2 pre-cleavage oxidations. Per glucose, the net products are 1 lactate, 1 ethanol, 1 CO_2, and 1 ATP. Thus, the efficiency is about half that of the EMP pathway.

Two types of phosphoketolases exist: the single-specificity enzyme that catalyzes the key reaction of the heterolactic pathway (EC 4.1.2.9) and a dual-specificity enzyme (XFP) that also can hydrolyze β-D-fructose 6-phosphate to D-erythrose 4-phosphate and acP (EC 4.1.2.22). The dual-function enzyme functions in the Bifidobacterium shunt (93–95). This shunt is used by several *Bifidobacterium* species and *Gardnerella vaginalis* (Fig. 7). Glucose is first activated and isomerized to β-D-fructose-6-phosphate, which is cleaved by the dual-specificity phosphoketolase to acP and D-erythrose 4-phosphate (EC 4.1.2.22). AckA converts the former to acetate and ATP. The latter reacts with another β-D-fructose-6-phosphate to ultimately form two molecules of xylulose-5-phosphate, which are cleaved by XFP to acP and D-glyceraldehyde-3-phosphate (EC 4.1.29) (96–98). The former is converted

to acetate and ATP, while the latter is converted to lactate, as described for the EMP pathway. Thus, the result is a mixed-acid fermentation (acetate and lactate). Because 2 glucose molecules (in the form of 2 molecules of fructose 6-phosphate) are required, the net products per glucose come to 1 acetate, ½ lactate, and 2 ATP.

CONCLUSION

I tell my medical students that they are walking, talking incubators; we cannot survive without our bacteria. I tell my graduate students that the metabolic chart is just a map. If one understands the symbols, then one can visit intriguing places. For example, knowledge of mixed-acid fermentation helps explain why short-chain fatty acids (SCFA) constitute about two-thirds of the colonic anion concentration, primarily as acetate, propionate, and butyrate. These SCFA are produced by diverse bacteria, are rapidly absorbed by the colonic mucosa, and provide the primary energy source for colonocytes, hepatic cells, fat cells, and muscle cells (90, 99–102). SCFA also perform functions of considerable significance to the health of the host, including protecting colonocytes against colitis ulcerosa, diverticulosis, and colorectal cancer (90, 91, 103–107). As microbiologists in a microbiome world, we must embrace metabolism. It is the gateway to a deeper understanding of pathogenesis.

ACKNOWLEDGMENTS

This work was supported by grants from the NIH and DOE: R01 AI108255, R21 DK097435 and DE-SC0012443.

CITATION

Wolfe AJ. 2015. Glycolysis for Microbiome Generation, Microbiol Spectrum 3(3):MBP-0014-2014.

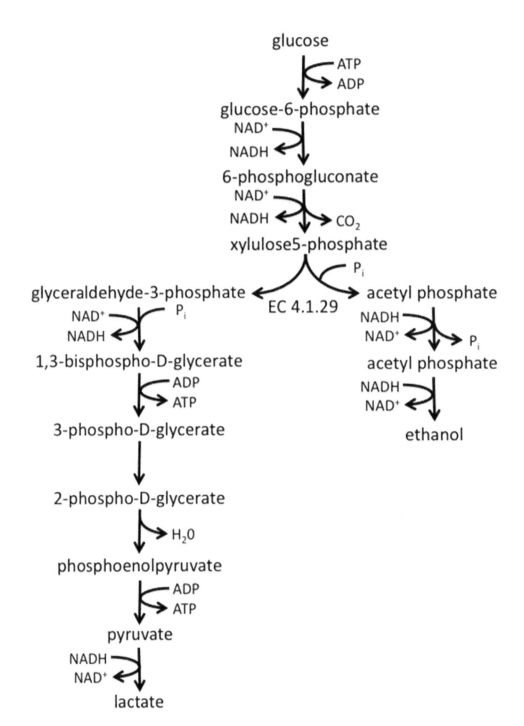

FIGURE 6 The heterolactic pathway. EC 4.1.29 is a phosphoketolase that cleaves a 5-carbon phosphosugar (xylulose 5-phosphate) into a 3-carbon phosphate (glyceraldehyde 3-phosphate) and a 2-carbon phosphate (acetyl phosphate). Note that all the NAD^+-consuming steps are balanced by NAD^+-producing steps. Because acetyl phosphate is used to recycle NAD^+, it is not used to generate ATP. doi:10.1128/microbiolspec.MBP-0014-2014.f6

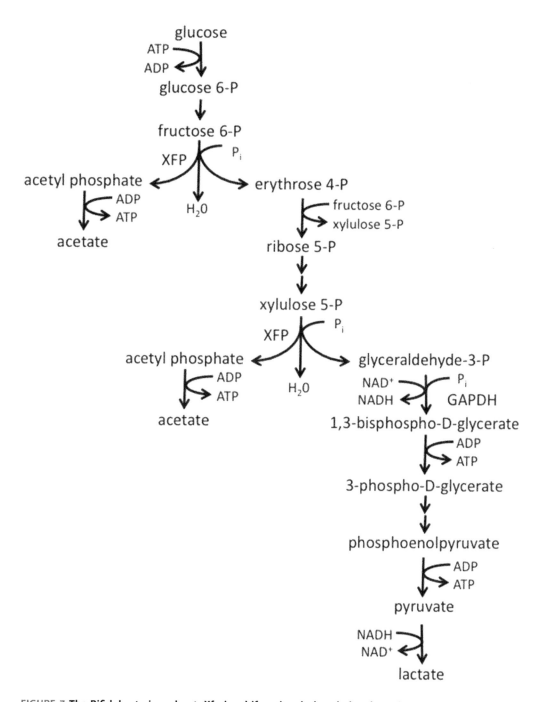

FIGURE 7 The Bifidobacterium shunt. Xfp is a bifunctional phosphoketolase. One activity (EC 4.1.2.22) cleaves a 6-carbon phosphosugar (fructose 6-phosphate) into a 4-carbon phosphate (erythrose 4-phosphate) and a 2-carbon phosphate (acetyl phosphate). A second activity (EC 4.1.29) cleaves a 5-carbon phosphosugar (xylulose 5-phosphate) into a 3-carbon phosphate (glyceraldehyde 3-phosphate) and a 2-carbon phosphate (acetyl phosphate). Note that all the NAD^+-consuming steps are balanced by NAD^+-producing steps. Acetyl phosphate is used to generate ATP. doi:10.1128/microbiolspec.MBP-0014-2014.f7

REFERENCES

1. **Wolfe AJ, Conley MP, Berg HC.** 1988. Acetyladenylate plays a role in controlling the direction of flagellar rotation. *Proc Natl Acad Sci U S A* **85**(18):6711–6715.

2. **Lipmann F.** 1954. Development of the acetylation problem, a personal account. *Science* **120**(3126):855–865.

3. **Bentley R.** 2000. From 'reactive C2 units' to acetyl coenzyme A: a long trail with an acetyl phosphate detour. *Trends Biochem Sci* **25** (6):302–305.

4. **Barak R, Eisenbach M.** 2001. Acetylation of the response regulator, CheY, is involved in bacterial chemotaxis. *Mol Microbiol* **40**(3): 731–743.

5. **Barak R, Prasad K, Shainskaya A, Wolfe AJ, Eisenbach M.** 2004. Acetylation of the chemotaxis response regulator cheY by acetyl-CoA synthetase purified from *Escherichia coli*. *J Mol Biol* **342**(2):383–401.

6. **Barak R, Welch M, Yanovsky A, Oosawa K, Eisenbach M.** 1992. Acetyladenylate or its derivative acetylates the chemotaxis protein CheY in vitro and increases its activity at the flagellar switch. *Biochemistry* **31**:10099–10107.

7. **Barak R, Yan J, Shainskaya A, Eisenbach M.** 2006. The chemotaxis response regulator CheY can catalyze its own acetylation. *J Mol Biol* **359**(2):251–265.

8. **Li R, Gu J, Chen Y-Y, Xiao C-L, Wang L-W, Zhang Z-P, Bi L-J, Wei H-P, Wang X-D, Deng J-Y, Zhang X-E.** 2010. CobB regulates *Escherichia coli* chemotaxis by deacetylating the response regulator CheY. *Mol Microbiol* **76** (5):1162–1174.

9. **Liarzi O, Barak R, Bronner V, Dines M, Sagi Y, Shainskaya A, Eisenbach M.** 2010. Acetylation represses the binding of CheY to its target proteins. *Mol Microbiol* **76**(4):932–943.

10. **Yan J, Barak R, Liarzi O, Shainskaya A, Eisenbach M.** 2008. In vivo acetylation of heY, a response regulator in chemotaxis of *Escherichia coli*. *JMol Biol* **376**(5):1260–1271.

11. **Lukat GS, McCleary WR, Stock AM, Stock JB.** 1992. Phosphorylation of bacterial response regulator proteins by low molecular weight phospho-donors. *Proc Natl Acad Sci U S A* **89**(2):718–722.

12. **McCleary WR, Stock JB.** 1994. Acetyl phosphate and the activation of two-component response regulators. *J Biol Chem* **269**(50): 31567–31572.

13. **Barak R, Abouhamad WN, Eisenbach M.** 1998. Both acetate kinase and acetyl coenzyme A synthetase are involved in acetate-stimulated change in the direction of flagellar rotation in *Escherichia coli*. *J Bacteriol* **180**(4):985–988.

14. **Barak R, Eisenbach M.** 2004. Co-regulation of acetylation and phosphorylation of CheY, a response regulator in chemotaxis of *Escherichia coli*. *J Mol Biol* **342**(2):375–381.

15. **Li R, Chen P, Gu J, Deng JY.** 2013. Acetylation reduces the ability of CheY to undergo autophosphorylation. *FEMS Microbiol Lett* **347**(1): 70–76.

16. **Ramakrishnan R, Schuster M, Bourret RB.** 1998. Acetylation at Lys-92 enhances signaling by the chemotaxis response regulator protein CheY. *Proc Natl Acad Sci U S A* **95**(9):4918–4923.

17. **Fraiberg M, Afanzar O, Cassidy CK, Gabashvili A, Schulten K, Levin Y, Eisenbach M.** 2015. CheY's acetylation sites responsible for generating clockwise flagellar rotation in *Escherichia coli*. *Mol Microbiol*, in press.

18. **Danese PN, Silhavy TJ.** 1998. CpxP, a stress-combative member of the Cpx regulon. *J Bacteriol* **180**(4):831–839.

19. **Fredericks CE, Shibata S, Aizawa S-I, Reimann SA, Wolfe AJ.** 2006. Acetyl phosphate-sensitive regulation of flagellar biogenesis and capsular biosynthesis depends on the Rcs phosphorelay. *Mol Microbiol* **61**(3):734–747.

20. **Hu LI, Chi BK, Kuhn ML, Filippova EV, Walker-Peddakotla AJ, Basell K, Becher D, Anderson WF, Antelmann H, Wolfe AJ.** 2013. Acetylation of the response regulator RcsB controls transcription from a small RNA promoter. *J Bacteriol* **195**(18):4174–4186.

21. **Lima BP, Antelmann H, Gronau K, Chi BK, Becher D, Brinsmade SR, Wolfe AJ.** 2011. Involvement of protein acetylation in glucose-induced transcription of a stress-responsive promoter. *Mol Microbiol* **81**:1190–1204.

22. **Lima BP, Thanh Huyen TT, Bassell K, Becher D, Antelmann H, Wolfe AJ.** 2012. Inhibition of acetyl phosphate-dependent transcription by an acetylatable lysine on RNA polymerase. *J Biol Chem* **287**(38):32147–32160.

23. **Feng J, Atkinson MR, McCleary W, Stock JB, Wanner BL, Ninfa AJ.** 1992. Role of phosphorylated metabolic intermediates in the regulation of glutamine synthetase synthesis in *Escherichia coli*. *J Bacteriol* **174**(19):6061–6070.

24. **Mitra A, Fay PA, Vendura KW, Alla Z, Carroll RK, Shaw LN, Riordan JT.** 2014. Sigma(N)-dependent control of acid resistance and the locus of enterocyte effacement in

enterohemorrhagic *Escherichia coli* is activated by acetyl phosphate in a manner requiring flagellar regulator FlhDC and the sigma(S) antagonist FliZ. *Microbiologyopen* **3**(4):497–512.

25. **Boll JM, Hendrixson DR.** 2011. A specificity determinant for phosphorylation in a response regulator prevents in vivo cross-talk and modification by acetyl phosphate. *Proc Natl Acad Sci U S A* **108**(50):20160–20165.

26. **Xu H, Caimano MJ, Lin T, He M, Radolf JD, Norris SJ, Gherardini F, Wolfe AJ, Yang XF.** 2010. Role of acetyl-phosphate in activation of the Rrp2-RpoN-RpoS pathway in *Borrelia burgdorferi*. *PLoS Pathog* **6**(9):e1001104.

27. **Liu J, Obi IR, Thanikkal EJ, Kieselbach T, Francis MS.** 2011. Phosphorylated CpxR restricts production of the RovA global regulator in *Yersinia pseudotuberculosis*. *PLoS One* **6**(8):e23314.

28. **Shin S, Park C.** 1995. Modulation of flagellar expression in *Escherichia coli* by acetyl phosphate and the osmoregulator OmpR. *J Bacteriol* **177**(16):4696–4702.

29. **Bouche S, Klauck E, Fischer D, Lucassen M, Jung K, Hengge-Aronis R.** 1998. Regulation of RssB-dependent proteolysis in *Escherichia coli*: a role for acetyl phosphate in a response regulator-controlled process. *Mol Microbiol* **27**(4):787–795.

30. **Gueriri I, Bay S, Dubrac S, Cyncynatus C, Msadek T.** 2008. The Pta-AckA pathway controlling acetyl phosphate levels and the phosphorylation state of the DegU orphan response regulator both play a role in regulating *Listeria monocytogenes* motility and chemotaxis. *Mol Microbiol* **70**(6):1342–1357.

31. **Lawhon SD, Maurer R, Suyemoto M, Altier C.** 2002. Intestinal short-chain fatty acids alter *Salmonella typhimurium* invasion gene expression and virulence through BarA/SirA. *Mol Microbiol* **46**(5):1451–1464.

32. **Bang IS, Kim BH, Foster JW, Park YK.** 2000. OmpR regulates the stationary-phase acid tolerance response of *Salmonella enterica* serovar typhimurium. *J Bacteriol* **182**(8):2245–2252.

33. **Heyde M, Laloi P, Portalier R.** 2000. Involvement of carbon source and acetyl phosphate in the external-pH-dependent expression of porin genes in *Escherichia coli*. *J Bacteriol* **182**(1):198–202.

34. **Matsubara M, Mizuno T.** 1999. EnvZ-independent phosphotransfer signaling pathway of the OmpR-mediated osmoregulatory expression of OmpC and OmpF in *Escherichia coli*. *Biosci Biotechnol Biochem* **63**(2):408–414.

35. **Pruss BM.** 1998. Acetyl phosphate and the phosphorylation of OmpR are involved in the regulation of the cell division rate in *Escherichia coli*. *Arch Microbiol* **170**(3):141–146.

36. **Wolfe AJ.** 2005. The acetate switch. *Microbiol Mol Biol Rev* **69**(1):12–50.

37. **Wolfe AJ.** 2010. Physiologically relevant small phosphodonors link metabolism to signal transduction. *Curr Opin Microbiol* **13**(2):204–209.

38. **Starai VJ, Escalante-Semerena JC.** 2004. Identification of the protein acetyltransferase (Pat) enzyme that acetylates acetyl-CoA synthetase in *Salmonella enterica*. *J Mol Biol* **340**(5):1005–1012.

39. **Starai VJ, Celic I, Cole RN, Boeke JD, Escalante-Semerena JC.** 2002. Sir2-dependent activation of acetyl-CoA synthetase by deacetylation of active lysine. *Science* **298**(5602):2390–2392.

40. **Blander G, Guarente L.** 2004. The SIR2 family of protein deacetylases. *Ann Rev Biochem* **73**(1):417–435.

41. **Greiss S, Gartner A.** 2009. Sirtuin/Sir2 phylogeny, evolutionary considerations and structural conservation. *Mol Cells* **28**(5):407–415.

42. **Yang XJ, Seto E.** 2008. The Rpd3/Hda1 family of lysine deacetylases: from bacteria and yeast to mice and men. *Nat Rev Mol Cell Biol* **9**(3):206–218.

43. **Vetting MW, LP SdC, Yu M, Hegde SS, Magnet S, Roderick SL, Blanchard JS.** 2005. Structure and functions of the GNAT superfamily of acetyltransferases. *Arch Biochem Biophys* **433**(1):212–226.

44. **Hildmann C, Riester D, Schwienhorst A.** 2007. Histone deacetylases: an important class of cellular regulators with a variety of functions, p 487–497. *In Applied microbiology and biotechnology*, **vol. 75**. Springer, Berlin/Heidelberg.

45. **Kuhn ML, Zemaitaitis B, Hu LI, Sahu A, Sorensen D, Minasov G, Lima BP, Scholle M, Mrksich M, Anderson WF, Gibson BW, Schilling B, Wolfe AJ.** 2014. Structural, kinetic and proteomic characterization of acetyl phosphate-dependent bacterial protein acetylation. *PLoS One* **9**(4):e94816.

46. **Weinert Brian T, Iesmantavicius V, Wagner Sebastian A, Scholz C, Gummesson B, Beli P, Nystrom T, Choudhary C.** 2013. Acetyl-phosphate is a critical determinant of lysine acetylation in *E. coli*. *Mol Cells* **51**(2):265–272.

47. **Beckham KS, Connolly JP, Ritchie JM, Wang D, Gawthorne JA, Tahoun A, Gally DL, Burgess K, Burchmore RJ, Smith BO, Beatson SA, Byron O, Wolfe AJ, Douce GR,**

Roe AJ. 2014. The metabolic enzyme AdhE controls the virulence of *Escherichia coli* O157:H7. *Mol Microbiol* **93**(1):199–211.

48. **Witchell TD, Eshghi A, Nally JE, Hof R, Boulanger MJ, Wunder EA Jr, Ko AI, Haake DA, Cameron CE.** 2014. Post-translational modification of LipL32 during *Leptospira interrogans* infection. *PLoS Negl Trop Dis* **8**(10):e3280.

49. **Ding T, Schloss PD.** 2014. Dynamics and associations of microbial community types across the human body. *Nature* **509**(7500):357–360.

50. **Kraal L, Abubucker S, Kota K, Fischbach MA, Mitreva M.** 2014. The prevalence of species and strains in the human microbiome: a resource for experimental efforts. *PLoS One* **9**(5):e97279.

51. **Aagaard K, Petrosino J, Keitel W, Watson M, Katancik J, Garcia N, Patel S, Cutting M, Madden T, Hamilton H, Harris E, Gevers D, Simone G, McInnes P, Versalovic J.** 2013. The Human Microbiome Project strategy for comprehensive sampling of the human microbiome and why it matters. *FASEB J* **27**(3):1012–1022.

52. **Tojo R, Suarez A, Clemente MG, de Los Reyes-Gavilan CG, Margolles A, Gueimonde M, Ruas-Madiedo P.** 2014. Intestinal microbiota in health and disease: Role of bifidobacteria in gut homeostasis. *World J Gastroenterol* **20**(41):15163–15176.

53. **Turroni F, Ventura M, Butto LF, Duranti S, O'Toole PW, Motherway MO, van Sinderen D.** 2014. Molecular dialogue between the human gut microbiota and the host: a *Lactobacillus* and *Bifidobacterium* perspective. *Cell Mol Life Sci* **71**(2):183–203.

54. **Shreiner AB, Kao JY, Young VB.** 2015. The gut microbiome in health and in disease. *Curr Opin Gastroenterol* **31**:69–75.

55. **Ravel J, Gajer P, Abdo Z, Schneider GM, Koenig SS, McCulle SL, Karlebach S, Gorle R, Russell J, Tacket CO, Brotman RM, Davis CC, Ault K, Peralta L, Forney LJ.** 2011. Vaginal microbiome of reproductive-age women. *Proc Natl Acad Sci U S A* **108**(Suppl 1):4680–4687.

56. **Ravel J, Brotman RM, Gajer P, Ma B, Nandy M, Fadrosh DW, Sakamoto J, Koenig SS, Fu L, Zhou X, Hickey RJ, Schwebke JR, Forney LJ.** 2013. Daily temporal dynamics of vaginal microbiota before, during and after episodes of bacterial vaginosis. *Microbiome* **1**(1):29.

57. **Gajer P, Brotman RM, Bai G, Sakamoto J, Schutte UM, Zhong X, Koenig SS, Fu L, Ma ZS, Zhou X, Abdo Z, Forney LJ, Ravel J.** 2012. Temporal dynamics of the human vaginal microbiota. *Sci Transl Med* **4**(132):132ra152.

58. **Hickey RJ, Zhou X, Pierson JD, Ravel J, Forney LJ.** 2012. Understanding vaginal microbiome complexity from an ecological perspective. *Transl Res* **160**(4):267–282.

59. **Ma B, Forney LJ, Ravel J.** 2012. Vaginal microbiome: rethinking health and disease. *Annu Rev Microbiol* **66**:371–389.

60. **Fouts DE, Pieper R, Szpakowski S, Pohl H, Knoblach S, Suh MJ, Huang ST, Ljungberg I, Sprague BM, Lucas SK, Torralba M, Nelson KE, Groah SL.** 2012. Integrated next-generation sequencing of 16S rDNA and meta-proteomics differentiate the healthy urine microbiome from asymptomatic bacteriuria in neuropathic bladder associated with spinal cord injury. *J Transl Med* **10**:174.

61. **Hilt EE, McKinley K, Pearce MM, Rosenfeld AB, Zilliox MJ, Mueller ER, Brubaker L, Gai X, Wolfe AJ, Schreckenberger PC.** 2014. Urine is not sterile: use of enhanced urine culture techniques to detect resident bacterial flora in the adult female bladder. *J Clin Microbiol* **52**(3):871–876.

62. **Khasriya R, Sathiananthamoorthy S, Ismail S, Kelsey M, Wilson M, Rohn JL, Malone-Lee J.** 2013. Spectrum of bacterial colonization associated with urothelial cells from patients with chronic lower urinary tract symptoms. *J Clin Microbiol* **51**(7):2054–2062.

63. **Pearce MM, Hilt EE, Rosenfeld AB, Zilliox MJ, Thomas-White K, Fok C, Kliethermes S, Schreckenberger PC, Brubaker L, Gai X, Wolfe AJ.** 2014. The female urinary microbiome: a comparison of women with and without urgency urinary incontinence. *MBio* **5**(4):e01283–01214.

64. **Wolfe AJ, Toh E, Shibata N, Rong R, Kenton K, Fitzgerald M, Mueller ER, Schreckenberger P, Dong Q, Nelson DE, Brubaker L.** 2012. Evidence of uncultivated bacteria in the adult female bladder. *J Clin Microbiol* **50**(4):1376–1383.

65. **Mayer C, Boos W.** 2005. Hexose/pentose and hexitol/pentitol metabolism. *In EcoSal Plus 2005*.

66. **Romeo T, Snoep J.** 2005. Glycolysis and flux control. *In EcoSal Plus 2005*.

67. **Peekhaus N, Conway T.** 1998. What's for dinner? Entner-Doudoroff metabolism in *Escherichia coli*. *J Bacteriol* **180**(14):3495–3502.

68. **Lessie TG, Phibbs PVJ.** 1984. Alternative pathways of carbohydrate utilization in pseudomonads. *Annu Rev Microbiol* **38**:359–388.

69. **Conway T.** 1992. The Entner-Doudoroff pathway: history, physiology and molecular biology. *FEMS Microbiol Rev* **9**(1):1–27.

70. **Crabtree H.** 1928. The carbohydrate metabolism of certain pathological overgrowths. *Biochem J* **22**:1289–1298.

71. **Warburg O, Wind F, Negelein E.** 1927. The metabolism of tumors in the body. *J Gen Physiol* **8**:519–530.

72. **Greiner EF, Guppy M, Brand K.** 1994. Glucose is essential for proliferation and the glycolytic enzyme induction that provokes a transition to glycolytic energy production. *J Biol Chem* **269**(50):31484–31490.

73. **Warburg O.** 1956. On the origin of cancer cells. *Science* **123**(3191):309–314.

74. **Mustea I, Muresian T.** 1967. Crabtree effect in some bacterial cultures. *Cancer Res* **20**:1499–1501.

75. **Doelle HW, Ewings KN, Hollywood NW.** 1982. Regulation of glucose metabolism in bacterial systems. *Adv Biochem Eng* **23**:1–35.

76. **Luli GW, Strohl WR.** 1990. Comparison of growth, acetate production, and acetate inhibition of *Escherichia coli* strains in batch and fed-batch fermentations. *Appl Environ Microbiol* **56**(4):1004–1011.

77. **Holms H.** 1990. Flux analysis and control of the central metabolic pathways in *Escherichia coli*. *FEMS Microbiol Rev* 1996, **19**(2):85–116.

78. **Holms WH.** 1986. The central metabolic pathways of *Escherichia coli*: relationship between flux and control at a branch point, efficiency of conversion to biomass, and excretion of acetate. *Curr Top Cell Regul* **28**:69–105.

79. **Krebs HA.** 1972. The Pasteur effect and the relations between respiration and fermentation. *Essays Biochem* **8**:1–34.

80. **Diaz-Ruiz R, Averet N, Araiza D, Pinson B, Uribe-Carvajal S, Devin A, Rigoulet M.** 2008. Mitochondrial oxidative phosphorylation is regulated by fructose 1,6-bisphosphate. A possible role in Crabtree effect induction? *J Biol Chem* **283**(40):26948–26955.

81. **Diaz-Ruiz R, Rigoulet M, Devin A.** 2011. The Warburg and Crabtree effects: On the origin of cancer cell energy metabolism and of yeast glucose repression. *Biochim Biophys Acta* **1807**(6):568–576.

82. **Cocaign-Bousquet M, Garrigues C, Loubiere P, Lindley ND.** 1996. Physiology of pyruvate metabolism in *Lactococcus lactis*. *Antonie Van Leeuwenhoek* **70**(2–4):253–267.

83. **Melchiorsen CR, Jensen NB, Christensen B, Vaever Jokumsen K, Villadsen J.** 2001. Dynamics of pyruvate metabolism in *Lactococcus lactis*. *Biotechnol Bioeng* **74**(4):271–279.

84. **Pessione E.** 2012. Lactic acid bacteria contribution to gut microbiota complexity: lights and shadows. *Front Cell Infect Microbiol* **2**:86.

85. **Dashko S, Zhou N, Compagno C, Piskur J.** 2014. Why, when, and how did yeast evolve alcoholic fermentation? *FEMS Yeast Res* **14**(6):826–832.

86. **Bock A, Sawers G.** 1996. Fermentation, p 262–282. *In* Neidhardt FC, Curtiss R III, Ingraham JL, Lin ECC, Low KB, Magasanik B, Reznikoff WS, Riley M, Schaechter M, E UH (ed), *Escherichia coli and Salmonella: cellular and molecular biology*, 2nd ed. ASM Press, Washington, D.C.

87. **van Hoek MJ, Merks RM.** 2012. Redox balance is key to explaining full vs. partial switching to low-yield metabolism. *BMC Syst Biol* **6**:22.

88. **Xiao Z, Xu P.** 2007. Acetoin metabolism in bacteria. *Crit Rev Microbiol* **33**(2):127–140.

89. **Biebl H, Menzel K, Zeng AP, Deckwer WD.** 1999. Microbial production of 1,3-propanediol. *Appl Microbiol Biotechnol* **52**(3):289–297.

90. **Macfarlane GT, Macfarlane S.** 2012. Bacteria, colonic fermentation, and gastrointestinal health. *J AOAC Int* **95**(1):50–60.

91. **Puertollano E, Kolida S, Yaqoob P.** 2014. Biological significance of short-chain fatty acid metabolism by the intestinal microbiome. *Curr Opin Clin Nutr Metab Care* **17**(2):139–144.

92. **Zaunmuller T, Eichert M, Richter H, Unden G.** 2006. Variations in the energy metabolism of biotechnologically relevant heterofermentative lactic acid bacteria during growth on sugars and organic acids. *Appl Microbiol Biotechnol* **72**(3):421–429.

93. **Yin X, Chambers JR, Barlow K, Park AS, Wheatcroft R.** 2005. The gene encoding xylulose-5-phosphate/fructose-6-phosphate phosphoketolase (xfp) is conserved among *Bifidobacterium* species within a more variable region of the genome and both are useful for strain identification. *FEMS Microbiol Lett* **246**(2):251–257.

94. **Sgorbati B, Lenaz G, Casalicchio F.** 1976. Purification and properties of two fructose-6-phosphate phosphoketolases in *Bifidobacterium*. *Antonie Van Leeuwenhoek* **42**(1–2):49–57.

95. **Grill JP, Crociani J, Ballongue J.** 1995. Characterization of fructose 6 phosphate phosphoketolases purified from *Bifidobacterium* species. *Curr Microbiol* **31**(1):49–54.

96. **Heath EC, Hurwitz J, Horecker BL, Ginsburg A.** 1958. Pentose fermentation by *Lactobacillus plantarum*. I. The cleavage of xylulose 5-phosphate by phosphoketolase. *J Biol Chem* **231**(2):1009–1029.

97. **Lee JM, Jeong DW, Koo OK, Kim MJ, Lee JH, Chang HC, Kim JH, Lee HJ.** 2005. Cloning and characterization of the gene encoding phosphoketolase in *Leuconostoc mesenteroides* isolated from kimchi. *Biotechnol Lett* **27**(12):853–858.

98. **Posthuma CC, Bader R, Engelmann R, Postma PW, Hengstenberg W, Pouwels PH.** 2002. Expression of the xylulose 5-phosphate phosphoketolase gene, *xpkA*, from *Lactobacillus pentosus* MD363 is induced by sugars that are fermented via the phosphoketolase pathway and is repressed by glucose mediated by CcpA and the mannose phosphoenolpyruvate phosphotransferase system. *Appl Environ Microbiol* **68**(2):831–837.

99. **Macfarlane S, McBain AJ, Macfarlane GT.** 1997. Consequences of biofilm and sessile growth in the large intestine. *Adv Dent Res* **11**(1):59–68.

100. **McNeil NI.** 1984. The contribution of the large intestine to energy supplies in man. *Am J Clin Nutr* **39**(2):338–342.

101. **Mortensen PB, Clausen MR.** 1996. Short-chain fatty acids in the human colon: relation to gastrointestinal health and disease. *Scand J Gastroenterol Suppl* **216**:132–148.

102. **Topping DL, Clifton PM.** 2001. Short-chain fatty acids and human colonic function: roles of resistant starch and nonstarch polysaccharides. *Physiol Rev* **81**(3):1031–1064.

103. **Wadolkowski EA, Laux DC, Cohen PS.** 1988. Colonization of the streptomycin-treated mouse large intestine by a human fecal *Escherichia coli* strain: role of growth in mucus. *Infect Immun* **56**(5):1030–1035.

104. **Batt RM, Rutgers HC, Sancak AA.** 1996. Enteric bacteria: friend or foe? *J Small Anim Pract* **37**(6):261–267.

105. **Wong JM, de Souza R, Kendall CW, Emam A, Jenkins DJ.** 2006. Colonic health: fermentation and short chain fatty acids. *J Clin Gastroenterol* **40**(3):235–243.

106. **Tan J, McKenzie C, Potamitis M, Thorburn AN, Mackay CR, Macia L.** 2014. The role of short-chain fatty acids in health and disease. *Adv Immunol* **121**:91–119.

107. **Russell WR, Hoyles L, Flint HJ, Dumas ME.** 2013. Colonic bacterial metabolites and human health. *Curr Opin Microbiol* **16**(3):246–254.

108. **Hu LI, Lima BP, Wolfe AJ.** 2010. Bacterial protein acetylation: the dawning of a new age. *Mol Microbiol* **77**:15–21.

Pathogenesis — Thoughts from the Front Line

2

JAMES P. NATARO[1]

HISTORY

Conventionally, the history of microbiology begins with the studies of Leeuwenhoek in the 17th century [this section is reviewwzed elegantly in Bulloch (1938)]. Using crude but ingenious microscopes fashioned by hand, this erstwhile haberdasher first revealed the unseen microbial world. It would take more than two centuries before the contributions of these microorganisms to the ecology of the biosphere would be revealed. Pasteur inaugurated the study of functional microbiology with his work on fermentation in the late 1850s. In 1863, he described the phenomenon of anaerobic microbial life and first coined the terms aerobic and anaerobic. Pasteur is of course best known for his many seminal contributions to microbial pathogenesis later in his life.

The first infectious disease that was clearly linked to microorganisms was sepsis, specifically anthrax. In 1872, Davaine demonstrated that the blood of animals with septicemia induced fatal infection when injected into a normal animal. Soon after, Dreyer first observed bacteria in the blood of animals with sepsis. Robert Koch brought his precise systematic eye to sepsis in 1878 with his publication *Etiology of Traumatic Infectious Diseases*,

[1]University of Virginia.
Metabolism and Bacterial Pathogenesis
Edited by Tyrrell Conway and Paul Cohen
© 2015 American Society for Microbiology, Washington, DC
doi:10.1128/microbiolspec.MBP-0012-2014

wherein he expostulated the famous Henle-Koch's postulates. Although Koch is rightly considered the father of pathogenic bacteriology, the contributions of his mentor Ferdinand Cohn are routinely neglected. Nevertheless, it is remarkable that the techniques of pure culture, Petri dishes, solid agar, and other basic methods of bacteriology had been passed down almost unchanged from Koch's Berlin laboratory. The ensuing century produced an explosion of knowledge comparable to the Italian Renaissance, elucidating microbial causes of human, animal and plant diseases.

Another great moment in the history of microbial pathogenesis occurred in a delicatessen in Honolulu, Hawaii in 1972 (2) when Herbert Boyer (University of California San Francisco) and Stanley Cohen (Stanford University) first envisioned the experiment that would spawn the field of genetic engineering. By introducing a kanamycin-resistance–encoding gene into a tetracycline-resistance–conferring plasmid, made possible by Boyer's restriction endonucleases, these pioneers inaugurated the age of molecular biology. Within less than a decade, bacterial pathogeneticists came to realize the great promise of genetic engineering.

In the late 1970s, pathogenic microbiologists, including Professor Harry Smith (University of Birmingham), were envisioning the great contribution that genetic engineering would make to pathogenesis research. Smith wrote in 1980 "proving that a particular microbial product is a determinant of an important facet of pathogenicity is often difficult because of the complexity of microbial interactions with host cell systems. One way out of the difficulty is to compare in the appropriate pathogenicity test isogenic strains which differ only in the gene piece which determines the production of a putative virulence determinant" (3). In 1988, Stanley Falkow, recognizing the explosion of genetic pathogenesis studies now made possible, articulated what he termed "molecular Koch's postulates" (4),

but which today might be more appropriately called Falkow's postulates (see below). Falkow proposed that studies implicating a virulence factor would require construction of a mutation in the factor in question, demonstration that the mutant lost virulence, followed by restoration of virulence when the gene was replaced. This has now become the roadmap for pathogenesis studies, and has resulted in the implication of thousands of genetic loci in infectious diseases.

The genomic revolution, in which whole genomes of organisms and meta-genomic characterization of microbial populations are readily available, provides another powerful tool to understand the nature of microbial assault. At the other end of the pathogen-host dyad, genetic manipulation and detailed molecular characterization of cellular events illuminates the effects of pathogenic microorganisms on the host.

With our greatly enhanced power to dissect the phenomena of pathogenesis, it has become increasingly important to explore the denotations of its language. Casadevall and Pirofski have contributed the most lucid linguistic scheme, enunciated as the "damage-response" framework (5–8). According to these authors, "the pathological outcome of the host-microorganism interaction is determined by the amount of damage to the host" (8). Damage is defined as "disruption in the normal homeostatic mechanisms of a host that alter the functioning of cells, tissues or organs...." (8). The damage-response framework provides a powerful substratum to support discussions on microbial pathogenesis.

Understanding pathogenesis induced by microorganisms offers the very real and partially fulfilled prospect of effective disease prevention and treatment. Yet, many challenges remain, and some high profile conquests are in danger of being overturned with the rise of antimicrobial resistance, and perhaps, microbial or social escape from vaccine protection.

THE MICROBE DOES NOT CARE IF YOU ARE SICK OR NOT

What does it mean to call something a microbial pathogen? If a pathogen is simply considered a microorganism that elicits disease, the definition is best applied on a case-by-case basis: the same pneumococcus that causes fatal meningitis in host A may be a benign co-traveler in host B, and in fact may have been benign in host A had certain precipitating events not occurred. By contrast, *Staphylococcus epidermidis* is nearly always a benign co-traveler, yet in a neonate or individual with an indwelling central venous catheter, the organism may readily elicit disease and be a pathogen. In this chapter, the term *pathogen* is reserved for a specific case of interaction between host and microbe in which the host suffers injury (damage); a bacterium is not a pathogen when there is no damage. Defining the threshold where subtle injury becomes pathogenesis is an uncertain endeavor: many individuals harbor subclinical inflammation of various organs with no frank manifestations and risk for adverse outcome only demonstrable at the population level (9, 10).

There is clearly variation in the intrinsic virulence of bacteria on humans. Some bacteria rarely encounter a mammalian host without causing overt disease. At the other end of the spectrum, some rarely cause disease except in the most debilitated hosts. Use of the terms *virulent pathogen* and *opportunistic pathogen* would appropriately describe these scenarios.

Some pathogens elicit clinical syndromes that promote the spread of the microorganism through the host community, presumably a direct effect of natural selection to increase the reproductive rate of the organism, i.e., the number of secondary cases occurring for each primary case (11) (reproduced in Equation 1). In fact, all of the variables in the reproductive number equation are acted upon by natural selection. Thus, a pathogen that is benefited by causing disease in a host population will forge a productive balance between transmissibility and virulence that will both promote its transmission today and assure a sufficient supply of susceptible hosts for transmission tomorrow. (This perspective is nowhere better articulated than in Zinsser's classic work *Rats, Lice and History* [recently discussed in Weissman (2005)]. These observations hold for pathogens whose biological niches entail engendering disease in a host or host population. The strange case of tuberculosis comes in mind, in which the microorganism must cause symptomatic disease (cough) in order to be transmitted, but its long-term persistence in the community is favored by a long-term latent state with recrudescence of transmissibility once additional susceptibles are available. The term *essential pathogen* would be an adequate description of this phenomenon.

We have as well many examples of incidental pathogens: pathogens that may be fully adapted to pathogenesis in non-human hosts or at sites that are unlikely to promote transmission and persistence. Lyme disease is an example of the former prospect: the pathogen's virulence factors promote propagation in small animals (13), but transmission to man is a dead end for its life cycle. In humans, meningitis is not likely to entail transmission to other hosts, and therefore the factors promoting this clinical entity have probably been adapted for other functions. *Staphylococcus aureus* is a common human pathogen capable of causing pyogenic infections at many body sites. There is evidence to suggest that the virulence factors of *S. aureus* may have initially evolved to promote purulent skin infection (14), which may dramatically facilitate transmission of the microorganism through a human population.

For enteric pathogens, recent data revealed that when the organism induces diarrhea, the abundance of excretion is dramatically higher than when the host remains asymptomatic (15–17). This suggests that diarrhea mathematically promotes propagation, aside from the fact that frequent, voluminous, difficult-to-contain stools provide a far more dangerous vehicle for transmission to other susceptible individuals. Interestingly, this author's data from volunteer studies reveal that infection that induces softer stools not meeting the clinical definition of diarrhea produce as high an abundance of pathogen as that found in watery stools (J. Nataro, unpublished). Thus, for the pathogen, disease is not the endpoint; rather transmission dynamics are acted upon by natural selection.

Lastly, there are microorganisms that are statistically associated with disease in a population, but such studies rarely provide insight into the cause of disease at the individual level. In these cases, the term *epidemiologic pathogen* might be appropriate when pathogenesis exists in the aggregate.

WHAT IS A VIRULENCE FACTOR?

The conventional definition of a virulence factor is something that directly elicits damage to the host in the setting of infection (5). However, there are instances when a factor executes microbial metabolic functions while in the host (perhaps exclusively), but the factor does not directly damage the host. To reframe the question: must a virulence factor be one specific for a pathogenic microorganism and/or directly damage the host, or may it be a contribution of any kind that enables the pathogen to cause disease? The highly mosaic and plastic nature of microbial genomes complicates this phenomenon further, as genes that normally effect damage in a pathogen are likely to be silent in the genome of a harmless commensal. But it does not stop there.

A generation ago, Peter Smith articulated four simplified steps that all pathogens must achieve to elicit productive disease: colonization, multiplication, immune evasion, and damage (3). The first three of these four functions must also be accomplished by any

successfully invading commensal species. Of course, like pathogens, commensals must achieve transmission from host to host to assure their perpetuation within the host population. So, if colonization factors, metabolic genes assuring successful competition with incumbent microbiota for nutrients, and ability to avoid rapid clearance by the host's immune system are part of any microbial invader's arsenal, it is perhaps better to limit consideration of virulence factors as factors that expressly mediate damage to host tissues, cells, and processes. However, what if damage to the host occurs secondary to immune processes that are selected because they promote establishment, but which elicit host damage and disease as a common side effect? Both pathogenic and non-pathogenic microorganisms must execute essential metabolic functions, and do so in the rarefied conditions of host tissues, thus genes involved in these metabolic pathways cannot be considered true virulence factors. An exception may be factors that confer specific metabolic advantages to the pathogenic state. Examples may be the exploitation of tryptophan availability by intracellular *Chlamydia* (18, 19) and the ingenious ability of *Salmonella* strains to exploit inflamed tissue for metabolic advantage (20). Another metabolic function that may be specific to pathogens is the interesting ability of Shiga-toxin–producing *Escherichia coli* strains to adapt to the microenvironment of the epithelial cell surface: the intestinal lumen and mucus layer are glycolytic (generally monosaccharide-utilizing microenvironments), whereas the epithelial brush border appears to be gluconeogenic (21). The ability of pathogenic *E. coli*, but not commensal *E. coli,* to utilize gluconeogenic sources adapts it to the site of disease. As remarkable is the ability of enteric pathogens to utilize ethanolamine (22), a byproduct of phosphatidylethanolamine degradation; the abundant ethanolamine of the GI lumen may not only provide an energy source but also a signal inducing expression of virulence genes (23). Indeed, the interplay between metabolism and virulence gene expression is a major focus of this book.

All microorganisms that persist in the host must subvert host defenses at their chosen site. For commensals, they may need to compete for scarce resources with other indigenous microbiota and/or host sequestering functions. Pathogens need to execute the same functions, but what may be unique to the pathogen is the need to occupy a succession of different environments (see Shigatoxin–producing bacteria above), and a succession of host defenses featuring spatio-temporal specificity. Thus, an enteric pathogen must overcome gastric acidity, latch on in spite of peristalsis, compete with indigenous microbiota, avoid secretory enzymes and other antimicrobial factors (e.g., defensins), penetrate the mucus layer, and resist a series of cellular and soluble antimicrobial defenses that increase in complexity, effectiveness, and specificity over time. If the organism induces inflammation, and especially if it invades and persists within the host tissues, an astonishing number of antimicrobial factors may need to be overcome. Thus, the immune evasion factors that are particular to the pathogenic stages of the microbial attack can justly be considered pathogenesis factors, but perhaps not virulence factors.

Fortunately, pathogeneticists have devoted far more attention to characterizing virulence factors and their contributions than to caviling at the term. However, it may be defensible to revert to the Greek root of virulence meaning "poison," and to consider applying the term solely to factors that engender direct harm.

NOT SO FAST

While some pathogenesis events may appear to be accidents of nature, not all may be accidental. Le Gall, et al., reported that the virulence factors of extra-intestinal *E. coli* may increase the fitness of the strain in the normal gut environment (24). Many pro-inflammatory virulence factors of *E. coli* may have evolved to offer the advantage of nitrate reductase-positive *E. coli* upon induction of nitric oxide synthase (25).

At first glance, uropathogenic *E. coli* may not appear to be adapted to confer a special advantage, certainly not in the setting of pyelonephritis. But uroepithelial adhesins may promote persistent carriage in the GI tract, and chronic persistent asymptomatic urinary excretion may provide the opportunity for long-term carriage and frequent transmission (26). It may be sufficient to assume that all that we see is the result of natural selection, while recognizing that the pressure towards selection of a trait or set of traits may be almost entirely obscure.

Everything that has evolved in bacteria must be subject to inter-individual variation, on which natural selection acts. One area where strain-to-strain variation has remained curiously understudied is in the area of genetic regulation itself. Individual studies have suggested that certain highly virulent strains exhibit high levels of virulence factor expression (27, 28). Comprehensive studies to unravel the diversity of regulomics will almost certainly illuminate important features of infectious disease epidemiology.

ALL PATHOGENESIS IS LOCAL, OR WHEN IS A PATHOGEN NOT A PATHOGEN?

A pathogen is an agent that damages a particular individual host. So, whereas all pathogenesis is local with regard to an individual host, pathogenesis may also be local with regard to an individual population harboring the microorganism. A recent large, multi-site diarrheal surveillance case-control study aimed to elucidate the burden and etiology of diarrhea among children less than five in sub-Saharan Africa and south Asia (the Global Enteric Multi-center Study, or GEMS) (29). In this study, four enteric pathogens were statistically associated with symptomatic diarrhea at all sites: rotavirus, cryptosporidium, *Shigella* spp., and heat-stable enterotoxin producing enterotoxigenic *E. coli*. Interestingly, many putative pathogens were associated with disease at only one or several sites. For example, typical enteropathogenic *E. coli* was significantly associated with diarrhea at only two sites (Kenya and Mozambique) (J. Nataro, unpublished observations). In contrast, *Giardia enterica* was found to be statistically protective against diarrhea at all sites (29). Both *Giardia* and enteropathogenic *E. coli* have been clearly implicated in outbreaks of diarrhea historically; so, why are these not consistent pathogens at their sites? Perhaps ascertainment captured individuals who had seen the pathogens before and were therefore immune. Perhaps both pathogenic and non-pathogenic subtypes exist, and better molecular diagnostics could capture only the truly virulent types. Perhaps at some sites, the populations of host and pathogen have co-evolved over many generations of co-existence, such that susceptible *H. sapiens* and virulent microbe were selected against. We do not know the reasons for these surprising epidemiologic observations [reviewed in Levine and Robins-Browne (2012)], but continued discussions and investigations are warranted regarding what constitutes a pathogen in these settings.

A natural sequelae to this discussion must address maladies that remain poorly defined. In the 1960s, Lindenbaum and colleagues described a gastro-intestinal syndrome among returning Peace Corps workers (31); the afflicted patients manifested malabsorption and increased intestinal permeability, with resolution of the disorder after several months living back in the U.S. Subsequently, several investigators working in developing countries found that many (often most) children reared in impoverished settings likewise manifested subclinical malabsorption and altered permeability, which some linked to poor growth and cognition [reviewed in Keusch, et al. (2012)]. This disorder, termed tropical, and later environmental, enteropathy, remains enigmatic. Its cause is almost certainly exposure to enteric microbes, but the culprits remain unknown. Are there new undiscovered pathogens for this occult entity? Could the malady be caused by overwhelming exposure to microorganisms that at lower levels do not produce any disorder (i.e., the "pathogens" are not pathogens in the traditional sense of a specific microbe causing a specific disease)? Perhaps the most interesting prospect is this: if environmental enteropathy afflicts the majority of individuals in a given population, is it even a disease, or can it be considered the natural state of man, perhaps even one that offers protection from some other ailment?

PATHOGENESIS: FRIEND OR FOE?

In the context of evolutionary adaptation, we recognize that some microorganisms clearly adapt to pathogenesis as a niche. *Shigella* spp. for example, are rarely found in asymptomatic individuals, who likely contribute just a small fraction of pathogen transmission. *B. anthracis* may be another organism that benefits immensely from the pathogenic lifestyle. But in other cases, disease may represent an evolutionary co-adaptation between the host and microbe, particularly when the pathogen has committed itself to propagation within the very host it afflicts. Sansonetti and colleagues illuminated this perspective in an interesting way (33). These investigators infected rabbits with *S. flexneri*, then treated a group with anti-interleukin–8 antibodies, to mitigate the local inflammatory response. As predicted, the treated group developed less colitis and appeared healthier, at least until they succumbed to *Shigella*-induced bacteremia (!). In this case, the aggressive local inflammatory response that results in clinical shigellosis may be the price the host pays to keep the microorganism confined to the lumen.

Moreover, watery diarrhea may itself be not just a clinical nuisance, but a means of clearing the pathogen from the lumen. Watery diarrhea induced by STEC and other bacterial pathogens is indeed mitigated on treatment with gastrointestinal anti-motility agents, but at the cost of more prolonged bloody diarrhea (34).

Helicobacter presents another variation on this theme. *H. sapiens* has apparently co-evolved with *Helicobacter*, the latter having emerged from Africa carried by its human host (35). In this situation, some mutual benefit must be realized by the two populations, with the bacterium's very existence dependent on the robustness of the host. Thus, *Helicobacter*-induced peptic ulcer disease and gastric carcinoma are puzzling. Blaser and colleagues, however, offer another option (36, 37): eradication of *Helicobacter* from populations of children in New York City apparently increased their incidence of bronchial asthma, a disease of immune dysregulation and inflammation. Thus, chronic carriage of *H. pylori* may harm the few who develop excess inflammation, but may benefit the many who develop appropriate levels of response to environmental allergens.

PATHOGENESIS IS SYSTEMS ENGINEERING

A system is a community of interacting agents, arranged in a network. The agents in the system may be homogeneous (e.g., a population of bacteria, or a population of users of a website), or heterogeneous (e.g., predators and prey). The systems engineer seeks first to understand the basic motifs of interaction. Motifs are in turn located within larger modules comprised of several homogeneous or hetergeneous motifs; a full system or network may comprise many modules. Long the purview of the engineer, systems thinking now pervades a vast array of endeavors. Marketers apply systems thinking to ascertain who visits a website or a store and what influences their behavior. The rise of social media is a prime example of complex and (eventually) predictable human systems behavior.

To paraphrase Theodosius Dobzhansky, nothing in biology makes sense except in the light of systems thinking. Whether we talk of populations of an individual species, their interactions with other species in an ecosystem, the behavior of cells in an organ, or the interactions of proteins and genes inside a single cell, the principles of systems engineering pervade all aspects of our field. Remarkably, however, biological studies in which the emerging understanding of systems behavior and modeling thereof are remarkably few, but can be highly illuminating. Papin, et al, have applied systems thinking to elucidate promising drug targets in *Leishmania* (38, 39). Hewlett, et al., have applied systems thinking to understand the effects of *Clostridium difficile* toxins on cellular proteins. The approach reveals the sites of toxin action and how these direct actions produce downstream physiologic derangements (40).

Among the most powerful applications of systems biology will be its ability to illuminate the actions of the human microbiota. In particular, we need to understand how the ecosystems of the microbiota take shape and how they are impacted by specific insults and perturbations. A thorough analysis of the human microbial ecosystem will allow us to understand what aspects need to be preserved to assure human health, and how best to accomplish this objective. Moreover, the complex interactions of pathogens with the microbiota will only be understood in the light of systems thinking; such analysis will undoubtedly result in novel insights and therapeutic targets.

CONCLUSION

All ages of scientific discovery are exciting. All provide new tools and new opportunities to illuminate the dark reaches of biologic science. If ours is more exciting than others, it is only because application of the most powerful techniques is aided by a scientific industry, companies whose existence in the marketplace depends on their ability to accelerate scientific discovery at an affordable cost. More exciting still is the emerging global will to deploy scientific advances to assist those most in need, and to consider the economic implications of development.

Finally, the broadest approach to human health must address what human disease really is: the inability to live life to its fullest extent. In this context, maladies quite indirectly attributable to microbial insult, such as obesity, must ultimately come under the scientist's careful eye.

$$\frac{R_0 = BN}{\alpha + \beta + \nu} \qquad (1)$$

Where R_0 is the number of secondary infections occurring as a result of each primary infection; B is the rate constant of infectious transfer of the microorganism; N is the density of the susceptible host population; α is the rate of pathogen-induced mortality; β the rate of pathogen-independent mortality; and ν is the rate of recovery (11).

REFERENCES

1. **Bulloch W.** 1938. *The History of Bacteriology.* Dover Publications, Inc., New York.
2. **Falkow S.** 2008. I never met a microbe I didn't like. *Nat Med* **14**:1053–1057.
3. **Smith M.** 1980. Introduction, p 11–16. *In* Achtman M, Makela PH, Skehel JJ, Koch MA (ed), *The molecular basis of microbial pathogenicity.* Verlag Chemie, Weinheim.
4. **Falkow S.** 1988. Molecular Koch's postulates applied to microbial pathogenicity. *Rev Infect Dis* **10**(Suppl 2):S274–S276.
5. **Casadevall A, Pirofski LA.** 2009. Virulence factors and their mechanisms of action: the view from a damage-response framework. *J Water Health* **7**(Suppl 1):S2–S18.
6. **Casadevall A, Pirofski LA.** 2014. What is a host? Incorporating the Microbiota into the 'Damage-Response Framework'. *Infect Immun* **83**(1):2–7.
7. **Pirofski LA, Casadevall A.** 2008. The damage-response framework of microbial pathogenesis and infectious diseases. *Adv Exp Med Biol* **635**:135–146.
8. **Casadevall A, Pirofski LA.** 2003. The damage-response framework of microbial pathogenesis. *Nat Rev Microbiol* **1**:17–24.
9. **Lathe R, Sapronova A, Kotelevtsev Y.** 2014. Atherosclerosis and Alzheimer—diseases with a common cause? Inflammation, oxysterols, vasculature. *BMC Geriatr* **14**:36.
10. **Janakiram NB, Rao CV.** 2014. The role of inflammation in colon cancer. *Adv Exp Med Biol* **816**:25–52.
11. **Levin BR.** 1996. The evolution and maintenance of virulence in microparasites. *Emerg Infect Dis* **2**:93–102.
12. **Weissmann G.** 2005. Rats, lice, and Zinsser. *Emerg Infect Dis* **11**:492–496.
13. **Radolf JD, Caimano MJ, Stevenson B, Hu LT.** 2012. Of ticks, mice and men: understanding the dual-host lifestyle of Lyme disease spirochaetes. *Nat Rev Microbiol* **10**:87–99.
14. **Chi CY, Lin CC, Liao IC, Yao YC, Shen FC, Liu CC, Lin CF.** 2014. Panton-Valentine leukocidin facilitates the escape of Staphylococcus aureus from human keratinocyte endosomes and induces apoptosis. *J Infect Dis* **209**:224–235.
15. **Lindsay BR, Chakraborty S, Harro C, Li S, Nataro JP, Sommerfelt H, Sack DA, Colin SO.** 2014. Quantitative PCR and culture evaluation for enterotoxigenic Escherichia coli (ETEC) associated diarrhea in volunteers. *FEMS Microbiol Lett* **352**:25–31.
16. **Phillips G, Lopman B, Tam CC, Iturriza-Gomara M, Brown D, Gray J.** 2009. Diagnosing norovirus-associated infectious intestinal disease using viral load. *BMC Infect Dis* **9**:63.
17. **Taniuchi M, Sobuz SU, Begum S, Platts-Mills JA, Liu J, Yang Z, Wang XQ, Petri WA Jr, Haque R, Houpt ER.** 2013. Etiology of diarrhea in Bangladeshi infants in the first year of life analyzed using molecular methods. *J Infect Dis* **208**:1794–1802.
18. **Bonner CA, Byrne GI, Jensen RA.** 2014. Chlamydia exploit the mammalian tryptophan-depletion defense strategy as a counterdefensive cue to trigger a survival state of persistence. *Front Cell Infect Microbiol* **4**:17.
19. **Xie G, Bonner CA, Jensen RA.** 2002. Dynamic diversity of the tryptophan pathway in chlamydiae: reductive evolution and a novel operon for tryptophan recapture. *Genome Biol* **3**(9):research00511–research005117.

20. Rivera-Chavez F, Winter SE, Lopez CA, Xavier MN, Winter MG, Nuccio SP, Russell JM, Laughlin RC, Lawhon SD, Sterzenbach T, Bevins CL, Tsolis RM, Harshey R, Adams LG, Baumler AJ. 2013. Salmonella uses energy taxis to benefit from intestinal inflammation. *PLoS Pathog* **9:**e1003267.

21. Njoroge JW, Nguyen Y, Curtis MM, Moreira CG, Sperandio V. 2012. Virulence meets metabolism: Cra and KdpE gene regulation in enterohemorrhagic *Escherichia coli*. *M Bio* **3:** e00280-12.

22. Bertin Y, Girardeau JP, Chaucheyras-Durand F, Lyan B, Pujos-Guillot E, Harel J, Martin C. 2011. Enterohaemorrhagic *Escherichia coli* gains a competitive advantage by using ethanolamine as a nitrogen source in the bovine intestinal content. *Environ Microbiol* **13:**365–377.

23. Kendall MM, Gruber CC, Parker CT, Sperandio V. 2012. Ethanolamine controls expression of genes encoding components involved in interkingdom signaling and virulence in enterohemorrhagic *Escherichia coli* O157:H7. *M Bio* **3**(3):pii: e00050-12.

24. Le GT, Clermont O, Gouriou S, Picard B, Nassif X, Denamur E, Tenaillon O. 2007. Extraintestinal virulence is a coincidental by-product of commensalism in B2 phylogenetic group *Escherichia coli* strains. *Mol Biol Evol* **24:**2373–2384.

25. Winter SE, Winter MG, Xavier MN, Thiennimitr P, Poon V, Keestra AM, Laughlin RC, Gomez G, Wu J, Lawhon SD, Popova IE, Parikh SJ, Adams LG, Tsolis RM, Stewart VJ, Baumler AJ. 2013. Host-derived nitrate boosts growth of *E. coli* in the inflamed gut. *Science* **339:**708–711.

26. Ragnarsdottir B, Svanborg C. 2012. Susceptibility to acute pyelonephritis or asymptomatic bacteriuria: host-pathogen interaction in urinary tract infections. *Pediatr Nephrol* **27:**2017–2029.

27. Abou Chakra CN, Pepin J, Sirard S, Valiquette L. 2014. Risk factors for recurrence, complications and mortality in *Clostridium difficile* infection: a systematic review. *PLoS ONE* **9:**e98400.

28. Abu-Ali GS, Ouellette LM, Henderson ST, Whittam TS, Manning SD. 2010. Differences in adherence and virulence gene expression between two outbreak strains of enterohaemorrhagic *Escherichia coli* O157:H7. *Microbiology* **156:**408–419.

29. Kotloff KL, Nataro JP, Blackwelder WC, Nasrin D, Farag TH, Panchalingam S, Wu Y, Sow SO, Sur D, Breiman RF, Faruque AS, Zaidi AK, Saha D, Alonso PL, Tamboura B, Sanogo D, Onwuchekwa U, Manna B, Ramamurthy T, Kanungo S, Ochieng JB, Omore R, Oundo JO, Hossain A, Das SK, Ahmed S, Qureshi S, Quadri F, Adegbola RA, Antonio M, Hossain MJ, Akinsola A, Mandomando I, Nhampossa T, Acacio S, Biswas K, O'Reilly CE, Mintz ED, Berkeley LY, Muhsen K, Sommerfelt H, Robins-Browne RM, Levine MM. 2013. Burden and aetiology of diarrhoeal disease in infants and young children in developing countries (the Global Enteric Multicenter Study, GEMS): a prospective, case-control study. *Lancet* **382:** 209–222.

30. Levine MM, Robins-Browne RM. 2012. Factors that explain excretion of enteric pathogens by persons without diarrhea. *Clin Infect Dis* **55**(Suppl 4):S303–S311.

31. Lindenbaum J, Kent TH, Sprinz H. 1966. Malabsorption and jejunitis in American Peace Corps volunteers in Pakistan. *Ann Intern Med* **65:**1201–1209.

32. Keusch GT, Denno DM, Black RE, Duggan C, Guerrant RL, Lavery JV, Nataro JP, Rosenberg IH, Ryan ET, Tarr PI, Ward H, Bhutta ZA, Coovadia H, Lima A, Ramakrishna B, Zaidi AK, Hay Burgess DC, Brewer T. 2014. Environmental enteric dysfunction: pathogenesis, diagnosis, and clinical consequences. *Clin Infect Dis* **59** (Suppl 4):S207–S212.

33. Sansonetti PJ, Arondel J, Huerre M, Harada A, Matsushima K. 1999. Interleukin-8 controls bacterial transepithelial translocation at the cost of epithelial destruction in experimental shigellosis. *Infect Immun* **67:**1471–1480.

34. Bell BP, Griffin PM, Lozano P, Christie DL, Kobayashi JM, Tarr PI. 1997. Predictors of hemolytic uremic syndrome in children during a large outbreak of *Escherichia coli* O157:H7 infections. *Pediatrics* **100:**E12.

35. Falush D, Wirth T, Linz B, Pritchard JK, Stephens M, Kidd M, Blaser MJ, Graham DY, Vacher S, Perez-Perez GI, Yamaoka Y, Megraud F, Otto K, Reichard U, Katzowitsch E, Wang X, Achtman M, Suerbaum S. 2003. Traces of human migrations in *Helicobacter pylori* populations. *Science* **299:**1582–1585.

36. Chen Y, Blaser MJ. 2008. *Helicobacter pylori* colonization is inversely associated with childhood asthma. *J Infect Dis* **198:**553–560.

37. Reibman J, Marmor M, Filner J, Fernandez-Beros ME, Rogers L, Perez-Perez GI, Blaser MJ. 2008. Asthma is inversely associated with *Helicobacter pylori* status in an urban population. *PLoS ONE* **3:**e4060.

38. **Chavali AK, Whittemore JD, Eddy JA, Williams KT, Papin JA.** 2008. Systems analysis of metabolism in the pathogenic trypanosomatid *Leishmania major. Mol Syst Biol* **4:**177.

39. **Chavali AK, Blazier AS, Tlaxca JL, Jensen PA, Pearson RD, Papin JA.** 2012. Metabolic network analysis predicts efficacy of FDA-approved drugs targeting the causative agent of a neglected tropical disease. *BMC Syst Biol* **6:**27.

40. **D'Auria KM, Kolling GL, Donato GM, Warren CA, Gray MC, Hewlett EL, Papin JA.** 2013. In vivo physiological and transcriptional profiling reveals host responses to *Clostridium difficile* toxin A and toxin B. *Infect Immun* **81:**3814–3824.

Metabolic Adaptations of Intracellullar Bacterial Pathogens and their Mammalian Host Cells during Infection ("Pathometabolism")

3

WOLFGANG EISENREICH,[1] JÜRGEN HEESEMANN,[2] THOMAS RUDEL,[3] and WERNER GOEBEL[2]

Metabolic adaptation reactions are common when prokaryotes interact with eukaryotic cells, especially when the bacteria are internalized by these host cells. Such adaptations lead to significant changes in the metabolism of both partners. While the final outcome may be sometimes beneficial (e.g., in case of insect endosymbiosis) or (mainly) neutral for the interacting partners (e.g., microbiota and their hosts) (1–3), it is usually detrimental in infections of mammalian cells by intracellular bacterial pathogens. In this encounter, a host cell-defense program is initiated, including antimicrobial metabolic reactions aimed to damage the invading pathogen and/or to withdraw essential nutrients, while the intracellular pathogen tries to deprive nutrients from the host cell and to counteract the antimicrobial reactions, resulting in damaging of the host cell. Our knowledge of the metabolic adaptation processes occurring during this liaison and the link between these metabolic changes and the pathogenicity is still rather fragmentary. For these complex metabolic interactions, we coin the term "pathometabolism". Studies of pathometabolism are not only central for a deeper understanding of bacterial infections caused by intracellular

[1]Lehrstuhl für Biochemie, Technische Universität München, Germany; [2]Max von Pettenkofer-Institut für Hygiene und Medizinische Mikrobiologie, Ludwig-Maximilians-Universität München, Germany; [3]Lehrstuhl für Mikrobiologie, Biozentrum Universität Würzburg, Germany.
Metabolism and Bacterial Pathogenesis
Edited by Tyrrell Conway and Paul Cohen
© 2015 American Society for Microbiology, Washington, DC
doi:10.1128/microbiolspec.MBP-0002-2014

bacterial pathogens, but may also provide promising bacterial and host cell targets for the development of novel antimicrobial therapeutic measures.

GENERAL CONSIDERATIONS

The Intracellular Bacterial Pathogens

Intracellular bacteria that may cause severe infections in humans (which are exclusively discussed here), are characterized by their ability to actively invade human and other mammalian cells (and eventually also cells of lower eukaryotic organisms), efficiently replicate intracellularly, and finally exit the infected cells and reinfect new host cells. "Facultative intracellular" bacteria may grow intracellularly within suitable host cells or extracellularly in various natural and artificial environments. The so-called "obligate intracellular" bacteria are thought to exclusively replicate within appropriate host cells. However, recent investigations show that even these bacterial pathogens may thrive in axenic media (4–5). Within the infected host cells intracellular bacteria may replicate in specifically modified pathogen-containing vacuoles (PCV) (6). Typical representatives of this group belong to the genera *Salmonella, Mycobacteria, Brucella, Legionella, Coxiella, and Chlamydia*. The other group of intracellular bacterial pathogens, including members of the genera *Shigella, Listeria, Rickettsia* and *Francisella*, escapes into the host cell's cytosol where these bacteria actively replicate (6).

All intracellular bacteria must import basic nutrients from the host cell, including one or more suitable carbon, nitrogen, and sulphur sources and metal ions. Some intracellular bacteria require, in addition, several amino acids, nucleotides, vitamins, fatty acids, and even adenosine triphosphate (ATP) and nicotinamide adenine dinucleotide (NAD). The cytosolically replicating intracellular bacteria obviously have direct access to these host nutrients that are either taken up or produced by the host cells. Pathogens residing in PCVs, on the other hand, must transport cytosolic host nutrients across the vacuolar membrane. Depending on the pathogen, this membrane may obtain nutrient transporters by fusion with endocytic vesicles, including lysosomes, (macro)pinosomes, or exocytic vesicles, that contain enzymes, transporters, and/or nutrients (7–13).

Although intracellular bacteria seem to be optimally adapted to a specific intracellular host cell niche, bacteria normally replicating in PCV, such as *Mycobacterium tuberculosis* or *Salmonella enterica* serovar Typhimurium (14–17), can enter the cytosol under certain conditions and replicate in this host cell compartment. This compartment switching may even represent an important step in infections, e.g., as part of the evasion process (16, 18). *Vice versa*, cytosolically replicating bacteria, such as *Listeria monocytogenes*, seem to be able to also replicate in specific vacuoles under special conditions (19). These events show that at least some intracellular bacteria are able to obtain suitable nutrients that support their active intracellular growth in different host cell niches.

For optimal adaption to the (often varying) metabolic state of host cells, intracellular bacterial pathogens have acquired, in some cases, additional metabolic genes from other bacteria through horizontal transfer (20–22) or have deleted metabolic genes (present in their non-pathogenic relatives) that may be harmful for intracellular replication (23–26). But in general, the metabolic capacity of intracellular bacterial pathogens differs extremely, ranging from the presence of all basic catabolic and anabolic pathways (e.g., *Salmonella* serovars, *Shigella* spp.) to the extensive loss of catabolic reactions and most biosynthetic pathways (e.g., *Rickettsia* spp., *Chlamydia* spp.). This high metabolic variability among the intracellular bacterial pathogens (Figs. 1 and 2)

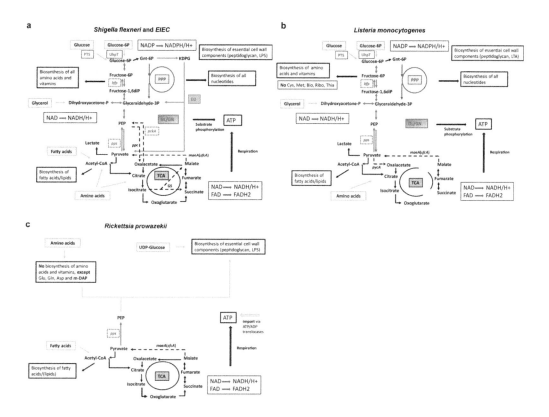

FIGURE 1 Metabolic potential of intracellular bacterial pathogens replicating in the cytosol of host cells. (a) *Shigella flexneri,* (b) *Listeria monocytogenes,* and (c) *Rickettsia prowazekii.* Solid arrows indicate reversible (double-headed arrows) and essentially irreversible (single-headed arrows) reactions of glycolysis and gluconeogenesis (GL/GN, red arrows), pentose phosphate pathway (PPP, blue arrows), Entner-Doudoroff pathway (ED, orange arrows), the tricarboxylic acid cycle (TCA). The irreversible reactions involved in GN are marked by dotted-framed boxes: fbp for phosphofructo-1,6-bisphosphatase, pps for PEP synthase, and pckA for PEP carboxykinase. Broken black arrows depict anaplerotic reactions: GS for glyoxylate shunt, pycA for pyruvate carboxylase, ppc for PEP carboxylase; pckA for PEP carboxykinase, and maeA (sfcA) for malate enzyme. Yellow boxes and arrows mark major carbon and energy substrates and black-framed boxes mark the biosynthesis of amino acids, vitamins, nucleotides, cell envelope components, and fatty acids/lipids as well as the major sites of ATP production. The major sites for the generation of NADH/H$^+$, NADPH/H$^+$ and FADH$_2$ are also shown. doi:10.1128/microbiolspec.MBP-0002-2014.f1

already suggests that their nutritional requirements and metabolic adaptation strategies to the host cells must also differ considerably. Mammalian cells obviously offer appropriate growth conditions in the cytosol and in vacuolar compartments for these metabolically highly diverse intracellular bacteria.

There seems to be a close coordination between the metabolism of the intracellular bacterial pathogens and the expression of virulence genes that encode the factors required for the various steps of the intracellular bacterial life cycle (i.e., invasion, proliferation in specific host cell niches, evasion and reinfection of new host cells, or by cell-to-cell spread) (27–29).

The metabolic activity of intracellular bacteria is also clearly dependent on the host cell metabolism, which in turn is subject to significant changes triggered during infection by the intracellular pathogen, e.g., by

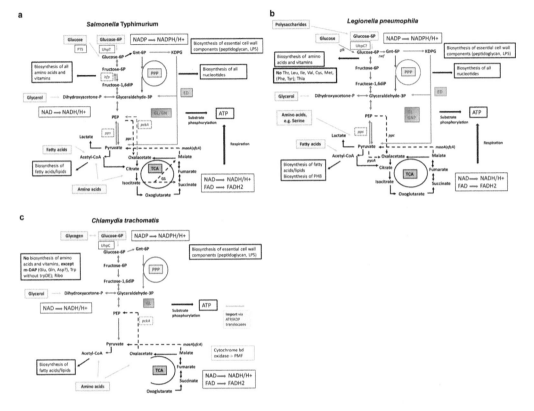

FIGURE 2 Metabolic potential of intracellular bacterial pathogens replicating in specialized vacuoles of the host cells. (a) *Salmonella enterica*, **(b)** *Legionella pneumophila*, **and (c)** *Chlamydia trachomatis*. See legend of Fig 1 for further explanations and abbreviations. doi:10.1128/microbiolspec.MBP-0002-2014.f2

the secretion of specific virulence factors and/or effector proteins (30) (see below).

Of course, the adequate supply of nutrients does not suffice to make the host cell an appropriate growth niche for an intracellular pathogen. Common and pathogen-specific metabolic reactions are also necessary to counteract the compartment-specific host cell defense mechanisms (see below) that otherwise would restrict the intracellular bacterial pathogen's ability to thrive (31–32). Common regulatory devices exist in mammalian cells that control both metabolic pathways and defense mechanisms (33). Their inactivation by the intracellular pathogens might coordinate acquisition of nutrients from the host cell and resistance to host defense mechanisms (see below).

Thus, common and indispensible properties of all intracellular bacteria that successfully proliferate within mammalian cells are the ability to take up and utilise appropriate host cell nutrients and to withstand the antimicrobial actions of the host cells. This ability is certainly not a common feature of all bacteria that eventually enter mammalian cells (34), but seems to be the evolutionary result of a fine-tuned interconnection between the host-cell-adapted intracellular metabolism of the pathogen, the metabolism-dependent expression of virulence genes and the infection-adapted metabolism of the host cell.

Although we are still far away from the understanding of this complex interconnection, some progress has been made (especially with some model pathogens) by the introduction of new techniques, such as comparative transcriptome, proteome and metabolome analyses, ^{13}C-isotopologue profiling, and determination of metabolic reactions and fluxes using nuclear magnetic resonance (NMR), mass spectroscopy (MS), secondary-ion mass spectroscopy, and Raman microspectrometry; for recent reviews, see (27–28, 30, 35–38).

Mammalian Cells Acting as Hosts for Intracellular Bacterial Pathogens

Phagocytes, especially dendritic cells (DC) and macrophages (MΦ), are often the first host cells encountered by intracellular bacterial pathogens during infection. These highly heterogeneous hematopoietic cells, found in nearly every tissue of the body (39), are characterized by a particularly high antimicrobial capacity, including production of reactive oxygen species (ROS) and reactive nitrogen intermediates (RNI). Synthesis of these antimicrobial agents is induced via the classical activation pathway (40–41) by microbial surface components (e.g., lipopolysaccharides [LPS] and other pathogen-associated molecular pattern [PAMP] molecules). These host cells also possess constitutive mechanisms to derive intracellular pathogens of required nutrients (42), including expression of the divalent cation transporter Nramp1, leading to iron depletion (43) and interferon (IFN)-γ-induced activation of indoleamine 2,3-dioxygenase which degrades tryptophan, leading to tryptophan starvation (44). As already stated above, successful intracellular pathogens must overcome (at least partially) these antimicrobial host cell activities.

The infected DC and MΦ may carry the intracellular bacteria to peripheral organs where they may invade new host cells, including epithelial cells, endothelial cells, hepatocytes, etc., either by cell-to-cell spreading (45) or by release from the carrier cells and re-infection of new host cells (16, 18). The mechanisms by which

intracellular bacteria, especially vacuolar pathogens, are released from primary infected cells are subject of extensive current research (16, 46–47). Active replication of intracellular bacteria represents a heavy metabolic burden for the infected cell, to which the host cell responds with general and specific metabolic changes. Compared to the progress concerning the metabolic adaptations of the intracellular pathogens, the knowledge of the metabolic host cell responses triggered by intracellular pathogens is still rather sketchy (30). There are, indeed, several obstacles that hamper the investigation of the metabolic host cell responses. A major problem is the nature of the model host cells, including mammalian cancer cell lines, protists, or invertebrate cells, which are typically applied to *in vitro* infection studies. Compared to the slowly proliferating primary host cells invaded by the bacterial pathogens under *in vivo* conditions (e.g., MΦ, epithelial cells, endothelial cells, etc.), the *in vitro*-used cancer cells have a highly altered carbon metabolism (48–49), which may lead to enhanced glucose and/or glutamine uptake, induced (aerobic) glycolysis and/or glutaminolysis, reduced tricarboxylic acid (TCA) cycle, reduced aerobic respiration, high lactate production, and increased anabolic activities (e.g., biosynthesis of lipids, proteins, and nucleotides). The reason for this metabolic dysregulation in the most frequently used mammalian cell lines are mutations in tumour suppressors (e.g., p53) or oncogenes (e.g., c-Myc, HIF-1) that function also as regulators of the metabolism in mammalian cells (see below) as well as mutations in different enzymes associated with the tricarboxylic acid (TCA) cycle (50–52). Intracellular bacteria will therefore encounter in these host cells completely different energy and nutrient conditions than in the target cells that they invade during infection. Possible activation of metabolic host cell reactions triggered by the intracellular bacteria may therefore not be detected in such model host cells, unless the metabolic changes are extremely pronounced. Protist and invertebrate infection models must also be critically assessed with respect to metabolic host responses since these cells often show different nutritional conditions and metabolic control mechanisms and, therefore, such metabolic data do not necessarily resemble the metabolic responses triggered by intracellular pathogens in mammalian cells.

Some Basic Aspects of the Mammalian Cell Metabolism

The metabolism of mammalian cells occurs in different compartments. The major catabolic reactions, essential for bioenergetic and biosynthetic processes, take place in the cytosol (glycolysis, oxidative pentose-phosphate pathway) and in mitochondria (tricarboxylic acid [TCA] cycle, glutaminolysis, β-oxidation). The inner mitochondrial membrane is also the site of the electron-transfer chain leading to oxidative phosphorylation. The anabolic reactions (gluconeogenesis, biosyntheses of amino acids, nucleotides, and fatty acids/lipids) occur mainly in the cytosol. There is an extensive exchange of certain metabolites between these two compartments followed by metabolic reactions in the cytosol, e.g., transport of citrate from mitochondria into the cytosol and conversion to oxaloacetate and acetyl-CoA by the cytosolic ATP-dependent citrate lyase, transport of malate and conversion to pyruvate and CO_2 by the cytosolic malic enzyme (Fig. 3a). Common cellular regulators, including tumor suppressors and oncogenes, are involved in the orchestration of the entire cell metabolism (53–54). A simplified overview on these metabolic processes and their regulation in mammalian cells is given in Figs. 3b and 3c.

In addition, several other membrane-surrounded vesicles are involved in the metabolic processes. The peroxisomes are metabolically linked to mitochondria as both organelles are involved in β-oxidation of fatty acids and implicated in the production of ROS. However, peroxisomes are not associated with energy production, but rather involved in detoxification processes. Other vesicles contain digestive enzymes (lysosomes, autophagosomes) or store enzymes, transporters, and other biomolecules for secretion to the cell's exterior (secretory vesicles).

Selected Intracellular Bacterial Pathogens

As already mentioned above, intracellular bacterial pathogens may differ extremely with respect to their metabolic properties. The pathogens that we discuss in the following belong to either the "vacuolar" or the "cytosolic" group. For each group, we selected three model bacteria with very different metabolic capacities (ranging from complete prototrophy to middle and high auxotrophy). The cytosolic group includes *Shigella flexneri* (*S. flexneri*) and the closely related enteroinvasive *E. coli* (EIEC), *Listeria monocytogenes* (*L. monocytogenes*), and *Rickettsia prowazekii* (*R. prowazekii*). The phagosomal group includes *Salmonella enterica* serovar Typhimurium (*S. typhimurium*) replicating in the *Salmonella*-containing vacuole (SCV), *Legionella pneumophila* (*L. pneumophila*) replicating in the *Legionella*-containing vacuole (LCV), and *Chlamydia trachomatis* (*C. trachomatis*) replicating in the so-called "inclusion".

Many relevant *in vitro* infection studies with the selected intracellular bacteria use established mammalian cell lines as host cells, and hence the following discussion is based to a considerable extent on results obtained with such cells. As outlined above, we are fully aware of the problems associated with the use of cancer cells when studying the intracellular metabolism of human bacterial pathogens (30) and especially the metabolic host responses to such infections. But we will emphasize the limits of these model systems and–wherever possible–compare the *in vitro* results with

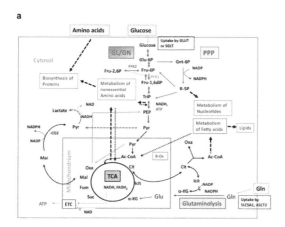

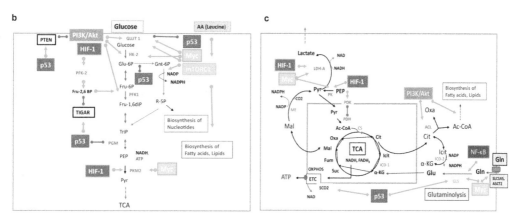

FIGURE 3 **Carbon metabolism and its regulation in mammalian cells. (a) Major carbon sources and their transporters are indicated by yellow boxes and arrows. Main catabolic and anabolic pathways, including glycolysis (GL, red solid arrows) with lactose formation (red broken arrows), pentose-phosphate pathway (PPP, blue arrows) in the cytosol, TCA cycle in the mitochondria and associated cytosolic reactions (black arrows), and glutaminolysis (purple arrows); catabolic breakdown of amino acids and fatty acids are indicated by dotted arrows. The reactions indicated by the green arrows are two anaplerotic reactions (catalysed by PCK and PYC). The starting points for the biosynthesis of the "non-essential" amino acids, nucleotides and fatty acids/lipids are schematically indicated by broken arrows. (b) Regulation of glucose uptake, glycolysis and pentose-phosphate pathway and (c) of the TCA cycle, glutaminolysis, aerobic respiration, and lactate production by general transcription factors, tumor suppressors, and oncogenes. Activation of the target enzymes (yellow boxes) are shown by green pointed arrows and inhibition by the red symbol. Explanations and abbreviations: Fru-2,6P: fructose 2,6-bisphosphate, the formation of which is catalysed by the fructokinase 2 (PFK2); Fru-2,6P activates fructokinase 1(PFK1); ACL: cytosolic ATP-dependent citrate lyase; ICD-2: cytosolic NADP-dependent isocitrate dehydrogenase; ME: cytosolic malate dehydrogenase (malic enzyme); GLS: glutaminase; HIF-1: hypoxia-inducible factor1; p53: tumor suppressor protein 53 encoded by the gene TP53; TIGAR: TP53-inducible glycolysis and apoptosis regulator; PTEN: phosphatase and tensin homolog; mTORC1: mammalian target of rapamycin complex 1; PI3K/Akt: phosphoinositide-dependent kinase-1/protein kinase B. For further details regarding the complex regulation circuit, see (30, 54) and further references cited there. doi:10.1128/microbiolspec.MBP-0002-2014.f3**

corresponding *in vivo* data, obtained in appropriate animal models.

Intracellular Bacterial Pathogens Replicating in the Cytosol

S. flexneri and EIEC

Metabolic adaptations of intracellular *S. flexneri* and EIEC

These two highly related enterobacteriaceae are typical food-borne intracellular pathogens causing a severe diarrheal disease in humans that is characterized by the invasion and destruction of the colonic mucosa (55). After invasion of intestinal epithelial cells, these bacteria replicate in the cytosol of the host cells (56). They produce numerous effector proteins, secreted by a type III secretion system (T3SS), that are important virulence factors essential for invasion, intracellular replication, and cell-to-cell spread (57).

S. flexneri and EIEC are facultative anaerobic, heterotrophic and prototrophic bacteria, equipped with all central catabolic pathways (glycolysis [GL], Entner-Doudoroff [ED] pathway, the pentose phosphate pathway [PPP], the tricarboxylic acid [TCA] cycle), and the main anaplerotic reactions, including the glyoxylate shunt (GS), the malic enzyme (encoded by *maeA*), phosphoenol-pyruvate (PEP) carboxy kinase (*pckA*), and PEP carboxylase (*ppc*). The biosynthetic pathways for all anabolic monomers (amino acids, vitamins and other co-factors, nucleotides, fatty acids, lipids) are functional (Fig. 1a). Accordingly, both pathogens are able to grow *in vitro* in minimal media supplemented with a single suitable carbon source under aerobic and anaerobic conditions. Glucose is a preferred carbon source for growth, taken up by *Shigella*/EIEC mainly by two PEP-dependent phosphotransferase systems (PTS); PtsG/Crr and ManXYZ (58). Glucose 6-phosphate (glucose-6P) is transported by the UhpT transporter (59). In both enterobacteria, the *uhpT* gene is under the control of a complex

two-component system (*uhpABC*) and carbon-catabolite repression (60). Transcription of *uhpT* is therefore very low in the presence of glucose. Both pathogens can catabolize various C_2-, C_3-, C_4-, and C_5-substrates and, hence, fatty acids, glycerol, pyruvate, lactate, and C_4-dicarboxylates may be used as carbon substrates under *in vitro* culture conditions.

Early information on the intracellular carbon metabolism of *Shigella* derives from (established) cell culture studies using differential gene-expression profiling, as well as from animal infection experiments applying mutants defective in specific catabolic or anabolic reactions (61–64). These studies suggested that as *S. flexneri* replicates in the cytosol, expression of the genes encoding the glucose transporters (*ptsG/crr* and *manXYZ*) were down-regulated, while those of *uhpT*-, *glpF*-, and *ugp*-encoding transporters for glucose-6P, glycerol, and glycerol 3-phosphate, respectively, were up-regulated. These genes are, however, subject to catabolite repression control and their up-regulation under intracellular conditions could simply reflect the lower glucose concentration within the host cells compared to the reference culture medium used. Indeed, a *uhpT* mutant showed wild-type replication in Henle cells (61). The glycolysis genes were down-regulated, while those for gluconeogenesis (*fbp* and *pps*) were up-regulated, suggesting gluconeogenic substrate(s) as carbon source for cytosolically replicating *Shigella*.

However, ^{13}C-profiling studies using external [U-^{13}C$_6$]glucose as major carbon source for Caco-2 cells infected with two different EIEC wild-type strains and mutants impaired in transport of glucose and glucose-6P transport strains, showed that glucose, but not glucose-6P, is the preferred carbon source for intracellular EIEC in these host cells (65). The ^{13}C-isotopologue profiles of amino acids deriving from intermediates of the glycolytic and TCA pathways suggested that glucose is catabolized through GL, PPP, and the TCA cycle in intracellularly replicating EIEC. However, in the absence of glucose, alternative carbon compounds, most likely C_3-compounds such as glycerol and/or glycerol-3P, were used instead (65). The exclusive consumption of these carbon sources requires gluconeogenesis.

A link between control of carbon metabolism and virulence gene expression by the carbon consumption regulators CsrA and Cra has been demonstrated (66). CsrA acted as posttranscriptional activator of GL and repressor of gluconeogenesis (67), whereas the transcriptional regulator Cra had the opposite effect (68). A *csrA* mutant showed decreased and a *cra* mutant increased invasion of *Shigella* in Henle cells, while both *csrA* and *cra* mutations seemed to inhibit the subsequent intracellular growth of *Shigella* (66). These data suggest a positive role of GL in invasion, possibly by affecting activity and expression, respectively, of the virulence gene regulators VirF and VirB through product(s) of these metabolic pathways (66). The inhibitory effect of *csrA* and *cra* mutations on intracellular growth suggests that both GL and gluconeogenesis are necessary and hence different carbon substrates may be used by *Shigella* in course of the intracellular replication cycle.

The choice of the carbon source seems to be also important for the capacity of *de novo* synthesis of anabolic monomers, such as amino acids and nucleotides, under intracellular conditions. An EIEC wild-type strain growing in Caco-2 epithelial cells and supplied with external [U-$^{13}C_6$]glucose as major carbon source (65) incorporated ^{13}C-label into all amino acids (albeit to a different extent). The *in vitro* and *in vivo* observed virulence attenuation of *aroB-D* (and also of *guaBA* and *thyA*) mutants of *S. flexneri*, also suggested that biosynthesis of aromatic amino acids (but also of nucleotides) is essential for full virulence of *Shigella* (64, 69–70). In contrast, a glucose-uptake mutant which–as stated above– consumes C_3-substrates, grew considerably slower than the wild-type strain in Caco-2 cells and incorporated ^{13}C-label only into Ala, Asp, Glu, Ser, Gly and Val. Apparently, all other amino acids were taken from the host under these conditions. Notably, the *de novo* synthesized amino acids (in contrast to the imported amino acids) required only one single step in their biosynthesis using a catabolic intermediate from the glycolytic or the TCA pathway, respectively, without ATP consumption.

Metabolic host cell responses triggered by intracellular *S. flexneri*

Shigella infection of established epithelial cells (HeLa and HCT116 cell lines) promotes p53 degradation by the calpain protease, which is activated by the virulence effector protein VirA (71). The effect of this process on necrosis and apoptosis of the *Shigella*-infected cells was extensively analyzed, but not its probable impact on host cell metabolism. However, the expected increase in glucose uptake and glycolysis due to the loss of p53 may not be seen (as mentioned above) in the cancer cell lines used in this study, because of the already enhanced glucose metabolism of these cells (30). Induction of glucose uptake in *Shigella*-infected primary intestinal epithelial cells due to loss of p53 would certainly favor the above described metabolism of intracellular *Shigella*.

Because of the narrow range of hosts that are susceptible to *Shigella* infection, robust data on *in vivo* metabolism of *Shigella* do not exist. A recent study that analyzed the proteome of *Shigella dysenteriae* isolated from the large bowel of infected gnotobiotic piglets came to the conclusion that the bacteria switch to an anaerobic metabolism accompanied by a strongly reduced TCA cycle (72). However, it remained unclear in this study whether the majority of the bacteria analyzed replicated within epithelial cells or extracellularly in the bowel.

Listeria monocytogenes

Metabolic adaptations of intracellular *L. monocytogenes*

L. monocytogenes is a Gram-positive bacterial pathogen that survives and replicates, like all other *Listeria* species, as a saprophyte in natural environments (73). But *L. monocytogenes* is also capable of causing serious systemic infections in susceptible individuals with the help of a number of virulence factors that are absent in the non-pathogenic *Listeria* species; for recent reviews see (74–75). All listerial species may take up and utilize amino acids, C_3-, C_4-, and C_5-carbon substrates, as well as many different carbohydrates (including glucose) via a large set of PEP-dependent transferase systems (PTS) (76). In contrast to the non-pathogenic species, *L. monocytogenes* can also import phosphorylated hexoses by the phosphate antiporter UhpT (Hpt). The gene for this transporter (*uhpT*) is under the control of PrfA, the major transcriptional regulator of most listerial virulence genes (77). Glucose and glucose-6P can be degraded via the intact glycolytic and pentose-phosphate pathway. The citrate cycle is, however, disrupted, lacking functional oxoglutarate dehydrogenase and malate dehydrogenase (78). Generation of oxaloacetate seems to occur exclusively by the ATP-dependent pyruvate carboxylase (PycA) (79). Asp is not taken up by *L. monocytogenes* and hence can not fill-in oxaloacetate (79). Gluconeogenesis starting from C_3- and C_4-components, but not from acetate (acetyl-CoA) is possible since PEP-synthetase (but not PEP carboxykinase) and phosphofructo-1,6-bisphosphatase (BFP) are present, and malic enzyme can convert malate to pyruvate (Fig. 1b). Indeed, *L. monocytogenes* is able to grow in defined minimal media (80–82) with a variety of carbohydrates and glycerol and, to a lesser extent, also with pyruvate and malate as sole carbon sources. In contrast, efficient growth with one or more amino acids as sole carbon source does not occur (83). Even with the above-mentioned growth-supporting carbon sources, proliferation of *L. monocytogenes* occurs only when a defined culture medium is supplemented with the three branched-chain amino acids (BCAA); Leu, Ile, and Val, the sulfur-containing amino acids, Met and Cys, and Arg together with the vitamins riboflavin, biotin, thiamine, and thioctic acid (80). The requirement of the latter vitamins is due to the lack of necessary genes for their biosynthesis. The need for the amino acids is more unexpected, as all genes encoding the enzymes for the biosynthesis of these amino acids are present (84) and expressed under appropriate conditions (85). We assume that shortage of essential biosynthetic intermediates might be the reason for the dependency on the BCAAs, Arg, Cys, and Met, i.e., limitation of pyruvate (necessary for the generation of oxaloacetate and acetyl-CoA, which are important for maintaining the truncated TCA cycle) and succinyl-CoA might be responsible for the failure of BCAA and Arg biosynthesis, respectively, while the missing sulfur-reductase leads to the lack of the necessary sulphide necessary for *de novo* Cys (and hence Met) biosynthesis.

There is an obvious link between the listerial metabolism and the expression of the PrfA-dependent virulence genes (27, 29, 86). Activity of PrfA, the major transcriptional regulator of the listerial virulence genes (87), is low in rich culture media or in presence of carbohydrates (especially glucose and cellobiose) in defined minimal media as major carbon and energy sources, but high in presence of glycerol (82, 88). The precise activation mechanism of PrfA is still poorly understood (89). In addition, the isoleucine response regulator CodY of Gram-positive bacteria (90) was shown to control *prfA* gene expression, thus linking synthesis of PrfA (and hence of virulence gene expression) to metabolic conditions, especially to the cellular BCAA concentration (86).

Until now, the intracellular metabolism of *L. monocytogenes* was mainly studied in established epithelial and macrophage cell lines and in primary murine bone marrow-derived macrophages (BMM) that were infected with *L. monocytogenes* wild-type strains or mutants defective in the transport or catabolism of different carbon sources (36, 85–86, 91–94). The conclusions drawn from these studies concerning the intracellular metabolism of *L. monocytogenes* are based on transcriptome, proteome, and ^{13}C-isotopologue analyses. Some interesting data concerning the listerial intracellular metabolism derived from recent *in vivo* studies (using a mouse-infection model), although the focus of these studies was primarily on other aspects of *L. monocytogenes* pathogenicity (95–96).

Together, these studies showed that the intracellular metabolism of *L. monocytogenes* relies on different carbon and nitrogen sources, which the host cell's cytosol provides in mostly limited and, apparently, host cell-specific concentrations (85, 91, 94). Host-derived glucose 6-phosphate (G6P) and glycerol seem to be preferential carbon sources when the listeriae grew intracellularly in epithelial cells and macrophages, as shown (a) by the ^{13}C-isotopologue distribution in certain amino acids in the presence of uniformly labeled ^{13}C-glucose (93), (b) by the upregulation of the genes encoding G6P transport (i.e., *uhpT*), as well as uptake and catabolism of glycerol (i.e., *glpF, glpK,* and *glpD*) (85–86, 91, 95–96), and (c) by the attenuation of intracellular replication of mutants defective in uptake or catabolism of these carbon substrates (36) (T. Fuchs, personal communication).

However, a mutant unable to grow with glycerol and G6P was still able to replicate–albeit at a reduced growth rate–in these epithelial and macrophage cells, and the ^{13}C-isotopologue analysis of the labeled amino acids suggests that either glucose or host cell-derived glucose catabolic intermediates (other than glycerol) can be used

by the intracellular listeriae under these conditions (T. Fuchs, personal communication). There is circumstantial evidence (deriving mainly from transcriptome and proteome data) that the oxidative part of the PPP rather than glycolysis is preferentially used for degradation of G6P, but also the non-oxidative part of PPP appears to be up-regulated. Expression of the glycolytic genes seems to be initially reduced but is resumed at a later stage in the intracellular replication cycle (85, 91–92). Whether the increased expression of the genes for the glycolytic enzymes in this state is necessary for catabolism of carbohydrates or for gluconeogenesis (when glycerol is used) remains an open question. The residual part of the disrupted TCA cycle (from oxaloacetate to oxoglutarate) is active as shown by the ^{13}C-isotopologue profiling data (93) and provides the intracellular *L. monocytogenes* probably not only with oxoglutarate (as intermediate for the synthesis of Glu and Gln) but also with reduced nicotinamide adenine dinucleotide (NADH) necessary for ATP production via aerobic respiration.

These data suggest that more than one carbon compound is used by *L. monocytogenes* when the bacteria grow in the cytosol of mammalian host cells. This finding raises the interesting question whether these carbon substrates are used successively depending on their available concentration within the host cell and/or their efficiency as carbon and energy source or whether the carbon substrates are co-metabolized at the same time, serving either as energy source (s) or as producers of intermediates for indispensible anabolic processes (see also discussion below).

As discussed above, *L. monocytogenes* has the genes encoding the biosynthetic pathways for all amino acids suggesting that *de novo* synthesis of all amino acids should be, in principle, possible. Yet, listerial amino acid and oligopeptide transporters were shown to be important for intracellular replication of *L. monocytogenes* (36, 97) and genes encoding these transporters were induced in intracellularly growing *L. monocytogenes* (85, 91). When a mixture of all ^{13}C-labeled amino acids (in presence of unlabeled glucose) was added to *L. monocytogenes*-infected epithelial cells, all labeled amino acids were taken up and incorporated into protein by the host cells and also–with the exception of Asp–by the intracellular listeriae. However, little turnover of these amino acids was observed, suggesting that the host amino acids were primarily used for protein biosynthesis and hardly catabolized by the intracellular listeriae (T. Fuchs, personal communication).

Ala, Asp, and Glu, the biosynthesis of which requires solely transamination of the central catabolic intermediates pyruvate, oxaloacetate, and oxoglutarate, respectively, were efficiently synthesized *de novo* by intracellular *L. monocytogenes*. At lower efficiency, *de novo* biosynthesis was also observed for Ser, Gly, Thr, and Val, which requires only a single catabolic intermediate (3-phosphoglycerate, oxaloacetate, and pyruvate, respectively) and a rather short biosynthetic pathway. All other amino acids, including His, Met, Leu, Ile, and the aromatic amino acids were not synthesized *de novo* under these conditions. This finding seems to contradict transcriptome data which indicated that transcription of the listerial genes encoding the enzymes for the biosynthesis of these latter amino acids is induced in intracellular *L. monocytogenes* (85–86, 91). Apparently, the biosynthetic pathways for these amino acids are derepressed on the transcriptional level at the low glucose level but either translation of the transcripts is inhibited or–more likely–essential substrates or critical co-factors (such as ATP and NAD) required for their biosynthesis are limited. This example shows again the need of more than one analytical method for obtaining reliable data on the intracellular metabolism of bacterial pathogens.

The few *in vivo* data on intracellular listerial metabolism, obtained mainly by transcriptome analysis of *L. monocytogenes* isolated from infected mice (95–96), were basically in line with the corresponding transcription data obtained by the *in vitro* cell culture experiments described above. This might be surprising in light of the dysregulated carbon metabolism of the cell lines used for the *in vitro* infection studies compared to the tightly regulated host cells which *L. monocytogenes* encounters during *in vivo* infection. However, a recent ^{13}C-profiling study which compared the metabolism of *L. monocytogenes* in primary bone marrow-derived macrophages (BMM) and in cells of the murine macrophage cell line J774A.1 (94) showed an almost identical intracellular listerial carbon metabolism in both host cell sytems.

Metabolic host cell responses triggered by intracellular *L. monocytogenes*

None of the genes involved in host cell metabolism was significantly altered in the human macrophage-like cell line THP-1 upon *L. monocytogenes* infection (98) which is likely due to the already enhanced carbon metabolism of the THP-1 cancer cells. Yet, a *L. monocytogenes* infection strongly changes the metabolism of primary host cells, especially when the bacteria are internalized (30, 94, 99). Comparative ^{13}C-isotopologue studies, using uniformly ^{13}C-labeled glucose or ^{13}C-labeled glutamine as carbon sources in infection assays of primary murine bone marrow-derived macrophages (BMM) and of murine macrophage-like J774A.1 cells (94), showed high induction of glucose uptake and of glycolysis in BMM upon infection, but not in the J774A.1 cells (which already exhibit dysregulated glucose transport and glycolysis). As already described above, the thereby enhanced glucose catabolism of the BMM apparently allows a similarly efficient proliferation of the intracellular listeriae in BMM as in J774A.1 cells. These *in vitro* data, obtained with primary host cells, are in accord with results deriving from *in vivo* studies which analyzed the transcript profiles of the intestinal epithelia of mice expressing humanized E-cadherin (99). These studies also revealed that transcription of most glycolysis genes is enhanced upon *L. monocytogenes* infection. Interestingly, the gene encoding host cell hexosekinase II (*hkII*) showed the highest increase (about 10-fold), which suggests that glucose taken up is mostly phosphorylated by this enzyme. As discussed above, the thereby generated glucose-6P is an important carbon substrate for intracellular *L. monocytogenes* in epithelial cells and macrophages (77, 93). These and other metabolic host responses (30, 99) appear to be triggered by the cytosolic bacteria since a non-invasive *inlA, inlB* double mutant of *L. monocytogenes* did not cause the upregulation of these metabolic genes. The listerial (virulence?) factors causing these changes of the host cell metabolism are still unknown.

Rickettsia prowazekii (R. prowazekii)

Metabolic adaptations of intracellular R. prowazekii

The genus *Rickettsia* comprises a variety of obligate intracellular pathogenic species (100). Among those, *R. prowazekii*, which is a member to the typhus group (TG) of *Rickettsia* species and transmitted from the human body louse to humans, belongs– with respect to metabolism–to the best-characterized *Rickettsia* species. This pathogen causes epidemic typhus (101) and multiplies in the cytosol of endothelial cells, the main host target cells (102). The pathogenic mechanisms, including rickettsial entry into metabolically active host cells, phagosomal escape, propelling of the bacteria by polymerized actin tails through the injected host cell, and into neighbouring cells were mainly studied in *R. conorii*, which belongs to the spotted fever group (SFG) of *Rickettsia* species. This type of intra- and intercellular motility resembles

that of *Shigella* and *L. monocytogenes* and was reviewed by Walker and Ismail (102). *R. prowazekii* does not form actin tails, suggesting differences in the intracellular life cycle. Identification of rickettsial virulence factors and their precise functions are complicated by the lack of an effective genetic manipulation system for rickettsiae.

This failure also hampers conclusions concerning the metabolism of *R. prowazekii* when replicating within the endothelial target cells. The annotation of its highly reduced genome, encoding only 835 proteins (103), suggests very limited catabolic and anabolic activities compared to those of *Shigella* and even of *L. monocytogenes*. The low intracellular growth rate of *R. prowazekii* (generation time about 10 h) (104) compared to growth rates of *Shigella* and *L. monocytogenes* (generation time about 1 h) seems to be consistent with this low metabolic capacity. Genes for the known glucose or glucose-6P-transporters, as well as for the enzymes of the glycolytic/gluconeogenic pathway and the pentose-phosphate pathway are missing, indicating that glucose or other carbohydrates cannot be used as carbon and energy sources and gluconeogenesis starting from gluconeogenic substrates is also not possible (Fig. 1c). *R. prowazekii* can take up, however, uridine 5'-diphosphoglucose (UDPG), which may serve as precursor for the biosynthesis of the sugar components required for the synthesis of the cell envelope components peptidoglycan and LPS (105). The rickettsial genome also carries the genes for the biosynthesis of meso-diaminopimelate (m-DAP) and for pyruvate phosphate dikinase required for PEP formation. Both, m-DAP and PEP are needed for peptidoglycan biosynthesis, and PEP in addition for the synthesis of 3-deoxy-D-mannooctulosonic acid (KDO) synthesis, a specific component of LPS. The rickettsial genome contains the genes encoding pyruvate dehydrogenase and all enzymes of the TCA cycle (Fig. 1c). The TCA enzymes are functional (106) suggesting that (at least in principle) C_2- (e.g., acetyl-CoA), C_3- (e.g. pyruvate), C_4- (e.g., malate or succinate), and C_5-catabolic intermediates (e.g., glutamate/ oxoglutarate), possibly imported from the host cell's cytosol, can be oxidized in the TCA cycle. The thereby produced NADH/H$^+$ and FADH$_2$ may be used as electron donors that are transferred to the existing carriers of the electron transfer chain for ATP production by aerobic respiration (106). *R. prowazekii* is also able to import ATP from the host via an ATP/ADP translocase (106). It is likely that this dual energy supply mechanisms reflect the adaptation of *R. rickettsia* to the metabolic activity of the host cell.

R. prowazekii has also lost most genes for the biosynthetic pathways of vitamins, (purine and pyrimidine) nucleotide mono-phosphates (but retained the genes encoding the enzymes for the conversion of the mononucleotides to all other ribo- and deoxyribonucleotides), and amino acids (besides m-DAP, only biosynthesis of Ala, Glu, and Asp is possible by transamination of the corresponding catabolic intermediates). Biosynthesis of fatty acids and (in part) of phospholipids can, however, function (107). The necessary sn-glycerol-3-phosphate which is not produced by *R. prowazekii* (due to the lack of the glycolytic and gluco-neogenic pathways) seems to be imported from the host cell, probably by several existing rickettsial triose phosphate transport systems (108). The existence of a large number of transporters for catabolic intermediates and anabolic monomers is indeed a hallmark of *R. rickettsiae* (105, 109–112). The hitherto poorly understood regulation of the rickettsial metabolite transporters in the course of the host cell infection is probably decisive for the balanced interaction of the host and the parasite since both partners compete, more or less, for the same nutrient pool. In this context, it has been proposed that the rickettsial transport systems might even be considered as metabolic virulence factors (108).

Metabolic host cell responses triggered by intracellular *R. prowazekii*

Rickettsiae trigger a variety of responses in endothelial cells, including induction of many cytokines and chemokines, enhanced production of ROS and RNI, and the induction of cyclo-oxygenase-2 (COX-2) leading to secretion of prostanoids; reviewed in references 100 and 113. Although most of these host responses are aimed to protect the host, some may favor the rickettsial survival and proliferation, e.g., NF-kB induction, which inhibits host cell death by apoptosis and thus maintains the intracellular niche for *R. prowazekii* (114). The rickettsial (virulence?) factors (including possible effectors secreted by the type IV secretion system (103)), responsible for the *R. prowazekii*-triggered host responses that may also affect the host cell metabolism, remain to be discovered.

Intracellular Bacterial Pathogens Replicating in Specialized Vacuoles

Salmonella enterica serovar Typhimurium (*S.* Typhimurium)

Metabolic adaptations of intracellular *S.* Typhimurium

The virulence mechanisms of *S.* Typhimurium have been extensively reviewed in the past and will not be further discussed here; for recent reviews, see references 115–119. *S.* Typhimurium has a very robust metabolic equipment (Fig. 2a) that allows these Gram-negative bacteria to grow (like *Shigella*) on many different glycolytic and gluconeogenic carbon sources and to produce all essential anabolic monomers (120). It is therefore not surprising that these bacteria are also able to proliferate within a variety of host cells (121) where they primarily proliferate in the Salmonella-containing vacuole (SCV). They are also able to efficiently replicate in the cytosol of epithelial cells (14, 17).

The intracellular metabolism of *S.* Typhimurium growing in cultured cells and model animals were studied using differential gene-expression profiling (DGEP), proteomics, and ^{13}C-isotopologue profiling (27, 122). The DGEP studies, which compared the transcripts of intracellular S. Typhimurium cultured in epithelial and macrophage cell lines (HeLa and J774, respectively) with those of *Salmonella* grown in rich LB medium, show up-regulation of the genes encoding enzymes for GL and ED and repression of most genes encoding the enzymes of the TCA cycle. Differential expression of these genes is more pronounced in HeLa than in J774 cells. From these studies, glucose, glucose-6P, and/or gluconate were suggested as probable carbon and energy source(s) for the intracellular metabolism of S. Typhimurium.

^{13}C-isotopologue profiling data showed that S. Typhimurium replicating in the SCV of Caco-2 cells (human intestinal cell line) uses glucose (but not glucose-6P) as preferential carbon substrate and incorporates ^{13}C-label from [U-^{13}C$_6$]glucose very efficiently into amino acids, suggesting an extensive *de novo* synthesis of these monomers when S. Typhimurium grows in the SCV. A mutant, impaired in glucose uptake, was, however, still able to replicate (albeit at a significantly reduced growth rate) in Caco-2 cells indicating that S. Typhimurium can switch in the absence of glucose to other nutrients as carbon and energy sources, most likely to C_3-substrates (65), but the use of C_2-substrate(s), deriving e.g., from fatty acids, could not be excluded by these studies.

Bowden and colleagues (123) analysed the replication of S. Typhimurium mutants, defective in glycolysis and glucose uptake, in J774 macrophages and also came to the conclusion that glucose represents the major carbon source and glycolysis is needed for efficient intracellular growth of S. Typhimurium in the SCV. The carbon metabolism of S. Typhimurium studied in systemically infected mice using S. Typhimurium wild-type and mutant strains (blocked in various steps of the primary carbon metabolism) seemed to depend on similar carbon substrates and

catabolic pathways as shown by the cell culture studies discussed above (124). The strong attenuation of mutants blocked in the biosynthesis of aromatic amino acids, including histidine and of purines (especially adenine), as well as the moderate attenuation of pyrimidine- and methionine-mutants observed in the infected mice, demonstrated the dependency of *in vivo*-growing S. Typhimurium on the biosynthesis of these anabolic monomers. The *in vitro* data supported this assumption: the DGEP studies also showed enhanced expression of the genes encoding enzymes for these anabolic pathways and, as already mentioned above, the ^{13}C-isotopologue profiling results (65) demonstrated significant *de novo* synthesis of amino acids in S. Typhimurium growing in Caco-2 cells. However, it should be emphasized that these *in vitro* data were obtained with bacteria actively growing in the SCV of their host cells and hence do not exclude the possibility that at later time points (when glucose becomes limited for the host cells), the expected change in the host cell physiology may lead to significant changes in the metabolism of the intracellular bacteria, possibly associated with the release of S. Typhimurium from the SCV into the cytosol and the later egress from the infected host cell (16, 18, 38).

Metabolic host cell responses triggered by intracellular S. Typhimurium

It is likely that effector proteins translocated by the TSS3 systems of S. Typhimurium–for a recent review, see (125)–act as modifiers of the host cell metabolism to support intracellular replication. As shown by Knodler and collegues (126), S. Typhimurium residing in the SCV translocates SopB which activates Akt, a central regulator of host cell carbon metabolism (127). This activation occurs within minutes after invasion and lasts for several hours. Among others, activated Akt phosphorylates Mdm2, a key regulator of p53 stability. Activated Mdm2 ubiquitinates p53 and thus promotes p53 degradation, which may in turn stimulate glucose flux via glycolysis. The Akt activation also leads to inhibition of apoptosis of the infected host cells (126), which is essential for *Salmonella* infection of epithelial cells and macrophages (128). AvrA, another *Salmonella* T3SS-translocated effector protein (129), might also be responsible, either directly or indirectly, for some of the specific *Salmonella*-induced metabolic host responses. As shown (130), AvrA leads to reduction of oxidative phosphorylation and to activation of mTOR, NF-κB, and p53 in intestinal *in vivo* infection. The activation of p53 is caused by AvrA-mediated p53 acetylation (131). While activated p53 will inhibit glycolysis, activated mTOR may stimulate amino acids uptake and catabolism. Thus, the AvrA-mediated activation of these regulators may again modify the carbon metabolism of *Salmonella*-infected host cells.

In a recent study, Lopez and colleagues showed that *S.* Typhimurium strains expressing the type III effector protein SopE (encoded by a lysogenic phage present in several epidemic strains) triggers the synthesis of iNOS in the host cells (22). The thereby produced nitric oxide is converted to nitrate in the host cell which acts as an efficient electron acceptor in anaerobic nitrate respiration. Under these respiratory conditions, *S.* Typhimurium is able to use the non-fermentable carbon source ethanolamine (132), a nutrient deriving from phosphatidylethanolamine that is present in the intestine (133). This metabolic capability generates a growth advantage for S. Typhimurium over the inherent anaerobic intestinal microbiota that is unable to respire ethanolamine.

Salmonella infection of murine macrophage-like cell lines (RAW264.7 and J774A.1) induced the expression of COX-2 (134–135). This induction depends on the *Salmonella* protein SpiC, a gene product encoded within SPI-2 and translocated into host cells by T3SS-2 (135). COX-2 encodes cyclooxygenase 2, a key enzyme involved in prostanoid synthesis, which leads to an increased production of PGE2 and PGI2, two prostanoids belonging to the eicosanoid hormones.

Legionella pneumophila (L. pneumophila)

Metabolic adaptations of the intracellular *L. pneumophila*

L. pneumophila, a Gram-negative bacterium, is the causative agent of Legionnaires' disease, a severe pulmonary disease (136). The natural hosts of this microorganism are apparently amoebae (137). By accident, *L. pneumophila* may also infect humans where it resides in alveolar macrophages. In these cells (and also in amoebae) the bacteria efficiently replicate within the so-called *Legionella*-containing vacuoles (LCV). The development of the LCV into a safe intracellular growth niche for *L. pneumophila* requires the remodeling of the LCV surface and the intimate interactions with ER-derived vesicles and organelles (mitochondria and ribosomes) (138). The close association with these vesicles and organelles may be important for the nutrient supply into the LCV. *L. pneumophila* actively replicates within the LCV for several hours after infection (replicative form, RF). Under nutrient limitation, the RF differentiates to the transmissive form (TF), which appears to be metabolically almost dormant, but expresses factors enabling the infection of new host cells. This biphasic lifestyle obviously requires specific metabolic adaptations in response to changing intracellular milieus (see below). The best-characterized virulence determinant is the Icm/Dot type IV secretion system, which may translocate about 300 effector proteins into the host cell cytosol (139–140). Several genes encoding effectors are highly expressed in *L. pneumophila* replicating in macrophages (141), indicating essential functions of these proteins for efficient replication within the LCV (142–143).

As for most other intracellular bacteria replicating in vacuoles, it is still rather unclear which nutrients are required for the intracellular metabolism of the RF and TF forms of *L. pneumophila*, how these nutrients are shuffled into the LCV, and which pathways are used to catabolize these compounds. The genome of *L. pneumophila* (144) contains all genes for the enzymes of the glycolytic pathway, the ED pathway, the TCA cycle, and the PPP, but typical PTS permeases for the import of glucose or other carbohydrates are missing (Fig. 2b). The genes *fbp* and *pck*, encoding fructose-1,6-bisphosphatase (FBP) and phosphoenolpyruvate carboxykinase (PCK), respectively, i.e., enzymes involved in gluconeogenesis, seem also to be absent. This is unexpected since gluconeogenesis should function as *L. pneumophila* is able to grow in defined culture media with gluconeogenic substrates (such as amino acids) as sole carbon sources (145). The genes for the biosynthetic pathways of Cys, Met, Thr, Val, Ile, and Leu are missing or incomplete in *L. pneumophila*, which is in accord with the lack of *de novo* synthesis of these amino acids under *in vitro* culture conditions (146).

Using *Acanthamoeba castellanii* as host for *L. pneumophila*, Brüggemann and collegues (147) could show that–in addition to genes involved in amino acid catabolism–genes encoding a putative sugar transporter, a putative glucokinase, and the enzymes of the ED pathway are upregulated in the actively growing (RF) bacteria, suggesting that–in addition to amino acids–hexoses (presumably glucose or glucose–6P) may be used as carbon substrates for intracellular growth of *L. pneumophila*. However, recent studies (148) feeding external ^{13}C-glucose to *A. castellanii* infected with (RF) *L. pneumophila* suggested that glucose is not a major nutrient for providing precursors for *de novo* synthesis of bacterial amino acids. Nevertheless, these data do not rule out that glucose-6P, rather than glucose, is an essential carbon substrate taken from the host cell. Indeed, *L. pneumophila* carries an

uhpC gene that is highly homologous to the *uhpC* gene of *Chlamydia pneumoniae*. The chlamydial UhpC has been shown to function as a glucose-6P transporter (149).

Amino acids (especially Ser, Thr, and Glu) may serve as major carbon, nitrogen, and energy sources for *L. pneumophila* when actively replicating in defined culture media or in macrophages (146, 150–152). The presence of genes for several other amino acid and peptide transporters, peptidases, and proteases in the *L. pneumophila* genome supports this finding. Indeed, the transcription profiles of *in vitro* grown, RF (compared to TF) *L. pneumophila* showed significant up-regulation of genes encoding amino acid transporters, amino acid-degrading enzymes (especially for serine and threonine), and several aminopeptidases and proteases (153). A recent transcriptome analysis (141), which compared the transcripts of intracellular *L. pneumophila* (replicating in human macrophage-like THP-1 cells) and of extracellular *L. pneumophila* (growing in AYE broth), also showed upregulation of many genes encoding transporters for amino acid and oligopeptides, as well as for enzymes involved in degradation of Lys, Arg, His, Thr, Glu, and Gln. Enhanced expression of genes encoding enzymes that are involved in glycerol catabolism was also observed, suggesting that not only amino acids but also glycerol could be used as a carbon source under intracellular growth conditions. The upregulation of the genes encoding phosphoenolpyruvate carboxylase and pyruvate carboxylase, respectively, supports this assumption as both enzymes connect glycerol catabolism with the TCA cycle by converting PEP and pyruvate, respectively, to oxaloacetate. This TCA intermediate is also a precursor in Asp and Lys biosynthesis. The genes for the biosynthesis of these two amino acids (as well as for His, Arg, and Pro) were also found to be induced. However, it remains to be elucidated whether the induction of the latter genes is really essential for intracellular growth of *L. pneumophila* or simply caused by de-repression due to the probably much lower levels of these amino acids within the host cells compared to the AYE medium.

The TF form of *L. pneumophila* contains substantial amounts of poly-3-hydroxybutyrate (PHB). PHB represents a bacterial energy and carbon storage compound and its degradation (leading to acetyl-CoA) in the TF stage may be important for maintaining the intracellular growth cycle of *L. pneumophila*. Indeed, a mutant defective in PHB metabolism showed a severe replication defect in amoebae and macrophages (154).

Together, these findings provide evidence that *L. pneumophila* uses–in addition to amino acids–carbohydrate(s), possibly including glucose-6P and glycerol as carbon sources for growth, at least in the late phase of intracellular growth. The carbohydrate (s) seem to be catabolized by ED and PPP rather than by GL (146). In line with this assumption is the finding that a Δ*zwf* mutant of *L. pneumophila* lacking glucose-6P dehydrogenase, an enzyme required for entering the ED pathway, is outcompeted by the isogenic wild-type strain after several rounds of infection in *A. castellanii*, although the replication rate of this mutant is not significantly altered (146). This indicates that carbohydrate(s) providing glucose-6P are necessary for the host-adapted intracellular metabolism of *L. pneumophila*, possibly for anabolic processes (e.g., for the synthesis of the sugar components of peptidoglycan and LPS). Indeed, a glucoamylase-like enzyme of *L. pneumophila* was shown to exhibit glycogen- and starch-degrading activity (155).

Metabolic host cell responses triggered by intracellular *L. pneumophila*

There are several lines of evidence indicating that *L. pneumophila* specifically manipulates the host cell's metabolism in favor of its proliferation. The amino acids transporter SLC1A5 (specific for neutral amino acids) was highly induced in the human monocyte cell line MM6 upon infection by *L. pneumophila* and was shown to be important for bacterial proliferation in these host cells (156). Preliminary evidence suggests that SLC1A5 is translocated to the LCV membrane thus supporting the uptake of cysteine, serine, glutamine, and other amino acids into the LCV. *L. pneumophila* effector proteins translocated by the T4SS into the host cells might be indirectly or even directly involved in the modification of metabolic processes of the infected host cells that support the intracellular metabolism of *L. pneumophila*, e.g., by provision of amino acids (157–161).

L. pneumophila-infected human macrophage-like U937 cells activated NF-κB signaling in a Dot/Icm-dependent manner (162) by the effector protein LnaB (163). NF-κB, well-known as an important regulator of eukaryotic cell survival and differentiation, is also linked to metabolic processes, e.g., glutamine metabolism (164). Putative Dot/Icm-dependent serine/threonine protein kinases and phosphatases may also modulate the activity of metabolic enzymes, as well as host-signaling pathways and possibly downstream metabolic processes (144, 165–166).

Chlamydia trachomatis (*C. trachomatis*)

Metabolic adaptations of intracellular *C. trachomatis*

The genus *Chlamydia* comprises obligate-intracellular, closely related, pathogenic species. The two major human pathogens, *C. trachomatis* and *C. pneumoniae*, cause infectious blinding trachoma or sexually transmitted disease and a community-acquired pneumonia, respectively (167–168). During these infections, the pathogens mainly grow in epithelial cells of the urogenital and the respiratory tract, respectively. After entry into host cells, the chlamydiae replicate in a specialized vacuole, termed "inclusion" (169) where they may exist in two morphologically and functionally distinct forms: the replicative, noninfectious, reticulate bodies (RB) and the elementary bodies (EB), which is infectious, but has a highly reduced metabolism (170–172). The residual metabolic activity of EB, including uptake and catabolism of glucose-6P, seems to be important for maintaining the EB infectivity of *C. trachomatis* (172). EB are released from the infected cells and may exist in a non-replicative extracellular state. The conversion of RB to EB occurs within the inclusion that–similar to the above described LCV–can fuse with ER- and Golgi-derived vesicles. This fusion leads to the acquisition of glycerophospholipids and sphingolipids from the vesicles, which increases the inclusion-membrane surface, and appears to be essential for intracellular growth of *C. trachomatis* (173–174) by assisting import of host-derived, high-energy metabolites and nutrients (which chlamydiae are unable to synthesize, see below) (175). Furthermore, the *Chlamydia* species may exist in a metabolically and morphologically aberrant RB-like persistent state (176).

C. trachomatis and *C. pneumoniae* have highly reduced genomes (1.2 Mb or less). Several genes involved in the central carbon and energy metabolism and most genes for anabolic pathways and regulatory factors are missing (177–178). However, these *Chlamydia* species carry the genes for all enzymes of GL and PPP, whereas the genes for the TCA cycle are incomplete in both pathogenic species (Fig. 2c), lacking the genes encoding the enzymes that catalyse the reactions from oxaloacetate to oxoglutarate. Similarly, as in rickettsiae, chlamydial genes (*adt*) encoding two ATP/ADP exchange proteins were identified (179–181), suggesting that ATP can be imported from host cells. The pathogenic *Chlamydia* spp. lack the genes for most cytochromes and flavoproteins of the electron-transfer chain and also for the F_1,F_0 -ATP synthase (catalyzing H^+-driven ATP synthesis), but retained *cydA* and *cydB*, encoding the cytochrome bd oxidase. The main bioenergetic function of this enzyme is the generation of a proton-motive force (PMF) by the vectorial charge transfer of protons. The genomes contain the genetic information for the subunits of a V-type ATPase that may likewise transport H^+ out of the bacterial cell at the expense of ATP. Hence, these reactions might be important for PMF-dependent transport processes (177).

The expression of *adt*, *cydA*, and *cydB*; *pyk*, *gap*, *pgk*, and *gnd*; *tal*,*fumC*, and *mdhC*, encoding glycolytic, PPP, and TCA enzymes, respectively, was demonstrated in *C. trachomatis*--infected Hep-2 cells in the active replicative phase, suggesting that during active growth the chlamydial energy requirement is met by ATP import from the host and by its own ATP production via the active glycolytic pathway. In persistent *C. trachomatis* bacteria (growing in human monocytes), only *adt1*, *cydB*, *fumC*, and *mdhC* transcripts were identified, suggesting that in this state ATP may also be taken primarily from the host cells (182).

On the other hand, proteome analyses detected the entire set of predicted glycolytic and PPP enzymes in *C. trachomatis* EB cells. These data suggest that the metabolic flux through these pathways, rather than *de novo* synthesis of the enzymes involved in these catabolic pathways, is induced when chlamydiae infect and actively replicate in the host cells (183). This hypothesis is also in line with the recent finding that some glucose catabolism may even occur in EB (5, 172). Furthermore, cell culture experiments using different carbon sources show that chlamydiae–unlike most free-living bacteria–do not alter the expression of genes involved in carbon metabolism in response to nutritional changes (184). Together, these data suggest that the expression of chlamydial genes encoding the enzymes involved in carbon catabolism is constitutive and little modified by the interaction of chlamydiae with the host cells, a conclusion that is consistent with the apparent lack of all global regulators known to control metabolic genes in bacteria.

What are the major carbon sources for the intracellular chlamydiae? As discussed above, ATP production by substrate-level phosphorylation seems to occur in RB during the active growth phase, suggesting the use of carbon substrates, such as carbohydrates including glycerol or glycerol-3P, that are oxidized in the ATP-delivering reactions of the lower part of the glycolytic pathway. Glucose transporters have not yet been identified in the pathogenic *Chlamydia* species. Indeed, ^{13}C-isotopologue studies using uniformly labeled ^{13}C-glucose provided to *C. trachomatis*-infected Caco-2 or HeLa cell cultures showed no significant glucose catabolism by the intracellular chlamydiae (A. Mehlitz, T. Rudel, W. Goebel, and W. Eisenreich, unpublished data). However, the *uhpC* gene present in all chlamydiae encodes a functional glucose-6P transporter (149), suggesting that imported glucose-6P may not be catabolized but rather serve as substrate for anabolic processes, e.g., biosynthesis of cell envelope components or production of glycogen. A chlamydial glycogen synthase (GlgA) has been identified which is secreted into the lumen of the inclusion and even into the cytosol of infected cells (185). *C. trachomatis* also has a functional glycogen phosphorylase and it has been suggested that glycogen might serve as a source for the generation of phosphorylated glucose at a later stage (183).

The protease CPAF shows a similar secretion pattern as GlgA (185). The action of this protease may be essential for providing host amino acids for chlamydial protein biosynthesis. Indeed, *C. trachomatis* seems to be totally dependent on the import of all amino acids from the host cell. However, with the exception of glutamate, the imported amino acids did not seem to be catabolized as shown by ^{13}C-isotopologue analysis (A. Mehlitz, personal communication). Glutamate appears to be channeled via oxoglutarate into the residual TCA cycle where it is converted to oxaloacetate providing NADH/H$^+$, FADH$_2$, and GTP. The

residual TCA cycle could be also fed by host-derived succinate or malate since a dicarboxylate translocator has been identified, even in EB of *C. trachomatis* (183). The production of oxaloacetate (as precursor of aspartate) might be important for the biosynthesis of meso-2,6-diaminopimelate (m-DAP). The m-DAP pathway is complete in chlamydiae (as in all peptidoglycan-containing bacteria).

Oxaloacetate may be also converted to PEP via PCK. The *pckA* gene encoding PCK is present on the chlamydial genome and the chlamydial PCK protein was identified by proteome analysis (183). This enzyme is essential for gluconeogenesis starting from TCA intermediates. Nevertheless, it remains questionable whether gluconeogenesis functions in chlamydiae, since a *fbp*-homologous gene, encoding the necessary FBP, has not been identified in *C. trachomatis*. However, the situation is similar in *L. pneumophila*, where an atypical FBP (186) must be present as gluconeogenesis should be functional since the bacteria are apparently able to grow with amino acids as the sole carbon source (145). In this context, it might be worthwhile to note that the reverse reaction leading to the formation of FBP by the glycolytic enzyme 6-phosphofructokinase (PFK) also seems to be catalyzed by atypical PFK(s) (187). Two enzymes (PFK-1 and PFK-2) are annotated in *C. trachomatis* (Kyoto Encyclopedia of Genes and Genomes [KEGG] GENES database) (206) with little homology to the classical ATP-PFKs from bacteria and eucarya and, to our knowledge, a biochemical PFK function has not been shown for either PFK-1 or PFK-2.

The capacity of chlamydiae for production of anabolic components is very limited. The two human-pathogenic *Chlamydia* species lack the genes for the biosynthesis of vitamins, purines, pyrimidines, and almost all amino acids. Hence, chlamydiae rely entirely on the import of these essential metabolites from the host cell. Several genes encoding transporters for purine and pyrimidine have been identified in the chlamydial genomes (177). In addition, chlamydiae possess a transporter for all four ribonucleoside triphosphates (188). Amino acids are mainly provided by the host cell through lysosomal degradation of external proteins (12) or through protein degradation by the chlamydial proteases, CPAF and cHtrA, that are secreted into the host cells cytosol (189–191). As expected, the chlamydiae possess several genes encoding amino acid transporters (177). The only amino acids that can be made de novo are those that simply need transamination of catabolic intermediates (Ala, Asp, and Glu) and meso-diaminopimelate (m-DAP) (177) (A. Mehlitz, personal communication). The entire gene set for the biosynthesis of m-DAP essential for peptidoglycan synthesis is present, whereas the subsequent gene for the diaminopimelate decarboxylase leading to Lys is missing. Indeed, all sequenced chlamydial genomes reveal the genetic information for peptidoglycan biosynthesis, although previous studies (192) could not detect peptidoglycan in chlamydial EB.

All aro genes necessary for the biosynthesis of chorismate–probably used for the synthesis of quinones–are also present. C. trachomatis, but not C. pneumonia, has also retained the genes trpA and trpB, encoding tryptophan synthetase (but not the other genes involved in tryptophan biosynthesis). The presence of this enzyme could allow the recycling of tryptophan precursors (especially indole) to tryptophan (193) and may constitute a selection advantage under conditions where host cell tryptophan is degraded, e.g., upon IFN-γ induction (194). Indeed, tryptophan starvation of Chlamydia-infected HEp-2 cells, mediated by IFN-γ treatment, blocks the conversion of the RB to the infectious EB form, which is normally induced by amino acid starvation. In addition to the de novo chlamydial fatty acid and (partial) phospholipid biosynthesis, host cell-derived phospholipids and sphingolipids required for the intracellular growth can also be trafficked and modified by chlamydiae (173, 195–196).

Metabolic host cell responses triggered by intracellular *C. trachomatis*

Chlamydial infections lead to increased glucose uptake and glucose flux through the glycolytic pathway of the infected host cells (197–198). Induction of the pentosephosphate shunt is also observed, which may be linked to the enhanced anabolic performance (increased protein and lipid synthesis) observed in the host cells during the proliferative phase of the chlamydiae. The remarkable up-regulation of many genes encoding transporters for nutrients needed for intracellular chlamydial growth (199) suggests increased import of these nutrients by the host cells and the possible recruitment of these nutrient transporters to the membrane surrounding the *Chlamydia*-containing inclusion. The pathogenic chlamydiae possess a type III secretion system (200–201) and various effector proteins essential for virulence (169, 202). But up to now, it is unknown whether some of these effectors may also trigger metabolic host responses observed in chlamydial infections.

CONCLUSIONS

The group of intracellular bacterial pathogens comprises microorganisms with highly different metabolic capacities. But, in each case, a successful intracellular pathogen adapts its intracellular metabolism to the metabolism of the infected host cell, triggers favorable metabolic host cell responses, and coordinates the expression of the genes encoding virulence factors needed for the intracellular bacterial life cycle, with its intracellular metabolism, in order to achieve efficient replication and to prevent premature death of the host cell. Common and pathogen-specific factors are apparently involved in this bilateral metabolic adaptation process (termed "pathometabolism"). Although our knowledge on the precise mechanisms responsible for these processes is still rather limited (27, 30), the available data suggest that the intracellular bacteria have apparently evolved individual strategies for this metabolic pathogen/host cell relationship, as outlined above for some selected examples. Since the final aim of this interaction is the optimal proliferation of the intracellular pathogen at the expense of the host cell, this interaction will be always parasitic.

Within the different individual strategies of the pathogens some common motifs are, however, recognizable, which concern in particular the handling of essential nutrients to insure extended survival of both partners. The use of more than one carbon source captured from the host cells by the intracellular bacteria seems to be a general mode for generating the necessary energy and metabolic intermediates to perform the indispensable bacterial biosynthesis of cell envelope components, of nucleic acids, and of proteins without harming the host cell too much. The choice and combination of the carbon sources used by the particular pathogen apparently depend on the metabolic capabilities of the pathogen and on the infection-adjusted host cell metabolism.

For those intracellular pathogens that possess redundant catabolic and anabolic capacities, the sequence of the preferentially used carbon sources may be adapted to the threshold concentration of the respective carbon substrate in the host cell. In this case each carbon substrate used may deliver both energy and intermediary metabolites, but the rate of the bacterial proliferation, and probably also the expression of virulence factors and the fate of the host cell, will depend on the choice of the actually used carbon source.

Most of the well-adapted intracellular pathogens have incomplete catabolic and, particularly, anabolic pathways and for these bacteria the simultaneous use of more than one carbon substrate might be necessary for producing enough energy and metabolic intermediates to guarantee bacterial proliferation. In this case, the availability of only a single carbon source may lead to a highly reduced metabolism providing just enough energy for the survival of the bacteria in a metabolically almost-dormant state in which the intracellular bacteria must still be at least able to carry out energy-requiring vital repair functions (in particular, DNA repair).

These co-metabolism strategies might also reduce the metabolic burden put on the host cell by the infection with intracellular pathogens and extend the host cell's lifespan–also to the advantage of the pathogen. Common and, in particular, pathogen-specific factors may support this metabolic pathogen/host cell relationship by inducing uptake and catabolism of basic nutrients (e.g., glucose or glutamine) by the host cell, and by silencing antimicrobial metabolic host cell reactions (30).

Another common strategic motif concerns the acquisition of host cell amino acids (and possibly other essential anabolic monomers) by the intracellular bacterial pathogens. Import of host cell amino acids is observed even when the intracellular bacteria are prototrophic for all amino acids. The extent of amino acid import seems to depend on the available carbon source, i.e., import is low and bacterial *de novo* amino acid synthesis is high when an energy-rich carbon compound, e.g., glucose, is used and *vice versa* when a less energy-rich carbon source, e.g., glycerol, is utilized. This strategy is consistent with the view that biosynthesis of most amino acids is energetically more costly for intracellular bacteria than import of these amino acids from the host cell and implies that the efficiency of intracellular bacterial replication will be significantly affected by the pool of free amino acids (and/or oligopeptides) in the host cell. This pool is mainly determined by the uptake of external amino acids or by stimulation of internal protein degradation (especially through the proteasomal machinery and autophagy), host cell processes that may be again induced by bacterial (virulence) factors (160, 203).

In this context, an interesting hypothesis has been proposed recently (38) postulating that Cys (the most limiting amino acid in mammalian cells) (204) or other limiting essential amino acids might function as nutritional rheostats controlling the proliferation of intracellular pathogens. Indeed, many intracellular pathogens are unable to synthesize Cys (including *L. monocytogenes, R. prowazeki, L. pneumophila, and C. trachomatis* among those discussed here) and hence completely rely on Cys supply from the host cell. Pathogen-specific strategies have been evolved to cope with the bottle neck of Cys supply (160, 205). Possibly, other vital anabolic monomers may play a similar role as Cys in controlling the intracellular metabolism of bacterial pathogens (152).

In summary, the ability of an intracellular bacterial pathogen to adapt its own metabolism to the intracellular milieu of the host cell, to modulate the host cell metabolism in favor of its own needs with the help of common and pathogen-specific (virulence) factors, and to adjust the expression of the necessary virulence genes to the intracellular metabolism, must be viewed to be central for the pathogenesis of infections by intracellular bacterial pathogens. For the complex metabolic interconnection between bacterial pathogens and host cells, we have introduced the term "pathometabolism."

ACKNOWLEDGMENTS

We thank A. Mehlitz for helpful discussions. We thank A. Mehlitz and T. Fuchs for allowing us to cite unpublished results. H. Hilbi is thanked for critical reading of the manuscript. Our own work, included in this chapter, was supported mainly by the Deutsche Forschungsgemeinschaft (e.g., SPP 1316, EI-384/5).

CITATION

Eisenreich W, Heesemann J, Rudel T, Goebel W. 2015. Metabolic adaptations of intracellullar bacterial pathogens and their mammalian host cells during infection ("pathometabolism"). Microbiol Spectrum 3(3):MBP-0002-2014.

REFERENCES

1. **Lupp C, Robertson ML, Wickham ME, Sekirov I, Champion OL, Gaynor EC, Finlay BB.** 2007. Host-mediated inflammation disrupts the intestinal microbiota and promotes the overgrowth of Enterobacteriaceae. *Cell Host Microbe* **2:**119–129.
2. **Stecher B, Robbiani R, Walker AW, Westendorf AM, Barthel M, Kremer M, Chaffron S, Macpherson AJ, Buer J, Parkhill J, Dougan G, von Mering C, Hardt WD.** 2007. *Salmonella enterica* serovar typhimurium exploits inflammation to compete with the intestinal microbiota. *PLoS Biol* **5:**2177–2189.
3. **Chaston J, Goodrich-Blair H.** 2010. Common trends in mutualism revealed by model associations between invertebrates and bacteria. *FEMS Microbiol Rev* **34:**41–58.
4. **Omsland A, Cockrell DC, Howe D, Fischer ER, Virtaneva K, Sturdevant DE, Porcella SF, Heinzen RA.** 2009. Host cell-free growth of the Q fever bacterium *Coxiella burnetii. Proc Natl Acad Sci U S A* **106:**4430–4434.
5. **Omsland A, Sager J, Nair V, Sturdevant DE, Hackstadt T.** 2012. Developmental stage-specific metabolic and transcriptional activity of *Chlamydia trachomatis* in an axenic medium. *Proc Natl Acad Sci U S A* **109:**19781–19785.
6. **Méresse S, Steele-Mortimer O, Moreno E, Desjardins M, Finlay B, Gorvel JP.** 1999. Controlling the maturation of pathogen-containing vacuoles: a matter of life and death. *Nat Cell Biol* **1:**E183–E188.
7. **Hackstadt T.** 1998. The diverse habitats of obligate intracellular parasites. *Curr Opin Microbiol* **1:**82–87.
8. **Beatty WL.** 2008. Late endocytic multi-vesicular bodies intersect the chlamydial inclusion in the absence of CD63. *Infect Immun* **76:**2872–2881.
9. **Kerr MC, Teasdale RD.** 2009. Defining macropinocytosis. *Traffic* **10:**364–371.
10. **Ma B, Xiang Y, An L.** 2011. Structural bases of physiological functions and roles of the vacuolar H(+)-ATPase. *Cell Signal* **23:**1244–1256.

11. **Lim JP, Gleeson PA.** 2011. Macropinocytosis: an endocytic pathway for internalising large gulps. *Immunol Cell Biol* **89:**836–843.

12. **Ouellette SP, Dorsey FC, Moshiach S, Cleveland JL, Carabeo RA.** 2011. *Chlamydia* species-dependent differences in the growth requirement for lysosomes. *PLoS One* **6:** e16783. doi:10.1371/journal.pone.0016783

13. **Liu Z, Zhang YW, Chang YS, Fang FD.** 2006. The role of cytoskeleton in glucose regulation. *Biochemistry (Mosc)* **71:**476–480.

14. **Beuzon CR, Salcedo SP, Holden DW.** 2002. Growth and killing of a *Salmonella enterica* serovar Typhimurium sifA mutant strain in the cytosol of different host cell lines. *Microbiology* **148:**2705–2715.

15. **van der Wel N, Hava D, Houben D, Fluitsma D, van Zon M, Pierson J, Brenner M, Peters PJ.** 2007. *M. tuberculosis* and *M. leprae* translocate from the phagolysosome to the cytosol in myeloid cells. *Cell* **129:**1287–1298.

16. **Knodler LA, Vallance BA, Celli J, Winfree S, Hansen B, Montero M, Steele-Mortimer O.** 2010. Dissemination of invasive *Salmonella* via bacterial-induced extrusion of mucosal epithelia. *Proc Natl Acad Sci U S A* **107:**17733–17738.

17. **Malik-Kale P, Winfree S, Steele-Mortimer O.** 2012. The bimodal lifestyle of intracellular *Salmonella* in epithelial cells: replication in the cytosol obscures defects in vacuolar replication. *PLoS One* **7:**e38732. doi:10.1371/journal.pone.0038732

18. **Grant AJ, Morgan FJ, McKinley TJ, Foster GL, Maskell DJ, Mastroeni P.** 2012. Attenuated *Salmonella* Typhimurium lacking the pathogenicity island-2 type 3 secretion system grow to high bacterial numbers inside phagocytes in mice. *PLoS Pathog* **8:**e1003070. doi:10.1371/journal. ppat.1003070

19. **Birmingham CL, Canadien V, Kaniuk NA, Steinberg BE, Higgins DE, Brumell JH.** 2008. Listeriolysin O allows *Listeria monocytogenes* replication in macrophage vacuoles. *Nature* **451:**350–354.

20. **Schmidt H, Hensel M.** 2004. Pathogenicity islands in bacterial pathogenesis. *Clin Microbiol Rev* **17:**14–56.

21. **Rohmer L, Hocquet D, Miller SI.** 2011. Are pathogenic bacteria just looking for food? Metabolism and microbial pathogenesis. *Trends Microbiol* **19:**341–348.

22. **Lopez CA, Winter SE, Rivera-Chavez F, Xavier MN, Poon V, Nuccio SP, Tsolis RM, Baumler AJ.** 2012. Phage-mediated acquisition of a type III secreted effector protein boosts growth of salmonella by nitrate respiration. *MBio* **3:**e00143-12. doi:10.1128/mBio.00143-12

23. **Maurelli AT, Fernández RE, Bloch CA, Rode CK, Fasano A.** 1998. "Black holes" and bacterial pathogenicity: a large genomic deletion that enhances the virulence of *Shigella* spp. and enteroinvasive *Escherichia coli*. *Proc Natl Acad Sci U S A* **95:**3943–3948.

24. **Ochman H, Moran NA.** 2001. Genes lost and genes found: evolution of bacterial pathogenesis and symbiosis. *Science* **292:**1096–1099.

25. **Rocco CJ, Escalante-Semerena JC.** 2010. In *Salmonella enterica*, 2-methylcitrate blocks gluconeogenesis. *J Bacteriol* **192:**771–778.

26. **Bliven KA, Maurelli AT.** 2012. Antivirulence genes: insights into pathogen evolution through gene loss. *Infect Immun* **80:**4061–4070.

27. **Eisenreich W, Dandekar T, Heesemann J, Goebel W.** 2010. Carbon metabolism of intracellular bacterial pathogens and possible links to virulence. *Nat Rev Microbiol* **8:**401–412.

28. **Fuchs TM, Eisenreich W, Heesemann J, Goebel W.** 2012. Metabolic adaptation of human pathogenic and related nonpathogenic bacteria to extra- and intracellular habitats. *FEMS Microbiol Rev* **36:**435–462.

29. **Aké FM, Joyet P, Deutscher J, Milohanic E.** 2011. Mutational analysis of glucose transport regulation and glucose-mediated virulence gene repression in *Listeria monocytogenes*. *Mol Microbiol* **81:**274–293.

30. **Eisenreich W, Heesemann J, Rudel T, Goebel W.** 2013. Metabolic host responses to infection by intracellular bacterial pathogens. *Front Cell Infect Microbiol* **3:**24.

31. **Creasey EA, Isberg RR.** 2012. The protein SdhA maintains the integrity of the *Legionella*-containing vacuole. *Proc Natl Acad Sci U S A* **109:**3481–3486.

32. **Aachoui Y, Leaf IA, Hagar JA, Fontana MF, Campos CG, Zak DE, Tan MH, Cotter PA, Vance RE, Aderem A, Miao EA.** 2013. Caspase-11 protects against bacteria that escape the vacuole. *Science* **339:**975–978.

33. **Pearce EL, Pearce EJ.** 2013. Metabolic pathways in immune cell activation and quiescence. *Immunity* **38:**633–643.

34. **Goetz M, Bubert A, Wang G, Chico-Calero I, Vazquez-Boland JA, Beck M, Slaghuis J, Szalay AA, Goebel W.** 2001. Microinjection and growth of bacteria in the cytosol of mammalian host cells. *Proc Natl Acad Sci U S A* **98:**12221–12226.

35. **Meibom KL, Charbit A.** 2010. *Francisella tularensis* metabolism and its relation to virulence. *Front Microbiol* **1:**140.

36. **Schauer K, Geginat G, Liang C, Goebel W, Dandekar T, Fuchs TM.** 2010. Deciphering

the intracellular metabolism of *Listeria monocytogenes* by mutant screening and modelling. *BMC Genomics* **11**:573.

37. **Fuchs TM, Eisenreich W, Kern T, Dandekar T.** 2012. Toward a systemic understanding of *Listeria monocytogenes* metabolism during infection. *Front Microbiol* **3**:23.

38. **Abu Kwaik Y, Bumann D.** 2013. Microbial quest for food *in vivo*: 'nutritional virulence' as an emerging paradigm. *Cell Microbiol* **15**:882–890.

39. **Gordon S, Taylor PR.** 2005. Monocyte and macrophage heterogeneity. *Nat Rev Immunol* **5**:953–964.

40. **Martinez FO, Sica A, Mantovani A, Locati M.** 2008. Macrophage activation and polarization. *Front Biosci* **13**:453–461.

41. **Odegaard JI, Chawla A.** 2011. Alternative macrophage activation and metabolism. *Annu Rev Pathol* **6**:275–297.

42. **Appelberg R.** 2006. Macrophage nutriprive antimicrobial mechanisms. *J Leukoc Biol* **79**:1117–1128.

43. **Cellier MF, Courville P, Campion C.** 2007. Nramp1 phagocyte intracellular metal withdrawal defense. *Microbes Infect* **9**:1662–1670.

44. **Taylor MW, Feng GS.** 1991. Relationship between interferon-gamma, indoleamine 2,3-dioxygenase, and tryptophan catabolism. *FASEB J* **5**:2516–2522.

45. **Gouin E, Welch MD, Cossart P.** 2005. Actin-based motility of intracellular pathogens. *Curr Opin Microbiol* **8**:35–45.

46. **Chen J, de Felipe KS, Clarke M, Lu H, Anderson OR, Segal G, Shuman HA.** 2004. *Legionella* effectors that promote nonlytic release from protozoa. *Science* **303**:1358–1361.

47. **Hagedorn M, Rohde KH, Russell DG, Soldati T.** 2009. Infection by tubercular mycobacteria is spread by nonlytic ejection from their amoeba hosts. *Science* **323**:1729–1733.

48. **Vander Heiden MG, Cantley LC, Thompson CB.** 2009. Understanding the Warburg effect: the metabolic requirements of cell proliferation. *Science* **324**:1029–1033.

49. **Smolková K, Ježek P.** 2012. The role of mitochondrial NADPH-dependent isocitrate dehydrogenase in cancer cells. *Int J Cell Biol* **2012**:273947.

50. **Dang L, Jin S, Su SM.** 2010. IDH mutations in glioma and acute myeloid leukemia. *Trends Mol Med* **16**:387–397.

51. **Chen JQ, Russo J.** 2012. Dysregulation of glucose transport, glycolysis, TCA cycle and glutaminolysis by oncogenes and tumor suppressors in cancer cells. *Biochim Biophys Acta* **1826**:370–384.

52. **Cardaci S, Ciriolo MR.** 2012. TCA cycle defects and cancer: when metabolism tunes redox state. *Int J Cell Biol* **2012**:161837.

53. **Jones RG, Thompson CB.** 2009. Tumor suppressors and cell metabolism: a recipe for cancer growth. *Genes Dev* **23**:537–548.

54. **Puzio-Kuter AM.** 2011. The role of p53 in metabolic regulation. *Genes Cancer* **2**:385–391.

55. **Parsot C.** 2005. *Shigella* spp. and enteroinvasive *Escherichia coli* pathogenicity factors. *FEMS Microbiol Lett* **252**:11–18.

56. **Bernardini ML, Mounier J, d'Hauteville H, Coquis-Rondon M, Sansonetti PJ.** 1989. Identification of icsA, a plasmid locus of *Shigella flexneri* that governs bacterial intra- and intercellular spread through interaction with F-actin. *Proc Natl Acad Sci U S A* **86**:3867–3871.

57. **Parsot C.** 2009. *Shigella* type III secretion effectors: how, where, when, for what purposes? *Curr Opin Microbiol* **12**:110–116.

58. **Barabote RD, Saier MH Jr.** 2005. Comparative genomic analyses of the bacterial phosphotransferase system. *Microbiol Mol Biol Rev* **69**:608–634.

59. **Island MD, Wei BY, Kadner RJ.** 1992. Structure and function of the uhp genes for the sugar phosphate transport system in *Escherichia coli* and *Salmonella typhimurium*. *J Bacteriol* **174**:2754–2762.

60. **Verhamme DT, Arents JC, Postma PW, Crielaard W, Hellingwerf KJ.** 2001. Glucose-6-phosphate-dependent phosphoryl flow through the Uhp two-component regulatory system. *Microbiology* **147**:3345–3352.

61. **Runyen-Janecky LJ, Payne SM.** 2002. Identification of chromosomal *Shigella flexneri* genes induced by the eukaryotic intracellular environment. *Infect Immun* **70**:4379–4388.

62. **Lucchini S, Liu H, Jin Q, Hinton JC, Yu J.** 2005. Transcriptional adaptation of *Shigella flexneri* during infection of macrophages and epithelial cells: insights into the strategies of a cytosolic bacterial pathogen. *Infect Immun* **73**:88–102.

63. **Hamilton S, Bongaerts RJ, Mulholland F, Cochrane B, Porter J, Lucchini S, Lappin-Scott HM, Hinton JC.** 2009. The transcriptional programme of *Salmonella enterica* serovar Typhimurium reveals a key role for tryptophan metabolism in biofilms. *BMC Genomics* **10**:599.

64. **Cersini A, Salvia AM, Bernardini ML.** 1998. Intracellular multiplication and virulence of *Shigella flexneri* auxotrophic mutants. *Infect Immun* **66**:549–557.

65. **Götz A, Eylert E, Eisenreich W, Goebel W.** 2010. Carbon metabolism of enterobacterial

human pathogens growing in epithelial colorectal adenocarcinoma (Caco-2) cells. *PLoS One* **5:** e10586. doi:10.1371/journal.pone.0010586

66. **Gore AL, Payne SM.** 2010. CsrA and Cra influence *Shigella flexneri* pathogenesis. *Infect Immun* **78:**4674–4682.

67. **Romeo T.** 1998. Global regulation by the small RNA-binding protein CsrA and the non-coding RNA molecule CsrB. *Mol Microbiol* **29:**1321–1330.

68. **Saier MH Jr, Ramseier TM.** 1996. The catabolite repressor/activator (Cra) protein of enteric bacteria. *J Bacteriol* **178:**3411–3417.

69. **Noriega FR, Losonsky G, Wang JY, Formal SB, Levine MM.** 1996. Further characterization of delta aroA delta virG *Shigella flexneri* 2a strain CVD 1203 as a mucosal *Shigella* vaccine and as a live-vector vaccine for delivering antigens of enterotoxigenic *Escherichia coli*. *Infect Immun* **64:**23–27.

70. **Cersini A, Martino MC, Martini I, Rossi G, Bernardini ML.** 2003. Analysis of virulence and inflammatory potential of *Shigella flexneri* purine biosynthesis mutants. *Infect Immun* **71:**7002–7013.

71. **Bergounioux J, Elisee R, Prunier AL, Donnadieu F, Sperandio B, Sansonetti P, Arbibe L.** 2012. Calpain activation by the *Shigella flexneri* effector VirA regulates key steps in the formation and life of the bacterium's epithelial niche. *Cell Host Microbe* **11:**240–252.

72. **Pieper R, Zhang Q, Parmar PP, Huang ST, Clark DJ, Alami H, Donohue-Rolfe A, Fleischmann RD, Peterson SN, Tzipori S.** 2009. The *Shigella dysenteriae* serotype 1 proteome, profiled in the host intestinal environment, reveals major metabolic modifications and increased expression of invasive proteins. *Proteomics* **9:**5029–5045.

73. **Velge P, Roche SM.** 2010. Variability of *Listeria monocytogenes* virulence: a result of the evolution between saprophytism and virulence? *Future Microbiol* **5:**1799–1821.

74. **Mostowy S, Cossart P.** 2012. Virulence factors that modulate the cell biology of Listeria infection and the host response. *Adv Immunol* **113:**19–32.

75. **Camejo A, Carvalho F, Reis O, Leitão E, Sousa S, Cabanes D.** 2011. The arsenal of virulence factors deployed by *Listeria monocytogenes* to promote its cell infection cycle. *Virulence* **2:**379–394.

76. **Stoll R, Goebel W.** 2010. The major PEP-phosphotransferase systems (PTSs) for glucose, mannose and cellobiose of *Listeria monocytogenes*, and their significance for extra- and intracellular growth. *Microbiology* **156:**1069–1083.

77. **Chico-Calero I, Suárez M, González-Zorn B, Scortti M, Slaghuis J, Goebel W, Vazquez-Boland JA.** 2002. Hpt, a bacterial homolog of the microsomal glucose-6-phosphate translocase, mediates rapid intracellular proliferation in *Listeria*. *Proc Natl Acad Sci U S A* **99:**431–436.

78. **Eisenreich W, Slaghuis J, Laupitz R, Bussemer J, Stritzker J, Schwarz C, Schwarz R, Dandekar T, Goebel W, Bacher A.** 2006. ^{13}C isotopologue perturbation studies of *Listeria monocytogenes* carbon metabolism and its modulation by the virulence regulator PrfA. *Proc Natl Acad Sci U S A* **103:**2040–2045.

79. **Schär J, Stoll R, Schauer K, Loeffler DI, Eylert E, Joseph B, Eisenreich W, Fuchs TM, Goebel W.** 2010. Pyruvate carboxylase plays a crucial role in carbon metabolism of extra- and intracellularly replicating *Listeria monocytogenes*. *J Bacteriol* **192:**1774–1784.

80. **Premaratne RJ, Lin WJ, Johnson EA.** 1991. Development of an improved chemically defined minimal medium for *Listeria monocytogenes*. *Appl Environ Microbiol* **57:**3046–3048.

81. **Tsai HN, Hodgson DA.** 2003. Development of a synthetic minimal medium for *Listeria monocytogenes*. *Appl Environ Microbiol* **69:** 6943–6945.

82. **Stoll R, Mertins S, Joseph B, Müller-Altrock S, Goebel W.** 2008. Modulation of PrfA activity in *Listeria monocytogenes* upon growth in different culture media. *Microbiology* **154:**3856–3876.

83. **Schneebeli R, Egli T.** 2013. A defined, glucose-limited mineral medium for the cultivation of *Listeria* spp. *Appl Environ Microbiol* **79:**2503–2511.

84. **Glaser P, Frangeul L, Buchrieser C, Rusniok C, Amend A, Baquero F, Berche P, Bloecker H, Brandt P, Chakraborty T, Charbit A, Chetouani F, Couvé E, de Daruvar A, Dehoux P, Domann E, Domínguez-Bernal G, Duchaud E, Durant L, Dussurget O, Entian KD, Fsihi H, García-del Portillo F, Garrido P, Gautier L, Goebel W, Gómez-López N, Hain T, Hauf J, Jackson D, Jones LM, Kaerst U, Kreft J, Kuhn M, Kunst F, Kurapkat G, Madueno E, Maitournam A, Vicente JM, Ng E, Nedjari H, Nordsiek G, Novella S, de Pablos B, Pérez-Diaz JC, Purcell R, Remmel B, Rose M, Schlueter T, Simoes N, Tierrez A, Vázquez-Boland JA, Voss H, Wehland J, Cossart P.** 2001. Comparative genomics of *Listeria* species. *Science* **294:**849–852.

85. **Joseph B, Przybilla K, Stühler C, Schauer K, Slaghuis J, Fuchs TM, Goebel W.** 2006. Identification of *Listeria monocytogenes* genes contributing to intracellular replication by expression profiling and mutant screening. *J Bacteriol* **188:**556–568.

86. **Lobel L, Sigal N, Borovok I, Ruppin E, Herskovits AA.** 2012. Integrative genomic analysis identifies isoleucine and CodY as regulators of *Listeria monocytogenes* virulence. *PLoS Genet* **8:**e1002887. doi:10.1371/journal.pgen.1002887

87. **Chakraborty T, Leimeister-Wächter M, Domann E, Hartl M, Goebel W, Nichterlein T, Notermans S.** 1992. Coordinate regulation of virulence genes in *Listeria monocytogenes* requires the product of the prfA gene. *J Bacteriol* **174:**568–574.

88. **Joseph B, Mertins S, Stoll R, Schär J, Umesha KR, Luo Q, Müller-Altrock S, Goebel W.** 2008. Glycerol metabolism and PrfA activity in *Listeria monocytogenes*. *J Bacteriol* **190:**5412–5430.

89. **de las Heras A, Cain RJ, Bielecka MK, Vázquez-Boland JA.** 2011. Regulation of *Listeria* virulence: PrfA master and commander. *Curr Opin Microbiol* **14:**118–127.

90. **Shivers RP, Sonenshein AL.** 2004. Activation of the *Bacillus subtilis* global regulator CodY by direct interaction with branched-chain amino acids. *Mol Microbiol* **53:**599–611.

91. **Chatterjee SS, Hossain H, Otten S, Kuenne C, Kuchmina K, Machata S, Domann E, Chakraborty T, Hain T.** 2006. Intracellular gene expression profile of *Listeria monocytogenes*. *Infect Immun* **74:**1323–1338.

92. **Donaldson JR, Nanduri B, Pittman JR, Givaruangsawat S, Burgess SC, Lawrence ML.** 2011. Proteomic expression profiles of virulent and avirulent strains of *Listeria monocytogenes* isolated from macrophages. *J Proteomics* **74:**1906–1917.

93. **Eylert E, Schär J, Mertins S, Stoll R, Bacher A, Goebel W, Eisenreich W.** 2008. Carbon metabolism of *Listeria monocytogenes* growing inside macrophages. *Mol Microbiol* **69:**1008–1017.

94. **Gillmaier N, Götz A, Schulz A, Eisenreich W, Goebel W.** 2012. Metabolic responses of primary and transformed cells to intracellular *Listeria monocytogenes*. *PLoS One* **7:**e52378. doi:10.1371/journal.pone.0052378

95. **Toledo-Arana A, Dussurget O, Nikitas G, Sesto N, Guet-Revillet H, Balestrino D, Loh E, Gripenland J, Tiensuu T, Vaitkevicius K, Barthelemy M, Vergassola M, Nahori MA, Soubigou G, Régnault B, Coppée JY, Lecuit M, Johansson J, Cossart P.** 2009. The *Listeria* transcriptional landscape from saprophytism to virulence. *Nature* **459:**950–956.

96. **Camejo A, Buchrieser C, Couvé E, Carvalho F, Reis O, Ferreira P, Sousa S, Cossart P, Cabanes D.** 2009. *In vivo* transcriptional profiling of *Listeria monocytogenes* and mutagenesis identify new virulence factors involved in infection. *PLoS Pathog* **5:**e1000449. doi:10.1371/journal.ppat.1000449

97. **Port GC, Freitag NE.** 2007. Identification of novel *Listeria monocytogenes* secreted virulence factors following mutational activation of the central virulence regulator, PrfA. *Infect Immun* **75:**5886–5897.

98. **Cohen P, Bouaboula M, Bellis M, Baron V, Jbilo O, Poinot-Chazel C, Galiègue S, Hadibi EH, Casellas P.** 2000. Monitoring cellular responses to *Listeria monocytogenes* with oligonucleotide arrays. *J Biol Chem* **275:**11181–11190.

99. **Lecuit M, Sonnenburg JL, Cossart P, Gordon JI.** 2007. Functional genomic studies of the intestinal response to a foodborne enteropathogen in a humanized gnotobiotic mouse model. *J Biol Chem* **282:**15065–15072.

100. **Valbuena G, Walker DH.** 2009. Infection of the endothelium by members of the order Rickettsiales. *Thromb Haemost* **102:**1071–1079.

101. **Cowan G.** 2000. Rickettsial diseases: the typhus group of fevers–a review. *Postgrad Med J* **76:**269–272.

102. **Walker DH, Ismail N.** 2008. Emerging and re-emerging rickettsioses: endothelial cell infection and early disease events. *Nat Rev Microbiol* **6:**375–386.

103. **Andersson SG, Zomorodipour A, Andersson JO, Sicheritz-Pontén T, Alsmark UC, Podowski RM, Näslund AK, Eriksson AS, Winkler HH, Kurland CG.** 1998. The genome sequence of *Rickettsia prowazekii* and the origin of mitochondria. *Nature* **396:**133–140.

104. **Winkler HH.** 1995. *Rickettsia prowazekii*, ribosomes and slow growth. *Trends Microbiol* **3:**196–198.

105. **Winkler HH, Daugherty RM.** 1986. Acquisition of glucose by *Rickettsia prowazekii* through the nucleotide intermediate uridine 5′-diphosphoglucose. *J Bacteriol* **167:**805–808.

106. **Zomorodipour A, Andersson SG.** 1999. Obligate intracellular parasites: *Rickettsia prowazekii* and *Chlamydia trachomatis*. *FEBS Lett* **452:**11–15.

107. **Fuxelius HH, Darby A, Min CK, Cho NH, Andersson SG.** 2007. The genomic and metabolic diversity of *Rickettsia*. *Res Microbiol* **158:**745–753.

108. **Frohlich KM, Audia JP.** 2013. Dual mechanisms of metabolite acquisition by the obligate intracytosolic pathogen *Rickettsia prowazekii* reveal novel aspects of triose phosphate transport. *J Bacteriol* **195:**3752–3760.

109. **Smith DK, Winkler HH.** 1977. Characterization of a lysine-specific active transport system in *Rickettsia prowazeki. J Bacteriol* **129:**1349–1355.

110. **Atkinson WH, Winkler HH.** 1989. Permeability of *Rickettsia prowazekii* to NAD. *J Bacteriol* **171:**761–766.

111. **Tucker AM, Winkler HH, Driskell LO, Wood DO.** 2003. S-adenosylmethionine transport in *Rickettsia prowazekii. J Bacteriol* **185:**3031–3035.

112. **Frohlich KM, Roberts RA, Housley NA, Audia JP.** 2010. *Rickettsia prowazekii* uses an sn-glycerol-3-phosphate dehydrogenase and a novel dihydroxyacetone phosphate transport system to supply triose phosphate for phospholipid biosynthesis. *J Bacteriol* **192:**4281–4288.

113. **Sahni SK, Rydkina E.** 2009. Host-cell interactions with pathogenic *Rickettsia* species. *Future Microbiol* **4:**323–339.

114. **Clifton DR, Goss RA, Sahni SK, van Antwerp D, Baggs RB, Marder VJ, Silverman DJ, Sporn LA.** 1998. NF-kappa B-dependent inhibition of apoptosis is essential for host cellsurvival during *Rickettsia rickettsii* infection. *Proc Natl Acad Sci U S A* **95:**4646–4651.

115. **Malik-Kale P, Jolly CE, Lathrop S, Winfree S, Luterbach C, Steele-Mortimer O.** 2011. *Salmonella* - at home in the host cell. *Front Microbiol* **2:**125.

116. **Swart AL, Hensel M.** 2012. Interactions of *Salmonella enterica* with dendritic cells. *Virulence* **3:**660–667.

117. **de Jong HK, Parry CM, van der Poll T, Wiersinga WJ.** 2012. Host-pathogen interaction in invasive Salmonellosis. *PLoS Pathog* **8:**e1002933. doi:10.1371/journal.ppat.1002933

118. **van der Heijden J, Finlay BB.** 2012. Type III effector-mediated processes in *Salmonella* infection. *Future Microbiol* **7:**685–703.

119. **Fàbrega A, Vila J.** 2013. *Salmonella enterica* serovar Typhimurium skills to succeed in the host: virulence and regulation. *Clin Microbiol Rev* **26:**308–341.

120. **McClelland M, Sanderson KE, Spieth J, Clifton SW, Latreille P, Courtney L, Porwollik S, Ali J, Dante M, Du F, Hou S, Layman D, Leonard S, Nguyen C, Scott K, Holmes A, Grewal N, Mulvaney E, Ryan E, Sun H, Florea L, Miller W, Stoneking T, Nhan M, Waterston R, Wilson RK.** 2001. Complete genome sequence of *Salmonella enterica* serovar Typhimurium LT2. *Nature* **413:**852–856.

121. **Haraga A, Ohlson MB, Miller SI.** 2008. Salmonellae interplay with host cells. *Nat Rev Microbiol* **6:**53–66.

122. **Bumann D.** 2009. System-level analysis of *Salmonella* metabolism during infection. *Curr Opin Microbiol* **12:**559–567.

123. **Bowden SD, Rowley G, Hinton JC, Thompson A.** 2009. Glucose and glycolysis are required for the successful infection of macrophages and mice by *Salmonella enterica* serovar typhimurium. *Infect Immun* **77:**3117–3126.

124. **Tchawa Yimga M, Leatham MP, Allen JH, Laux DC, Conway T, Cohen PS.** 2006. Role of gluconeogenesis and the tricarboxylic acid cycle in the virulence of *Salmonella enterica* serovar Typhimurium in BALB/c mice. *Infect Immun* **74:**1130–1140.

125. **Moest TP, Méresse S.** 2013. Salmonella T3SSs: successful mission of the secret(ion) agents. *Curr Opin Microbiol* **16:**38–44.

126. **Knodler LA, Finlay BB, Steele-Mortimer O.** 2005. The *Salmonella* effector protein SopB protects epithelial cells from apoptosis by sustained activation of Akt. *J Biol Chem* **280:**9058–9064.

127. **Yin C, Qie S, Sang N.** 2012. Carbon source metabolism and its regulation in cancer cells. *Crit Rev Eukaryot Gene Expr* **22:**17–35.

128. **Kuijl C, Savage ND, Marsman M, Tuin AW, Janssen L, Egan DA, Ketema M, van den Nieuwendijk R, van den Eeden SJ, Geluk A, Poot A, van der Marel G, Beijersbergen RL, Overkleeft H, Ottenhoff TH, Neefjes J.** 2007. Intracellular bacterial growth is controlled by a kinase network around PKB/AKT1. *Nature* **450:**725–730.

129. **Hardt WD, Gálan JE.** 1997. A secreted *Salmonella* protein with homology to an avirulence determinant of plant pathogenic bacteria. *Proc Natl Acad Sci U S A* **94:**9887–9892.

130. **Liu X, Lu R, Xia Y, Wu S, Sun J.** 2010. Eukaryotic signaling pathways targeted by *Salmonella* effector protein AvrA in intestinal infection *in vivo. BMC Microbiol* **10:**326.

131. **Wu S, Ye Z, Liu X, Zhao Y, Xia Y, Steiner A, Petrof EO, Claud EC, Sun J.** 2010. *Salmonella typhimurium* infection increases p53 acetylation in intestinal epithelial cells. *Am J Physiol Gastrointest Liver Physiol* **298:**G784–G794.

132. **Thiennimitr P, Winter SE, Winter MG, Xavier MN, Tolstikov V, Huseby DL, Sterzenbach T, Tsolis RM, Roth JR, Bäumler AJ.** 2011. Intestinal inflammation allows

Salmonella to use ethanolamine to compete with the microbiota. *Proc Natl Acad Sci U S A* **108:**17480–17485.

133. **Bertin Y, Girardeau JP, Chaucheyras-Durand F, Lyan B, Pujos-Guillot E, Harel J, Martin C.** 2011. Enterohaemorrhagic *Escherichia coli* gains a competitive advantage by using ethanolamine as a nitrogen source in the bovine intestinal content. *Environ Microbiol* **13:**365–377.

134. **Shi L, Chowdhury SM, Smallwood HS, Yoon H, Mottaz-Brewer HM, Norbeck AD, McDermott JE, Clauss TR, Heffron F, Smith RD, Adkins JN.** 2009. Proteomic investigation of the time course responses of RAW 264.7 macrophages to infection with *Salmonella enterica. Infect Immun* **77:**3227–3233.

135. **Uchiya K, Nikai T.** 2004. *Salmonella enterica* serovar Typhimurium infection induces cyclooxygenase 2 expression in macrophages: involvement of *Salmonella* pathogenicity island 2. *Infect Immun* **72:**6860–6869.

136. **Fraser DW, Tsai TR, Orenstein W, Parkin WE, Beecham HJ, Sharrar RG, Harris J, Mallison GF, Martin SM, McDade JE, Shepard CC, Brachman PS.** 1977. Legionnaires' disease: description of an epidemic of pneumonia. *N Engl J Med* **297:**1189–1197.

137. **Xu L, Luo ZQ.** 2013. Cell biology of infection by *Legionella pneumophila. Microbes Infect* **15:**157–167.

138. **Tilney LG, Harb OS, Connelly PS, Robinson CG, Roy CR.** 2001. How the parasitic bacterium *Legionella pneumophila* modifies its phagosome and transforms it into rough ER: implications for conversion of plasma membrane to the ER membrane. *J Cell Sci* **114:**4637–4650.

139. **Heidtman M, Chen EJ, Moy MY, Isberg RR.** 2009. Large-scale identification of *Legionella pneumophila* Dot/Icm substrates that modulate host cell vesicle trafficking pathways. *Cell Microbiol* **11:**230–248.

140. **Zhu W, Banga S, Tan Y, Zheng C, Stephenson R, Gately J, Luo ZQ.** 2011. Comprehensive identification of protein substrates of the Dot/Icm type IV transporter of *Legionella pneumophila. PLoS One* **6:**e17638. doi:10.1371/journal.pone.0017638

141. **Faucher SP, Mueller CA, Shuman HA.** 2011. *Legionella pneumophila* transcriptome during intracellular multiplication in human macrophages. *Front Microbiol* **2:**60.

142. **Isberg RR, O'Connor TJ, Heidtman M.** 2009. The *Legionella pneumophila* replication vacuole: making a cosy niche inside host cells. *Nat Rev Microbiol* **7:**13–24.

143. **Xu L, Shen X, Bryan A, Banga S, Swanson MS, Luo ZQ.** 2010. Inhibition of host vacuolar H$^+$-ATPase activity by a *Legionella pneumophila* effector. *PLoS Pathog* **6:**e1000822. doi:10.1371/journal.ppat.1000822

144. **Chien M, Morozova I, Shi S, Sheng H, Chen J, Gomez SM, Asamani G, Hill K, Nuara J, Feder M, Rineer J, Greenberg JJ, Steshenko V, Park SH, Zhao B, Teplitskaya E, Edwards JR, Pampou S, Georghiou A, Chou IC, Iannuccilli W, Ulz ME, Kim DH, Geringer-Sameth A, Goldsberry C, Morozov P, Fischer SG, Segal G, Qu X, Rzhetsky A, Zhang P, Cayanis E, De Jong PJ, Ju J, Kalachikov S, Shuman HA, Russo JJ.** 2004. The genomic sequence of the accidental pathogen *Legionella pneumophila. Science* **305:**1966–1968.

145. **Warren WJ, Miller RD.** 1979. Growth of Legionnaires disease bacterium (*Legionella pneumophila*) in chemically defined medium. *J Clin Microbiol* **10:**50–55.

146. **Eylert E, Herrmann V, Jules M, Gillmaier N, Lautner M, Buchrieser C, Eisenreich W, Heuner K.** 2010. Isotopologue profiling of *Legionella pneumophila*: role of serine and glucose as carbon substrates. *J Biol Chem* **285:**22232–22243.

147. **Brüggemann H, Cazalet C, Buchrieser C.** 2006. Adaptation of *Legionella pneumophila* to the host environment: role of protein secretion, effectors and eukaryotic-like proteins. *Curr Opin Microbiol* **9:**86–94.

148. **Heuner K, Eisenreich W.** 2013. The intracellular metabolism of legionella by isotopologue profiling. *Methods Mol Biol* **954:**163–181.

149. **Schwöppe C, Winkler HH, Neuhaus HE.** 2002. Properties of the glucose-6-phosphate transporter from *Chlamydia pneumoniae* (HPTcp) and the glucose-6-phosphate sensor from *Escherichia coli* (UhpC). *J Bacteriol* **184:**2108–2115.

150. **George JR, Pine L, Reeves MW, Harrell WK.** 1980. Amino acid requirements of *Legionella pneumophila. J Clin Microbiol* **11:**286–291.

151. **Tesh MJ, Morse SA, Miller RD.** 1983. Intermediary metabolism in *Legionella pneumophila*: utilization of amino acids and other compounds as energy sources. *J Bacteriol* **154:**1104–1109.

152. **Sauer JD, Bachman MA, Swanson MS.** 2005. The phagosomal transporter A couples threonine acquisition to differentiation and replication of *Legionella pneumophila* in macrophages. *Proc Natl Acad Sci U S A* **102:**9924–9929.

153. **Brüggemann H, Hagman A, Jules M, Sismeiro O, Dillies MA, Gouyette C, Kunst F, Steinert M, Heuner K, Coppée JY, Buchrieser C.** 2006. Virulence strategies for

infecting phagocytes deduced from the *in vivo* transcriptional program of *Legionella pneumophila*. *Cell Microbiol* **8**:1228–1240.

154. **Aurass P, Pless B, Rydzewski K, Holland G, Bannert N, Flieger A.** 2009. bdhA-patD operon as a virulence determinant, revealed by a novel large-scale approach for identification of *Legionella pneumophila* mutants defective for amoeba infection. *Appl Environ Microbiol* **75**:4506–4515.

155. **Herrmann V, Eidner A, Rydzewski K, Blädel I, Jules M, Buchrieser C, Eisenreich W, Heuner K.** 2011. GamA is a eukaryotic-like glucoamylase responsible for glycogen- and starch-degrading activity of *Legionella pneumophila*. *Int J Med Microbiol* **301**:133–139.

156. **Wieland H, Ullrich S, Lang F, Neumeister B.** 2005. Intracellular multiplication of *Legionella pneumophila* depends on host cell amino acid transporter SLC1A5. *Mol Microbiol* **55**:1528–1537.

157. **Otto GP, Wu MY, Clarke M, Lu H, Anderson OR, Hilbi H, Shuman HA, Kessin RH.** 2004. Macroautophagy is dispensable for intracellular replication of *Legionella pneumophila* in *Dictyostelium discoideum*. *Mol Microbiol* **51**:63–72.

158. **Amer AO, Swanson MS.** 2005. Autophagy is an immediate macrophage response to *Legionella pneumophila*. *Cell Microbiol* **7**:765–778.

159. **Tung SM, Unal C, Ley A, Peña C, Tunggal B, Noegel AA, Krut O, Steinert M, Eichinger L.** 2010. Loss of Dictyostelium ATG9 results in a pleiotropic phenotype affecting growth, development, phagocytosis and clearance and replication of *Legionella pneumophila*. *Cell Microbiol* **12**:765–780.

160. **Price CT, Al-Quadan T, Santic M, Rosenshine I, Abu Kwaik Y.** 2011. Host proteasomal degradation generates amino acids essential for intracellular bacterial growth. *Science* **334**:1553–1557.

161. **Hubber A, Kubori T, Nagai H.** 2013. Modulation of the ubiquitination machinery by *Legionella*. *Curr Top Microbiol Immunol* **376**:227–247.

162. **Losick VP, Isberg RR.** 2006. NF-kappaB translocation prevents host cell death after low-dose challenge by *Legionella pneumophila*. *J Exp Med* **203**:2177–2189.

163. **Losick VP, Haenssler E, Moy MY, Isberg RR.** 2010. LnaB: a *Legionella pneumophila* activator of NF-kappaB. *Cell Microbiol* **12**:1083–1097.

164. **Rathore MG, Saumet A, Rossi JF, de Bettignies C, Tempé D, Lecellier CH, Villalba M.** 2012. The NF-kappaB member p65 controls glutamine metabolism through miR-23a. *Int J Biochem Cell Biol* **44**:1448–1456.

165. **Li Z, Dugan AS, Bloomfield G, Skelton J, Ivens A, Losick V, Isberg RR.** 2009. The amoebal MAP kinase response to *Legionella pneumophila* is regulated by DupA. *Cell Host Microbe* **6**:253–267.

166. **Hervet E, Charpentier X, Vianney A, Lazzaroni JC, Gilbert C, Atlan D, Doublet P.** 2011. Protein kinase LegK2 is a type IV secretion system effector involved in endoplasmic reticulum recruitment and intracellular replication of *Legionella pneumophila*. *Infect Immun* **79**:1936–1950.

167. **Wright HR, Turner A, Taylor HR.** 2008. Trachoma. *Lancet* **371**:1945–1954.

168. **Blasi F, Tarsia P, Aliberti S.** 2009. *Chlamydophila pneumoniae*. *Clin Microbiol Infect* **15**:29–35.

169. **Betts HJ, Wolf K, Fields KA.** 2009. Effector protein modulation of host cells: examples in the *Chlamydia* spp. arsenal. *Curr Opin Microbiol* **12**:81–87.

170. **Campbell LA, Kuo CC.** 2004. *Chlamydia pneumoniae*--an infectious risk factor for atherosclerosis? *Nat Rev Microbiol* **2**:23–32.

171. **Haider S, Wagner M, Schmid MC, Sixt BS, Christian JG, Hacker G, Pichler P, Mechtler K, Müller A, Baranyi C, Toenshoff ER, Montanaro J, Horn M.** 2010. Raman microspectroscopy reveals long-term extracellular activity of Chlamydiae. *Mol Microbiol* **77**:687–700.

172. **Sixt BS, Siegl A, Müller C, Watzka M, Wultsch A, Tziotis D, Montanaro J, Richter A, Schmitt-Kopplin P, Horn M.** 2013. Metabolic features of *Protochlamydia amoebophila* elementary bodies--a link between activity and infectivity in Chlamydiae. *PLoS Pathog* **9**: e1003553.

173. **van Ooij C, Kalman L, van Ijzendoorn, Nishijima M, Hanada K, Mostov K, Engel JN.** 2000. Host cell-derived sphingolipids are required for the intracellular growth of *Chlamydia trachomatis*. *Cell Microbiol* **2**:627–637.

174. **Robertson DK, Gu L, Rowe RK, Beatty WL.** 2009. Inclusion biogenesis and reactivation of persistent *Chlamydia trachomatis* requires host cell sphingolipid biosynthesis. *PLoS Pathog* **5**:e1000664. doi:10.1371/journal.ppat.1000664

175. **Wyrick PB.** 2000. Intracellular survival by *Chlamydia*. *Cell Microbiol* **2**:275–282.

176. **Schoborg RV.** 2011. *Chlamydia* persistence – a tool to dissect chlamydia--host interactions. *Microbes Infect* **13**:649–662.

177. **Stephens RS, Kalman S, Lammel C, Fan J, Marathe R, Aravind L, Mitchell W, Olinger L, Tatusov RL, Zhao Q, Koonin EV, Davis RW.** 1998. Genome sequence of an obligate

intracellular pathogen of humans: *Chlamydia trachomatis*. *Science* **282:**754–759.

178. **Kalman S, Mitchell W, Marathe R, Lammel C, Fan J, Hyman RW, Olinger L, Grimwood J, Davis RW, Stephens RS.** 1999. Comparative genomes of *Chlamydia pneumoniae* and *C. trachomatis*. *Nat Genet* **21:**385–389.

179. **Hatch TP, Al-Hossainy E, Silverman JA.** 1982. Adenine nucleotide and lysine transport in *Chlamydia psittaci*. *J Bacteriol* **150:**662–670.

180. **Wyllie S, Ashley RH, Longbottom D, Herring AJ.** 1998. The major outer membrane protein of *Chlamydia psittaci* functions as a porin-like ion channel. *Infect Immun* **66:**5202–5207.

181. **Trentmann O, Horn M, van Scheltinga AC, Neuhaus HE, Haferkamp I.** 2007. Enlightening energy parasitism by analysis of an ATP/ADP transporter from chlamydiae. *PLoS Biol* **5:**e231. doi:10.1371/journal.pbio.0050231

182. **Gérard HC, Freise J, Wang Z, Roberts G, Rudy D, Krauss-Opatz B, Köhler L, Zeidler H, Schumacher HR, Whittum-Hudson JA, Hudson AP.** 2002. *Chlamydia trachomatis* genes whose products are related to energy metabolism are expressed differentially in active vs. persistent infection. *Microbes Infect* **4:**13–22.

183. **Skipp P, Robinson J, O'Connor CD, Clarke IN.** 2005. Shotgun proteomic analysis of *Chlamydia trachomatis*. *Proteomics* **5:**1558–1573.

184. **Iliffe-Lee ER, McClarty G.** 2000. Regulation of carbon metabolism in *Chlamydia trachomatis*. *Mol Microbiol* **38:**20–30.

185. **Lu C, Lei L, Peng B, Tang L, Ding H, Gong S, Li Z, Wu Y, Zhong G.** 2013. *Chlamydia trachomatis* GlgA is secreted into host cell cytoplasm. *PLoS One* **8:**e68764. doi:10.1371/journal.pone.0068764

186. **Meurice G, Deborde C, Jacob D, Falentin H, Boyaval P, Dimova D.** 2004. In silico exploration of the fructose-6-phosphate phosphorylation step in glycolysis: genomic evidence of the coexistence of an atypical ATP-dependent along with a PPi-dependent phosphofructokinase in *Propionibacterium freudenreichii* subsp. shermanii. *In Silico Biol* **4:**517–528.

187. **Siebers B, Klenk HP, Hensel R.** 1998. PPi-dependent phosphofructokinase from *Thermoproteus tenax*, an archaeal descendant of an ancient line in phosphofructokinase evolution. *J Bacteriol* **180:**2137–2143.

188. **Tjaden J, Winkler HH, Schwoppe C, Van Der Laan M, Mohlmann T, Neuhaus HE.** 1999. Two nucleotide transport proteins in *Chlamydia trachomatis*, one for net nucleoside triphosphate uptake and the other for transport of energy. *J Bacteriol* **181:**1196–1202.

189. **Zhong G, Fan P, Ji H, Dong F, Huang Y.** 2001. Identification of a chlamydial protease-like activity factor responsible for the degradation of host transcription factors. *J Exp Med* **193:**935–942.

190. **Wu X, Lei L, Gong S, Chen D, Flores R, Zhong G.** 2011. The chlamydial periplasmic stress response serine protease cHtrA is secreted into host cell cytosol. *BMC Microbiol* **11:**87.

191. **Chen AL, Johnson KA, Lee JK, Sütterlin C, Tan M.** 2012. CPAF: a Chlamydial protease in search of an authentic substrate. *PLoS Pathog* **8:**e1002842. doi:10.1371/journal.ppat.1002842

192. **Fox A, Rogers JC, Gilbart J, Morgan S, Davis CH, Knight S, Wyrick PB.** 1990. Muramic acid is not detectable in *Chlamydia psittaci* or *Chlamydia trachomatis* by gas chromatography-mass spectrometry. *Infect Immun* **58:**835–837.

193. **Wood H, Fehlner-Gardner C, Berry J, Fischer E, Graham B, Hackstadt T, Roshick C, McClarty G.** 2003. Regulation of tryptophan synthase gene expression in *Chlamydia trachomatis*. *Mol Microbiol* **49:**1347–1359.

194. **Pfefferkorn ER.** 1984. Interferon gamma blocks the growth of *Toxoplasma gondii* in human fibroblasts by inducing the host cells to degrade tryptophan. *Proc Natl Acad Sci U S A* **81:**908–912.

195. **Wylie JL, Hatch GM, McClarty G.** 1997. Host cell phospholipids are trafficked to and then modified by *Chlamydia trachomatis*. *J Bacteriol* **179:**7233–7242.

196. **Beatty WL.** 2007. Lysosome repair enables host cell survival and bacterial persistence following *Chlamydia trachomatis* infection. *Cell Microbiol* **9:**2141–2152.

197. **Rupp J, Gieffers J, Klinger M, van Zandbergen G, Wrase R, Maass M, Solbach W, Deiwick J, Hellwig-Burgel T.** 2007. *Chlamydia pneumoniae* directly interferes with HIF-1alpha stabilization in human host cells. *Cell Microbiol* **9:**2181–2191.

198. **Werth N, Beerlage C, Rosenberger C, Yazdi AS, Edelmann M, Amr A, Bernhardt W, von Eiff C, Becker K, Schafer A, Peschel A, Kempf VA.** 2010. Activation of hypoxia inducible factor 1 is a general phenomenon in infections with human pathogens. *PLoS One* **5:**e11576. doi:10.1371/journal.pone.0011576

199. **Natividad A, Freeman TC, Jeffries D, Burton MJ, Mabey DC, Bailey RL, Holland MJ.** 2010. Human conjunctival transcriptome analysis reveals the prominence of innate defense in *Chlamydia trachomatis* infection. *Infect Immun* **78:**4895–4911.

200. **Fields KA, Mead DJ, Dooley CA, Hackstadt T.** 2003. *Chlamydia trachomatis* type III secretion: evidence for a functional apparatus during early-cycle development. *Mol Microbiol* **48:**671–683.

201. **Peters J, Wilson DP, Myers G, Timms P, Bavoil PM.** 2007. Type III secretion a la *Chlamydia. Trends Microbiol* **15:**241–251.

202. **Valdivia RH.** 2008. *Chlamydia* effector proteins and new insights into chlamydial cellular microbiology. *Curr Opin Microbiol* **11:**53–59.

203. **Niu H, Xiong Q, Yamamoto A, Hayashi-Nishino M, Rikihisa Y.** 2012. Autophagosomes induced by a bacterial Beclin 1 binding protein facilitate obligatory intracellular infection. *Proc Natl Acad Sci U S A* **109:**20800–20807.

204. **Young VR.** 1994. Adult amino acid requirements: the case for a major revision in current recommendations. *J Nutr* **124:**1517S–1523S.

205. **Alkhuder K, Meibom KL, Dubail I, Dupuis M, Charbit A.** 2009. Glutathione provides a source of cysteine essential for intracellular multiplication of *Francisella tularensis. PLoS Pathog* **5:**e1000284. doi:10.1371/journal.ppat.1000284

206. **Kanehisa Laboratories.** 2014. *KEGG Genes Database.* http://www.genome.jp/kegg/genes.html

Small RNAs Regulate Primary and Secondary Metabolism in Gram-negative Bacteria

4

MAKSYM BOBROVSKYY,[1] CARIN K. VANDERPOOL,[1] and
GREGORY R. RICHARDS[2]

MECHANISMS OF REGULATION BY SRNAS

Base-Pairing–Dependent sRNA Regulation

Characteristics of base-pairing sRNA regulation

A diverse palette of novel small regulatory RNAs (sRNAs) has been identified in bacteria in recent years, and many play important roles in regulating gene expression and adaptation to constantly changing physiological and metabolic needs. To date, more than a hundred sRNAs have been identified and characterized in gram-negative bacteria, with crucial and sometimes global regulatory functions that can easily rival protein regulators. The most studied and widely dispersed class of sRNAs in gram-negative bacteria act post-transcriptionally by base pairing to target mRNAs in order to effect positive or negative regulatory outcomes.

Base-pairing sRNAs are broadly assigned into two main groups based on their location on the chromosome relative to the genes they regulate. sRNAs encoded at loci distinct from their regulated genes are said to be *trans*-encoded, and are characterized by limited and imperfect base

[1]Department of Microbiology, University of Illinois, Urbana, IL 61801; [2]Biological Sciences Department, University of Wisconsin-Parkside, Kenosha, WI 53141.
Metabolism and Bacterial Pathogenesis
Edited by Tyrrell Conway and Paul Cohen
© 2015 American Society for Microbiology, Washington, DC
doi:10.1128/microbiolspec.MBP-0009-2014

pairing interactions with the transcripts they target. *Trans*-encoded sRNAs are usually 50 nucleotides (nt) to 300 nt long and are expressed in response to specific physiological, metabolic, and/or stress signals such as iron limitation (1), sugar–phosphate overload (2), oxidative stress (3), growth-phase transition (4), anaerobiasis (5) and many others (6–8). sRNAs encoded on the DNA strand directly opposite their targets (antisense) are known as *cis*-encoded sRNAs and share a region of perfect complementarity with the genes they regulate. *Cis*-encoded sRNAs vary greatly in size, ranging from 100 nt to 7000 nt in length (9–14). *Cis*-encoded sRNAs on plasmids, phages, and transposons were among the first regulatory RNAs identified. Later, sRNAs encoded in *cis* were found on the chromosomes of many bacteria. Several of them are encoded antisense to transposase genes and inhibit transposition when expressed (11, 15–18). Other *cis*-encoded sRNAs act as antitoxins, regulating expression of small, hydrophobic proteins are toxic to bacterial cells if they accumulate to high levels (10, 19, 20). Together, *trans*- and *cis*-encoded sRNAs are now known to regulate countless mRNA targets and to play a role in most global regulatory responses in bacteria.

Both *trans*- and *cis*-encoded sRNAs base pair with target mRNAs and regulate their expression at the level of translation and/or mRNA stability (6, 21). sRNA–mRNA complex formation often involves rearrangements of the secondary structure of each individual interacting RNA molecule (6). For *trans*-encoded sRNAs, RNA stability and secondary structure rearrangements that allow annealing with mRNA targets often depend on a conserved and abundant RNA chaperone Hfq, described in more detail below. In contrast, *cis*-encoded sRNAs usually do not rely on Hfq for stability or to facilitate base pairing, potentially due to their extensive complementarity with their target transcripts.

Mechanisms of activation by base pairing sRNAs

Post-transcriptional activation of gene expression by sRNAs is achieved via stimulation of translation or stabilization of the target transcript. Leader sequences in the 5′ untranslated regions (UTRs) of some mRNAs form secondary structures that occlude ribosome binding sites (RBS) or translational enhancer elements. The binding of sRNAs to complementary sequences within such 5′ UTRs can prevent formation of the translation-inhibitory structure and allow enhanced translation of the mRNA (22–24). Post-transcriptional regulation of the stationary phase sigma factor (σ^S), encoded by *rpoS*, is the paradigm for positive regulation by sRNAs in enteric bacteria, such as *E. coli*. Three Hfq-dependent sRNAs, DsrA, RprA, and ArcZ, activate translation of *rpoS* under conditions of low-temperature, osmotic shock, and aerobic stationary phase growth, respectively (25–27). The long 5′ UTR of *rpoS* mRNA folds into a stem-loop structure that inhibits translation of the message (28, 29). Each of the three sRNAs can base pair with sequences in the *rpoS* 5′ UTR to prevent formation of the inhibitory hairpin structure and allow ribosomes to access the RBS (25–27, 30, 31).

Recently, examples of translation-independent activation of gene expression by sRNAs have been reported. In these cases, the sRNA does not directly stimulate translation, but rather stabilizes the mRNA (32–34). For example, the sRNA SgrS, the master regulator of the glucose-phosphate stress response, stabilizes *yigL* mRNA, coding for a haloacid dehalogenase (HAD)-like sugar phosphatase (33, 35). A dicistronic mRNA, *pldB-yigL* is processed by RNase E, cleaving the transcript within the *pldB* coding sequence (∼200 nt upstream of *yigL* start codon). In the absence of SgrS, processed fragments are further degraded. When present, SgrS base pairs at a site downstream of the processing site (within the 3′ region of *pldB*) and prevents further RNase-E–mediated degradation, thus stabilizing the 'pldB-yigL mRNA. In essence, SgrS competes with RNase E and stabilizes the yigL mRNA, allowing more rounds of translation and increased production of YigL protein (33). Another sRNA, RydC, post-transcriptionally activates cfa, a gene encoding cyclopropane fatty acid synthase, by a similar mechanism. RydC stabilizes cfa mRNA by interfering with the RNase E-dependent degradation (34). It is likely that even more examples of stability-dependent activation by sRNAs will be emerging because ribonucleolytic decay can be readily interfered with through sRNA-mediated occlusion of processing sites recognized by various endoribonucleases.

Mechanisms of repression by base-pairing sRNAs

Negative regulation of gene expression by sRNAs is highly prevalent in bacteria and has been well documented. The canonical mechanism of repression involves sRNA interference with translation initiation of the target mRNA. The translation initiation region (TIR) of the mRNA may consist of several cis-elements important for translation, namely the RBS (36), translation enhancer element (37, 38), ribosome standby site (39) and leader open reading frame (ORF) (40), any of which may be targeted for inhibition by sRNAs.

Many sRNAs base pair within the leader sequence (5′ UTR) of the mRNA target and inhibit translation by blocking the RBS. Established boundaries of the ribosome footprint are 15 nt upstream and 20 nt downstream of the start codon (41, 42), and sRNA pairing within this region results in inhibition of translation initiation. For example, the sRNA SgrS, which will be described in more detail below, regulates ptsG mRNA by pairing at the RBS and blocking ribosome access, successfully inhibiting synthesis of the PtsG protein (36). However, sRNAs binding outside of this region have been reported to inhibit translation in Escherichia coli and Salmonella (37, 39, 43, 44). RBS occlusion can occur via sRNA binding

at a distance via the Hfq protein (45). Alternatively, some sRNAs target enhancer elements to reduce translation initiation (46). The best-characterized example of this is the sRNA GcvB (discussed more below), which inhibits translation of several mRNAs by base pairing upstream of the RBS, at C/A-rich regions that were shown to enhance translation (37, 38). Repression by sRNA occlusion of other elements in the translation initiation region, such as putative ribosome standby sites (39) or leader ORFs (47), are not well characterized, but have been described as mechanisms of post-transcriptional control (46).

Although translation inhibition alone is often sufficient to achieve gene silencing (36), the sRNA-mRNA duplexes formed to cause translational repression are also typically targeted for degradation (48–50). Few cases are known where sRNA acts solely by either translational repression (51) or by directly promoting degradation (52). In E. coli and related organisms, duplex decay is mediated by recruiting the endoribonuclease RNase E, a major component of the degradosome often involved in trans-encoded sRNA-mediated regulation (12, 48–50) or RNase III, an endoribonuclease that recognizes double-stranded RNA as substrate for processing (53, 54). Together, translation inhibition and duplex destabilization provide an effective and irreversible system for post-transcriptional repression of gene expression in bacteria.

Transcription inhibition by cis-encoded sRNAs

Transcriptional regulation by cis-encoded sRNAs can occur via transcription interference or attenuation. Transcription interference takes place when transcription from one gene is inhibited by transcription from the cis-encoded, convergently oriented gene on the opposite strand. In contrast to post-transcriptional regulatory mechanisms and transcription attenuation, transcription interference does not require direct interaction between the sRNA and a cognate mRNA. Rather, premature termination of transcription can occur when convergently replicating RNA polymerases collide, resulting in dissociation, backtracking, or stalling of one or both polymerase complexes (55, 56). Mathematical models predict additional mechanisms of interference where elongating RNA polymerase can collide with a separate polymerase open complex at the sensitive promoter and prevent it from proceeding to elongation (57), or where the promoter of one gene is blocked by the pausing of the polymerase bound to the gene on the opposite strand (58). A transcription interference mechanism has been suggested for the sRNA-encoded antisense to the *ubiG-mccBA* operon involved in methionine to cysteine conversion in *Clostridium acetobutylicum* (59). Transcription attenuation, on the other hand, requires base pairing between the sRNA and an mRNA target. Attenuation occurs when the sRNA binds to the elongating mRNA and induces secondary structural rearrangements that inhibit the formation of an antiterminator structure and favor the formation of a terminator hairpin (60, 61). In *Vibrio anguillarum* species, a non-coding sRNA RNAβ interferes with transcription of *angR* by pairing within the intergenic region of *fatA-angR* mRNA, which results in premature transcription termination (60).

Base-Pairing–Independent sRNA Regulation

Protein-binding sRNAs

While the vast majority of characterized sRNAs utilize base pairing as the means of regulating gene expression, a few sRNAs are known to bind and modulate protein activity. A subset of these sRNAs modifies activity and abundance of the proteins that bind nucleic acid substrates by mimicking structures of their cognate targets. A well-characterized example of this category of sRNAs is 6S RNA, and it is highly conserved in a wide variety of bacteria, including gram-negative bacteria like *E. coli*, *Pseudomonas aeruginosa* and *Haemophilus influenzae* (62, 63). *E. coli* 6S RNA preferentially binds and inhibits RNA polymerase holoenzyme containing the housekeeping sigma factor, σ^{70} (62, 64). In 6S RNA, there is a highly conserved secondary structure that consists of a double-stranded RNA hairpin with a bubble that resembles the open complex of DNA promoters. This mimicry enables 6S RNA to fit into the active site of the RNA polymerase, preventing it from initiating transcription (63). Expression of 6S RNA increases in the stationary phase of growth and is highest in late-stationary phase, when it inhibits transcription from a subset of σ^{70} promoters (62, 65–67). While 6S RNA sequesters σ^{70} and inhibits expression from some σ^{70}-dependent promoters, expression of the alternative sigma factor σ^{S} is upregulated and induces transcription of genes necessary for stationary phase growth (65). In this way, 6S RNA is thought to help mediate a switch in sigma factor usage in late stationary phase.

Another class of protein-binding sRNAs is exemplified by transfer-messenger RNA (tmRNA) involved in *trans*-translation, a mechanism to rescue stalled ribosomes in bacteria (68, 69). Defective mRNAs that lack the termination signals allowing ribosome release cause stalling of the ribosomes and results in a general loss of translational efficiency, and accumulation of aberrant mRNA and nascent peptides (70, 71). The tmRNA can be charged with alanine and contains a short mRNA reading frame (72–74). tmRNA enters the ribosomal A-site on stalled ribosomes, and translation then proceeds using the tmRNA as a template. Translation terminates at the tmRNA-encoded stop codon, and the defective mRNA and "tagged" polypeptide are released from the ribosome. The tagged polypeptide is targeted for degradation and the ribosome is free to participate in further rounds of translation (68–70, 75). Yet another distinct class of protein-binding RNAs includes two sRNAs, CsrB and CsrC, found in *E. coli* and related bacteria (76). These sRNAs act by binding to and sequestering the translational regulator CsrA, which regulates a large number of genes related to carbon storage and metabolism (77). The CsrB/C family of sRNAs will be discussed in more detail below. These select examples provide just a glimpse into the variety of mechanisms by which bacterial sRNAs can modulate activity, sequester, or provide structural scaffolding to target proteins. It is conceivable that a much greater variety exists and will be uncovered by further searches for novel protein-binding sRNAs.

Protein Factors Globally Involved in sRNA Regulation

Hfq: A major facilitator of sRNA regulation

As previously mentioned, Hfq is a widely conserved and highly abundant (reported estimates vary from ~400 (78) to 10,000 (79, 80) copies per *E. coli* cell) RNA chaperone protein involved in regulation by sRNAs in bacteria (81). Hfq (host factor for Qβ) was first identified as a bacterial host factor involved in the replication of the Qβ phage in *E. coli* (82). It binds the 3′ end of the viral plus-strand RNA and is required for minus-strand synthesis by Qβ replicase (83). A much broader function for Hfq was recognized when its inactivation in *E. coli* and *Salmonella* resulted in pleiotropic growth and virulence phenotypes (84–86), in addition to a general deregulation of ~20% of *Salmonella* genome (87). Of interest was that *E. coli* *hfq* mutants adjusted poorly to environmental stresses such as stationary phase growth, osmotic imbalance, oxidation and others (88–90). These defective stress phenotypes were later found to be due to Hfq's vital role in the sRNA regulatory network responding to a variety of stresses and growth conditions. Hfq regulates its own expression at the post-transcriptional level by binding within the *hfq* mRNA leader sequence, which

interferes with translation initiation (91). This phenomenon is particularly of note because normally sRNAs associated with Hfq determine the specificity of regulation; in this case, Hfq seems to act alone to alter translation of *hfq* mRNA. Hfq may in fact act alone on a number of other mRNA targets as evidenced by a recent study showing that Hfq binds and inhibits translation of *cirA* mRNA, encoding a siderophore transporter (and colicin Ia receptor), in the absence of an sRNA (92).

A great deal has been learned about the function of Hfq by studying its structure. Hfq is a member of the Sm-like (LSm) protein family, which is conserved in all domains of life. A distinct protein fold called "LSm-domain" and the ability to interact with nucleic acids characterize proteins in this family. The LSm-domain structure is partially conserved in bacterial Hfq monomers and consists of five antiparallel β-strands forming a β-sheet, with an α-helix stacked on top (93). Hfq monomers in *E. coli* and related species assemble into homohexameric rings. Both forms likely exist in equilibrium within the cell, with hexamers being most active in RNA binding and annealing (94). In *E. coli*, the Hfq ring contains sRNA-binding sites on the so-called proximal (side of the ring and inner rim of the central cavity containing α-helixes) and lateral (outer rim of the ring) faces, and an mRNA-binding region on the distal (surface opposite to the proximal side containing β-sheets) face of the ring (95–99). Hfq was shown to preferentially recognize a common sRNA structure: an A/U rich sequence followed by a Rho-independent terminator hairpin and a 3′ poly-U tail (100–102). Similarly, an Hfq-binding element called the $(ARN)_x$ motif (where A is adenine, R is a purine, and N is any nucleotide) is found in the leader sequences of many sRNA-regulated mRNAs. It consists of an A-rich sequence flanked by highly structured regions shown to specifically interact with Hfq. Recognition of $(ARN)_x$-containing mRNAs occurs through an A-R-E motif (where A is the adenosine specificity site, R is the purine selectivity site, and E is the non-discriminatory entrance/exit site) present on each Hfq monomer. A total of six A-R-E motifs in the Hfq hexamer allow simultaneous binding of up to six ARN repeats (95, 103–105). Functionally, Hfq binds and protects sRNAs from nucleolytic decay (51, 106) and facilitates their base pairing to target mRNAs (51, 107). Binding of Hfq to both sRNA and mRNA increases their local concentrations, stimulates structural remodeling to facilitate pairing, and increases annealing rates of cognate pairs (108–111). While Hfq is quite abundant, several recent studies have demonstrated that there is competition among sRNAs for binding of Hfq (112–114). Since many sRNAs compete for binding, the structure of Hfq is highly optimized to allow for cooperative binding of several RNA molecules. In addition, a rapid RNA dissociation rate aids in the cycling of many different combinations of sRNAs and mRNAs in order to maximize chances of a productive interaction (112). While such features provide for robust regulatory response, it was recently shown that abnormally high levels of one sRNA could effectively disrupt signaling through other sRNAs *in vivo* (113). Overexpression of one sRNA decreases stability and accumulation of other sRNAs, as well as reduces their binding to Hfq, which causes perturbations in the regulation of target mRNAs and changes the outcome of gene regulation in the cell (114). While competition between sRNAs for Hfq binding is concentration dependent (108), different sRNAs compete with different efficiencies (114), likely determined by the intrinsic properties of each sRNA, further complicating RNA regulatory networks and their study.

In addition to binding RNAs, Hfq is also known to directly interact with a number of protein factors such as RNA polymerase (115), Rho factor (116), poly-A polymerase I (117), RNase E (118) and PNPase (119, 120). Of particular interest are Hfq associations with the components of the degradosome: RNase E and PNPase, and the role of Hfq in recruiting these proteins during sRNA-mediated regulation of target transcripts (121). Hfq likely serves as a scaffold for sRNA–mRNA interactions and a platform for recruiting protein factors important for a robust regulatory outcome.

RNase E and the RNA degradosome in sRNA-mediated target decay

Many base-pairing sRNAs modulate stability of their mRNA targets by recruiting ribonucleases to the duplex. RNase E is a single-strand–specific endoribonuclease involved in the regulatory mechanisms of *trans*-encoded sRNAs (48–50), and also plays important roles in RNA processing and bulk mRNA turnover in *E. coli* and related organisms (122, 123). The active enzyme is a homotetramer composed of catalytic, RNA binding and protein binding domains. The C-terminal protein-binding domain of RNase E serves as a scaffold on which other components, namely polynucleotide phosphorylase (PNPase), RNase helicase B (RhlB), and the glycolytic enzyme enolase, assemble to form a large RNA degrading complex called the degradosome (123). Importantly, the catalytic domain and C-terminal scaffold domain are both necessary for sRNA-targeted mRNA degradation.

Hfq interacts with RNase E in a large ribonucleoprotein complex. Exposed A/U-rich single-stranded RNA regions are preferentially recognized by RNase E, which is stimulated by the 5′-terminal monophosphate of the sRNA to cleave its RNA substrates (12, 53, 122, 124, 125). Regulatory sRNAs might stimulate degradation of their target mRNAs by promoting structural remodeling that exposes A/U-rich recognition sequences and/or by providing a 5′-teminal monophosphate for to stimulate RNase E-mediated cleavage. Although not experimentally demonstrated, helicase (RhlB) associated with the degradosome might facilitate unwinding of the sRNA-mRNA duplex or clearing the secondary structure that occludes RNase E recognition sequences. Additionally, sRNA-mediated translation inhibition of mRNAs frees these mRNAs from ribosomes, which otherwise protect mRNAs from RNase E-mediated cleavage (126–129).

A novel ribonuclease, YbeY, and its potential role in sRNA regulation

YbeY is a novel bacterial exoribonuclease that was recently implicated in regulation by sRNAs. The *ybeY* gene is highly conserved and found in nearly every sequenced bacterial species (130, 131). YbeY is a 17-kDa exoribonuclease that bears a close resemblance to eukaryotic Argonaute (AGO) proteins, which are principal components of the RNA-induced silencing complex (132, 133). YbeY specifically binds single-stranded RNA and together with other ribonucleases like RNase R and PNPase, it mediates processing required for the maturation of rRNAs (131, 134). YbeY also functions in controlling the quality of the 70S subunit after it has been assembled, to prevent protein mistranslation (133).

An alternative and less understood function of YbeY is in sRNA-mediated regulation. Gram-negative *Sinorhizobium meliloti ybeY* mutants exhibit a pleiotropic phenotype of poor growth under a variety of stress conditions, very similar to *S. meliloti hfq* mutants (135). Moreover, altered levels of at least nine sRNAs and their respective target mRNAs was seen in both *S. meliloti ybeY* and *hfq* mutants (135). Studies performed in *E. coli* suggest that YbeY modulates levels of many sRNAs and their cognate targets under certain stress conditions, independent of sRNA requirements for Hfq (136). Another recent study demonstrated that YbeY, besides its role in rRNA maturation, also affects regulation of virulence and stress response-associated sRNAs in *Vibrio cholerae* (131). Taken together, these studies suggest that YbeY plays a general and important role in sRNA regulatory processes. However, current understanding of the basis for regulation by this novel RNase is limited, and additional research efforts are needed.

REGULATION OF PRIMARY METABOLISM BY SRNAS

sRNAs Regulating Carbon Transport and Metabolism

SgrS and the response to sugar–phosphate accumulation in *Escherichia coli* and *Salmonella enterica* serovar Typhimurium

The Hfq-binding sRNA SgrS operates at the interface of carbon metabolism and stress. It is indispensable for regulating the response of *E. coli* to metabolic stress caused by sugar–phosphates (like glucose-6-phosphate [G6P]), which inhibit growth through an unknown mechanism when they accumulate to high intracellular levels (137). This accumulation prompts the transcription factor SgrR (2, 138–141) to activate *sgrS* transcription. Both *sgrR* and *sgrS* are crucial for recovery from glucose-phosphate stress, as demonstrated by the fact that *E. coli sgrR* and *sgrS* mutant growth is severely inhibited (compared to wild-type cell growth) during stress (36, 141).

Along with Spot 42 and RyhB (both described below), molecular mechanisms of regulation by SgrS are perhaps the best characterized among sRNAs that regulate metabolism. It has been known for nearly a decade that SgrS rescues cell growth under stress conditions by preventing accumulation of sugar–phosphates. SgrS acts in large part by negatively regulating *ptsG*, which encodes the major glucose phosphoenolpyruvate phosphotransferase system (PTS) transporter EIICB^{Glc}. SgrS acts at a posttranscriptional level by specifically base pairing with *ptsG* mRNA, blocking the RBS and inhibiting translation (36, 141, 142). The *ptsG*-SgrS complex is subsequently degraded by the RNase E degradosome (36, 50, 142), though degradation is not required for regulation of *ptsG*. In addition to inhibiting translation of *ptsG*, SgrS is one of a small number of sRNAs that also encodes a small protein, SgrT, with a complementary role in the response to glucose-phosphate stress. While the exact mechanism has yet to be characterized, SgrT is known to act at the level of protein activity, inhibiting further EIICB^{Glc}-mediated uptake of sugar–phosphates (143).

While inhibition of EIICB^{Glc} synthesis and activity has long seemed the primary *raison d'être* of the SgrR/ST stress response, in recent years it has become clear that SgrR/S regulation of additional targets is also important. For example, SgrS negatively regulates expression of a second PTS transporter, EIIABCD^{Man}, which is encoded by *manXYZ* and transports sugars including mannose and glucose. SgrS exhibits a novel binding mechanism for *manXYZ*, distinct from that of *ptsG*. One SgrS binding site on *manXYZ* mRNA is within the *manX* coding region, not at the RBS (144). Further work identified a second, distinct binding site in the intergenic region between *manX* and *manY* that is required to fully inhibit translation of *manY* and *manZ* (translation of the latter two genes is coupled) (145). While binding at each site individually is sufficient for translational repression of the corresponding coding sequence, binding at both sites is required for SgrS-mediated degradation of *manXYZ* mRNA (145). Even more recently, the first positively regulated target of SgrS was identified: *yigL*, which encodes a sugar–phosphatase that dephosphorylates accumulated sugar–phosphates so that sugars can be pumped out of cells (33). YigL plays a role in recovery from stress, as *yigL* mutants are defective in growth during stress (33). SgrS acts by a novel mechanism on this third target: it stabilizes the *yigL* transcript by preventing its degradation via RNase E (33). Intriguingly, *yigL* is necessary for full virulence of the enteric insect pathogen *Xenorhabdus nematophila* (146), and *yigL* and/or other players in the glucose-phosphate stress response could promote the pathogenesis of other gram-negative bacteria. For example, in *Salmonella* Typhimurium, SgrS inhibits translation of *sopD* mRNA (147), which encodes a secreted effector protein required for full virulence (148). This regulation is exceptionally specific, as the highly similar *sopD2* virulence effector, also required for virulence, is not regulated by SgrS due to a sequence difference in a single base pair (147).

In addition to genes regulated by SgrS during the glucose-phosphate stress response, SgrR regulates transcription of additional genes, two of which have been described (2, 149). The first, *setA*, encodes an efflux pump that can transport a wide array of sugars (150). While *setA* contributes to rescue of growth during stress, it has been shown that it is not a major efflux transporter of glucose during stress; rather, SetA is speculated to carry out efflux of other, uncharacterized metabolites that contribute to stress (149). The second SgrR target, *alaC* (formerly *yfdZ*), encodes a glutamic-pyruvic transaminase that contributes to alanine synthesis (151). While no function for AlaC in glucose-phosphate stress has been established, it can lead to production of pyruvate (151), which could conceivably affect glycolytic metabolism during stress (2, 152). Moreover, genomic analyses (Balasubramanian, Bobrovskyy, Richards, and Vanderpool, unpublished data) suggest additional regulatory targets of SgrR and SgrS may be involved in central carbon, amino acid, and phosphate metabolism. Suggestions of broader metabolic roles for the glucose-phosphate stress response are supported by the recent finding that induction of the phosphate starvation (Pho) regulon can partially rescue the growth defects of *sgrS* and *sgrR* mutants during glucose-phosphate stress (153). Another recent study found that growth defects during stress are worse in nutrient-poor minimal media than in rich media, and SgrS regulation of different targets is required. In minimal media, regulation of additional targets (beyond those required in rich media) is required to rescue cells from stress, and addition of amino acids can improve the growth of an *sgrS* mutant during stress. (154).

Intriguingly, bioinformatics and genetic studies have revealed that the protein and RNA components of the glucose–phosphate stress response are not uniformly distributed across different species of enterics, or even among strains of *E. coli*. For example, based on bioinformatics analysis, some species like *S.* Typhimurium and *Klebsiella pneumoniae* have both *sgrS* and *sgrT*, while in others like *Yersinia pestis* and *Photorhabdus luminescens, sgrT* was absent (155). Likewise, *E. coli* strains K12 and CFT073 possess both *sgrS* and *sgrT*, while O157:H7 has only *sgrS*, suggesting variations in the regulatory response to glucose-phosphate stress based on the two *sgrS* functions (base pairing with mRNAs and producing SgrT) (155). These differences have been shown to have some physiological relevance: when expressed ectopically, *sgrS* homologs from *E. coli* and *Y. pestis* require SgrS base pairing activity to rescue growth of an *E. coli* mutant during stress. On the other hand, base pairing-deficient *sgrS* homologs of *K. pneumoniae, Pectobacterium carotovora,* and *S.* Typhimurium could all rescue growth when only SgrT was functional (156). The coordination of these two distinct functions has been most extensively studied in *S.* Typhimurium, where it appears the base pairing and SgrT production functions of SgrS may have a prioritized hierarchy. Mutant alleles of *sgrS* that did not produce SgrT performed base-pairing–dependent regulation of *ptsG* mRNA similarly to wild-type SgrS (157). In contrast, mutations in the base pairing region of SgrS led to an increase in SgrT production (as measured by Western blotting with an anti-SgrT antibody) (157). Moreover, whereas SgrS is produced immediately on induction of stress (141, 157), SgrT levels did not increase until later (~40 minutes after induction). Taken together, these results suggest that in *S.* Typhimurium, SgrS-mediated inhibition of *ptsG* mRNA translation is the immediate mediator of glucose-phosphate stress rescue, with SgrT inhibition of EIICB^{Glc} activity

playing a supporting role later in the stress response (157).

Despite the well-characterized mechanisms of SgrS regulation of mRNA targets, the metabolic cause of stress, and therefore the biological role of the stress response, is not entirely understood. In other words, it is not inherently clear why the accumulation of sugar–phosphates is stressful to the cell. In the laboratory, stress is induced artificially in one of two ways: first, by addition of glucose analogs like α-methyl glucoside (αMG), which is transported by EIICB^{Glc} and accumulates as α-methyl glucoside-6-phosphate (αMG6P) because it cannot be metabolized (141); second, by introducing mutations that block glycolysis (such as *pgi*, which encodes phosphoglucose isomerase), resulting in accumulation of G6P (158, 159). Both methods result not only in accumulation of sugar–phosphates, but occur as a consequence of inhibiting the glycolytic pathway. Therefore, it has been posited that stress could be due either to toxicity resulting from high levels of sugar–phosphates (36, 159) or to depletion of glycolytic or other related metabolic intermediates (152).

Current evidence most strongly supports a role of metabolic depletion as the cause of stress (158–160), although the two causes are not necessarily mutually exclusive. One such piece of evidence is the fact that stress caused by mutation of glycolytic genes can be abrogated by addition of glycolytic intermediates downstream of the block in the glycolytic pathway. For example, stress in *pgi* mutants (which accumulate G6P) is rescued by addition of fructose-6-phosphate, which presumably allows glycolysis to continue, despite the presence of another sugar–phosphate (158, 159). Stress can similarly be induced by other glycolytic mutations such as *pfk* (which leads to accumulation of fructose-6-phosphate) and *fda* (which leads to fructose-1,6 bisphosphate accumulation), and stress can likewise be alleviated by supplying other downstream glycolytic compounds (158, 161). Perhaps the most telling evidence to date was reported by a recent study that showed addition of glycolytic compounds, such as G6P, also rescues an *sgrS* mutant from stress caused by αMG (160). Significantly, G6P and other compounds rescued growth of an *sgrS* mutant during stress even when αMG6P was still present in cells (160). Taken together, these studies strongly support a role of glycolytic intermediate depletion, and not toxicity of accumulated sugar–phosphates, as the underlying metabolic cause of glucose-phosphate stress. This appears to be true regardless of whether stress is induced by mutational block or a nonmetabolizable glucose analog; in either case, bypassing the metabolic block (e.g., by adding back those intermediates) aids in recovery from stress (158–160).

While glycolytic depletion contributes to glucose-phosphate stress, the offending molecule has not yet been identified. Nevertheless, some findings implicate later glycolytic intermediates in stress. In contrast to upstream glycolytic intermediates, adding pyruvate during αMG stress is lethal to an *sgrS* mutant, and ectopic expression of PpsA, the phosphoenolpyruvate (PEP) synthetase that converts pyruvate to PEP, rescues an *sgrS* mutant from this fate (160). This finding suggests that the balance of PEP-pyruvate levels in the cell may play an important role in the stress response. This would not be surprising, as PEP directly connects glycolysis and PTS transport. The glycolytic block during glucose-phosphate stress is expected to decrease the intracellular PEP concentration, since PEP is used to drive sugar transport via the PTS and the block prevents PEP regeneration through metabolism. Thus, glucose-phosphate stress-associated growth inhibition could conceivably be caused by PEP depletion (152).

Escherichia coli Spot 42 regulates carbon metabolism and catabolite repression

Spot 42 is one of the first identified and best-characterized bacterial sRNAs. While growth defects resulting from Spot 42 overexpression have been appreciated for decades, it was in 2002 that a regulatory mechanism was first characterized (162). Even more recently, it has been recognized that Spot 42 plays a much larger, global role in regulating metabolism.

Spot 42 was first found to regulate expression of *galK*, which, along with *galE* and *galT*, is part of an operon that encodes products required for catabolism of the sugar galactose. Spot 42 was demonstrated to act by the relatively common sRNA mechanism of inhibiting *galK* mRNA translation by base pairing with and obstructing the RBS (162). However, Spot 42 has little to no effect on the translation of *galE* and *galT*, and is thus said to mediate discoordinate regulation of the operon. This differential regulation is thought to have a particular physiological function dependent on the availability of certain carbon sources (162). When galactose is provided as the carbon source, Spot 42 expression is inhibited and the entire *gal* operon is expressed since all its products are required for galactose catabolism (162). When the preferred carbon source glucose is present, cells engage in catabolite repression, inhibiting expression of genes encoding catabolic pathways for other, less preferred carbon sources. However, GalT and GalE must still be expressed in the presence of glucose because they have a second function in synthesizing precursors for lipopolysaccharide biosynthesis (163), while GalK is not required for this function. Thus, when expressed in the presence of glucose, Spot 42 inhibits translation of nonessential *galK* while permitting needed expression of the *galE* and *galT* mRNA (162).

Since the initial characterization of its role in galactose utilization, the Spot 42 regulon recently has expanded to include an astounding array of targets involved in carbon and secondary metabolism, catabolite repression and redox metabolism (164). This discovery is consistent with previous regulatory and phenotypic connections of Spot 42 to other aspects of cell physiology.

For example, when ectopically overexpressed from a multicopy plasmid, Spot 42 leads to a variety of growth defects, including decreased colony size on rich media, a growth lag when subcultured from minimal into rich media, and increased generation time when grown with various carbon sources (165). Clearly, such phenotypes must be attributed to regulatory effects beyond those on galactose utilization. An additional connection of Spot 42 to carbon catabolite repression has long been known: CRP (cAMP receptor protein) repressed transcription of Spot 42. This regulation is consistent with the role of Spot 42 in galactose utilization. CRP is most active in the absence of glucose (when its co-activating molecule cAMP is highest), when Spot 42 is not required for discoordinate regulation of the *gal* operon (166, 167). In contrast, cells grown with glucose (i.e., when cAMP levels are low and CRP is inactive) or in the absence CRP have increased Spot 42 levels and, therefore, discoordinate expression of the *gal* mRNA (162, 163, 168, 169).

Beyond these initial observations, a new view of Spot 42 as a global regulator of multiple aspects of metabolism was discovered through microarray analysis (164). When cells are grown with glucose as a carbon source, Spot 42 is produced and decreases expression of transcripts encoding a wide variety of functions, including central and secondary metabolism (e.g., *gltA*, which encodes citrate synthase, and *maeA*, which contributes to malate catabolism), transport and catabolism of alternative carbon sources (e.g., *srlA*, which encodes part of the glucitol-sorbitol transporter, and *fucI*, which aids in fucose catabolism), and redox metabolism (e.g., *sthA*, which aids in the oxidation of NADPH) (164). Gene expression changes for these potential Spot 42 targets were confirmed with *lacZ* translational fusions, and Spot 42 was found to inhibit translation of several targets via specific base pairing (164). Reinforcing the connection of Spot 42 to catabolite repression, many of these targets are also known to be activated by CRP during

growth without glucose. While wild-type Spot 42 expression during growth with glucose inhibits target expression, a mutant lacking Spot 42 displays increased (or "leaky") expression of *srlA* and *fucI* (164). Based on these observations, Beisel and Storz proposed that Spot 42 and CRP form a regulatory multioutput feed-forward loop in which both protein and sRNA differentially regulate (increase and decrease, respectively) the same targets, while at the same time CRP represses transcription of Spot 42 (164). Therefore, during growth in glucose, target gene transcription is reduced (due to low CRP activity), and expression is decreased even further post-transcriptionally by Spot 42. The Spot 42-mediated reduction in leaky expression of genes not required during growth in glucose could in theory increase metabolic efficiency and energy conservation energy use (164).

Pseudomonas aeruginosa CrcZ regulates carbon catabolite repression

CrcZ, a regulator of an atypical type of carbon catabolite repression system in *Pseudomonas aeruginosa*, is representative of the diversity of regulatory roles sRNAs play in gram-negative bacterial metabolism. While recent findings have made the precise regulatory role of this sRNA unclear (as discussed below), CrcZ does appear play a part in regulating carbon source utilization. Compared to wild-type *P. aeruginosa*, *crcZ* mutants exhibit growth defects on carbon sources like fructose, gluconic acid, and glycerol, as well as amino acids like D-alanine and L-lysine (170). *Pseudomonas* catabolite repression differs from the *E. coli* paradigm because glucose is not the preferred carbon source. Rather, TCA cycle intermediates such as citrate and succinate are preferentially used over glucose, other sugars, and amino acids (171–174). In *Pseudomonas* species, carbon catabolite repression is not mediated by CRP, but rather through the protein Crc (175–177). Crc was thought to act by binding and repressing translation of mRNA targets related to alternative carbon source transport and catabolism, thus allowing use of preferred carbon substrates (178–180). However, studies by the Bläsi lab have challenged this notion, noting that apparent RNA binding properties of purified Crc were due to the contamination of preparations with the RNA-binding protein Hfq; Crc failed to bind RNA upon further purification (181, 182). Moreover, structural and in vitro activity analyses of Crc have revealed that it does not display any RNA-binding activity (181). Recent findings suggest that Crc acts by supporting regulation mediated by Hfq, which is a major regulator of carbon catabolite repression in *P. aeruginosa* (183). For example, Hfq represses translation of target mRNA *amiE* by binding to A-rich RBS sequences. Crc was shown to be required for complete regulation of *amiE* (and other carbon catabolite repression targets) by Hfq (183). Although its mechanism of action remains uncertain, it is clear that Crc exerts effects on gene expression in various *Pseudomonas* species, as *crc* mutants exhibit changes in expression of a wide array of genes related to carbon catabolism (e.g., catabolite repression of some carboxylic acids), type IV pili formation, and resistance to stress (175, 178, 184–186).

While an RNA-binding mechanism for the Crc protein is no longer certain, the sRNA CrcZ does appear to regulate the activity of Crc itself. When *Pseudomonas* species are grown in the presence of non-preferred carbon sources (e.g., mannitol), the two-component system CbrA–CbrB and the alternative sigma factor RpoN help activate transcription of *crcZ* (170). CrcZ then inhibits Crc protein activity, which permits expression of genes required for alternative carbon source utilization (170). In contrast, *crcZ* expression is low on preferred carbon sources like succinate, allowing Crc to inhibit expression of genes for alternative carbon source catabolism. CrcZ was recently demonstrated to bind strongly to Hfq itself, and appears to sequester Hfq and prevent

regulation by Hfq on other targets (183). In sum, a novel sRNA (CrcZ) exerts an analogous effect on carbon catabolism as Spot 42 in *E. coli*, underscoring the commonalities in regulation of carbon source utilization by sRNAs in gram-negative bacteria.

Unlike *P. aeruginosa*, which only possesses CrcZ, at least two *Pseudomonas* species (*P. putida* and *P. syringae*) have an additional sRNA regulator of carbon catabolism (CrcY and CrcX, respectively) (187, 188). CrcY of *P. putida* is functionally redundant with CrcZ, as mutation of both genes is required to abolish sRNA regulation of Crc-mediated catabolite repression (187).

sRNAs Regulating Amino Sugar Metabolism

Escherichia coli GlmY and GlmZ sRNAs discoordinately regulate amino sugar metabolism

In a manner reminiscent of Spot 42, the sRNAs GlmY and GlmZ engage in discoordinate post-transcriptional regulation of genes encoded on a single transcript. GlmY/Z regulate expression of *glmUS*, which encode important enzymes needed, respectively, for synthesis of the amino-sugars uridine diphosphate-*N*-acetylglucosamine (UDP-GlcNAc) and glucosamine-6-phosphate (GlcN-6-P), both of which are precursors for the synthesis of cell wall peptidoglycan and lipopolysaccharide (189–193). Discoordinate expression is required because while GlmU is needed constitutively to synthesize UDP-GlcNAc, GlmS is only needed to synthesize GlcN-6-P if an external source of amino sugars is unavailable (in which case fructose-6-phosphate is rerouted from glycolysis to GlmS) (189–193).

The *glmUS* mRNA is unstable because it is targeted by RNase E, which cleaves the transcript at the site of the *glmU* stop codon (189, 190). In the absence of GlcN-6-P (i.e., when GlmS is needed to synthesize GlcN-6-P), the sRNA GlmZ helps to stabilize the *glmS* transcript and allow its translation by binding the 5′ UTR and unmasking the RBS, which is otherwise obfuscated by an internal hairpin structure that inhibits translation (191–193). However, GlmZ activity is controlled by the adaptor protein RapZ, which promotes RNase E-mediated degradation of GlmZ (194). RNase E removes the 3′ region of GlmZ, which is required for base pairing with the *glmS* transcript (193, 194). The second sRNA, GlmY, counteracts degradation of GlmZ by RNase E by acting as a "decoy." GlmY is very similar in sequence to GlmZ, and it stabilizes GlmZ indirectly, by binding to and sequestering RapZ and preventing its targeting of GlmZ to RNase E (191, 193, 194). In the absence of GlcN-6-P, GlmY levels increase, stabilizing full-length GlmZ when its action is required. In sum, the two sRNAs are part of a regulatory hierarchy with one metabolic goal: increase expression of the *glmS* mRNA and thus synthesis of the amino sugar GlcN-6-P under conditions where exogenous GlcN is not available.

sRNAs Regulating Amino Acid Metabolism

Escherichia coli GcvB and amino acid transport and metabolism

GcvB, a global regulator of amino acid metabolism and transport in *E. coli* and other gram-negative bacteria, exhibits two unusual features for a trans-acting sRNA. First, base pairing contributes to but is not the only mechanism of regulation by GcvB. GcvB was named as such because its expression is in *E. coli* regulated by GcvR and GcvA, transcriptional regulators of genes involved in glycine cleavage (195). However, GcvB was found not to regulate expression of glycine cleavage genes. Rather it was first identified as an inhibitor of expression of *oppA* and *dppA*, encoding periplasmic binding proteins that are part of, respectively, oligopeptide and dipeptide transport systems. *gcvB* mutants exhibit increased, constitutive expression of OppA and DppA, even during growth in rich medium (*i.e.*, when their expression is normally repressed) (195). Intriguingly, while

GcvB regulates translation of *oppA* and *dppA* transcripts at least in part through specific base pairing, it does not fully explain regulation of these targets (195, 196). While some mutations in predicted base-pairing regions of GcvB affected *oppA* and *dppA* translation, others did not affect expression (196). Also, where effects on expression were observed, compensatory or complementary mutational changes in *oppA* and *dppA* mRNAs restored regulation of *oppA* but not *dppA* by GcvB (196). Thus, while base pairing undoubtedly contributes to translational inhibition by GcvB, it is apparent that a more complex regulatory mechanism exists or that other factors are involved.

A second unusual feature of GcvB was identified in its regulation of another target, *cycA*, which encodes a transporter of the amino acids glycine, serine, and D-alanine. While CrcZ and CrcY of pseudomonads exemplify two different sRNAs with functional redundancy, GcvB regulation of *cycA* demonstrates redundancy *within* the same sRNA molecule (197, 198). Two separate regions of GcvB exhibit complementarity with *cycA* mRNA, and either region is sufficient to repress *cycA* translation; only mutations in both base-pairing regions of *gcvB* resulted in significant loss of *cycA* regulation (198).

GcvB is highly conserved among enteric bacteria and is also found in other gram-negative bacteria such as members of the Vibrionaceae (199–201). Interestingly, as with Spot 42, genomic studies have recently extended the regulatory function of GcvB to a global scale in *S.* Typhimurium, with GcvB potentially exerting a posttranscriptional regulatory effect on as much as 1% of the genome (approximately 45 genes) (202). Contributing to this broad regulatory ability is a highly conserved, G/U-rich region (199, 202). Experiments using complementarity predictions and *gfp* reporter fusions revealed that the GcvB regulon in *S.* Typhimurium extends to more than 20 direct targets, some of which are previously unrecognized transcripts involved in amino acid and cofactor transport and metabolism. These targets include *serA* (which encodes the enzyme of the first step in L-serine synthesis), *gdhA* (encoding an NADP glutamate dehydrogenase), *tppB* (encoding a putative tripeptide transporter), *lrp* (encoding the global regulator of amino acid metabolism and other functions), and *ndk* (encoding a nucleoside diphosphate kinase) (202).

Escherichia coli RybA regulates aromatic acid synthesis during peroxide stress

The sRNA RybA was identified in 2001 by Wassarman, et al., but a regulatory function was not characterized until more than a decade later (203, 204). RybA is noteworthy in part because it regulates amino acid metabolism, but only under particular environmental conditions. Specifically, it regulates expression of genes involved in biosynthesis of aromatic amino acids. However, as with SgrS, this metabolic regulation has only been observed under conditions of stress (in this case, resulting from peroxide or cold shock) (204). In a *rybA* mutant grown under peroxide stress, levels of several mRNAs encoding functions involved in aromatic metabolism (such as shikimate kinase-encoding *aroL*, chorismate mutase-encoding *tyrA*, and *yliL*, a gene of unknown function that overlaps the *rybA* gene) are significantly increased (204). The regulation of these genes by RybA requires TyrR, a transcriptional regulator of many genes required for aromatic amino acid transport and biosynthesis. Moreover, many of the potential RybA targets are also regulated by TyrR (204). The mechanism of RybA regulation has not been determined, but given their overlapping regulons, RybA is posited to work by regulating synthesis or activity of TyrR (204). Although not yet clear, Gerstle et al., suggest that RybA inhibition of aromatic amino acid biosynthesis might connect to peroxide stress by affecting available levels of chorismate, which is needed for ubiquinone and enterobactin synthesis. Ubiquinone helps protect the cell membrane during oxidative stress, while enterobactin is an iron siderophore whose production could have a negative impact on cells under oxidative stress (204).

Aar regulates amino acid metabolism in *Acinetobacter baylyi*

While the vast majority of sRNA work has focused on enterics (particularly *E. coli* and *S.* Typhimurium), some studies have begun to shed light on sRNA regulation of metabolism in other gram-negatives, including the gammaproteobacterium *Acinetobacter baylyi* (order Pseudomonales). Identified by bioinformatics, the sRNA Aar is located in the intergenic region between *trpS* and *sucD*. Based on Northern blot analysis, overexpression of Aar caused an increase in transcript levels of seven genes related to functions such as branched chain amino acid metabolism (*ilvI*, *fadA*, and *leuC*), pyruvate metabolism (*ppc*), and nitrogen fixation (*glnE*) (205). Aar levels vary depending on nutritional and environmental conditions (205), but the regulatory mechanisms controlling Aar synthesis have not been uncovered. While mechanisms of Aar-mediated regulation of target genes have likewise not yet been determined, Hfq appears to affect the stability and/or processing of Aar, as a second, smaller transcript is present in an *hfq* mutant when Aar is detected by Northern blot (205).

sRNAs Regulating Metal Homeostasis

Regulation of iron homeostasis by *Escherichia coli* RyhB

RyhB is one of a handful of sRNAs for which a definitive metabolic function has been extensively characterized: it helps *E. coli* regulate iron homeostasis, in large part by inhibiting translation of mRNAs encoding non-essential iron-containing enzymes under iron starvation conditions in order to free up iron for essential cellular functions. RyhB also increases expression of targets involved in synthesizing siderophores, low molecular weight iron chelators that help cells scavenge iron during growth in iron-depleted environments. RyhB has long been known to be essential to iron homeostasis; *ryhB* mutants have a much longer lag phase compared to wild-type *E. coli* when transferred to iron-limited media (48).

Iron is an essential and limiting nutrient for organisms growing in many natural environments. Acquisition of sufficient iron is particularly difficult for microbial cells growing in aerobic and neutral aqueous environments, where metabolizable forms of iron can be in short supply (206, 207). The same is true for many pathogens growing in host organisms, because many hosts sequester their iron stores to inhibit growth of invading microbes. Both RyhB and the ferric uptake regulator (Fur) transcription factor play important, complementary roles in regulating expression of *E. coli* genes under iron starvation conditions. In fact, RyhB was initially described over a decade ago as a regulatory target of Fur (48). Fur is the major regulator responsible for sensing and responding to changes in iron availability, not just in *E. coli* but other gram-negative species such as *Pseudomonas*, *Bordetella*, and *Yersinia*. Fur is actually a transcriptional repressor of genes needed during iron starvation. When iron concentrations are high, Fur binds to the corepressor ferrous (Fe^{2+}) iron and inhibits transcription of a regulon of more than 50 genes, for example many that are required for synthesis of high affinity iron acquisition systems as well as *ryhB*. Under iron-limiting conditions (i.e., lower Fe^{2+} corepressor availability), the inactive Fur apoenzyme derepresses genes in the Fur regulon, including *ryhB*. RyhB then inhibits expression of an impressively wide array of targets both directly and indirectly; a genomic microarray analysis examining the effects of ectopically expressed RyhB identified 56 genes as potential targets of RyhB regulation (1, 208).

Transcripts that are confirmed targets directly inhibited by RyhB include many encoding non-essential, iron-requiring proteins. For example, RyhB directly inhibits translation of *sdhCDAB* (encoding the iron-containing TCA cycle enzyme succinate dehydrogenase) and *sodB* (encoding the iron-dependent superoxide dismutase) mRNAs (48, 209, 210). RyhB acts through direct base pairing interactions with sequences in the RBS regions of *sdhD* and *sodB*, inhibiting translation and resulting in degradation of the mRNAs. RyhB likewise inhibits translation of several other mRNAs encoding nonessential, iron-using enzymes, including TCA cycle members fumarase (*fumA*), fumarate reductase (*frdABCD*), and aconitase (*acnA* and *acnB*) and respiratory proteins encoded by *nuo* and *fdo* (48, 208). RyhB-dependent inhibition of the synthesis of these iron-containing proteins frees up iron for essential iron-utilizing enzymes, an effect referred to as "iron sparing."

RyhB has a positive effect on expression of other targets, both indirectly and directly. For example, RyhB indirectly increases expression of genes involved in the biosynthesis of the siderophore enterobactin through an indirect mechanism. Iron-depleted conditions lead to derepression of transcription of the *entCEBAH* genes required for enterobactin synthesis (211). However, *ryhB* is also required for normal regulation of enterobactin biosynthesis, as evidenced by the fact that *ryhB* mutants produce less enterobactin compared to wild-type *E. coli* (212). Reduced production of enterobactin in *ryhB* mutants is caused by at least two separate effects: 1) a still-uncharacterized mechanism causes a decrease in *entCEBAH* transcription, and 2) RyhB indirectly increases levels of L-serine (required for enterobactin biosynthesis) by directly repressing translation of *cysE* mRNA (212). CysE consumes L-serine to synthesize cysteine, reducing the amount of L-serine available for enterobactin biosynthesis. Thus, it has been suggested that RyhB-dependent inhibition of *cysE* frees up L-serine for use in enterobactin synthesis during iron starvation (212). Supporting this notion, enterobactin production in the *ryhB* mutant was increased by either mutation of *cysE* or by supplying an extracellular source of serine (212).

Direct, positive regulation by RyhB has so far been reported for only one target, *shiA* mRNA, which encodes a permease required to bring shikimate, an enterobactin precursor, into the cell (24). The 5′ UTR of *shiA* mRNA contains a secondary structure that occludes the RBS, repressing translation (24). Under iron-depleted conditions, RyhB acts by base pairing with the 5′ UTR, freeing up the RBS to allow translation and stabilizing the *shiA* transcript (24). Taken together, these findings demonstrate that RyhB acts through multiple direct and indirect mechanisms to enhance siderophore biosynthesis (and thus iron scavenging) under iron-limited conditions.

PrrF1 and PrrF2 regulate iron homeostasis in *Pseudomonas aeruginosa*

As with *E. coli* RyhB, the sRNAs PrrF1 and PrrF2 (named for <u>P</u>seudomonas <u>r</u>egulatory <u>R</u>NA involving iron) play an essential role in the response of *Pseudomonas aeruginosa* to iron starvation (213–215). Remarkably, despite similar regulatory effects, the sequence of the PrrF sRNAs is not similar to that of RyhB (214). Bioinformatic analyses based on sequence similarity have uncovered potential RyhB orthologs within the *Enterobacteriaceae* family but not in other gram-negative bacteria (48). The lack of RyhB homologs spurred a bioinformatic search for sRNAs potentially involved in *P. aeruginosa* iron homeostasis. This search identified PrrF1 and PrrF2, which are encoded in tandem in an intergenic region (214). The two sRNAs share more than 95% identity with each other and have been demonstrated to play RyhB-like roles in *P. aeruginosa* iron scavenging and homeostasis (214). Both the *prrF1* and *prrF2* genes have Fur-binding elements located upstream, and their expression is induced under iron-starved conditions (214). While the exact regulatory mechanisms have not been determined, deletion of both *prrF* genes results in increased mRNA levels of several potential target transcripts, and

predicted base-pairing interactions have been identified. These mRNA targets include nonessential iron-containing enzymes encoded by *sdh* and *sodB* (both also *E. coli* RyhB targets), as well as *bfrB*, which encodes a subunit of the iron storage protein bacterioferritin mRNAs (214). While other potential roles of the PrrF sRNAs have not yet been characterized, their expression is repressed by both Fur and heme (215). Intriguingly, a longer transcript, termed PrrH, has recently been described and consists of both PrrF1, PrrF2, and their intergenic region (215). PrrH appears to regulate targets of PrrF1 and PrrF2 (including *sdhD*), and also may inhibit expression of novel targets including two (*nirL* and *thiE*) involved in the biosynthesis of heme (215).

REGULATION OF SECONDARY METABOLISM AND VIRULENCE BY SRNAS

Carbon Storage, Motility, Biofilm Formation, and Membrane Composition

sRNAs link metabolism to motility and biofilm formation in *E. coli* and other enterics

In addition to extensive roles in regulating primary metabolism, sRNAs have been shown in *E. coli* and related species to control complex regulatory circuits coordinating metabolism with behaviors such as motility and biofilm formation. Three noteworthy targets of sRNAs are major regulators of these processes: FlhDC, CsrA, and CsgD. *flhDC* encodes the master transcriptional activator of flagellar biosynthesis. The global regulator CsrA (carbon storage regulator) is a key coordinator of transitions between motile and sessile (biofilm) lifestyles, regulating expression of genes involved in carbon storage and production of flagella and biofilm matrix, among others. The LuxR superfamily transcription factor CsgD likewise regulates expression of genes involved in biofilm formation, flagellar production, and curli fimbriae synthesis and assembly. We will focus on the regulation of carbon storage, motility, and biofilm formation by the sRNAs CsrB and CsrC, McaS, and OmrA and OmrB.

Regulation of carbon storage and motility by sRNAs CsrB and CsrC

The global protein regulator CsrA, central to modulating metabolic and motile to sessile transitions in *E. coli* (77, 216), is itself regulated by two sRNAs called CsrB and CsrC. CsrA is an RNA-binding regulatory protein that represses translation of certain genes, including the operon *pgaABCD*, which encodes products necessary for the synthesis of the polysaccharide poly-β-1,6-N-acetyl-d-glucosamine (PGA), a component of the extracellular matrix in many gram-negatives that mediates cell-to-cell and cell-to-surface adhesion in biofilms (217, 218). CsrA represses *pgaA* translation by binding to the *pgaA* mRNA leader sequence, and hindering ribosome access to the RBS as well as destabilizing the *pgaABCD* transcript (219, 220). Autoregulation of *csrA* mRNA occurs through the same translation inhibition mechanism (221). In addition, CsrA positively regulates expression of some genes involved in active growth and the planktonic motile lifestyle, such as the *flhDC* operon (219, 220, 222). Activation by CsrA is less understood, but in the case of *flhDC* occurs via CsrA protection of *flhDC* mRNA from cleavage by RNase E (223). CsrA-mediated stabilization of *flhDC* mRNA results in increased translation of the message (222) (Fig. 1). Independent of the regulatory outcome, CsrA-regulated transcripts each contain between one and six CsrA binding sites (224) consisting of a GGA motif located in the loop of a hairpin structure (225). Structural studies demonstrated that CsrA and CsrA homologs act as homodimers with RNA binding regions on opposite sides of the dimer (226).

In *E. coli*, two homologous sRNAs, CsrB and CsrC, possess CsrA binding sites that allow them to bind and sequester CsrA, thus preventing it from binding to its mRNA targets. CsrB is a 336-nt sRNA that was first identified in *E. coli* in 1997, when it was co-purified with CsrA (227). Remarkably, CsrB contains 22 predicted GGA motifs that can bind and sequester ~9 CsrA homodimers simultaneously (227). The 245-nt CsrC sRNA, was identified as a homolog of CsrB and sequesters CsrA in a similar fashion, although CsrC contains only nine GGA motifs for CsrA binding (228).

CsrB and CsrC sRNAs are regulated at the levels of synthesis and stability. Transcription of *csrB* and *csrC* is activated by the two-component system BarA–UvrY. The BarA sensor kinase and UvrY response regulator are activated by formate and acetate (229), thus connecting CsrB and CsrC synthesis to metabolic changes in the cell (230). UvrY induces CsrB and CsrC at the onset of stationary phase (228, 230). The protein CsrD controls the stability of CsrB and CsrC by increasing their rate of RNase E-mediated degradation (231), although the underlying mechanism is not understood.

Orthologs of CsrA and its antagonizing sRNAs (CsrB orthologs) are found in other gram-negative bacteria (231–234). RsmA, a CsrA ortholog in *Pseudomonas* species, regulates both conserved and novel mRNA targets involved in carbon storage, quorum sensing, virulence, and antifungal metabolite synthesis (76, 235, 236). RsmA is sequestered by three functionally redundant sRNAs: RsmX, RsmY, and RsmZ. Two sRNAs, RsmY and RsmZ, are broadly distributed among pseudomonads, while RsmX is only found in a few species (236, 237). All three sRNAs are ~120 nt in length and contain between 6 and 8 GGA motifs (236, 238, 239). Synthesis of RsmX, RsmY, and RsmZ is controlled by the two-component system GacS–GacA responsible for pathogenicity and production of antimicrobial compounds in pseudomonads (240, 241). GacS–GacA shares sequence similarity

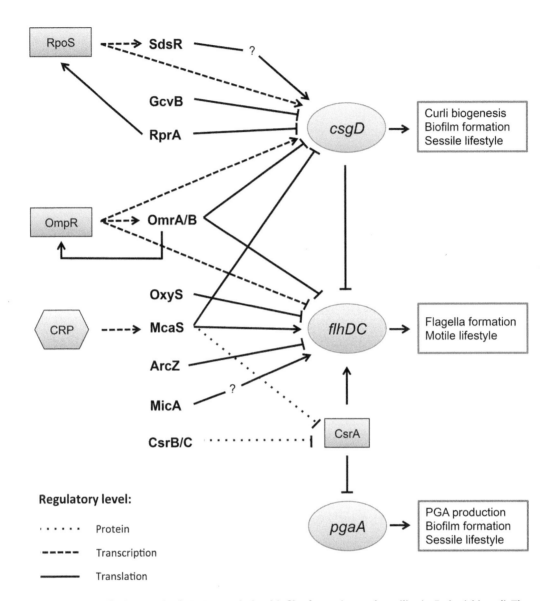

FIGURE 1 Simplified network of sRNAs regulating biofilm formation and motility in *Escherichia coli*. The regulatory network shows a set of sRNAs (in bold) and relevant protein factors (grey boxes) controlling *csgD* (blue circle), *flhDC* (red circle) and *pgaA* (violet circle) at the transcriptional (dashed lines), translational (solid lines) and protein (dotted lines) levels. Other factors known to regulate *csgD*, *flhDC*, and *pgaA* were omitted for clarity. Arrows indicate activating interactions, and lines with blunt ends indicate inhibitory interactions. doi:10.1128/microbiolspec.MBP-0009-2014.f1

with the *E. coli* BarA–UvrY and performs the similar function of inducing synthesis of sRNAs that sequester a global post-transcriptional regulator of secondary carbon metabolism.

Regulation of *E. coli* motility and biofilm formation by the multifunctional sRNA McaS

An *E. coli* sRNA, McaS, was recently found to regulate motility and PGA synthesis by sequestering CsrA in a manner similar to CsrB and CsrC (242). In addition to its ability to titrate the CsrA protein, McaS also possesses a second function: it regulates mRNA targets outside the CsrB–C regulon, including those required for flagella and curli fimbriae biosynthesis, using a canonical base pairing mechanism (243).

McaS is a 95-nt sRNA expressed in *E. coli* in stationary phase, reaching its highest levels in mid-stationary phase, and decreasing significantly in late stationary phase. McaS requires Hfq, both for stability and to exert its regulatory effects. Whether there is competition between Hfq and CsrA for binding to McaS is not clear; this is the first documented case where an Hfq-dependent sRNA also sequesters a protein. Transcription of *mcaS* is activated by CRP when glucose is limiting (243).

The secondary structure of McaS resembles that of CsrB and consists of three hairpin structures with highly conserved GGA motifs in the single-stranded loops. A recent study determined that McaS sRNA activates *pgaA* translation by titrating CsrA protein so it can no longer bind and inhibit *pgaA* translation (242) (Fig. 1). McaS also relieves repression of other CsrA targets such as *ydeH*, *ycdT*, and *glgC*, which enhances biofilm formation and glycogen accumulation (242q).

In addition to its protein-binding function, McaS has been shown to post-transcriptionally regulate two mRNA targets, *csgD* and *flhDC*, via base pairing interactions. CsgD activates expression of the *csgBAC* operon coding for structural components, assembly and export of curli fimbriae (244). CsgD also activates transcription of *adrA*, which encodes a diguanylate cyclase that synthesizes cyclic di-GMP, a second messenger that stimulates cellulose production and biofilm formation (245, 246). In addition to initiating curli production, CsgD also directly represses transcription of two flagellar operons (247, 248). McaS represses *csgD* by binding to a 16-nt region of the leader sequence of *csgD* mRNA, resulting

in translation inhibition (243). Unlike the majority of sRNAs, McaS pairs far upstream of the *csgD* RBS and start codon and thus could not interfere directly with ribosome binding, suggesting an alternate mechanism of regulation (243).

While McaS inhibits curli formation via repression of CsgD synthesis, it exerts a positive effect on flagellar synthesis by increasing translation of *flhDC*. The 198-nt leader sequence of the *flhDC* mRNA is predicted to form a secondary structure that sequesters the RBS and reduces translation. Cooperative base pairing of McaS at two distinct sites in the *flhDC* leader is thought to increase translation by preventing formation of this secondary structure (243).

The proposed physiological role for McaS is in modulating the transition from a motile to a sessile lifestyle as cells progress through stationary phase. Early in the stationary phase as bacteria seek nutrients whose levels are diminishing, McaS enhances motility by activating synthesis of the master flagellar regulators FlhDC. However, during mid-stationary phase when McaS levels reach their peak, McaS antagonizes CsrA activity so that CsrA-regulated genes such as *pgaA* and *glgC* are depressed, allowing the transition to a sessile lifestyle in a biofilm (242).

Many layers of transcriptional and post-transcriptional control of *csgD* and *flhDC* make the overall regulation very complex (Fig. 1). The global transcription factor CRP controls transcription of *mcaS* and its targets, *flhDC* and *csgD*. The CRP-McaS-*flhDC* circuit forms a feed-forward loop where CRP activates *flhDC* directly by binding to its promoter and indirectly through activation of *mcaS* transcription (243) (Fig. 1). In contrast, the CRP-McaS-*csgD* regulatory circuit forms an "incoherent" feedforward loop, since CRP activates *csgD* and *mcaS* transcription while McaS represses *csgD* translation (Fig. 1). Regulation of *csgD* mRNA is further complicated by the involvement of the sRNAs OmrA and OmrB (discussed below), as well as

GcvB and RprA, all of which are predicted to bind the same region of *csgD* mRNA as McaS (243). Likewise, translational control of *flhDC* is complex, with repression by OmrA, OmrB, ArcZ and OxyS sRNAs (249).

Regulation of *E. coli* outer membrane proteins and curli fimbriae by the sRNAs OmrA and OmrB

In addition to McaS, two homologous *E. coli* sRNAs, OmrA and OmrB, exhibit functionally redundant regulation of motility and biofilm formation, as well as of distinct targets that encode outer membrane (OM) proteins. OmrA and OmrB (OmpR regulated sRNAs A and B) are, respectively, 88 nt and 82 nt in length and encoded tandemly in the *aas-galR* intergenic region (250). This arrangement is retained in most but not all Enterobacteriaceae (250). The first 21 nt and last 35 nt of OmrA and OmrB sRNAs are nearly identical and well conserved among enteric bacteria (250). The highly conserved 5′-ends of OmrA and OmrB are responsible for base pairing with all known mRNA targets of these sRNAs (44, 249–251). Transcription of *omrA* and *omrB* is activated by the EnvZ-OmpR two-component system, in which EnvZ is a kinase that senses high osmolarity and changes in OM composition, leading to activation of the response regulator OmpR (251). Stability of OmrA and OmrB is dependent on Hfq (203, 250–252).

OmrA and OmrB regulate *csgD* and *flhD* mRNAs, inhibiting translation and destabilizing both transcripts (44, 249). As with McaS, pairing of OmrA and OmrB with *csgD* occurs substantially upstream (~50 nt) from the RBS, suggesting an indirect mechanism of translation inhibition (44). In addition to these targets, OmrA and OmrB repress translation of mRNAs encoding the OM protease OmpT and three TonB-dependent OM receptors CirA, FepA, and FecA (250). OmrA and OmrB base pair near the RBS and start codon, resulting in translation inhibition and subsequent RNase E-dependent degradation of these target mRNAs (250).

Interestingly, OmrA and OmrB also indirectly regulate their own transcription by base pairing with *ompR* mRNA and inhibiting its translation (251). Thus, OmrA and OmrB form a feedforward loop by inhibiting *csgD* both directly (by base pairing) and indirectly by inhibiting OmpR, a transcriptional activator of *csgD* (Fig. 1). In contrast, an incoherent feed-forward loop is formed by OmrA and OmrB direct inhibition of *flhDC* and indirect activation of the same, via *ompR* repression (Fig. 1). Together, the two redundant sRNAs OmrA and OmrB modulate the composition of the outer membrane and regulate production of higher order structures such as flagella and curli fimbriae in *E. coli* and other enteric gram-negative bacteria.

Quorum Sensing

Regulation of quorum sensing by Qrr sRNAs in *Vibrio harveyi* and *Vibrio cholerae*

Homologous, redundant Qrr (quorum regulatory RNA) sRNAs are central to quorum sensing in *Vibrio* species. Quorum sensing is a mechanism by which bacteria indirectly sense population density (or diffusion properties of their environment) (253, 254) via sensing concentrations of self-produced secondary metabolites known as autoinducers (Fig. 2) (255, 256). The quorum-sensing and signal-transduction system causes changes in gene expression resulting in changes in cell physiology or behavior that give individual cells or the population an advantage under specific conditions (255, 257). In gram-negative bacteria, the autoinducers (AIs) produced are typically acyl-homoserine lactones (AHLs) or molecules derived from S-adenosylmethionine (SAM). The marine bacterium and human pathogen *V. cholerae* produces two different AI signaling molecules synthesized from SAM: CAI-1 produced by the CqsA enzyme and AI-2 made by the highly conserved LuxS (255, 257). CAI-1 and AI-2 are recognized by their respective receptors found in the bacterial cell membrane (Fig. 2). Both receptors function as kinases in the absence of ligand binding, which results in phosphorylation of the response regulator LuxO through the phosphorelay protein LuxU. Phosphorylated LuxO (LuxO~P), together with the alternative sigma factor σ^{54}, activates transcription of four homologous sRNAs, called Qrr1-4 (258). Therefore, Qrr1-4 sRNAs are not made at high cell density when AIs are abundant, but are produced in high amounts at low cell density when signal molecules are scarce (Fig. 2).

The Qrr sRNAs are highly conserved Hfq-dependent sRNAs found in most sequenced *Vibrio* species (258). In *V. cholerae*, four Qrr sRNAs act redundantly to post-transcriptionally activate the master regulator of low cell density genes, AphA, by pairing with the *aphA* mRNA leader sequence to promote translation. In addition, with the help of Hfq, Qrr1-4 base pair in the 5' UTR of the *hapR* mRNA, encoding a master regulator of high cell density genes, to repress its translation (259–261). Analogous to the *V. cholerae* quorum-sensing system, *V. harveyi* encodes five Qrr homologs (Qrr1-5), that when induced at low cell density, activate AphA (260), and repress HapR ortholog LuxR (261, 262) (Fig. 2).

By inversely regulating AphA and HapR expression, Qrr RNAs indirectly affect expression of hundreds of genes involved in modulating *V. cholerae* physiology and behaviors such as virulence, biofilm formation, and cell defenses. For example, at low cell density, AphA activates production of ToxT, which in turn regulates a multitude of virulence genes, whereas HapR repression allows expression of genes for biofilm formation. Recently, Qrr1-4 were shown to repress expression of Type VI secretion system (T6SS) genes by a negative feed-forward mechanism. The sRNAs indirectly inhibit T6SS by repressing the T6SS activator HapR. Simultaneously, Qrr1-4 repress a subset of T6SS genes through direct base pairing interactions with mRNAs (263).

It has been noted that feedback loops and autoregulation are common features among many components of this system. For example, in addition to being inversely regulated by Qrr1-4 sRNAs, AphA and HapR repress each other's transcription (260), while HapR also autoregulates its own expression. Besides autorepression, LuxO~P is negatively regulated by Qrr1-4 at the post-transcriptional level (264). Such redundancy in Qrr sRNA control, together with complex feedback loops at numerous branch points establish a highly responsive regulatory system that ensures precise timing of behaviors controlled by quorum sensing (260, 261, 264–266) (Fig. 2). More specifically, this system enables *V. cholerae* to form biofilms and replicate at low cell density within the host, until bacteria accumulate to high cell density, at which point Qrr sRNAs inhibit virulence and biofilm genes to facilitate dispersal back into the environment.

Oxidative Stress and Acid Resistance

A function of RaoN sRNA in mediating *Salmonella* intramacrophage survival

RaoN is an sRNA that was recently discovered in *Salmonella* to confer resistance to oxidative stress and survival during infection (267). *Salmonella* pathogenesis initiates with penetrating the gut barrier through the M cells found in the Peyer's patches of the intestine. The bacteria then infect phagocytes, and after surviving harsh conditions of nutrient limitation, reactive oxygen and nitrogen species produced by phagocytes, can disseminate to cause a systemic infection (268). *Salmonella* virulence depends on the stress response and invasion systems encoded by the genes within the 11 pathogenicity islands (SPIs) of the *Salmonella* genome (269, 270).

SPI-11 contains genes important for *Salmonella* survival in macrophages (267, 271, 272) and was recently shown to encode an sRNA, RaoN. RaoN is ~200 nt-long, is stabilized by Hfq, and is encoded in the intergenic region between SPI-11 genes *cspH* and *envE* (267). It was first identified

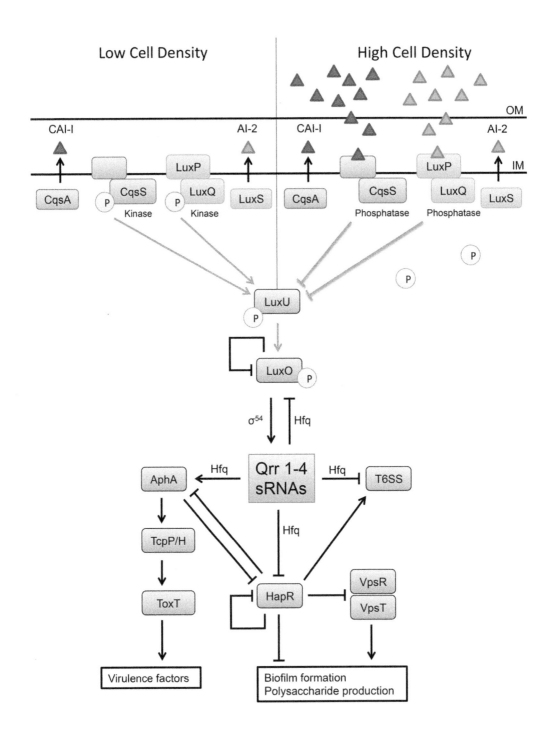

in a global transcriptomic analysis of cells growing under the conditions of nutrient limitation (minimal medium) and oxidative stress (hydrogen peroxide) that mimic the hostile environment within the macrophage. RaoN transcription is highly induced by this combined treatment, and although less, expression is still observed in response to oxidative stress alone (267). Disruption of *raoN* results in high susceptibility to the combined starvation and oxidative stress, and importantly, reduced survival within murine macrophages. RaoN confers stress resistance at least in part by regulating expression of *ldhA*, encoding lactate dehydrogenase A, which regenerates NAD$^+$ via NADH oxidation when converting pyruvate to lactate. LdhA maintains glycolysis and confers resistance to oxidative stress; however, its overexpression is toxic. RaoN negatively regulates *ldhA*, which seems to maintain appropriate levels of lactate dehydrogenase suited for optimal resistance to oxidative stress inside the macrophage (267). However, regulation of *ldhA* only partially accounts for RaoN-mediated intramacrophage survival, strongly suggesting additional contributions of other unknown targets of this sRNA (267).

GadY sRNA and its role in regulating acid response genes in *E. coli*

GadY sRNA, originally identified as IS183 (273), controls expression of genes involved in glutamate-dependent acid resistance in *E. coli* (9, 274, 275). Acid resistance is an important property of *E. coli* (and other enteric bacteria) that allows it to survive the acidity of the mammalian gastrointestinal tract (276, 277). For example, acid-resistant properties of enterohemorrhagic *E. coli* serotype O157:H7 are responsible for its low infectious dose, playing a major role in its virulence (278–280). One of the systems responsible for extreme acid tolerance in *E. coli* involves expression of two decarboxylases, namely GadA and GadB, that use glutamate as their substrate to produce γ-aminobutyric acid. This reaction consumes intracellular protons to raise internal pH. An additional component, the antiporter GadC, exchanges γ-aminobutyric acid for extracellular glutamate to replenish its concentrations in the cell. Expression of this glutamate-dependent acid resistance system is regulated by a LuxR-family regulator GadE (281), and poorly understood AraC-like GadX and GadW regulators (282). GadX transcriptionally activates GadA and GadB, whereas GadW has an opposing effect, acting as a repressor of the decarboxylases' expression (283).

In the heart of the glutamate-dependent acid resistance system is the Hfq-dependent GadY, encoded within the ~371 nt *gadX-gadW* intergenic region and oriented divergently in *cis* to the 3′ UTR of the *gadX* gene (9, 284). GadY is conserved in organisms

---

FIGURE 2 Quorum sensing systems of *Vibrio cholerae*. V. cholerae uses histidine kinases CqsS and LuxPQ to sense autoinducers (AIs) CAI-1 (violet triangle) and AI-2 (orange triangle) respectively. Receptors function as kinases at low cell density (LCD), when concentrations of CAI-1 and AI-2, which are produced by CqsA and LuxS, respectively, are low. This stimulates σ^{54}-dependent activation of *qrr* gene expression through LuxU and LuxO phosphorylation cascade. The Qrr 1-4 sRNAs (red square), facilitated by Hfq, activate *aphA*, which stimulates expression of *toxT*, a known activator of the major virulence factors. Additionally Qrr 1-4 repress *hapR*, which leads to polysaccharide production and biofilm formation. Qrr 1-4 also negatively regulate genes for biogenesis of type VI secretion system (T6SS). In contrast, at high cell density (HCD) receptors sense the presence of AIs and function as phosphatases that stimulate dephosphorylation of LuxU, resulting in cessation of *qrr* expression. In the absence of Qrr sRNAs, *hapR* expression increases, which leads to the inhibition of biofilm formation and shut down of virulence factor production, while stimulating T6SS biosynthesis. Lines with arrowheads indicate activating interactions, and lines with blunt ends indicate inhibitory interactions. doi:10.1128/microbiolspec.MBP-0009-2014.f2

that possess the *gadX-gadW* operon, such as closely related *E. coli* and *Shigella*, but is absent from most other enteric bacteria (9). Expression of GadY is dependent on *rpoS* (σ^S) and highest during stationary phase when bacteria are growing under acidic conditions (9, 275). Forming a regulatory-feedback loop, GadX is also required for GadY maximal activity (275). GadX binds in the GadY/GadW overlapping promoter region, which positively affects transcription of GadY, but inhibits GadW (275).

Once expressed, GadY positively regulates *gadX* and *gadW* at the post-transcriptional level, by pairing with the complementary sequence of the 3′ UTR of *gadX* within the *gadXW* polycistronic mRNA. This interaction stimulates RNase III processing downstream from the *gadX* stop codon, which results in stable *gadX* and *gadW* mRNA products (9, 274, 275). Several species of *gadY* with sizes 105 nt, 90 nt, and 59 nt were observed and are believed to be the result of endonucleolytic processing, the mechanistic details of which are not known. Still, even the shortest 59-nt GadY fragment was able to promote formation of a stable *gadX* mRNA (9). Additionally, it has been noted that GadY-initiated processing of *gadXW* occurs in the absence of RNase III ribonuclease, at the different nucleotides that are adjacent to the RNase III cleavage sites (274). The redundant ribonuclease has not been identified (274); however, some evidence points towards the involvement of RNase E in this process (284). In conclusion, GadY sRNA positively affects GadX and GadW synthesis at the post-transcriptional level, which then regulate *gadA* and *gadBC* to improve *E. coli* adaptation to acid stress and stationary phase growth.

CONCLUSIONS

It has become abundantly clear in the last 15 years that bacterial sRNAs have pivotal roles in controlling primary and secondary metabolism. For some sRNAs of gram-negative bacteria, molecular mechanisms of regulation have been well characterized (e.g., SgrS, CsrB/C, Qrr1-4), while for others a metabolic function has been ascribed but the underlying regulatory mechanism remains elusive (e.g., Aar, RaoN). For very few sRNAs (e.g., Spot 42 and RyhB) have both aspects—mechanism and role in cellular metabolism—been definitively described. In the future, a more holistic approach integrating mechanism with metabolic function—as well as an understanding of how different sRNAs coordinate variegated aspects of cellular physiology and metabolism among each other—will help illuminate regulation by sRNAs. Moreover, the majority of work has been done in *E. coli* (and, to a lesser extent, *S. enterica*). Undoubtedly, with advances in available genomic, transcriptomic, and bioinformatics approaches, a vast diversity of metabolic regulation by sRNAs in other gram-negative bacteria awaits further characterization.

CITATION

Bobrovskyy M, Vanderpool CK, Richards GR. 2015. Small RNAs regulate primary and secondary metabolism in gram-negative bacteria. Microbiol Spectrum 3(3):MBP-0009-2014.

REFERENCES

1. **Masse E, Salvail H, Desnoyers G, Arguin M.** 2007. Small RNAs controlling iron metabolism. *Cur Opin Microbiol* **10**:140–145.
2. **Vanderpool CK, Gottesman S.** 2007. The novel transcription factor SgrR coordinates the response to glucose-phosphate stress. *J Bacteriol* **189**:2238–2248.
3. **Altuvia S, Weinstein-Fischer D, Zhang A, Postow L, Storz G.** 1997. A small, stable RNA induced by oxidative stress: role as a pleiotropic regulator and antimutator. *Cell* **90**:43–53.
4. **Frohlich KS, Papenfort K, Berger AA, Vogel J.** 2012. A conserved RpoS-dependent small RNA controls the synthesis of major porin OmpD. *Nucleic Acids Res* **40**:3623–3640.

5. **Boysen A, Moller-Jensen J, Kallipolitis B, Valentin-Hansen P, Overgaard M.** 2010. Translational regulation of gene expression by an anaerobically induced small non-coding RNA in *Escherichia coli. J Biol Chem* **285:** 10690–10702.

6. **Storz G, Vogel J, Wassarman KM.** 2011. Regulation by small RNAs in bacteria: expanding frontiers. *Mol Cell* **43:**880–891.

7. **Richards GR, Vanderpool CK.** 2011. Molecular call and response: the physiology of bacterial small RNAs. *Biochim Biophys Acta* **1809:**525–531.

8. **Hoe CH, Raabe CA, Rozhdestvensky TS, Tang TH.** 2013. Bacterial sRNAs: regulation in stress. *Int J Med Microbiol* **303:**217–229.

9. **Opdyke JA, Kang JG, Storz G.** 2004. GadY, a small-RNA regulator of acid response genes in *Escherichia coli. J Bacteriol* **186:**6698–6705.

10. **Kawano M, Aravind L, Storz G.** 2007. An antisense RNA controls synthesis of an SOS-induced toxin evolved from an antitoxin. *Mol Microbiol* **64:**738–754.

11. **Toledo-Arana A, Dussurget O, Nikitas G, Sesto N, Guet-Revillet H, Balestrino D, Loh E, Gripenland J, Tiensuu T, Vaitkevicius K, Barthelemy M, Vergassola M, Nahori MA, Soubigou G, Regnault B, Coppee JY, Lecuit M, Johansson J, Cossart P.** 2009. The *Listeria* transcriptional landscape from saprophytism to virulence. *Nature* **459:**950–956.

12. **Lee E-J, Groisman E.** 2010. An antisense RNA that governs the expression kinetics of a multifunctional virulence gene. *Mol Microbiol* **76:** 1020–1033.

13. **Stazic D, Lindell D, Steglich C.** 2011. Antisense RNA protects mRNA from RNase E degradation by RNA-RNA duplex formation during phage infection. *Nucleic Acids Res* **39:** 4890–4899.

14. **Wurtzel O, Sesto N, Mellin JR, Karunker I, Edelheit S, Becavin C, Archambaud C, Cossart P, Sorek R.** 2012. Comparative transcriptomics of pathogenic and non-pathogenic *Listeria* species. *Mol Sys Biol* **8:**583.

15. **Simons R, Kleckner N.** 1983. Translational control of IS10 transposition. *Cell* **34:**683–691.

16. **Landt S, Abeliuk E, McGrath P, Lesley J, McAdams H, Shapiro L.** 2008. Small noncoding RNAs in *Caulobacter crescentus. Mol Microbiol* **68:**600–614.

17. **Padalon-Brauch G, Hershberg R, Elgrably-Weiss M, Baruch K, Rosenshine I, Margalit H, Altuvia S.** 2008. Small RNAs encoded within genetic islands of *Salmonella* typhimurium show host-induced expression and role in virulence. *Nucleic Acids Res* **36:**1913–1927.

18. **Waters JL, Salyers AA.** 2012. The small RNA RteR inhibits transfer of the Bacteroides conjugative transposon CTnDOT. *J Bacteriol* **194:** 5228–5236.

19. **Fozo E, Kawano M, Fontaine F, Kaya Y, Mendieta K, Jones K, Ocampo A, Rudd K, Storz G.** 2008. Repression of small toxic protein synthesis by the Sib and OhsC small RNAs. *Mol Microbiol* **70:**1076–1093.

20. **Fozo E, Makarova K, Shabalina S, Yutin N, Koonin E, Storz G.** 2010. Abundance of type I toxin-antitoxin systems in bacteria: searches for new candidates and discovery of novel families. *Nucleic Acids Res* **38:**3743–3759.

21. **Georg J, Hess W.** 2011. cis-antisense RNA, another level of gene regulation in bacteria. *Microbiol Mol Biol Rev* **75:**286–300.

22. **Morfeldt E, Taylor D, von Gabain A, Arvidson S.** 1995. Activation of alpha-toxin translation in *Staphylococcus aureus* by the trans-encoded antisense RNA, RNAIII. *EMBO J* **14:**4569–4577.

23. **Lease RA, Cusick ME, Belfort M.** 1998. Riboregulation in *Escherichia coli*: DsrA RNA acts by RNA:RNA interactions at multiple loci. *Proc Natl Acad Sci U S A* **95:**12456–12461.

24. **Prevost K, Salvail H, Desnoyers G, Jacques JF, Phaneuf E, Masse E.** 2007. The small RNA RyhB activates the translation of shiA mRNA encoding a permease of shikimate, a compound involved in siderophore synthesis. *Mol Microbiol* **64:**1260–1273.

25. **Sledjeski DD, Gupta A, Gottesman S.** 1996. The small RNA, DsrA, is essential for the low temperature expression of RpoS during exponential growth in *Escherichia coli. EMBO J* **15:**3993–4000.

26. **Majdalani N, Chen S, Murrow J, St John K, Gottesman S.** 2001. Regulation of *RpoS* by a novel small RNA: the characterization of RprA. *Mol Microbiol* **39:**1382–1394.

27. **Mandin P, Gottesman S.** 2010. Integrating anaerobic/aerobic sensing and the general stress response through the ArcZ small RNA. *EMBO J* **29:**3094–3107.

28. **Brown L, Elliott T.** 1997. Mutations that increase expression of the *rpoS* gene and decrease its dependence on hfq function in *Salmonella* typhimurium. *J Bacteriol* **179:**656–662.

29. **Cunning C, Brown L, Elliott T.** 1998. Promoter substitution and deletion analysis of upstream region required for rpoS translational regulation. *J Bacteriol* **180:**4564–4570.

30. **Majdalani N, Cunning C, Sledjeski D, Elliott T, Gottesman S.** 1998. DsrA RNA regulates translation of RpoS message by an anti-antisense

mechanism, independent of its action as an antisilencer of transcription. *Proc Natl Acad Sci U S A* **95:**12462–12467.

31. **Majdalani N, Hernandez D, Gottesman S.** 2002. Regulation and mode of action of the second small RNA activator of RpoS translation, RprA. *Mol Microbiol* **46:**813–826.

32. **Mackie G.** 1998. Ribonuclease E is a 5′-end-dependent endonuclease. *Nature* **395:** 720–723.

33. **Papenfort K, Sun Y, Miyakoshi M, Vanderpool CK, Vogel J.** 2013. Small RNA-mediated activation of sugar phosphatase mRNA regulates glucose homeostasis. *Cell* **153:**426–437.

34. **Frohlich KS, Papenfort K, Fekete A, Vogel J.** 2013. A small RNA activates CFA synthase by isoform-specific mRNA stabilization. *EMBO J* **32:**2963–2979.

35. **Koonin EV, Tatusov RL.** 1994. Computer analysis of bacterial haloacid dehalogenases defines a large superfamily of hydrolases with diverse specificity. Application of an iterative approach to database search. *J Mol Biol* **244:** 125–132.

36. **Morita T, Mochizuki Y, Aiba H.** 2006. Translational repression is sufficient for gene silencing by bacterial small noncoding RNAs in the absence of mRNA destruction. *Proc Natl Acad Sci U S A* **103:**4858–4863.

37. **Sharma CM, Darfeuille F, Plantinga TH, Vogel J.** 2007. A small RNA regulates multiple ABC transporter mRNAs by targeting C/A-rich elements inside and upstream of ribosome-binding sites. *Genes Dev* **21:**2804–2817.

38. **Yang Q, Figueroa-Bossi N, Bossi L.** 2014. Translation enhancing ACA motifs and their silencing by a bacterial small regulatory RNA. *PLoS Genet* **10:**e1004026.

39. **Darfeuille F, Unoson C, Vogel J, Wagner EG.** 2007. An antisense RNA inhibits translation by competing with standby ribosomes. *Mol Cell* **26:**381–392.

40. **Vecerek B, Moll I, Blasi U.** 2007. Control of Fur synthesis by the non-coding RNA RyhB and iron-responsive decoding. *EMBO J* **26:** 965–975.

41. **Beyer D, Skripkin E, Wadzack J, Nierhaus KH.** 1994. How the ribosome moves along the mRNA during protein synthesis. *J Biol Chem* **269:**30713–30717.

42. **Huttenhofer A, Noller HF.** 1994. Footprinting mRNA-ribosome complexes with chemical probes. *EMBO J* **13:**3892–3901.

43. **Bouvier M, Sharma CM, Mika F, Nierhaus KH, Vogel J.** 2008. Small RNA binding to 5′ mRNA coding region inhibits translational initiation. *Mol Cell* **32:**827–837.

44. **Holmqvist E, Reimegard J, Sterk M, Grantcharova N, Romling U, Wagner EG.** 2010. Two antisense RNAs target the transcriptional regulator CsgD to inhibit curli synthesis. *EMBO J* **29:**1840–1850.

45. **Desnoyers G, Masse E.** 2012. Noncanonical repression of translation initiation through small RNA recruitment of the RNA chaperone Hfq. *Genes Devel* **26:**726–739.

46. **Desnoyers G, Bouchard MP, Masse E.** 2013. New insights into small RNA-dependent translational regulation in prokaryotes. *Trends Genet* **29:**92–98.

47. **Sonnleitner E, Gonzalez N, Sorger-Domenigg T, Heeb S, Richter AS, Backofen R, Williams P, Huttenhofer A, Haas D, Blasi U.** 2011. The small RNA PhrS stimulates synthesis of the *Pseudomonas aeruginosa* quinolone signal. *Mol Microbiol* **80:**868–885.

48. **Masse E, Gottesman S.** 2002. A small RNA regulates the expression of genes involved in iron metabolism in *Escherichia coli. Proc Natl Acad Sci U S A* **99:**4620–4625.

49. **Masse E, Escorcia FE, Gottesman S.** 2003. Coupled degradation of a small regulatory RNA and its mRNA targets in *Escherichia coli. Genes Devel* **17:**2374–2383.

50. **Morita T, Maki K, Aiba H.** 2005. RNase E-based ribonucleoprotein complexes: mechanical basis of mRNA destabilization mediated by bacterial noncoding RNAs. *Genes Devel* **19:**2176–2186.

51. **Moller T, Franch T, Udesen C, Gerdes K, Valentin-Hansen P.** 2002. Spot 42 RNA mediates discoordinate expression of the *E. coli* galactose operon. *Genes Devel* **16:**1696–1706.

52. **Pfeiffer V, Papenfort K, Lucchini S, Hinton J, Vogel Jr.** 2009. Coding sequence targeting by MicC RNA reveals bacterial mRNA silencing downstream of translational initiation. *Nat Struct Mol Biol* **16:**840–846.

53. **Carpousis AJ, Luisi BF, McDowall KJ.** 2009. Endonucleolytic initiation of mRNA decay in *Escherichia coli. Prog Mol Biol Transl Sci* **85:** 91–135.

54. **Viegas SC, Silva IJ, Saramago M, Domingues S, Arraiano CM.** 2011. Regulation of the small regulatory RNA MicA by ribonuclease III: a target-dependent pathway. *Nucleic Acids Res* **39:**2918–2930.

55. **Sneppen K, Dodd I, Shearwin K, Palmer A, Schubert R, Callen B, Egan J.** 2005. A mathematical model for transcriptional interference by RNA polymerase traffic in *Escherichia coli. J Mol Biol* **346:**399–409.

56. **Crampton N, Bonass W, Kirkham J, Rivetti C, Thomson N.** 2006. Collision events between

RNA polymerases in convergent transcription studied by atomic force microscopy. *Nucleic Acids Res* **34:**5416–5425.

57. **Callen B, Shearwin K, Egan J.** 2004. Transcriptional interference between convergent promoters caused by elongation over the promoter. *Mol Cell* **14:**647–656.

58. **Palmer A, Ahlgren-Berg A, Egan J, Dodd I, Shearwin K.** 2009. Potent transcriptional interference by pausing of RNA polymerases over a downstream promoter. *Mol Cell* **34:**545–555.

59. **Andre G, Even S, Putzer H, Burguiere P, Croux C, Danchin A, Martin-Verstraete I, Soutourina O.** 2008. S-box and T-box riboswitches and antisense RNA control a sulfur metabolic operon of *Clostridium acetobutylicum*. *Nucleic Acids Res* **36:**5955–5969.

60. **Stork M, Di Lorenzo M, Welch T, Crosa J.** 2007. Transcription termination within the iron transport-biosynthesis operon of *Vibrio anguillarum* requires an antisense RNA. *J Bacteriol* **189:**3479–3488.

61. **Giangrossi M, Prosseda G, Tran C, Brandi A, Colonna B, Falconi M.** 2010. A novel antisense RNA regulates at transcriptional level the virulence gene icsA of *Shigella flexneri*. *Nucleic Acids Res* **38:**3362–3375.

62. **Wassarman K, Storz G.** 2000. 6S RNA regulates *E. coli* RNA polymerase activity. *Cell* **101:**613–623.

63. **Barrick J, Sudarsan N, Weinberg Z, Ruzzo W, Breaker R.** 2005. 6S RNA is a widespread regulator of eubacterial RNA polymerase that resembles an open promoter. *RNA* **11:**774–784.

64. **Trotochaud A, Wassarman K.** 2005. A highly conserved 6S RNA structure is required for regulation of transcription. *Nat Struct Mol Biol* **12:**313–319.

65. **Trotochaud A, Wassarman K.** 2004. 6S RNA function enhances long-term cell survival. *J Bacteriol* **186:**4978–4985.

66. **Cavanagh AT, Klocko AD, Liu X, Wassarman KM.** 2008. Promoter specificity for 6S RNA regulation of transcription is determined by core promoter sequences and competition for region 4.2 of sigma70. *Mol Microbiol* **67:**1242–1256.

67. **Klocko AD, Wassarman KM.** 2009. 6S RNA binding to Esigma(70) requires a positively charged surface of sigma(70) region 4.2. *Mol Microbiol* **73:**152–164.

68. **Dulebohn D, Choy J, Sundermeier T, Okan N, Karzai A.** 2007. Trans-translation: the tmRNA-mediated surveillance mechanism for ribosome rescue, directed protein degradation, and nonstop mRNA decay. *Biochemistry* **46:** 4681–4693.

69. **Shpanchenko O, Golovin A, Bugaeva E, Isaksson L, Dontsova O.** 2010. Structural aspects of trans-translation. *IUBMB Life* **62:**120–124.

70. **Keiler KC, Waller PR, Sauer RT.** 1996. Role of a peptide tagging system in degradation of proteins synthesized from damaged messenger RNA. *Science* **271:**990–993.

71. **Withey JH, Friedman DI.** 2003. A salvage pathway for protein structures: tmRNA and trans-translation. *Annu Rev Microbiol* **57:**101–123.

72. **Williams KP, Bartel DP.** 1996. Phylogenetic analysis of tmRNA secondary structure. *RNA* **2:**1306–1310.

73. **Felden B, Himeno H, Muto A, Atkins JF, Gesteland RF.** 1996. Structural organization of *Escherichia coli* tmRNA. *Biochimie* **78:**979–983.

74. **Komine Y, Kitabatake M, Yokogawa T, Nishikawa K, Inokuchi H.** 1994. A tRNA-like structure is present in 10Sa RNA, a small stable RNA from *Escherichia coli*. *Proc Natl Acad Sci U S A* **91:**9223–9227.

75. **Karzai AW, Roche ED, Sauer RT.** 2000. The SsrA-SmpB system for protein tagging, directed degradation and ribosome rescue. *Nat Struct Biol* **7:**449–455.

76. **Babitzke P, Romeo T.** 2007. CsrB sRNA family: sequestration of RNA-binding regulatory proteins. *Curr Opin Microbiol* **10:**156–163.

77. **Timmermans J, Van Melderen L.** 2010. Posttranscriptional global regulation by CsrA in bacteria. *Cell Mol Life Sci* **67:**2897–2908.

78. **Carmichael GG, Weber K, Niveleau A, Wahba AJ.** 1975. The host factor required for RNA phage Qbeta RNA replication in vitro. Intracellular location, quantitation, and purification by polyadenylate-cellulose chromatography. *J Biol Chem* **250:**3607–3612.

79. **Kajitani M, Kato A, Wada A, Inokuchi Y, Ishihama A.** 1994. Regulation of the *Escherichia coli* hfq gene encoding the host factor for phage Q beta. *J Bacteriol* **176:**531–534.

80. **Ali Azam T, Iwata A, Nishimura A, Ueda S, Ishihama A.** 1999. Growth phase-dependent variation in protein composition of the *Escherichia coli* nucleoid. *J Bacteriol* **181:**6361–6370.

81. **Weichenrieder O.** 2014. RNA binding by Hfq and ring-forming (L)Sm proteins: A trade-off between optimal sequence readout and RNA backbone conformation. *RNA Biol* **11(5):**537–549.

82. **Franze de Fernandez MT, Eoyang L, August JT.** 1968. Factor fraction required for the synthesis of bacteriophage Qbeta-RNA. *Nature* **219:**588–590.

83. **Franze de Fernandez MT, Hayward WS, August JT.** 1972. Bacterial proteins required for replication of phage Q ribonucleic acid. Pruification and properties of host factor I, a ribonucleic acid-binding protein. *J Biol Chem* **247:**824–831.

84. **Tsui HC, Leung HC, Winkler ME.** 1994. Characterization of broadly pleiotropic phenotypes caused by an hfq insertion mutation in *Escherichia coli* K-12. *Mol Microbiol* **13:**35–49.

85. **Sittka A, Pfeiffer V, Tedin K, Vogel J.** 2007. The RNA chaperone Hfq is essential for the virulence of *Salmonella* typhimurium. *Mol Microbiol* **63:**193–217.

86. **Hayashi-Nishino M, Fukushima A, Nishino K.** 2012. Impact of hfq on the intrinsic drug resistance of *Salmonella enterica* serovar typhimurium. *Front Microbiol* **3:**205.

87. **Sittka A, Lucchini S, Papenfort K, Sharma CM, Rolle K, Binnewies TT, Hinton JC, Vogel J.** 2008. Deep sequencing analysis of small noncoding RNA and mRNA targets of the global post-transcriptional regulator, Hfq. *PLoS Genet* **4:**e1000163.

88. **Muffler A, Traulsen DD, Fischer D, Lange R, Hengge-Aronis R.** 1997. The RNA-binding protein HF-I plays a global regulatory role which is largely, but not exclusively, due to its role in expression of the sigmaS subunit of RNA polymerase in *Escherichia coli*. *J Bacteriol* **179:**297–300.

89. **Tsui HC, Feng G, Winkler ME.** 1997. Negative regulation of mutS and mutH repair gene expression by the Hfq and RpoS global regulators of *Escherichia coli* K-12. *J Bacteriol* **179:**7476–7487.

90. **Vytvytska O, Moll I, Kaberdin VR, von Gabain A, Blasi U.** 2000. Hfq (HF1) stimulates ompA mRNA decay by interfering with ribosome binding. *Genes Dev* **14:**1109–1118.

91. **Vecerek B, Moll I, Blasi U.** 2005. Translational autocontrol of the *Escherichia coli* hfq RNA chaperone gene. *RNA* **11:**976–984.

92. **Salvail H, Caron MP, Belanger J, Masse E.** 2013. Antagonistic functions between the RNA chaperone Hfq and an sRNA regulate sensitivity to the antibiotic colicin. *EMBO J* **32:**2764–2778.

93. **Sauer E.** 2013. Structure and RNA-binding properties of the bacterial LSm protein Hfq. *RNA Biol* **10:**610–618.

94. **Panja S, Woodson SA.** 2012. Hexamer to monomer equilibrium of E. coli Hfq in solution and its impact on RNA annealing. *J Mol Biol* **417:**406–412.

95. **Link TM, Valentin-Hansen P, Brennan RG.** 2009. Structure of *Escherichia coli* Hfq bound to polyriboadenylate RNA. *Proc Natl Acad Sci U S A* **106:**19292–19297.

96. **Mikulecky PJ, Kaw MK, Brescia CC, Takach JC, Sledjeski DD, Feig AL.** 2004. *Escherichia coli* Hfq has distinct interaction surfaces for DsrA, rpoS and poly(A) RNAs. *Nat Struct Mol Biol* **11:**1206–1214.

97. **Zhang A, Schu DJ, Tjaden BC, Storz G, Gottesman S.** 2013. Mutations in Interaction Surfaces Differentially Impact E. coli Hfq Association with Small RNAs and Their mRNA Targets. *J Mol Biol* **425**(19):3678–3697.

98. **Sauer E, Schmidt S, Weichenrieder O.** 2012. Small RNA binding to the lateral surface of Hfq hexamers and structural rearrangements upon mRNA target recognition. *Proc Natl Acad Sci U S A* **109:**9396–9401.

99. **Robinson KE, Orans J, Kovach AR, Link TM, Brennan RG.** 2014. Mapping Hfq-RNA interaction surfaces using tryptophan fluorescence quenching. *Nucleic Acids Res* **42:**2736–2749.

100. **Valentin-Hansen P, Eriksen M, Udesen C.** 2004. The bacterial Sm-like protein Hfq: a key player in RNA transactions. *Mol Microbiol* **51:**1525–1533.

101. **Ishikawa H, Otaka H, Maki K, Morita T, Aiba H.** 2012. The functional Hfq-binding module of bacterial sRNAs consists of a double or single hairpin preceded by a U-rich sequence and followed by a 3′ poly(U) tail. *RNA* **18:**1062–1074.

102. **Otaka H, Ishikawa H, Morita T, Aiba H.** 2011. PolyU tail of rho-independent terminator of bacterial small RNAs is essential for Hfq action. *Proc Natl Acad Sci U S A* **108**(32):13059–13064.

103. **Soper TJ, Woodson SA.** 2008. The rpoS mRNA leader recruits Hfq to facilitate annealing with DsrA sRNA. *RNA* **14:**1907–1917.

104. **Salim NN, Faner MA, Philip JA, Feig AL.** 2012. Requirement of upstream Hfq-binding (ARN)x elements in glmS and the Hfq C-terminal region for GlmS upregulation by sRNAs GlmZ and GlmY. *Nucleic Acids Res* **40:**8021–8032.

105. **Salim NN, Feig AL.** 2010. An upstream Hfq binding site in the fhlA mRNA leader region facilitates the OxyS-fhlA interaction. *PLoS One* **5**(9):e13028.

106. **Sledjeski DD, Whitman C, Zhang A.** 2001. Hfq is necessary for regulation by the untranslated RNA DsrA. *J Bacteriol* **183:**1997–2005.

107. **Zhang A, Wassarman KM, Ortega J, Steven AC, Storz G.** 2002. The Sm-like Hfq protein increases OxyS RNA interaction with target mRNAs. *Molec Cell* **9:**11–22.

108. **Fender A, Elf J, Hampel K, Zimmermann B, Wagner EG.** 2010. RNAs actively cycle on the

Sm-like protein Hfq. *Genes Develop* **24**:2621–2626.

109. **Hopkins JF, Panja S, Woodson SA.** 2011. Rapid binding and release of Hfq from ternary complexes during RNA annealing. *Nucleic Acids Res* **39**:5193–5202.

110. **Maki K, Morita T, Otaka H, Aiba H.** 2010. A minimal base-pairing region of a bacterial small RNA SgrS required for translational repression of ptsG mRNA. *Molecular Microbiol* **76**:782–792.

111. **Soper T, Mandin P, Majdalani N, Gottesman S, Woodson SA.** 2010. Positive regulation by small RNAs and the role of Hfq. *Proc Natl Acad Sci U S A* **107**:9602–9607.

112. **Adamson DN, Lim HN.** 2011. Essential requirements for robust signaling in Hfq dependent small RNA networks. *PLoS Comput Biol* **7**:e1002138.

113. **Hussein R, Lim HN.** 2011. Disruption of small RNA signaling caused by competition for Hfq. *Proc Natl Acad Sci U S A* **108**:1110–1115.

114. **Moon K, Gottesman S.** 2011. Competition among Hfq-binding small RNAs in *Escherichia coli*. *Mol Microbiol* **82**:1545–1562.

115. **Sukhodolets MV, Garges S.** 2003. Interaction of *Escherichia coli* RNA polymerase with the ribosomal protein S1 and the Sm-like ATPase Hfq. *Biochemistry* **42**:8022–8034.

116. **Rabhi M, Espeli O, Schwartz A, Cayrol B, Rahmouni AR, Arluison V, Boudvillain M.** 2011. The Sm-like RNA chaperone Hfq mediates transcription antitermination at Rho-dependent terminators. *EMBO J* **30**:2805–2816.

117. **Hajnsdorf E, Regnier P.** 2000. Host factor Hfq of *Escherichia coli* stimulates elongation of poly(A) tails by poly(A) polymerase I. *Proc Natl Acad Sci U S A* **97**:1501–1505.

118. **Ikeda Y, Yagi M, Morita T, Aiba H.** 2011. Hfq binding at RhlB-recognition region of RNase E is crucial for the rapid degradation of target mRNAs mediated by sRNAs in *Escherichia coli*. *Mol Microbiol* **79**:419–432.

119. **De Lay N, Gottesman S.** 2011. Role of polynucleotide phosphorylase in sRNA function in *Escherichia coli*. *RNA* **17**:1172–1189.

120. **Mohanty BK, Maples VF, Kushner SR.** 2004. The Sm-like protein Hfq regulates polyadenylation dependent mRNA decay in *Escherichia coli*. *Mol Microbiol* **54**:905–920.

121. **Bandyra KJ, Luisi BF.** 2013. Licensing and due process in the turnover of bacterial RNA. *RNA Biol* **10**:627–635.

122. **Belasco JG.** 2010. All things must pass: contrasts and commonalities in eukaryotic and bacterial mRNA decay. Nature reviews. *Mol Cell Biol* **11**:467–478.

123. **Carpousis AJ.** 2007. The RNA degradosome of *Escherichia coli*: an mRNA-degrading machine assembled on RNase E. *Ann Rev Microbiol* **61**:71–87.

124. **Bandyra KJ, Said N, Pfeiffer V, Gorna MW, Vogel J, Luisi BF.** 2012. The seed region of a small RNA drives the controlled destruction of the target mRNA by the endoribonuclease RNase E. *Mol Cell* **47**:943–953.

125. **Jiang X, Belasco JG.** 2004. Catalytic activation of multimeric RNase E and RNase G by 5′-monophosphorylated RNA. *Proc Natl Acad Sci U S A* **101**:9211–9216.

126. **Dreyfus M.** 2009. Killer and protective ribosomes. *Prog Mol Biol Transl Sci* **85**:423–466.

127. **Joyce SA, Dreyfus M.** 1998. In the absence of translation, RNase E can bypass 5′ mRNA stabilizers in *Escherichia coli*. *J Mol Biol* **282**:241–254.

128. **Deana A, Belasco JG.** 2005. Lost in translation: the influence of ribosomes on bacterial mRNA decay. *Genes Dev* **19**:2526–2533.

129. **Prevost K, Desnoyers G, Jacques JF, Lavoie F, Masse E.** 2011. Small RNA-induced mRNA degradation achieved through both translation block and activated cleavage. *Genes Dev* **25**:385–396.

130. **Davies BW, Walker GC.** 2008. A highly conserved protein of unknown function is required by *Sinorhizobium meliloti* for symbiosis and environmental stress protection. *J Bacteriol* **190**:1118–1123.

131. **Vercruysse M, Kohrer C, Davies BW, Arnold MF, Mekalanos JJ, RajBhandary UL, Walker GC.** 2014. The Highly Conserved Bacterial RNase YbeY Is Essential in *Vibrio cholerae*, Playing a Critical Role in Virulence, Stress Regulation, and RNA Processing. *PLoS Pathog* **10**:e1004175.

132. **Meister G.** 2013. Argonaute proteins: functional insights and emerging roles. *Nat Rev Genet* **14**:447–459.

133. **Jacob AI, Kohrer C, Davies BW, RajBhandary UL, Walker GC.** 2013. Conserved bacterial RNase YbeY plays key roles in 70S ribosome quality control and 16S rRNA maturation. *Mol Cell* **49**:427–438.

134. **Davies BW, Kohrer C, Jacob AI, Simmons LA, Zhu J, Aleman LM, Rajbhandary UL, Walker GC.** 2010. Role of *Escherichia coli* YbeY, a highly conserved protein, in rRNA processing. *Mol Microbiol* **78**:506–518.

135. **Pandey SP, Minesinger BK, Kumar J, Walker GC.** 2011. A highly conserved protein of unknown function in *Sinorhizobium meliloti* affects sRNA regulation similar to Hfq. *Nucleic Acids Res* **39**:4691–4708.

136. **Pandey SP, Winkler JA, Li H, Camacho DM, Collins JJ, Walker GC.** 2014. Central role for RNase YbeY in Hfq-dependent and Hfq-independent small-RNA regulation in bacteria. *BMC Genomics* **15:**121.

137. **Englesberg E, Anderson RL, Weinberg R, Lee N, Hoffee P, Huttenhauer G, Boyer H.** 1962. L-Arabinose-sensitive, L-ribulose 5-phosphate 4-epimerase-deficient mutants of *Escherichia coli. J Bacteriol* **84:**137–146.

138. **Horler RS, Vanderpool CK.** 2009. Homologs of the small RNA SgrS are broadly distributed in enteric bacteria but have diverged in size and sequence. *Nucleic Acids Res* **37:**5465–5476.

139. **Martinez-Hackert E, Stock AM.** 1997. Structural relationships in the OmpR family of winged-helix transcription factors. *J Mol Biol* **269:**301–312.

140. **Tam R, Saier MH Jr.** 1993. Structural, functional, and evolutionary relationships among extracellular solute-binding receptors of bacteria. *Microbiol Rev* **57:**320–346.

141. **Vanderpool CK, Gottesman S.** 2004. Involvement of a novel transcriptional activator and small RNA in post-transcriptional regulation of the glucose phosphoenolpyruvate phosphotransferase system. *Mol Microbiol* **54:**1076–1089.

142. **Kawamoto H, Koide Y, Morita T, Aiba H.** 2006. Base-pairing requirement for RNA silencing by a bacterial small RNA and acceleration of duplex formation by Hfq. *Mol Microbiol* **61:**1013–1022.

143. **Wadler C, Vanderpool C.** 2007. A dual function for a bacterial small RNA: SgrS performs base pairing-dependent regulation and encodes a functional polypeptide. *Proc Natl Acad Sci U S A* **104:**20454–20459.

144. **Rice J, Vanderpool C.** 2011. The small RNA SgrS controls sugar-phosphate accumulation by regulating multiple PTS genes. *Nucleic Acids Res* **39:**3806–3819.

145. **Rice JB, Balasubramanian D, Vanderpool CK.** 2012. Small RNA binding-site multiplicity involved in translational regulation of a polycistronic mRNA. *Proc Natl Acad Sci U S A* **109:**e2691–e2698.

146. **Richards GR, Vivas EI, Andersen AW, Rivera-Santos D, Gilmore S, Suen G, Goodrich-Blair H.** 2009. Isolation and characterization of *Xenorhabdus nematophila* transposon insertion mutants defective in lipase activity against Tween. *J Bacteriol* **191:**5325–5331.

147. **Papenfort K, Podkaminski D, Hinton JC, Vogel J.** 2012. The ancestral SgrS RNA discriminates horizontally acquired *Salmonella* mRNAs through a single G-U wobble pair. *Proc Natl Acad Sci U S A* **109:**e757–e764.

148. **Jiang X, Rossanese OW, Brown NF, Kujat-Choy S, Galan JE, Finlay BB, Brumell JH.** 2004. The related effector proteins SopD and SopD2 from *Salmonella enterica* serovar Typhimurium contribute to virulence during systemic infection of mice. *Mol Microbiol* **54:**1186–1198.

149. **Sun Y, Vanderpool CK.** 2011. Regulation and function of *Escherichia coli* sugar efflux transporter A (SetA) during glucose-phosphate stress. *J Bacteriol* **193:**143–153.

150. **Liu JY, Miller PF, Willard J, Olson ER.** 1999. Functional and biochemical characterization of *Escherichia coli* sugar efflux transporters. *J Biol Chem* **274:**22977–22984.

151. **Kim SH, Schneider BL, Reitzer L.** 2010. Genetics and regulation of the major enzymes of alanine synthesis in *Escherichia coli. J Bacteriol* **192:**5304–5311.

152. **Vanderpool CK.** 2007. Physiological consequences of small RNA-mediated regulation of glucose-phosphate stress. *Curr Opin Microbiol* **10:**146–151.

153. **Richards GR, Vanderpool CK.** 2012. Induction of the Pho regulon suppresses the growth defect of an *Escherichia coli sgrS* mutant, connecting phosphate metabolism to the glucose-phosphate stress response. *J Bacteriol* **194:**2520–2530.

154. **Sun Y, Vanderpool CK.** 2013. Physiological consequences of multiple-target regulation by the small RNA SgrS in *Escherichia coli. J Bacteriol* **195:**4804–4815.

155. **Horler R, Vanderpool C.** 2009. Homologs of the small RNA SgrS are broadly distributed in enteric bacteria but have diverged in size and sequence. *Nucleic Acids Res* **37:**5465–5476.

156. **Wadler C, Vanderpool C.** 2009. Characterization of homologs of the small RNA SgrS reveals diversity in function. *Nucleic Acids Res* **37:**5477–5485.

157. **Balasubramanian D, Vanderpool CK.** 2013. Deciphering the interplay between two independent functions of the small RNA regulator SgrS in *Salmonella. J Bacteriol* **195:**4620–4630.

158. **Kimata K, Tanaka Y, Inada T, Aiba H.** 2001. Expression of the glucose transporter gene, ptsG, is regulated at the mRNA degradation step in response to glycolytic flux in *Escherichia coli. EMBO J* **20:**3587–3595.

159. **Morita T, El-Kazzaz W, Tanaka Y, Inada T, Aiba H.** 2003. Accumulation of glucose 6-phosphate or fructose 6-phosphate is responsible for destabilization of glucose transporter

mRNA in *Escherichia coli*. *J Biol Chem* **278**: 15608–15614.

160. **Richards GR, Patel MV, Lloyd CR, Vanderpool CK.** 2013. Depletion of glycolytic intermediates plays a key role in glucose-phosphate stress in *Escherichia coli*. *J Bacteriol* **195**:4816–4825.

161. **Morita T, El-Kazzaz W, Tanaka Y, Inada T, Aiba H.** 2003. Accumulation of glucose 6-phosphate or fructose 6-phosphate is responsible for destabilization of glucose transporter mRNA in *Escherichia coli*. *J Biol Chem* **278**: 15608–15614.

162. **Møller T, Franch T, Udesen C, Gerdes K, Valentin-Hansen P.** 2002. Spot 42 RNA mediates discoordinate expression of the *E. coli* galactose operon. *Genes Develop* **16**:1696–1706.

163. **Adhya S.** 1987. The galactose operon, p 1503–1512. *In* Neidhardt FC, Ingraham JL, Curtiss R (ed), *Escherichia coli and Salmonella typhimurium: Cellular and molecular biology*. ASM Press, Washington, D.C.

164. **Beisel CL, Storz G.** 2011. The base-pairing RNA spot 42 participates in a multioutput feedforward loop to help enact catabolite repression in *Escherichia coli*. *Mol Cell* **41**:286–297.

165. **Rice PW, Dahlberg JE.** 1982. A gene between *polA* and *glnA* retards growth of *Escherichia coli* when present in multiple copies: physiological effects of the gene for spot 42 RNA. *J Bacteriol* **152**:1196–1210.

166. **Polayes DA, Rice PW, Garner MM, Dahlberg JE.** 1988. Cyclic AMP-cyclic AMP receptor protein as a repressor of transcription of the *spf* gene of *Escherichia coli*. *J Bacteriol* **170**:3110–3114.

167. **Sahagan B, Dahlberg JE.** 1979. A small, unstable RNA molecule of *Escherichia coli*: Spot 42 RNA. *J Mol Biol* **131**:593–605.

168. **Joseph E, Danchin A, Ullmann A.** 1981. Regulation of galactose operon expression: glucose effects and role of cyclic adenosine 3′,5′-monophosphate. *J Bacteriol* **146**:149–154.

169. **Queen C, Rosenberg M.** 1981. Differential translation efficiency explains discoordinate expression of the galactose operon. *Cell* **25**:241–249.

170. **Sonnleitner E, Abdou L, Haas D.** 2009. Small RNA as global regulator of carbon catabolite repression in *Pseudomonas aeruginosa*. *Proc Natl Acad Sci U S A* **106**:21866–21871.

171. **Liu P.** 1952. Utilization of carbohydrates by *Pseudomonas aeruginosa*. *J Bacteriol* **64**:773–781.

172. **Smyth PF, Clarke PH.** 1975. Catabolite repression of *Pseudomonas aeruginosa* amidase: The effect of carbon source on amidase synthesis. *J Gen Microbiol* **90**:81–90.

173. **Collier DN, Hager PW, Phibbs PV.** 1996. Catabolite repression control in the pseudomonads. *Res Microbiol* **147**:551–561.

174. **Rojo F, Dinamarca MA.** 2004. Catabolite repression and physiological control, p 365–387. *In* Ramos JL (ed), *Pseudomonas: virulence and gene regulation*, **vol. 2**. Kluwer Academic/Plenum, New York.

175. **MacGregor CH, Arora SK, Hager PW, Dail MB, Phibbs PVJ.** 1996. The nucleotide sequence of the *Pseudomonas aeruginosa pyrE-crc-rph* region and the purification of the *crc* gene product. *J Bacteriol* **178**:5627–5635.

176. **MacGregor CH, Wolff JA, Arora SK, Phibbs PVJ.** 1991. Cloning of a catabolite repression control (*crc*) gene from *Pseudomonas aeruginosa*, expression of the gene in *Escherichia coli*, and identification of the gene product in *Pseudomonas aeruginosa*. *J Bacteriol* **173**:7204–7212.

177. **Rojo F.** 2010. Carbon catabolite repression in *Pseudomonas*: optimizing metabolic versatility and interactions with the environment. *FEMS Microbiol Rev* **34**:658–684.

178. **Moreno R, Martinez-Gomariz M, Yuste L, Gil C, Rojo F.** 2009. The *Pseudomonas putida* Crc global regulator controls the hierarchical assimilation of amino acids in a complete medium: evidence from proteomic and genomic analyses. *Proteomics* **9**:2910–2928.

179. **Moreno R, Rojo F.** 2008. The target for the *Pseudomonas putida* Crc global regulator in the benzoate degradation pathway is the BenR transcriptional regulator. *J Bacteriol* **190**:1539–1545.

180. **Moreno R, Ruiz-Manzano A, Yuste L, Rojo F.** 2007. The *Pseudomonas putida* Crc global regulator is an RNA binding protein that inhibits translation of the AlkS transcriptional regulator. *Mol Microbiol* **64**:665–675.

181. **Milojevic T, Grishkovskaya I, Sonnleitner E, Djinovic-Carugo K, Bläsi U.** 2013. The *Pseudomonas aeruginosa* catabolite repression control protein Crc is devoid of RNA binding activity. *PLoS ONE* **8**:e64609.

182. **Milojevic T, Sonnleitner E, Romeo A, Djinović-Carugo K, Bläsi U.** 2013. False positive RNA binding activities after Ni-affinity purification from *Escherichia coli*. *RNA Biol* **10**:1066–1069.

183. **Sonnleitner E, Bläsi U.** 2014. Regulation of Hfq by the RNA CrcZ in *Pseudomonas aeruginosa* carbon catabolite repression. *PLoS Genet* **10**:e1004440.

184. **Wolff JA, MacGregor CH, Eisenberg RC, Phibbs PV.** 1991. Isolation and characterization of catabolite repression control mutants of *Pseudomonas aeruginosa* PAO. *J Bacteriol* **173**:4700–4706.

185. **Amador CI, Canosa I, Govantes F, Santero E.** 2010. Lack of CbrB in *Pseudomonas putida*

affects not only amino acids metabolism but also different stress responses and biofilm development. *Environ Microbiol* **12:**1748–1761.

186. **O'Toole GA, Gibbs KA, Hager PW, Phibbs PVJ, Kolter R.** 2000. The global carbon metabolism regulator Crc is a component of a signal transduction pathway required for biofilm development by *Pseudomonas aeruginosa. J Bacteriol* **182:**425–431.

187. **Moreno R, Fonseca P, Rojo F.** 2012. Two small RNAs, CrcY and CrcZ, act in concert to sequester the Crc global regulator in *Pseudomonas putida,* modulating catabolite repression. *Mol Microbiol* **83:**24–40.

188. **Filiatrault MJ, Stodghill PV, Wilson J, Butcher BG, Chen H, al e.** 2013. CrcZ and CrcX regulate carbon source utilization in *Pseudomonas syringae* pathovar tomato strain DC3000. *RNA Biol* **10:**245–255.

189. **Joanny G, Le Derout J, Brechemier-Baey D, Labas V, Vinh J, Régnier P, Hajnsdorf E.** 2007. Polyadenylation of a functional mRNA controls gene expression in *Escherichia coli. Nucleic Acids Res* **35:**2494–2502.

190. **Kalamorz F, Reichenbach B, Marz W, Rak B, Gorke B.** 2007. Feedback control of glucosamine-6-phosphate synthase GlmS expression depends on the small RNA GlmZ and involves the novel protein YhbJ in *Escherichia coli. Mol Microbiol* **65:**1518–1533.

191. **Reichenbach B, Maes A, Kalamorz F, Hajnsdorf E, Gorke B.** 2008. The small RNA GlmY acts upstream of the sRNA GlmZ in the activation of *glmS* expression and is subject to regulation by polyadenylation in *Escherichia coli. Nucleic Acids Res.* **36:**2570–2580.

192. **Urban JH, Papenfort K, Thomsen J, Schmitz RA, Vogel J.** 2007. A conserved small RNA promotes discoordinate expression of the glmUS operon mRNA to activate GlmS synthesis. *Journal Mol Biol* **373:**521–528.

193. **Urban JH, Vogel J.** 2008. Two seemingly homologous noncoding RNAs act hierarchically to activate glmS mRNA translation. *PLoS Biol* **6:**e64.

194. **Göpel Y, Papenfort K, Reichenbach B, Vogel J, Görke B.** 2013. Targeted decay of a regulatory small RNA by an adaptor protein for RNase E and counteraction by an anti-adaptor RNA. *Genes Devel* **27:**552–564.

195. **Urbanowski ML, Stauffer LT, Stauffer GV.** 2000. The *gcvB* gene encodes a small untranslated RNA involved in expression of the dipeptide and oligopeptide transport systems in *Escherichia coli. Mol Microbiol* **37:**856–868.

196. **Pulvermacher SC, Stauffer LT, Stauffer GV.** 2008. The role of the small regulatory RNA GcvB in GcvB/mRNA posttranscriptional regulation of *oppA* and *dppA* in *Escherichia coli. FEMS Microbiol Lett* **281:**42–50.

197. **Pulvermacher SC, Stauffer LT, Stauffer GV.** 2009. Role of the sRNA GcvB in regulation of *cycA* in *Escherichia coli. Microbiology* **155:**106–114.

198. **Stauffer LT, Stauffer GV.** 2012. The *Escherichia coli* GcvB sRNA uses genetic redundancy to control *cycA* expression. *ISRN Microbiol* **2012.**

199. **Sharma C, Darfeuille F, Plantinga T, Vogel J.** 2007. A small RNA regulates multiple ABC transporter mRNAs by targeting C/A-rich elements inside and upstream of ribosome-binding sites. *Genes Devel* **21:**2804–2817.

200. **McArthur SD, Pulvermacher SC, Stauffer GV.** 2006. The *Yersinia pestis gcvB* gene encodes two small regulatory RNA molecules. *BMC Microbiol* **6:**52.

201. **Silveira AC, Robertson KL, Lin B, Wang Z, Vora GJ, Vasconcelos AT, Thompson FL.** 2010. Identification of non-coding RNAs in environmental vibrios. *Microbiology* **156:**2452–2458.

202. **Sharma CM, Papenfort K, Pernitzsch SR, Mollenkopf HJ, Hinton JC, Vogel J.** 2011. Pervasive post-transcriptional control of genes involved in amino acid metabolism by the Hfq-dependent GcvB small RNA. *Mol Microbiol* **81** (5):1144–1165.

203. **Wassarman KM, Repoila F, Rosenow C, Storz G, Gottesman S.** 2001. Identification of novel small RNAs using comparative genomics and microarrays. *Genes Dev* **15:**1637–1651.

204. **Gerstle K, Klätschke K, Hahn U, Piganeau N.** 2012. The small RNA RybA regulates key-genes in the biosynthesis of aromatic amino acids under peroxide stress in *E. coli. RNA Biol* **9:**458–468.

205. **Schilling D, Findeiss S, Richter AS, Taylor JA, Gerischer U.** 2010. The small RNA Aar in *Acinetobacter baylyi:* a putative regulator of amino acid metabolism. *Arch Microbiol* **192:**691–702.

206. **Andrews SC, Robinson AK, Rodriguez-Quinones F.** 2003. Bacterial iron homeostasis. *FEMS Microbiol Rev* **27:**215–237.

207. **Schaible UE, Kaufmann SH.** 2004. Iron and microbial infection. *Nat Rev Microbiol* **2:**946–953.

208. **Masse E, Vanderpool CK, Gottesman S.** 2005. Effect of RyhB small RNA on global iron use in *Escherichia coli. J Bacteriol* **187:**6962–6971.

209. **Geissmann, Touati D.** 2004. Hfq, a new chaperoning role: binding to messanger RNA determines access for small RNA regulator. *EMBO J* **23:**396–405.

210. **Vecerek B, Moll I, Afonyushkin T, Kaberdin V, Blasi U.** 2003. Interaction of the RNA chaperone Hfq with mRNAs: direct and indirect roles of Hfq in iron metabolism of *Escherichia coli*. *Molec Microbiol* **50:**897–909.

211. **Hantke K.** 1981. Regulation of ferric iron transport in *Escherichia coli* K12: isolation of a constitutive mutant. *Mol Gen Genet* **182:**288–292.

212. **Salvail H, Lanthier-Bourbonnais P, Sobota JM, Caza M, Benjamin JA, Mendieta ME, Lepine F, Dozois CM, Imlay J, Masse E.** 2010. A small RNA promotes siderophore production through transcriptional and metabolic remodeling. *Proc Natl Acad Sci U S A* **107:**15223–15228.

213. **Ochsner UA, Wilderman PJ, Vasil AI, Vasil ML.** 2002. GeneChip expression analysis of the iron starvation response in *Pseudomonas aeruginosa*: identification of novel pyoverdine biosynthesis genes. *Mol Microbiol* **45:**1277–1287.

214. **Wilderman PJ, Sowa NA, FitzGerald DJ, FitzGerald PC, Gottesman S, Ochsner UA, Vasil ML.** 2004. Identification of tandem duplicate regulatory small RNAs in *Pseudomonas aeruginosa* involved in iron homeostasis. *Proc Natl Acad Sci U S A* **101:**9792–9797.

215. **Oglesby-Sherrouse AG, Vasil ML.** 2010. Characterization of a heme-regulated noncoding RNA encoded by the *prrF* locus of *Pseudomonas aeruginosa*. *PLoS ONE* **5:**e9930.

216. **Romeo T, Vakulskas CA, Babitzke P.** 2013. Post-transcriptional regulation on a global scale: form and function of Csr/Rsm systems. *Environ Microbiol* **15:**313–324.

217. **Wang X, Preston JF 3rd, Romeo T.** 2004. The pgaABCD locus of *Escherichia coli* promotes the synthesis of a polysaccharide adhesin required for biofilm formation. *J Bacteriol* **186:**2724–2734.

218. **Itoh Y, Wang X, Hinnebusch BJ, Preston JF 3rd, Romeo T.** 2005. Depolymerization of beta-1,6-N-acetyl-D-glucosamine disrupts the integrity of diverse bacterial biofilms. *J Bacteriol* **187:**382–387.

219. **Jackson DW, Suzuki K, Oakford L, Simecka JW, Hart ME, Romeo T.** 2002. Biofilm formation and dispersal under the influence of the global regulator CsrA of *Escherichia coli*. *J Bacteriol* **184:**290–301.

220. **Wang X, Dubey AK, Suzuki K, Baker CS, Babitzke P, Romeo T.** 2005. CsrA post-transcriptionally represses pgaABCD, responsible for synthesis of a biofilm polysaccharide adhesin of *Escherichia coli*. *Mol Microbiol* **56:**1648–1663.

221. **Yakhnin H, Yakhnin AV, Baker CS, Sineva E, Berezin I, Romeo T, Babitzke P.** 2011. Complex regulation of the global regulatory gene csrA: CsrA-mediated translational repression, transcription from five promoters by Esigma (7)(0) and Esigma(S), and indirect transcriptional activation by CsrA. *Mol Microbiol* **81:**689–704.

222. **Wei BL, Brun-Zinkernagel AM, Simecka JW, Pruss BM, Babitzke P, Romeo T.** 2001. Positive regulation of motility and flhDC expression by the RNA-binding protein CsrA of *Escherichia coli*. *Mol Microbiol* **40:**245–256.

223. **Yakhnin AV, Baker CS, Vakulskas CA, Yakhnin H, Berezin I, Romeo T, Babitzke P.** 2013. CsrA activates flhDC expression by protecting flhDC mRNA from RNase E-mediated cleavage. *Mol Microbiol* **87:**851–866.

224. **Mercante J, Edwards AN, Dubey AK, Babitzke P, Romeo T.** 2009. Molecular geometry of CsrA (RsmA) binding to RNA and its implications for regulated expression. *J Mol Biol* **392:**511–528.

225. **Dubey AK, Baker CS, Romeo T, Babitzke P.** 2005. RNA sequence and secondary structure participate in high-affinity CsrA-RNA interaction. *RNA* **11:**1579–1587.

226. **Schubert M, Lapouge K, Duss O, Oberstrass FC, Jelesarov I, Haas D, Allain FH.** 2007. Molecular basis of messenger RNA recognition by the specific bacterial repressing clamp RsmA/CsrA. *Nat Struct Mol Biol* **14:**807–813.

227. **Liu MY, Gui G, Wei B, Preston JF 3rd, Oakford L, Yuksel U, Giedroc DP, Romeo T.** 1997. The RNA molecule CsrB binds to the global regulatory protein CsrA and antagonizes its activity in *Escherichia coli*. *J Biol Chem* **272:**17502–17510.

228. **Weilbacher T, Suzuki K, Dubey AK, Wang X, Gudapaty S, Morozov I, Baker CS, Georgellis D, Babitzke P, Romeo T.** 2003. A novel sRNA component of the carbon storage regulatory system of *Escherichia coli*. *Mol Microbiol* **48:**657–670.

229. **Chavez RG, Alvarez AF, Romeo T, Georgellis D.** 2010. The physiological stimulus for the BarA sensor kinase. *J Bacteriol* **192:**2009–2012.

230. **Suzuki K, Wang X, Weilbacher T, Pernestig AK, Melefors O, Georgellis D, Babitzke P, Romeo T.** 2002. Regulatory circuitry of the CsrA/CsrB and BarA/UvrY systems of *Escherichia coli*. *J Bacteriol* **184:**5130–5140.

231. **Suzuki K, Babitzke P, Kushner SR, Romeo T.** 2006. Identification of a novel regulatory protein (CsrD) that targets the global regulatory RNAs CsrB and CsrC for degradation by RNase E. *Genes Dev* **20:**2605–2617.

232. **Fields JA, Thompson SA.** 2008. *Campylobacter jejuni* CsrA mediates oxidative stress responses, biofilm formation, and host cell invasion. *J Bacteriol* **190:**3411–3416.

233. **Jonas K, Edwards AN, Ahmad I, Romeo T, Romling U, Melefors O.** 2010. Complex regulatory network encompassing the Csr, c-di-GMP and motility systems of *Salmonella* Typhimurium. *Environ Microbiol* **12:**524–540.

234. **Jones MK, Warner EB, Oliver JD.** 2008. csrA inhibits the formation of biofilms by *Vibrio vulnificus. Appl Environ Microbiol* **74:**7064–7066.

235. **Blumer C, Haas D.** 2000. Iron regulation of the hcnABC genes encoding hydrogen cyanide synthase depends on the anaerobic regulator ANR rather than on the global activator GacA in *Pseudomonas fluorescens* CHA0. *Microbiology* **146**(Pt 10):2417–2424.

236. **Kay E, Dubuis C, Haas D.** 2005. Three small RNAs jointly ensure secondary metabolism and biocontrol in *Pseudomonas fluorescens* CHA0. *Proc Natl Acad Sci U S A* **102:**17136–17141.

237. **Moll S, Schneider DJ, Stodghill P, Myers CR, Cartinhour SW, Filiatrault MJ.** 2010. Construction of an rsmX co-variance model and identification of five rsmX non-coding RNAs in *Pseudomonas syringae* pv. tomato DC3000. *RNA Biol* **7:**508–516.

238. **Heeb S, Blumer C, Haas D.** 2002. Regulatory RNA as mediator in GacA/RsmA-dependent global control of exoproduct formation in *Pseudomonas fluorescens* CHA0. *J Bacteriol* **184:**1046–1056.

239. **Valverde C, Heeb S, Keel C, Haas D.** 2003. RsmY, a small regulatory RNA, is required in concert with RsmZ for GacA-dependent expression of biocontrol traits in *Pseudomonas fluorescens* CHA0. *Mol Microbiol* **50:**1361–1379.

240. **Laville J, Voisard C, Keel C, Maurhofer M, Defago G, Haas D.** 1992. Global control in *Pseudomonas fluorescens* mediating antibiotic synthesis and suppression of black root rot of tobacco. *Proc Natl Acad Sci U S A* **89:**1562–1566.

241. **Zuber S, Carruthers F, Keel C, Mattart A, Blumer C, Pessi G, Gigot-Bonnefoy C, Schnider-Keel U, Heeb S, Reimmann C, Haas D.** 2003. GacS sensor domains pertinent to the regulation of exoproduct formation and to the biocontrol potential of *Pseudomonas fluorescens* CHA0. *Mol Plant Microbe Interact* **16:**634–644.

242. **Jorgensen MG, Thomason MK, Havelund J, Valentin-Hansen P, Storz G.** 2013. Dual function of the McaS small RNA in controlling biofilm formation. *Genes Dev* **27:**1132–1145.

243. **Thomason MK, Fontaine F, De Lay N, Storz G.** 2012. A small RNA that regulates motility and biofilm formation in response to changes in nutrient availability in *Escherichia coli. Mol Microbiol* **84:**17–35.

244. **Hammar M, Arnqvist A, Bian Z, Olsen A, Normark S.** 1995. Expression of two csg operons is required for production of fibronectin- and congo red-binding curli polymers in *Escherichia coli* K-12. *Mol Microbiol* **18:**661–670.

245. **Pesavento C, Becker G, Sommerfeldt N, Possling A, Tschowri N, Mehlis A, Hengge R.** 2008. Inverse regulatory coordination of motility and curli-mediated adhesion in *Escherichia coli. Genes Dev* **22:**2434–2446.

246. **Zogaj X, Nimtz M, Rohde M, Bokranz W, Romling U.** 2001. The multicellular morphotypes of Salmonella typhimurium and Escherichia coli produce cellulose as the second component of the extracellular matrix. *Mol Microbiol* **39:**1452–1463.

247. **Ogasawara H, Yamamoto K, Ishihama A.** 2011. Role of the biofilm master regulator CsgD in cross-regulation between biofilm formation and flagellar synthesis. *J Bacteriol* **193:**2587–2597.

248. **Wolfe AJ, Visick KL.** 2008. Get the message out: cyclic-Di-GMP regulates multiple levels of flagellum-based motility. *J Bacteriol* **190:**463–475.

249. **De Lay N, Gottesman S.** 2012. A complex network of small non-coding RNAs regulate motility in *Escherichia coli. Mol Microbiol* **86:**524–538.

250. **Guillier M, Gottesman S.** 2006. Remodelling of the *Escherichia coli* outer membrane by two small regulatory RNAs. *Mol Microbiol* **59:**231–247.

251. **Guillier M, Gottesman S.** 2008. The 5′ end of two redundant sRNAs is involved in the regulation of multiple targets, including their own regulator. *Nucleic Acids Res* **36:**6781–6794.

252. **Zhang A, Wassarman KM, Rosenow C, Tjaden BC, Storz G, Gottesman S.** 2003. Global analysis of small RNA and mRNA targets of Hfq. *Mol Microbiol* **50:**1111–1124.

253. **Redfield RJ.** 2002. Is quorum sensing a side effect of diffusion sensing? *Trends Microbiol* **10:**365–370.

254. **West SA, Winzer K, Gardner A, Diggle SP.** 2012. Quorum sensing and the confusion about diffusion. *Trends Microbiol* **20:**586–594.

255. **Ng WL, Bassler BL.** 2009. Bacterial quorum-sensing network architectures. *Ann Rev Genet* **43:**197–222.

256. **Waters CM, Bassler BL.** 2005. Quorum sensing: cell-to-cell communication in bacteria. *Ann Rev Cell Devel Bio* **21:**319–346.

257. **Rutherford ST, Bassler BL.** 2012. Bacterial quorum sensing: its role in virulence and possibilities for its control. *Cold Spring Harb Perspect Med* **2**(11):a012427.

258. **Lenz DH, Mok KC, Lilley BN, Kulkarni RV, Wingreen NS, Bassler BL.** 2004. The small RNA chaperone Hfq and multiple small RNAs control quorum sensing in *Vibrio harveyi* and *Vibrio cholerae. Cell* **118:**69–82.

259. **Bardill JP, Zhao X, Hammer BK.** 2011. The *Vibrio cholerae* quorum sensing response is mediated by Hfq-dependent sRNA/mRNA base pairing interactions. *Mol Microbiol* **80:**1381–1394.

260. **Rutherford ST, van Kessel JC, Shao Y, Bassler BL.** 2011. AphA and LuxR/HapR reciprocally control quorum sensing in vibrios. *Genes Dev* **25:**397–408.

261. **Shao Y, Bassler BL.** 2012. Quorum-sensing non-coding small RNAs use unique pairing regions to differentially control mRNA targets. *Mol Microbiol* **83:**599–611.

262. **Tu KC, Bassler BL.** 2007. Multiple small RNAs act additively to integrate sensory information and control quorum sensing in *Vibrio harveyi. Genes Dev* **21:**221–233.

263. **Shao Y, Bassler BL.** 2014. Quorum regulatory small RNAs repress type VI secretion in *Vibrio cholerae. Mol Microbiol* **92:**921–930.

264. **Svenningsen SL, Tu KC, Bassler BL.** 2009. Gene dosage compensation calibrates four regulatory RNAs to control *Vibrio cholerae* quorum sensing. *EMBO J* **28:**429–439.

265. **Lin W, Kovacikova G, Skorupski K.** 2005. Requirements for *Vibrio cholerae* HapR binding and transcriptional repression at the hapR promoter are distinct from those at the aphA promoter. *J Bacteriol* **187:**3013–3019.

266. **Svenningsen SL, Waters CM, Bassler BL.** 2008. A negative feedback loop involving small RNAs accelerates *Vibrio cholerae*'s transition out of quorum-sensing mode. *Genes Dev* **22:**226–238.

267. **Lee YH, Kim S, Helmann JD, Kim BH, Park YK.** 2013. RaoN, a small RNA encoded within *Salmonella* pathogenicity island-11, confers resistance to macrophage-induced stress. *Microbiology* **159:**1366–1378.

268. **de Jong HK, Parry CM, van der Poll T, Wiersinga WJ.** 2012. Host-pathogen interaction in invasive Salmonellosis. *PLoS Pathog* **8:**e1002933.

269. **Fabrega A, Vila J.** 2013. *Salmonella enterica* serovar Typhimurium skills to succeed in the host: virulence and regulation. *Clin Microbiol Rev* **26:**308–341.

270. **Sabbagh SC, Forest CG, Lepage C, Leclerc JM, Daigle F.** 2010. So similar, yet so different: uncovering distinctive features in the genomes of *Salmonella enterica* serovars Typhimurium and Typhi. *FEMS Microbiol Lett* **305:**1–13.

271. **Gunn JS, Alpuche-Aranda CM, Loomis WP, Belden WJ, Miller SI.** 1995. Characterization of the *Salmonella* typhimurium pagC/pagD chromosomal region. *J Bacteriol* **177:**5040–5047.

272. **Miller SI, Kukral AM, Mekalanos JJ.** 1989. A two-component regulatory system (phoP phoQ) controls *Salmonella* typhimurium virulence. *Proc Natl Acad Sci U S A* **86:**5054–5058.

273. **Chen S, Lesnik EA, Hall TA, Sampath R, Griffey RH, Ecker DJ, Blyn LB.** 2002. A bioinformatics based approach to discover small RNA genes in the *Escherichia coli* genome. *Bio Systems* **65:**157–177.

274. **Opdyke JA, Fozo EM, Hemm MR, Storz G.** 2011. RNase III participates in GadY-dependent cleavage of the gadX-gadW mRNA. *J Mol Biol* **406:**29–43.

275. **Tramonti A, De Canio M, De Biase D.** 2008. GadX/GadW-dependent regulation of the *Escherichia coli* acid fitness island: transcriptional control at the gadY-gadW divergent promoters and identification of four novel 42 bp GadX/GadW-specific binding sites. *Mol Microbiol* **70:**965–982.

276. **Giannella RA, Broitman SA, Zamcheck N.** 1972. Gastric acid barrier to ingested microorganisms in man: studies in vivo and in vitro. *Gut* **13:**251–256.

277. **Giannella RA, Broitman SA, Zamcheck N.** 1973. Influence of gastric acidity on bacterial and parasitic enteric infections. A perspective. *Ann Intern Med* **78:**271–276.

278. **Arnold KW, Kaspar CW.** 1995. Starvation- and stationary-phase-induced acid tolerance in Escherichia coli O157:H7. *Appl Environ Microbiol* **61:**2037–2039.

279. **Benjamin MM, Datta AR.** 1995. Acid tolerance of enterohemorrhagic *Escherichia coli. Appl Environ Microbiol* **61:**1669–1672.

280. **Conner DE, Kotrola JS.** 1995. Growth and survival of *Escherichia coli* O157:H7 under acidic conditions. *Appl Environ Microbiol* **61:**382–385.

281. **Ma Z, Gong S, Richard H, Tucker DL, Conway T, Foster JW.** 2003. GadE (YhiE) activates glutamate decarboxylase-dependent acid resistance in *Escherichia coli* K-12. *Mol Microbiol* **49:**1309–1320.

282. **Ma Z, Richard H, Tucker DL, Conway T, Foster JW.** 2002. Collaborative regulation of *Escherichia coli* glutamate-dependent acid resistance by two AraC-like regulators, GadX and GadW (YhiW). *J Bacteriol* **184**:7001–7012.

283. **Tramonti A, De Canio M, Delany I, Scarlato V, De Biase D.** 2006. Mechanisms of transcription activation exerted by GadX and GadW at the gadA and gadBC gene promoters of the glutamate-based acid resistance system in *Escherichia coli. J Bacteriol* **188**:8118–8127.

284. **Takada A, Umitsuki G, Nagai K, Wachi M.** 2007. RNase E is required for induction of the glutamate-dependent acid resistance system in *Escherichia coli. Biosci Biotechnol Biochem* **71**:158–164.

Sialic acid and N-acetylglucosamine Regulate type 1 Fimbriae Synthesis

5

IAN C. BLOMFIELD[1]

INTRODUCTION

Type 1 fimbriae of *Escherichia coli*, a member of the chaperon-usher family of bacterial adhesins, are synthesised by the majority of strains of the bacterium. Although frequently produced by commensal strains, the adhesin is nevertheless a virulence factor in extraintestinal pathogenic *E. coli* (ExPEC). The role of the adhesin in pathogenesis is best understood in uropathogenic *E. coli* (UPEC). Host attachment and particularly invasion by type 1 fimbriate bacteria activates inflammatory pathways, with TLR4 signalling playing a predominant role (1). In a mouse model of cystitis, type 1 fimbriation not only enhances UPEC adherence to oligomannosides of uroplakin 1a on the surface of superficial umbrella cells of the bladder urothelium, but is both necessary and sufficient for their invasion (2, 3). Moreover, more surprisingly, the adhesin plays a role in the formation of transient intracellular bacterial communities (IBCs) within the cytoplasm of urothelial cells as part of UPEC cycles of invasion (4).

The expression of type 1 fimbriation is controlled by phase variation at the transcriptional level, a mode of gene regulation in which bacteria switch reversibly between fimbriate and afimbriate phases. Phase variation has

[1]School of Biosciences, University of Kent, Kent, UK.

Metabolism and Bacterial Pathogenesis
Edited by Tyrrell Conway and Paul Cohen
© 2015 American Society for Microbiology, Washington, DC
doi:10.1128/microbiolspec.MBP-0015-2014

been widely considered a mechanism enabling immune evasion, although alternative explanations for the phenomenon do exist (5). Notwithstanding the apparently random nature of phase variation, switching of type 1 fimbrial expression is nevertheless controlled by a range of environmental signals that include the amino sugars sialic acid and N-acetylglucosamine (GlcNAc). Sialic acid plays a pivotal role in innate immunity, including signalling by the toll-like receptors. Here how sialic acid and GlcNAc control type 1 fimbriation is described and the potential significance of this regulatory response is discussed.

The Regulation of Type 1 Fimbriation

The genes required for the expression of type 1 fimbriation are situated at around 98 minutes on the genome of *E. coli* K-12. The adhesin's expression is controlled by the inversion of a short (314 bp) element of DNA (*fimS*) that contains the promoter required for the transcription of the operon that encodes the fimbrial structural genes (*fimAICDFGH*) (6). An example of site-specific recombination, the inversion produces reversible switching (phase variation) in the adhesin's expression. Inversion of *fimS* is catalysed by the tyrosine recombinases FimB and FimE, which are encoded by genes *fimB* and *fimE* that lie upstream of *fimS* in the on or fimbriate orientation (7, 8). While FimB catalyses inversion in either direction at a relatively low frequency ($\sim$10^{-3} per cell per generation), FimE is capable of promoting on-to-off inversion with both high frequency (up to 0.7 per cell per generation) and specificity (9–11). Thus at equilibrium the relative activities of FimB and FimE favour the afimbriate phase, with FimB playing a key role in determining the proportion of fimbriate bacteria. Additional recombinases that can also catalyse the inversion of *fimS*, while not present in *E. coli* K-12, have been identified elsewhere on the genomes of some pathogenic strains of the bacterium (12, 13).

In addition to the recombinases, integration host factor (IHF) and the leucine-responsive regulatory protein (Lrp) also play a direct role in *fimS* inversion (14, 15). The invertible element is relatively short ($\sim$314 bp) and the asymmetric binding of IHF and Lrp to sites within *fimS* presumably allows the DNA bending necessary to bring the two sites of recombination (IRL and IRR) into a juxtaposition for synapses and strand exchange (16). Phase switching, while apparently random, is nevertheless controlled by a range of environmental signals that affect either recombinase gene expression or the inversion of *fimS* directly. For example, the branched-chain amino acids and alanine stimulate the *fim* inversion by promoting the formation of an Lrp nucleoprotein complex that enhances recombination over an alternative complex that does not (11, 17). The inversion of *fimS* is also affected by temperature, pH, osmolarity and oxygen availability among other signals (11, 18–20). *fimB* expression and hence type 1 fimbriation is inhibited selectively by the amino sugars N-acetylglucosamine (GlcNAc) and sialic acid (21, 22).

Sialic Acids: Their Role in Host Immunity and *E. coli* Metabolism

Sialic acids, a family of naturally-occurring nine-carbon keto amino sugars, are found almost exclusively in animals of the deuterostome lineage. When present in the glycan chains of glycoproteins and glycolipids they occupy the terminal position, providing cell-surface receptors for both host and microbial lectins alike. Sialic acids play a critical and complex role in the immune system, inhibiting the innate immune response by engaging Siglecs (Sialic acid-binding immunoglobulin-like lectins) on immune cells and by preventing the activation of complement by recruiting factor H (23). More rarely, sialic acids are present on bacterial and protozoan animal pathogen cell surfaces, where they enhance virulence by enabling the microorganisms to masquerade as self and hence suppress host defences.

While the ability of pathogens to decorate their surfaces with sialic acid is comparatively rare, many more bacteria that occupy an animal niche can utilise sialic acid as a carbon, nitrogen and energy source (24). However, among this group only a subset of bacteria possess sialidases able to liberate free sialic acid from glycan chains. Those that are able to metabolise sialic acid, but which do not produce a sialidase, must rely upon other microorganisms, or the host itself, to release sialic acid from glycoconjugates before it can be used. Both *E. coli* and *S. typhimurium*, which fall into this latter category, appear to utilise sialic acid in the course of infection (25, 26). For example, UPEC mutants unable to metabolise sialic acid are attenuated for growth in a murine model of bacteremia (25). These results, while open to alternative interpretations as discussed below, nevertheless imply that sialic acid, presumably released by host sialidases, is available to *E. coli* in the bloodstream. While sialic acid metabolism does not appear to enhance UPEC virulence in the murine model of cystitis, the genes for its utilisation are induced in human urine (27). Free sialic acid levels are among a small group of compounds that are increased in the urine of cystitis patients (28). Sialic acid also appears to be available to *E. coli* in the urinary tract.

In *E. coli*, the metabolism of the most abundant form of sialic acid (N-acetylneuraminic acid or Neu5Ac) to N-acetylglucosamine-6-phosphate (GlcNAc-6-P) requires the *nanATEK* operon (29). To enter glycolysis, GlcNAc-6-P is converted to fructose-6-phosphate by the sequential action of N-acetylglucosamine-6-phosphate deacetylase (NagA), generating GlcN-6-P, and glucosamine-6-phosphate deaminase (NagB) (Fig. 1) (30). However, GlcN-6-P can also be converted to GlcN-1-P, the precursor of UDP-GlcNAc required for both peptidoglycan and LPS biosynthesis, by phosphoglucosamine

mutase (GlmM). Thus, sialic acid metabolism provides intermediates for both central metabolism and cell wall biosynthesis. The uptake of Neu5Ac across the cytoplasmic membrane, which occurs without modification, requires NanT, while its catabolism to GlcNAc-6-P requires N-acetylneuraminate aldolase (NanA), which generates N-acetylmannosamine and pyruvate, together with N-acetylmannosamine kinase (NanK) and N-acetylmannosamine-6-phosphate-2-epimerase (NanE) (24, 31).

The expression of the *nanATEK* operon is repressed by the transcriptional regulator NanR, which is induced by sialic acid, and activated by CRP (cAMP receptor protein) (32, 33). The *nagAB* operon is repressed by NagC and induced by GlcNAc-6-P (34, 35). Accumulation of intracellular sialic acid in a *nanA* mutant is growth inhibitory, indicating that the interaction of sialic acid with one or more additional targets is detrimental to the cell (29). On the other hand, sialic acid also inhibits the growth of a *nanR* mutant, but not a *nanR nagC* double mutant, suggesting that excess GlcNAc-6-P is also detrimental (36). Thus, it seems that both the uptake of sialic acid, and the rate of its catabolism, require careful regulation to avoid growth inhibition. Given the toxicity of sialic acid in the absence of NanA, it is possible that this effect, rather than the loss of sialic acid as a nutrient, accounts for the attenuated growth of a *nanA* mutant of UPEC in the bloodstream (24, 25).

The Regulation of *fimB* and the *nanCMS* Expression by Sialic Acid and GlcNAc

fimB is separated from the divergently transcribed *nanCMS* operon (formally *yjhATS* respectively) by a large (>1.4-kbp) intergenic region (Fig. 2). Both *fimB* and the *nanCMS* operon have long 5′ untranslated regions (5′-UTR) and share a set of cis-acting regulatory elements that are positioned close to the *nan* operon promoter (21, 22, 37). Moreover, while *fimA* expression is

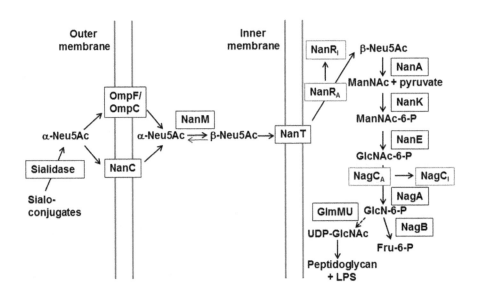

FIGURE 1 A model for the sialic acid Neu5Ac utilisation pathway of *E. coli*. Sialic acids, released from glycoconjugates in the α-anomer, cross the outer membrane by diffusion either through the sialic acid-selective channel NanC or the general porins OmpF and OmpC. The porins are presumably non-selective for the anomeric forms of Neu5Ac, so that any of the β-anomer formed spontaneously would also be taken up via this route (not shown). Periplasmic mutarotase NanM catalyses the rapid equilibrium of the α-anomer with the thermodynamically more stable β-anomer, before uptake by the inner membrane transporter NanT. Cytoplasmic Neu5Ac induces the *nanATEK* operon by converting NanR from the active transcriptional inhibitor (NanR$_A$) to the inactive form (NanR$_I$). Conversion of Neu5Ac to GlcNAc-6-P is catalysed by N-acetylneuraminate aldolase (NanA), which generates N-acetylmannosamine and pyruvate, together with N-acetylmannosamine kinase (NanK) and N-acetylmannosamine-6-phosphate-2-epimerase (NanE). To enter glycolysis, GlcNAc-6-P is converted to fructose-6-phosphate by N-acetylglucosamine-6-phosphate deacetylase (NagA), generating GlcN-6-P, followed by glucosamine-6-phosphate deaminase (NagB). The *nagAB* operon is induced by GlcNAc-6-P, which inactivates the transcriptional repressor NagC. GlcN-6-P can also be converted to UDP-GlcNAc, required for both peptidoglycan and LPS biosynthesis, by the sequential action of phosphoglucosamine mutase (GlmM) and the bifunctional glucosamine-1-phosphate acetyltransferase and N-acetylglucosamine-1-phosphate uridyltransferase (GlmU). Proteins are shown in boxes (enzymes and porins in black-bordered boxes, while transcriptional regulators are bordered by red). The broken arrow linking GlcN-6-P to UDP-GlcNAc indicates that more than one step is involved. doi:10.1128/microbiolspec.MBP-0015-2014.f1

suppressed by the small regulatory RNAs (sRNA) OmrA and OmrB, *nanA* expression is enhanced by these factors (38). Thus, the expression of type 1 fimbriation and sialic acid metabolism are coordinated at a number of levels.

Initially it was determined that a 252 bp deletion (termed Δ2), more than 500 bp upstream of the *fimB* transcriptional start site, decreased *fimB* expression and FimB recombination five- and eighteen-fold respectively (21). Substitution mutagenesis of the sequences within Δ2 localised two specific regions (termed region 1 and region 2 or NagC1 originally, but renamed O_{NR} and O_{NCI}) that stimulate *fimB* expression. Mutations in O_{NR} have a lesser effect than those in O_{NCI}, while combination of both produced the greatest effect. O_{NR} (underlined) falls within a 27 bp sequence (5′-ACCTTTATACCTGTTATACCAGATCAA) conserved upstream of both the *nanATEK* and *yjhBC* operons. This sequence, which includes the operator site for NanR, over-

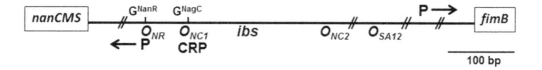

FIGURE 2 The organisation of the 1.4 Kbp *fimB-nanCMS* **intergenic region. The location of binding sites for NanR (O_{NR}), NagC (O_{NC1} and O_{NC2}), IHF (*ibs*) and SlyA (ON_{SA12}) and Dam methylation sites GATC^{NanR} (G^{NanR}) and GATC^{NagC} (G^{NagC}) are indicated. The transcription start sites for the** *nanCMS* **and** *fimB* **promoters (P), and the direction of transcription (arrows), are shown. The sections delineated by sloping parallel lines indicate where sections are omitted to allow the regulatory region shared by** *fimB* **and** *nanCMS* **(O_{NR} to O_{NC2}) to be shown to scale. The diagram is drawn to the scale indicated in base pairs (bp). doi:10.1128/microbiolspec.MBP-0015-2014.f2**

laps the known -10 region of both the *nanATEK* and *nanCMS* promoters (32, 37).

O_{NR} encompasses, and O_{NC1} lies immediately next to, Dam methylation sites (termed GATC^{NanR} and GATC^{NagC}, respectively) that are unmodified in a subpopulation of cells (21, 22). The pattern of protection and methylation indicates that the simultaneous protection of both sites occurs rarely, if at all. Whereas a mutation of *nanR* produces a loss of methylation protection at GATC^{NanR} specifically, Neu$_5$Ac inhibits protection at both GATC^{NanR} and GATC^{NagC} (22). Further analysis shows that O_{NC1} binds NagC and that exogenous GlcNAc inhibits both *fimB* expression and FimB recombination. As expected, NagC is required for methylation protection at GATC^{NagC}, and the addition of GlcNAc to growth medium, which generates GlcNAc-6-P on uptake, inhibits the methylation protection of GATC^{NagC} selectively. Since, as noted above, Neu5Ac catabolism generates GlcNAc-6-P, the initially perplexing observation that Neu5Ac inhibits methylation protection at both sites is readily explained. Taken together, these observations support a model in which *fimB* expression is activated alternately by NanR and NagC, with GlcNAc inhibiting one, and Neu$_5$Ac inhibiting both, of these pathways. In contrast, expression of the *nanCMS* operon is inhibited by these factors (37). As noted above, O_{NR} overlaps the -10 region of the *nanCMS* operon promoter. NagC binding O_{NC1} prevents CRP binding to an overlapping

operator site and hence inhibits activation of the *nanCMS* promoter.

NagC sites characteristically function in pairs, but although NagC binds to a second element (O_{NC2}; originally termed NagC2) situated 212 bp closer to *fimB* than O_{NC1}, the cooperative binding of NagC to O_{NC1} and O_{NC2} was not apparent in early in vitro studies (32). Furthermore a 304-bp deletion that included O_{NC2} (Δ3) did not diminish *fimB* expression. However, further work showed that a point mutation in O_{NC2} both diminishes *fimB* expression and results in a loss of the methylation protection of GATC^{NagC} (39). These apparent discrepancies were resolved by showing that NanR compensates for the loss of NagC activation of *fimB* in the Δ3 deletion mutant background. Furthermore, IHF was determined to bind to a site (*ibs*) midway between O_{NC1} and O_{NC2} to facilitate the cooperative binding of NagC to its two operator sites.

Further work demonstrates that *fimB* expression is also enhanced by SlyA, which binds to operator sites situated much closer to the *fimB* promoter than either NanR or NagC do (40). Although the effects of NanR, NagC and SlyA on *fimB* expression are all diminished in the absence of the abundant nucleiod associated protein H-NS, neither NanR nor NagC require SlyA to activate *fimB* expression. How NagC-IHF and NanR are able to activate *fimB* expression from binding sites situated so far upstream of the *fimB* promoter remains unclear. What is clear, however, is that the dual control of *fimB* by NanR and NagC increases the effect of sialic acid on *fimB* expression, while enabling it to inhibition by GlcNAc too.

The Function of *nanCMS* Operon

Although Neu5Ac can cross the outer membrane by diffusion through the general porins OmpF and OmpC, NanC is a sialic acid specific channel (37, 41). Growth on Neu5Ac is unaffected in a *nanC* mutant, but the activity of NanC suggests that the uptake of sialic acid through OmpF and OmpC is suboptimal in some circumstances. The pyranose ring of free sialic acid spontaneously alternates between alpha and beta anomers, with >90% of molecules existing in the thermodynamically more stable beta-anomeric form in solution. In contrast, sialic acid is trapped in the alpha-anomeric form in sialoconjugates and it is in this configuration that it is released by sialidases (24). The second member of the *nanC* operon, *nanM*, encodes a periplasmic sialic acid mutarotase that accelerates the normally slow equilibrium between the anomeric forms of Neu5Ac (42). This activity presumably allows *E. coli* to exploit sialic acid released from glycoconjugates more rapidly than would otherwise be possible. The final member of the *nanC* operon, *nanS*, encodes a sialyl esterase needed to convert Neu5,9Ac$_2$ to Neu5Ac to permit *E. coli* to use the 9-O acetylated sialic acid for growth (43, 44). NanS is necessary for Neu5,9Ac$_2$ to induce expression of the NanR-regulated operons, suggesting that either NanT is unable to transport Neu5,9Ac$_2$ into the cytoplasm, or that Neu5,9Ac$_2$ is unable to induce NanR following its uptake (33). Given that NanS seems to lack a signal sequence the later possibility appears to be more likely. While mutation of *nanS* abolishes growth on Neu5,9Ac$_2$ as the sole carbon source, mutations in *nanC* and *nanM* diminish it too (our unpublished data). This raises the possibility that NanC and NanM also play a role in Neu5,9Ac$_2$ metabolism, although further

work is required to rule out polarity as an explanation for the effects of the *nanC* and *nanM* mutations.

The Regulation of Type 1 Fimbriation by Sialic Acid and its Potential Implications for the Host-parasite Relationship

It is well documented that the interaction of type 1 fimbriate bacteria with host cells is pro-inflammatory, with activation of TLR4 signalling being key in UPEC pathogenesis (1). The release of sialic acid from the host surface is a pivotal step in the activation of inflammatory pathways by freeing Siglecs from cis inhibition by adjacent glycoconjugates (23). We have proposed that *E. coli* recognises free sialic acid as an indicator of increased inflammation, providing the raison d'être for its effect on *fimB* expression (21). As the release of sialic acid from host cell surfaces can enhance the binding of type 1 fimbriate *E. coli* (45), this could be particularly important to prevent an escalating level of bacterial-host attachment and invasion leading to excessive inflammation and bacterial elimination. Moreover, while it has been known for some time that TLR4 signalling enables host sialidase Neu1 up-regulation (46), it has only more recently been demonstrated that this overcomes an inhibitory TLR-Siglec interaction that acts as a break on the inflammatory response (47). Thus, a direct link has now been established between the innate immune signalling pathway that type 1 fimbriate bacteria activate and the release of sialic acid from the host cell surface.

Notwithstanding the comments above, it is also apparent that the ability of some UPEC strains to respond adaptively to sialic acid in the course of infection in the mouse model of cystitis is limited (48). UPEC infection of the mouse bladder is enhanced when the host is coinfected with Group B *Streptococcus*, even though the latter is eliminated rapidly from the urinary tract. This effect requires Siglec-E (the murine equivalent of Siglec-9) trans engagement with the Streptococcal sialic acid capsule (49). The effect is attenuated in mice lacking an intact TLR4 signalling pathway. These observations demonstrate that the release of Siglec-sialic acid binding does have a detrimental effect on UPEC in this model system. The observations also imply that UPEC's ability to detect and respond appropriately to sialic acid was at best suboptimal in these experiments. One potential caveat in this work is that the UPEC strain used, UT189, does itself produce a sialic acid capsule (50). Both the effect of the K1 capsule produced by this strain on Siglec signalling, and the bacterium's ability to recognise exogenous sialic acid, may differ from other UPEC isolates and it would be interesting to know if other, non-sialylated UPEC strains behave differently to UT189.

CONCLUDING REMARKS

As a signal of host immune function, free sialic acid is a potentially attractive candidate for bacteria, including *E. coli*. Sialic acid is restricted to animals and their pathogens in which it plays a crucial role in controlling innate immunity. Furthermore, while *E. coli* lacks a sialidase, and so is unable to release the amino sugar from host glycoconjugates, host sialidases liberate free monomers upon immune activation. The demonstration that type 1 fimbriation, which allows both host attachment and invasion, is inhibited by sialic acid is therefore intriguing. It seems possible that other virulence factors are also controlled by sialic acid. Following this line of reasoning, it would also be interesting to know if sialic acid induces the expression of anti-inflammatory factors produced by UPEC, but absent from *E. coli* K-12 (51, 52).

For *E. coli* to be able to recognise free sialic acid as a reliable indicator of host immune activation it would seem important for the bacterium to be able to distinguish somehow sialic acid released by the host from that liberated by microbial sialidases. In this

respect, host sialidases are generally better able to catalyse the release of O-acetylation sialic acid than their microbial counterparts (24). The presence of sialyl esterase (NanS) in *E. coli* and the close association between *fimB* and the *nanCMS* operon that encodes it may enhance the bacterium's ability to achieve this. Further work is also required to determine the functions of both NanC and NanM and the effect that the loss of both these proteins and NanS have on the interaction of *E. coli* with its animal hosts.

CITATION

Blomfield IC. 2015. Sialic acid and N-acetyl-glucosamine Regulate type 1 Fimbriae Synthesis, Microbiol Spectrum 3(3):MBP-0015-2014.

REFERENCES

1. **Hannan TJ, Totsika M, Mansfield KJ, Moore KH, Schembri MA, Hultgren SJ.** 2012. Host-Pathogen Checkpoints and Population Bottlenecks in Persistent and Intracellular Uropathogenic *E. coli* Bladder Infection. *FEMS Microbiol Rev* **36:**616–648.
2. **Martinez JJ, Mulvey MA, Schilling JD, Pinkner JS, Hultgren SJ.** 2000. Type 1 pilus-mediated bacterial invasion of bladder epithelial cells. *EMBO J* **19:**2803–2812.
3. **Zhou G, Mo WJ, Sebbel P, Min G, Neubert TA, Glockshuber R, Wu XR, Sun TT, Kong XP.** 2001. Uroplakin Ia is the urothelial receptor for uropathogenic *Escherichia coli*: evidence from in vitro FimH binding. *J Cell Sci* **114:**4095–4103.
4. **Wright KJ, Seed PC, Hultgren SJ.** 2007. Development of intracellular bacterial communities of uropathogenic *Escherichia coli* depends on type 1 pili. *Cell Microbiol* **9:**2230–2241.
5. **van der Woude MW.** 2006. Re-examining the role and random nature of phase variation. *FEMS Microbiol Lett* **254:**190–197.
6. **Abraham JM, Freitag CS, Clements JR, Eisenstein BI.** 1985. An invertible element of DNA controls phase variation of type 1 fimbriae of *Escherichia coli. Proc Natl Acad Sci USA* **82:**5724–5727.
7. **Klemm P.** 1986. Two regulatory *fim* genes, *fimB* and *fimE*, control the phase variation of type 1 fimbriae in *Escherichia coli. EMBO J* **5:**1389–1393.

8. **Gally DL, Leathart J, Blomfield IC.** 1996. Interaction of FimB and FimE with the *fim* switch that controls the phase variation of type 1 fimbriae in *Escherichia coli* K-12. *Mol Microbiol* **21:**725–738.

9. **Blomfield IC, McClain MS, Eberhardt KJ, Eisenstein BI.** 1993. Type 1 fimbriation and *fimE* mutants of *Escherichia coli* K-12. *J Bacteriol* **173:**5298–5307.

10. **McClain MS, Blomfield IC, Eberhardt KJ, Eisenstein BI.** 1993. Inversion-independent phase variation of type 1 fimbriae in *Escherichia coli*. *J Bacteriol* **175:**4335–4344.

11. **Gally DL, Bogan JA, Eisenstein BI, Blomfield IC.** 1993. Environmental regulation of the *fim* switch controlling type 1 fimbrial phase variation in *Escherichia coli* K-12: Effects of temperature and media. *J Bacteriol* **175:**6186–6193.

12. **Bryan A, Roesch P, Davis L, Moritz R, Pellet S, Welch RA.** 2006. Regulation of type 1 fimbriae by unlinked FimB- and FimE-like recombinases in uropathogenic *Escherichia coli* strain CFT073. *Infect Immun* **74:**1072–1083.

13. **Xie Y, Yao Y, Kolisnychenko V, Teng CH, Kim KS.** 2006. HbiF regulates type 1 fimbriation independently of FimB and FimE. *Infect Immun* **74:**4039–4047.

14. **Dorman CJ, Higgins CF.** 1987. Fimbrial phase variation in *Escherichia coli*: dependence on integration host factor and homologies with other site-specific recombinases. *J Bacteriol* **169:**3840–3843.

15. **Gally DL, Rucker TJ, Blomfield IC.** 1994. The leucine-responsive regulatory protein binds to the fim switch to control phase variation of type 1 fimbrial expression in Escherichia coli K-12. *J Bacteriol* **176:**5665–5672.

16. **Blomfield IC, Kulasekera DH, Eisenstein BI.** 1997. Integration host factor stimulates both FimB- and FimE-mediated site-specific DNA inversion that controls phase variation of type 1 fimbriae expression in *Escherichia coli*. *Mol Microbiol* **23:**705–717.

17. **Roesch PL, Blomfield IC.** 1998. Leucine alters the interaction of the leucine-responsive regulatory protein (Lrp) with the *fim* switch to stimulate site-specific recombination in *Escherichia coli*. *Mol Microbiol* **27:**751–761.

18. **Schwan WR, Lee JL, Lenard FA, Matthews BT, Beck MT.** 2002. Osmolarity and pH growth conditions regulate *fim* gene transcription and type 1 pilus expression in uropathogenic *Escherichia coli*. *Infect Immun* **70:**1391–1402.

19. **Aberg A, Shingler V, Balsalobre C.** 2006. (p)ppGpp regulates type 1 fimbriation of *Escherichia coli* by modulating the expression of the site-specific recombinase FimB. *Mol Microbiol* **60:**1520–1533.

20. **Floyd KA, Moore JL, Eberly AR, Good JA, Shaffer CL, Zaver H, Almqvist F, Skaar EP, Caprioli RM, Hadjifrangiskou M.** 2015. Adhesive figer stratification in uropathogenic *Escherichia coli* biofilms unveils oxygen-mediated control of type 1 pili. *PLoS Pathog* **11**(3):e1004697. doi:10.1371/journal.ppat.1004697. eCollection 2015 Mar.

21. **El-Labany S, Sohanpal BK, Lahooti M, Akerman R, Blomfield IC.** 2003. Distant *cis*-active sequences and sialic acid control the expression of *fimB* in *Escherichia coli* K-12. *Mol Microbiol* **49:**1109–1118.

22. **Sohanpal BK, El-labany S, Lahooti M, Plumbridge JA, Blomfield IC.** 2004. Integrated regulatory responses of *fimB* to *N*-acetylneuraminic (sialic) acid and GlcNAc in *Escherichia coli* K-12. *Proc Natl Acad Sci USA* **101:**16322–16327.

23. **Varki A, Gagneux P.** 2012. Multifarious roles of sialic acids in immunity. *Ann NY Acad Sci* **1253:**16–36.

24. **Vimr ER.** 2013. Unified theory of bacterial sialometabolism; how and why bacteria metabolize sialic acids. *ISRN Microbiol* **2013:**816713.10.1155/816713.

25. **Smith SN, Hagan EC, Lane MC, Mobley HL.** 2010. Dissemination and systemic colonization of uropathogenic *Escherichia coli* in a murine model of bacteremia. *Mbio 1.* pii: e00262–10. doi:10.1128/mBio.00262-10.

26. **Ng KM, Ferreyra JA, Higginbottom SK, Lynch JB, Kashyap PC, Gopinath S, Naidu N, Choudhury B, Weimer BC, Monack DM, Sonnenburg JL.** 2013. Microbiota-liberated host sugars facilitate post-antibiotic expansion of enteric pathogens. *Nature* **502:**96–99.

27. **Alteri CJ, Smith SN, Mobley HL.** 2009. Fitness of *Escherichia coli* during urinary tract infection requires gluconeogenesis and the TCA cycle. *PLoS Pathog* **5**(5):e1000448.

28. **Lv H, Hung CS, Chaturvedi KS, Hooton TM, Henderson JP.** 2011. Development of an integrated metabolic profiling approach for infectious diseases research. *Analyst* **136:**4752–4763.

29. **Vimr ER, Troy FA.** 1985. Identification of an inducible catabolic system for sialic acid (nan) in *Escherichia coli*. *J Bacteriol* **164:**845–853.

30. **Plumbridge J, Vimr ER.** 1999. Convergent pathways for utilization of the amino sugars *N*-acetylglucosamine, *N*-acetylmannosamine, and *N*-acetylneuraminic acid by *Escherichia coli*. *J Bacteriol* **181:**47–54.

31. **Martinez J, Steenbergen S, Vimr E.** 1995. Derived structure of the putative sialic acid

transporter from *Escherichia coli* predicts a novel sugar permease domain. *J Bacteriol* **177:**6005–6010.

32. **Kalivoda KA, Steenbergen SM, Vimr ER, Plumbridge J.** 2003. Regulation of sialic acid catabolism by the DNA binding protein NanR in *Escherichia coli*. *J Bacteriol* **185:**4806–4815.

33. **Kalivoda KA, Steenbergen SM, Vimr ER.** 2013. Control of the *Escherichia coli* sialoregulon by transcriptional repressor NanR. *J Bacteriol* **195:**4689–4701.

34. **Plumbridge JA.** 1991. Repression and induction of the *nag* regulon of *Escherichia coli* K12: the roles of *nagC* and *nagA* in maintenance of the uninduced state. *Mol Microbiol* **5:**2053–2062.

35. **Plumbridge JA, Kolb A.** 1991. CAP and Nag repressor binding to the regulatory regions of the *nagE-B* and *manX* genes of *E. coli*. *J Mol Biol* **217:**661–679.

36. **Chu D, Roobol J, Blomfield IC.** 2008. A theoretical interpretation of the transient sialic acid toxicity of a *nanR* mutant of *Escherichia coli*. *J Mol Biol* **375:**875–889.

37. **Condemine G, Berrier C, Plumbridge J, Ghazi A.** 2005. Function and expression of an N-acetylneuraminic acid-inducible outer membrane channel in *Escherichia coli*. *J Bacteriol* **187:**1959–1965.

38. **Guillier M, Gottesman S.** 2006. Remodelling of the *Escherichia coli* outer membrane by two small regulatory RNAs. *Mol Micro* **59:**231–247.

39. **Sohanpal BK, Friar S, Roobol J, Plumbridge JA, Blomfield IC.** 2007. Multiple co-regulatory elements and IHF are necessary for the control of *fimB* expression in response to sialic acid and N-acetylglucosamine in *Escherichia coli* K-12. *Mol Microbiol* **35:**1279–1288.

40. **McVicker G, Sun L, Sohanpal BK, Gashi K, Williamson RA, Plumbridge J, Blomfield IC.** 2011. SlyA protein activates *fimB* gene expression and type 1 fimbriation in *Escherichia coli* K-12. *J Biol Chem* **286:**32026–32035.

41. **Giri J, Tang JM, Wirth C, Peneff CM, Eisenberg B.** 2012. Single-channel measurements of an N-acetylneuraminic acid-inducible outer membrane channel in *Escherichia coli*. 2012. *Eur Biophys J* **41:**259–271.

42. **Severi E, Müller A, Potts JR, Leech A, Williamson D, Wilson KS, Thomas GH.** 2008. Sialic acid mutarotation is catalysed by the *Escherichia coli* β-propeller protein YjhT. *J Biol Chem* **283:**4841–4849.

43. **Steenbergen SM, Jirik JL, Vimr ER.** 2009. YjhS (NanS) is required for *Escherichia coli* to gron on 9-O-acetylated N-acetylneuraminic acid. *J Bacteriol* **191:**7134–7139.

44. **Rangarajan ES, Ruane KM, Proteau A, Schrag JD, Valladares R, Gonzalez CF, Gilbert M, Yakunin AF, Cygler M.** 2011. Structural and enzymatic characterization of NanS (YjhS), a 9-O-Acetyl N-acetylneuraminic acid esterase from Escherichia coli O157:H7. *Protein Sci* **20:**1208–1219.

45. **Davis CP, Avots-Avotins AE, Fader RC.** 1981. Evidence for a bladder cell glycolipid receptor for *Escherichia coli* and the effect of neuraminic acid and colominic acid on adherence. *Infect Immun* **34:**944–948.

46. **Amith SR, Jayanth P, Franchuk S, Siddiqui S, Seyrantepe V, Gee K, Basta S, Beyaert R, Pshezhetsky AV, Szewczuk MR.** 2009. Dependence of pathogen molecule-induced toll-like receptor activation and cell function on Neu1 sialidase. *Glycoconj J* **26:**1197–1212.

47. **Chen GY, Brown NK, Wu W, Khedri Z, Yu H, Chen X, van de Vlekkert D, D'Azzo A, Zheng P, Liu Y.** 2014. Broad and direct interaction between TLR and Siglec families of pattern recognition receptors and its regulation of Neu1. *Elife* Sep 3;3e04066. doi:10.7554/eLife.04066.

48. **Kline KA, Schwartz DJ, Gilbert NM, Hultgren SJ, Lewis AL.** 2012. Immune modulation by group B *Streptococcus* influences host susceptibility to urinary tract infection by uropathogenic *Escherichia coli*. *Infect Immun* **80:**4186–4194.

49. **Weiman S, Uchiyama S, Lin FY, Chaffin D, Varki A, Nizet V, Lewis AL.** 2010. O-acetylation of sialic acid on group B *Streproccocus* inhibits neutrophil suppression and virulence. *Biochem J* **428:**163–168.

50. **Anderson GG, Goller CC, Justice S, Hultgren SJ, Seed PC.** 2010. Polysaccharide capsule and sialic acid-mediated regulation promote biofilm-like intracellular bacterial communities during cystitis. *Infect Immun* **78:**963–975.

51. **Hunstad DA, Justice SS, Hung CS, Lauer SR, Hultgren SJ.** 2005. Suppression of bladder epithelial cytokine response by uropathogenic *Escherichia coli*. *Infect Immun* **73:**3999–4006.

52. **Billips BK, Forrestal SG, Rycyk MT, Johnson JR, Klumpp DJ, Schaeffer AJ.** 2007. Modulation of host innate immune response in the bladder by uropathogenic *Echerichia coli*. *Infect Immun* **75:**5353–5360.

Trigger Enzymes: Coordination of Metabolism and Virulence Gene Expression

6

FABIAN M. COMMICHAU[1] and JÖRG STÜLKE[1]

INTRODUCTION

As for all other organisms, life of bacterial pathogens has been subject to a selective pressure to grow and multiply. For these bacteria, the host is a huge source of nutrients, and it is their primary aim to utilize these nutrients rather than to cause damage to the host. In consequence, metabolism is intimately linked to virulence of these organisms (1–3).

This close relation between metabolism and virulence is reflected by the tight regulation of pathogenicity determinants by signals derived from both host and pathogen metabolism. In this chapter, we will review recent discoveries on a particular class of enzymes, the trigger enzymes. These proteins have a second regulatory function that is governed by substrate availability but is not biochemically linked to their enzymatic activity. Their common property is the transduction of signals derived from metabolism to the control of gene expression (4).

To be effective in the competition with both their fellow bacteria and the host cells, bacteria need to be able to survey their environments and to respond to even subtle changes. Of particular importance are changes of the physico-chemical conditions, such as the temperature or concentrations of

[1]Department of General Microbiology, Georg-August-University Göttingen, Grisebachstr. 8, 37077 Göttingen, Germany.

Metabolism and Bacterial Pathogenesis
Edited by Tyrrell Conway and Paul Cohen
© 2015 American Society for Microbiology, Washington, DC
doi:10.1128/microbiolspec.MBP-0010-2014

particular ions and the availability of nutrients. In order to respond to such changes, bacteria have developed intricate regulatory networks and may devote as much as 10% of their coding-genome capacity to regulatory proteins (5). Due to the extreme instability of bacterial mRNAs (6), most regulation in these organisms takes place at the level of transcription. Such regulation is highly efficient with respect to the consumed resources and effective with respect to the short time required to achieve the required output of the regulation.

Transcriptional regulation is most often exerted by DNA-binding regulatory proteins, sometimes also by RNA-binding proteins or RNA-binding metabolites and RNAs. A major question is how the regulators perceive their particular signal. Three major mechanisms of signal perception can be distinguished: (i) Many DNA-binding proteins respond to the presence of small molecule messengers, such as the *E. coli* Lac repressor and the Crp activator protein that are controlled by the direct binding of the inducer allolactose and the cofactor cyclic AMP, respectively. (ii) A second large group of bacterial transcription factors is controlled by protein phosphorylation. This has been most intensively studied for the widespread two-component regulatory systems and for a class of RNA- or DNA-binding transcription factors that are controlled by phosphotransferase system-mediated phosphorylation events (7, 8). (iii) Finally, accumulating evidence demonstrates the control of transcription factors' activity by interactions with other, often sensory, proteins. Such mechanisms have been intensively studied in the control of nitrogen metabolism in Gram-positive bacteria or in the control of sigma factors by their dedicated antagonists and anti-antagonists (9, 10).

While the basic concepts of the control of gene expression have essentially been elucidated in the analysis of the model organisms *Escherichia coli* and *Bacillus subtilis*, much less is known about the control of gene expression in pathogens. Importantly, pathogens are able to recognize the host, and they can sense specific nutrients in the host. These regulatory host-pathogen interactions involve responses to temperature increases, to nutrient availability, and to specific signal molecules and proteins derived from the host. Since the colonization of a human host corresponds to a significant increase in temperature as compared to natural environments, it is not surprising that temperature is a key signal to indicate to the bacteria that they have reached a potential host. Interestingly, temperature changes can be detected both by RNAs and proteins. RNA thermometers that allow expression of particular genes only at a specific temperature are widespread in pathogens (see ref. 10 for a review), and have been intensively studied for the master regulators of virulence in *Yersinia* species and *Listeria monocytogenes*, LcrF and PrfA, respectively, as well as for several virulence genes of *Neisseria meningitidis* (12–14). Temperature-sensing proteins have been described in both *Yersinia* sp and in *L. monocytogenes*. In both species, thermoregulated proteolysis of a transcription factor or of an antagonist of a transcription repressor, respectively, controls the expression of virulence genes (15, 16). Similarly, the presence of host-derived nutrients is an important signal to control virulence gene expression. This relation has been intensively investigated for the control of the transcription activators PrfA and AtxA from *L. monocytogenes* and the causative agent of anthrax, *Bacillus anthracis*, respectively (17, 18).

A field that has so far been largely unexplored in the study of bacterial pathogens is the role of enzymes in the control of gene expression and other cellular activities of pathogenic bacteria. This is mainly due to the fact that these enzymes have a primary function in cellular metabolism, e.g., as an enzyme in the tricarboxylic acid (TCA) cycle. However, under specific conditions, they exert an additional activity. Thus, these enzymes belong to the so-called moonlighting proteins (19–21). Moonlighting of enzymes has been implicated in the control of cell size by carbon source availability in both *E. coli* and *B. subtilis* (22–24). Moreover, acetaldehyde-CoA dehydrogenase is implicated in the control of virulence in *E. coli* O157:H7 (25). In this chapter, we will present the current state of research on trigger enzymes that have a second activity in controlling gene expression with a specific focus on pathogenic bacteria, whenever this is possible.

Trigger Enzymes that Control Gene Expression by Direct DNA Binding

Due to the short half-life of bacterial mRNA, regulation of transcription initiation is the major mechanism to control gene expression in these organisms. A few enzymes have acquired DNA-binding domains that allow them to act directly as transcription factor in response to the availability of the substrate. The best-studied examples of this class of trigger enzymes are the proline dehydrogenase PutA and the biotin-protein ligase BirA.

Proline degradation requires uptake of the amino acid and its subsequent oxidation and decarboxylation to yield glutamate. In eukaryotes and Gram-positive bacteria, the two latter functions are catalyzed by distinct enzymes. In contrast, most Gram-negative bacteria possess bifunctional enzymes that catalyze both reactions. Interestingly, several pathogenic proteobacteria, including the enterobacteria, *Bordetella*, and *Pseudomonas*, possess trifunctional proline dehydrogenases that not only harbor the two consecutive enzymatic activities, but that also contain a ribbon-helix-helix DNA-binding domain at their N-terminus (26–28) (Fig. 1). In the presence of proline, the reduced form of these trifunctional enzymes is bound to the membrane and exerts its catalytic activity. In contrast, the oxidized enzymes that prevail in the absence of the substrate are released from the membrane, and are now able to bind an operator site in the intergenic region of the *putP-putA* divergon to repress the expression of both the permease (*putP*) and the dehydrogenase (*putA*) genes (29–31). In

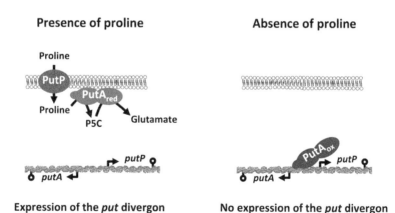

Presence of proline   **Absence of proline**

Expression of the *put* divergon   No expression of the *put* divergon

FIGURE 1 **The localization of the PutA protein within the cell determines its role in proline catabolism. In the presence of exogenous proline, the trifunctional PutA enzyme catalyzes the two-step conversion of proline to glutamate, which may serve as a carbon and nitrogen source. This catabolically active, reduced form of PutA (PutA$_{red}$) localizes to the membrane. The *put* divergon, encoding the proline transporter PutP and the PutA trigger enzyme, respectively, is expressed in the presence of proline. In the absence of proline, the oxidized PutA protein (PutA$_{ox}$) binds to the intergenic region of the *putA* and *putP* genes to repress their transcription. P5C, Δ1-pyrroline-5-carboxylate. doi:10.1128/microbiolspec. MBP-0010-2014.f1**

consequence, PutA acts both as an enzyme and a transcription repressor.

Biotin is an essential component of several proteins involved in carboxylation reactions. Biotin is covalently attached to its target proteins by the biotin-protein ligase BirA (32). In addition to its catalytic activity, BirA is also able to bind specific DNA sequences and to act as transcriptional repressor of a small regulon involved in biotin acquisition (33). The DNA-binding activity of BirA depends on its interaction with its corepressor biotinoyl-5′-AMP as a signal for sufficient levels of biotin that results in repression of biotin biosynthetic and transport genes (34). While the principles of BirA-mediated signal transduction are well understood, the molecular details are still a matter of debate. Specifically, there are experimental data that support and contradict, respectively, the idea of a role for the biotinylated protein AccB as an interaction partner of BirA that controls its regulatory capacity (35, 36). Recently, it was shown that the wing of the DNA-binding winged helix-turn-helix motif in *E. coli* BirA is not only essential for DNA binding, but also for the enzymatic activity of the protein, suggesting an intimate interdependence of the two functionalities. In contrast, the N-terminal DNA-binding domain can be deleted from *B. subtilis* BirA, and the truncated protein retains its biotin protein-ligase activity, but has lost the ability to bind DNA. Thus, the molecular details of BirA activity seem to differ between the organisms (37, 38). The ability of BirA to act as a trigger enzyme and to control the expression of genes involved in biotin acquisition is observed in a wide variety of bacteria and archaea, and it has been suggested that BirA is an evolutionary ancient DNA-binding transcription factor (33).

Several reports suggest that DNA-binding trigger enzymes are wide-spread in bacteria. Similar to PutA and BirA, the nicotinamide mononucleotide adenylyltransferase NadR and the DNA alkyltransferase AdaA contain DNA-binding domains that allow these proteins to regulate the genes for nicotinamide adenine dinucleotide (NAD) biosynthesis and uptake and those for the adaptive response to alkylative DNA damage (39–43). The enterobacterial *carAB* operon encoding the two subunits of carbamoyl phosphate synthetase is highly interesting since it is subject to regulation by three different trigger enzymes: The UMP kinase PyrH represses the expression of the operon in the presence of excess pyrimidines (44). A similar trigger enzyme activity was also suggested for the UMP kinase of *Mycobacterium tuberculosis* (45). In addition to PyrH, the aminopeptidase PepA represses the expression of the *carAB* operon (46, 47). The signal triggering this latter regulation is unknown, but it is tempting to speculate that the regulation by PepA allows integration of *carAB* into the amino acid regulatory network since carbamoyl phosphate is also a precursor for the biosynthesis of arginine. Finally, the repressor PurR represses the *carAB* operon in the absence of phosphoribosyl pyrophosphate possibly in order to balance the purine and pyrimidine biosynthetic capacities in the cell (44). Interestingly, PurR is an example for a transcription factor that contains a metabolite-binding domain that has lost enzymatic activity (see below, *Trigger enzymes as evolutionary link between enzymes and regulators*) (48). An interesting feedback regulation by a trigger enzyme was observed in *E. coli*: the alanyl-tRNA synthetase binds an operator site in its own promoter region to repress gene expression in the presence of the cognate amino acid alanine (49). A recent and highly unusual addition to the list of DNA-binding trigger enzymes was made by the discovery that the *B. subtilis* tRNA^{Ile}-lysidine synthetase TilS and the hypoxanthine phosphoribosyltransferase HprT form a complex that binds and activates the promoter of the *ftsH* gene (50). However, the signal that provokes the formation of the DNA-binding TilS-HprT complex as well as the physiological relevance of this regulation are subject of further investigation.

Trigger Enzymes that Control Gene Expression by RNA Binding

In contrast to DNA, RNAs can adopt a variety of different structures, and they can bind and interact with virtually any molecule, including metabolites, proteins and other RNAs (51–53). Thus, it is not surprising that proteins with very different structures are able to bind RNA (54). Recent proteome-wide studies in yeast revealed that many metabolic enzymes bind specific RNAs and are implicated in the control of gene expression (55, 56).

There is one unique example of a trigger enzyme with RNA-binding activity in organisms from bacteria to man, i.e., aconitase (57, 58). This enzyme catalyzes the reversible conversion of citrate to isocitrate in the tricarboxylic acid (TCA) cycle and requires an iron-sulfur cluster for activity (59). Three factors make the aconitase an excellent candidate for being a trigger enzyme: First, due to its important metabolic function it is a very abundant enzyme in many organisms. Second, the iron-sulfur cluster is rather unstable and disassembles under condition of iron limitation (60). Third, iron is an essential nutrient for most organisms with the notable exception of the causative agent of syphilis, *Treponema pallidum* (61). However, the common ion salts are highly insoluble making them difficult to acquire by cells. Indeed, iron is usually the growth-limiting nutrient for pathogenic bacteria and they have developed a variety of strategies to get access to iron in the host organisms (62). Thus, aconitase is an ideal sensor for the iron supply of the cell. Under conditions of iron-limitation, the TCA cycle cannot operate due to the loss of the iron-sulfur center and the resulting inactivity of aconitase. However, aconitase takes over a second function, and becomes an RNA-binding protein. In this way, the enzyme can help to overcome iron limitation by binding so-called iron-responsive elements (IREs) in the mRNAs of genes involved in iron homeostasis. This was first shown for the human enzyme, which was actually initially characterized as RNA-binding protein (iron-responsive element-binding protein, IRP) and subsequently found to be similar to mitochondrial aconitase and to exhibit aconitase activity (63). The analysis of aconitases from the model bacteria *E. coli* and *Bacillus subtilis* revealed that these enzymes do also have dual activities as enzymes and IRE-binding proteins (64, 65). These activities were also observed for the enzymes of pathogens such as *M. tuberculosis*, *Helicobacter pylori*, and *Salmonella enterica* (66–68). Moreover, the evidence suggests that the aconitases of the human and plant pathogenic bacteria *Legionella pneumophila*, *Staphylococcus epidermidis*, *Staphylococcus aureus*, *Pseudomonas aeruginosa*, and *Xanthomonas campestris* also act as RNA-binding trigger enzymes (69–73).

The determination of the structures of IRE-bound aconitase and the enzymatically active protein permitted an understanding of the molecular basis for the two mutually exclusive activities of aconitase (74–76). In the presence of iron, the enzyme has a compact conformation with the iron-sulfur cluster as a ligand. In the absence of iron or under conditions of oxidative stress, the iron-sulfur cluster disassembles and the free (apo–) aconitase adopts a more open conformation with respect to two domains that are located outside of the core of the protein. This opening allows the binding of the IRE, with sequence-specific contacts of a C in the central bulge of the IRE and an AGU triplet in the terminal loop. These contacts involve the two different domains of apo-aconitase. It is worth noting that only a few bases are conserved in the iron-responsive elements, and that these conserved bases are brought into the right position by secondary structure elements (76). Recent studies suggest that it is the structure of the IRE RNA, rather than its actual sequence, that is important for recognition and binding by aconitase. However, the presence of IREs with different sequences and resulting different binding affinities for aconitase allows establishing a regulatory hierarchy (77, 78).

Based on the overall sequence similarity of the eukaryotic and *B. subtilis* aconitases and IREs (65), it can be assumed that the bacterial enzyme follows the same mechanism as outlined above for the mammalian enzyme. *B. subtilis* aconitase binds IREs in the untranslated regions of the *qoxD* and *feuAB* mRNAs. These genes encode the iron-containing protein cytochrome aa_3 oxidase and an iron-uptake system, respectively (65). *B. subtilis* mutants lacking aconitase are defective in sporulation. This was attributed to the reduced accumulation of the *gerE* mRNA encoding a transcription regulator of late sporulation genes. Indeed, a secondary structure similar to the IRE can be formed by the untranslated region of the *gerE* mRNA, and the aconitase binds this sequence. Thus, aconitase might affect sporulation by stabilizing the mRNA of the GerE transcription factor (79). The substrate of aconitase, citrate, is not only the first metabolite of the TCA cycle but also a chelator of iron ions. If aconitase is inactive due to the unavailability of iron, this may result in the accumulation of citrate and subsequently in the chelation of iron ions making them thus even scarcer (80). Recently it has been shown that aconitase does also bind a poorly conserved IRE in the 5′ untranslated region of the *citZ* mRNA encoding citrate synthase. This binding results in the destabilization of the mRNA and results in reduced accumulation of citrate synthase and thus in reducing citrate synthesis (81). This "feedback regulation" at the level of mRNA stability is in excellent agreement with previous observations of mutations inactivating the *citZ* gene in mutants defective in aconitase in *B. subtilis*, *Corynebacterium glutamicum*, and *Streptomyces coelicolor* (80, 82, 83). Both the control of *citZ* expression and the acquisition of *citZ* suppressor mutants prevent the accumulation of the iron-chelator citrate in the cell.

E. coli possesses two aconitases, of which aconitase B (AcnB) is the main enzyme involved in the TCA cycle. In contrast to the monomeric enzymes of eukaryotes and *B. subtilis*, AcnB forms dimers. An analysis of the requirements of AcnB for RNA binding revealed that it is independent of the iron-sulfur cluster in the active centre of the enzyme. However, while the dimeric protein is endowed with catalytic activity as aconitase, the monomer prevails is the absence of the iron-sulfur center, and this monomeric form binds IRE RNAs. Moreover, the arrangement of the RNA in the protein seems to be different from that observed in the mammalian aconitase-IRE complex (84). This is in good agreement with the observation that the RNA elements recognized by *E. coli* aconitase B differ from the classical IREs (64).

The role of aconitase as a trigger enzyme in pathogens has been studied only recently. In *S. enterica*, loss of the iron-sulfur cluster due to the presence of reactive oxygen species converts aconitase to a RNA-binding protein. By destabilizing the mRNA of the FtsH protease, aconitase indirectly enhances the expression of the flagellum protein FliC, and thus motility (85). Similarly, reactive oxygen species result in loss of the iron-sulfur cluster in the aconitase of *H. pylori* and thus in RNA-binding activity. In this species, RNA-binding to the *pgdA* mRNA encoding peptidoglycan deacetylase results in stabilization of this transcript and thus in increased cell wall modification. This helps the bacteria in becoming more resistant to lysozyme and in evading the host's immune system (67). A role for the RNA-binding form of aconitase in the protection against oxidative stress has recently also been discovered for the enzyme of *Streptomyces viridochromogenes*. This protein is involved in triggering herbicide production to help the bacteria in the competition with plants in their environment (86).

Given the key importance of iron acquisition for pathogens, bacterial aconitase might also be regarded as a potential target for antimicrobial drugs. However, such screens have to take into account that aconitase does also exist in the host cells.

There is a large variety of potential RNA structures that might be able to bind diverse ligands and, indeed, many eukaryotic enzymes have been shown to moonlight in RNA interactions (55, 56). Thus, many bacterial enzymes might be trigger enzymes involved in protein-RNA interactions. Indeed, distinct regulatory activities upon RNA binding have been described for several bacterial enzymes.

Further examples for RNA-binding trigger enzymes include the *E. coli* threonyl-tRNA synthetase that negatively controls its own expression by binding to its mRNA (87). Moreover, the riboflavin kinase RibR from *B. subtilis* binds the leader region of the riboflavin biosynthetic *rib* operon and is required for the riboregulation by the FMN-sensing riboswitch of the operon (88). Finally, the minor uracil phosphoribosyltransferase PyrR from *B. subtilis* acts as transcriptional antiterminator of the *pyr* operon for the biosynthesis of pyrimidine nucleotides by binding its RNA target in the absence of pyrimidine bases (89–91). It is tempting to speculate that novel examples of RNA-binding trigger enzymes await their discovery.

Trigger Enzymes that Control Gene Expression by Covalent Modification of Transcription Factors

The trigger enzymes discussed above exert their regulatory activity by direct binding to nucleic acids, either DNA or RNA, and thus controlling gene expression. In addition, trigger enzymes may also act by modulating the activity of transcription factors, either by covalent modification or by regulatory protein-protein interactions.

A class of transcription factors is controlled by phosphorylation exerted by trigger enzymes that have an enzymatic activity as permeases and phosphotransferases. These enzymes are part of the bacterial phosphoenolpyruvate:sugar phosphotransferase

system (PTS). In this system, a cytoplasmic protein, called Enzyme I, autophosphorylates at the expense of phosphoenolpyruvate and transfers the phosphate group to a second cytoplasmic protein, HPr. HPr, in turn, phosphorylates the A domain of the sugar-specific permease. This domain may be part of a membrane-bound multi-domain protein or occur as a single cytoplasmic protein. From the A domain of the permease (also referred to as Enzyme IIA) the phosphate group is then transferred to a second sugar-specific phosphotransferase domain of the permease (Enzyme IIB), and from there ultimately to the incoming sugar that enters the cell via the membrane-integral Enzyme IIC permease domain (for review, see ref. 92). In the absence of their substrates, the sugar-specific permease domains IIA and IIB accumulate in the phosphorylated form. Thus, the phosphorylation state of the PTS proteins is informative, and precisely this information can be used to modulate the activity of transcription factors that control the expression of the sugar-specific permeases and the enzymes required for the utilization of their sugar substrates (93, 94).

The regulators controlled by Enzyme II-catalyzed phosphorylation all contain so-called PTS regulation domains (PRDs). These domains are found in DNA-binding transcriptional activators and RNA-binding antitermination proteins, and they always occur in duplicate in a protein (95). Many bacteria contain antitermination proteins that control the expression of β-glucoside utilization. It has been proposed that the absence of these sugars is an indication for the human pathogen *L. monocytogenes*, that it does not thrive in soil or in the rhizosphere, and that it triggers the switch to pathogenicity (96, 97). The BglG and LicT antiterminators from *E. coli* and *B. subtilis*, respectively, are the paradigm of PRD-containing regulators that are subject to control via phosphorylation (see ref. 94 for review). In the absence of the sugar substrate of the system (the β-glucoside salicin), the corresponding Enzyme II of the

PTS is phosphorylated, and the Enzyme IIB (EIIB) domain can phosphorylate and thereby inactivate the cognate antiterminator protein BglG. If β-glucosides become available, the phosphate group is drained from BglG to the incoming substrate resulting in activation of BglG and subsequent positive regulation of *bgl* gene expression for β-glucoside utilization (98–100; see Fig. 2). Another layer of control is exerted by the HPr protein of the PTS, which also phosphorylates BglG. This phosphorylation occurs in the absence of preferred carbon sources and results in activation of BglG (or LicT). The two antagonistically acting phosphorylation events take place at distinct sites: While EIIB-dependent inactivation occurs in PRD-1, PRD-2 is the site of positive regulation by HPr (101). An interesting problem with respect to the duplicated PRDs and quite similar phosphorylation sites in both domains is the achievement of regulatory specificity of HPr and EIIB for PRD-2 and PRD-1, respectively. For the glucose-specific antiterminator GlcT from *B. subtilis*, it has been shown that the position of the PRD in the protein determines whether it can be phosphorylated by HPr or not (102). While many Gram-negative bacteria contain only one regulatory system involving a trigger enzyme of this class and a PRD-containing regulator (i.e., the β-glucoside utilization system), Gram-positive bacteria often contain several paralogous systems. In the model organism *B. subtilis*, the four paralogous systems are involved in the utilization of

Presence of β-glucosides

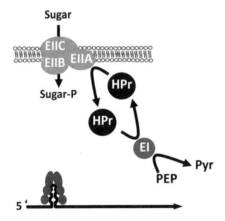

- **BglG is active**
- **Transcription antitermination of the *bgl* operon occurs**

Absence of β-glucosides

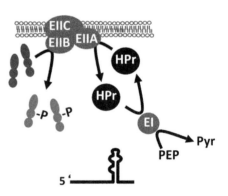

- **BglG is inactivated by EIIB**
- **Transcription termination of the *bgl* operon occurs**

FIGURE 2 The β-glucoside permease controls the activity of the transcription antiterminator protein BglG in response to β-glucoside availability. In the presence of β-glucosides, the sugar is taken up by the β-glucoside permease of the PTS and concomitantly phosphorylated. The phosphoryl group is derived from phosphoenolpyruvate (PEP) and transferred via the phosphocarriers Enzyme I (EI) and HPr to the EIIB component of the β-glucoside permease. Under these conditions, the transcription-antiterminator protein BglG binds a stem-loop structure of the *bgl* mRNA, thereby preventing the formation of a terminator structure, and the transcription of the *bgl* mRNA can continue. Inactivation of the antiterminator protein BglG occurs in the absence of β-glucosides. BglG receives a phosphoryl goup from the β-glucoside permease and is now unable to bind the *bgl* mRNA. The formation of a termination structure occurs and the transcription of the *bgl* operon is aborted. Pyr, pyruvate. doi:10.1128/microbiolspec.MBP-0010-2014.f2

β-glucosides, glucose, and sucrose (93). Each system is made up of a specific trigger enzyme (the EIIB domain of the sugar permease), a PRD-containing transcriptional antiterminator, and the characteristic RNA sequences to which the antiterminators bind to allow expression of their target operons. Again, the interaction of conserved elements raises the problem of specificity in signal transduction. A series of genetic and biochemical studies have shown that this specificity is achieved at all levels—protein-RNA, protein-protein, and protein-sugar interaction (103–105).

Since the sugar permeases that phosphorylate the PRD-containing regulators are membrane-bound proteins (106, 107), the localization of the PRD-regulators has recently attracted much attention (see ref. 8 for review). For *E. coli* BglG, it has been observed that the protein is sequestered at the membrane when phosphorylated by the β-glucoside permease. In contrast, the active BglG protein is evenly distributed in the cytosol (106, 108). For LicT from *B. subtilis*, no sequestration at the membrane was observed. In contrast, this protein is evenly distributed in the cytoplasm but it rapidly relocalizes to the part of the cytosol that also harbors the ribosomes upon activation by EIIB-dependent dephosphorylation of the PRD-1 (107). Contradicting results have been reported for the PRD-containing transcription activator MtlR from *B. subtilis*: Bouraoui et al. (109) report that the protein requires direct activation by the membrane-bound mannitol permease MtlA whereas Heravi and Altenbuchner (110) found dephosphorylation of MtlR as sufficient for full activity and no requirement for direct interaction with MtlA. This issue clearly requires further investigation.

Importantly, a similar mechanism of control by an Enzyme II trigger enzyme was also proposed for the regulation of the central regulator of *Bacillus anthracis* virulence genes, AtxA. It has been suggested that the phosphorylation state of AtxA reports the availability of host-specific carbon sources.

Unfortunately, the permease responsible for this control has not yet been discovered (111).

Trigger Enzymes that Control Gene Expression by Regulatory Protein-Protein Interactions with Transcription Factors

Many trigger enzymes are not endowed with nucleic acid-binding activity nor can they modify other proteins. Such trigger enzymes exert their regulatory activity by direct protein-protein interactions with transcription factors. Since the initial introduction of the concept of trigger enzymes, several new examples of this class have been discovered. In addition, a subclass of trigger enzymes, the trigger transporters, has been recognized (4, 112, 113).

Control by regulator-binding trigger enzymes has been most intensively studied in the regulation of nitrogen metabolism in Gram-positive bacteria. Two trigger enzymes, the glutamine synthetase (GS) and the glutamate dehydrogenase (GDH), control the expression of genes for both core-nitrogen metabolism (uptake of ammonia and synthesis of glutamate and glutamine) as well as those genes for the utilization of alternative nitrogen sources. Since the role and mechanisms of regulation by these enzymes has recently been extensively reviewed (4, 9), we will only briefly present some novel findings here. Importantly, the structures of the active and the feedback-inhibited GS of *B. subtilis* have been solved. Massive structural rearrangements upon feedback inhibition provide the basis for the selective interaction of the feedback-inhibited form with the transcription factors TnrA and GlnR (114). Another interesting observation is the reciprocity in the regulatory activity between GS and TnrA: feedback-inhibited GS binds TnrA and prevents it from binding to its DNA targets. On the other hand, the interaction with TnrA also inhibits the catalytic activity of glutamine synthetase (118). Importantly, the GS acts as a trigger enzyme by interacting with the transcription factor GlnR, also in a large variety of low-GC Gram-positive bacteria, including important pathogens such as *Streptococcus pneumoniae*, and both the GS and members of the GlnR regulon contribute to virulence of the bacteria (116–118).

Another example for a trigger enzyme that controls gene expression by directly modulating the activity of a transcription factor was recently discovered in *B. subtilis* and in *S. aureus*. Both *O*-acetylserine-thiol-lyases CysK and CysM from *B. subtilis* and *S. aureus*, respectively, can form a complex with CymR, the master regulator of cysteine metabolism (119–122) (Fig. 3). In the absence of exogenous cysteine, the serine acetyltransferase CysE catalyses the acetylation of serine to *O*-acetyl-serine (OAS), which is then converted to sulphide and cysteine by the OAS-thiol-lyase CysK. CysE and CysK together form a bienzyme complex that is called cysteine synthase (123). If cysteine is provided by the environment, it has been suggested that this amino acid prevents the formation of the bienzyme CysE-CysK complex (121). The free OAS-thiol-lyase CysK then forms a complex with the transcription factor CymR and stimulates its DNA-binding activity, which in turn results in repression and activation of several genes that are involved

in cysteine formation and sulfur utilization, respectively (119, 120). In analogy to the interaction between the glutamine synthetase and the transcription factor GlnR, the OAS-thiol-lyase CysK acts like a chaperone that helps to stabilize the CymR-DNA complex (121, 124). The regulatory interaction between the OAS-thiol-lyase and CymR enables the bacteria to adapt their sulfur metabolism to the availability of exogenous cysteine and other sulfur sources. It is interesting to note that the gene sets that are directly regulated by CymR differ in *B. subtilis* and in the opportunistic pathogen *S. aureus*. The different composition of the CymR regulons in *B. subtilis* and in *S. aureus* seems to reflect the nutritional conditions the bacteria may encounter in their habitats (122). For instance, CymR directly regulates the genes that are involved in the utilization of sulfur sources of human origin such as taurine and homocysteine.

Very recently, a novel, highly interesting trigger enzyme has been discovered in *E. coli*. This enzyme, YdaM, is a diguanylate cyclase that produces cyclic di-GMP. This protein can directly interact with the transcription activator MlrA and stimulates its DNA-binding activity. Thus, YdaM exerts a positive effect on *E. coli* biofilm formation (125).

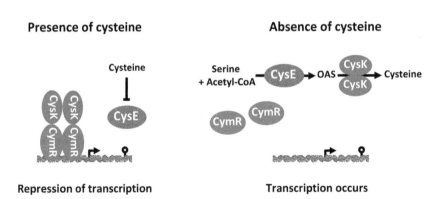

Presence of cysteine   **Absence of cysteine**

Repression of transcription   **Transcription occurs**

FIGURE 3 Control of CymR DNA-binding activity by CysK. In the presence of cysteine, the acetyltransferase CysE is inhibited and the *O*-acetyl-serine (OAS)-thiol-lyase CysK forms a complex with the transcription factor CymR. The protein complex binds to the CymR-regulated genes and prevents transcription. At low cysteine levels, the OAS-thiol-lyase converts serine and acetyl-CoA to OAS, which serves as the substrate for CysK to produce cysteine. doi:10.1128/microbiolspec.MBP-0010-2014.f3

As mentioned above, a specific class of trigger enzymes acts as bifunctional transporter and signal-transduction protein. Such so-called trigger transporters have been identified only recently, but their relevance is increasingly recognized (112, 126, 127). Many Gram-positive bacteria can synthesize antimicrobial peptides, and in turn they are also able to defend themselves against these compounds. Resistance is often achieved by ABC-transporter-mediated export of these peptides. The genes for several of these transporters are controlled by two-component signal transduction systems. Interestingly, the sensor kinases of this class do not contain large helices that span into the extracellular space; they are therefore referred to as intramembrane-sensing kinases. These kinases receive the signal that the antimicrobial peptide is present from dedicated regulatory ABC transporters that sense the substrate and interact with their cognate kinases in order to activate them. This mechanism of control has been intensively studied for the defense against bacitracin in *B. subtilis*, *Enterococcus faecalis*, and *S. aureus* (128–130). Moreover, bioinformatic analyses have revealed that this mechanism of genomically and functionally linked ABC transporters that serve as trigger enzymes to transduce the information on the availability of antimicrobial compounds to two-component sensor kinases is widespread in bacteria (113). Indeed, such a regulatory mechanism that depends on the interaction between the VraG ABC transporter subunit and the GraS sensor kinase has also been observed for the control of resistance towards cationic antimicrobial peptides in *S. aureus* (131).

Trigger Enzymes as Evolutionary Link Between Enzymes and Regulators

As presented above, trigger enzymes belong to a special class of moonlighting proteins that are both active in metabolism and in controlling gene expression (4). Trigger enzymes regulate gene expression in many different ways and we can find different evolutionary stages in nature (Fig. 4). Trigger enzymes like the glutamate dehydrogenase RocG and the glutamine synthetase GlnA from *B. subtilis* have acquired the ability to control the DNA-binding activity of the

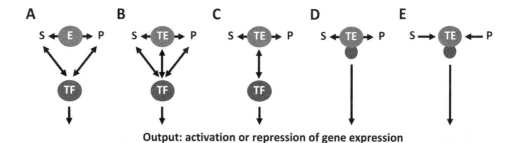

Output: activation or repression of gene expression

FIGURE 4 **Evolutionary stages of (trigger) enzymes A. Conventional enzymes (E), such as the β-galactosidase LacZ, catalyze metabolic reactions without controlling gene expression through modulating the activity of a transcription factor (TF). B. Bifunctional trigger enzymes (TEs) such as the glutamate dehydrogenase (GDH) from *B. subtilis* can control the activity of TFs by a direct protein-protein interaction. It has been suggested that the metabolites that are converted by the GDH also directly modulate the activity of the GDH-controlled TF, GltC. C. TEs like the glutamine synthetase (GS) from *B. subtilis* control the activities of TFs that do not response to metabolites. D. TFs such as the trifunctional PutA enzyme may have acquired a DNA-binding motif, which allows the enzyme to regulate gene expression depending of the metabolic state of the cell. E. TFs like BzdR from *B. japonicum* may be composed of a DNA-binding domain and an enzymatic domain that has lost its catalytic activity during evolution. These TE sense metabolites without converting them. S, substrate; P, product. doi:10.1128/microbiolspec.MBP-0010-2014.f4**

TABLE 1 A compilation of trigger enzymes in bacteria

Organism	Enzyme, catalyzed reaction	Effector/Signal	Target	Reference
DNA-binding transcription factors				
E. coli	AlaS, aminoacyl tRNA synthetase	Alanine	AlaS promoter	49
E. coli, Salmonella typhi, Klebsiella pneumoniae, Yersinia pestis, Legionella pneumophila, B. subtilis	BirA, biotin-protein ligase	Biotin	Biotin biosynthetic genes	33
B. subtilis	PurR, phosphoribosyltransferase	?	Purine biosynthetic genes	48
E. coli, S. typhimurium	PutA, bifunctional proline dehydrogenase and Δ^1-pyrroline-5-carboxylate dehydrogenase	Proline	Promoter region of the put operon	30
E. coli, S. typhimurium	NadR, nicotinamide mononucleotide adenylyl-transferase and ribosylnicotinamide kinase	NMN/NAD$^+$	Promoter region of the NAD$^+$-biosynthetic genes, modulation of PnuC activity?	40
B. subtilis	AdaA, DNA alkyltransferase	Alkylated DNA	adaA operon, alkA gene	41
E. coli	Ada, DNA alkyltransferase	Alkylated DNA	ada, aidB, alkA genes	147
B. subtilis	TilS, tRNA$^{\text{Ile2}}$ lysidine synthetaseHprT, hypoxanthine-guanine phosphoribosyltransferase	?	TilS-HprT complex stimulates ftsH transcription	50
E. coli	DcuB, fumarate/ succinate antiporter	?	DcuS sensor kinase	148
E. coli S. typhimurium	PyrH, UMP kinase	Uracil	carAB operon	44
E. coli S. typhimurium	PepA, aminopeptidase	?	carAB operon	46
RNA-binding transcription factors				
B. subtilis	CitB, aconitase	Iron	Iron-responsive elements	65
B. subtilis	PyrR, uracil phosphoribosyltransferase	Pyrimidines	pyr mRNA	149
E. coli, S. typhimurium	AcnA, AcnB, aconitase	ROS	Iron-responsive elements	64
E. coli	ThrS, threonyl-tRNA synthetase	?	thrS mRNA	90
H. pylori	AcnB, aconitase	ROS	Iron-responsive elements	66
M. tuberculosis	Acn, aconitase	Iron	Iron-responsive elements	67
S. viridochromogenes	AcnA, aconitase	Iron	Iron-responsive elements	86
Covalent modification of a transcription regulator				
E. coli	BglF, β-glucoside permease (EIIB subunit)	Salicin	BglG	100
B. subtilis	PtsG, glucose permease (EIIB subunit)	Glucose	GlcT	150
B. subtilis	LevE, fructose permease (EIIB subunit)	Fructose	LevR	151

(Continued on next page)

TABLE 1 **A compilation of trigger enzymes in bacteria** *(Continued)*

Organism	Enzyme, catalyzed reaction	Effector/Signal	Target	Reference
B. subtilis	LicB, lichenan-specific permease (EIIB subunit)	Oligo-β-glucosides	LicR	152
B. subtilis	ManP, mannose-specific permease (EIIB domain)	Mannose	ManR	153
B. subtilis	MtlA, mannitol-specific permease (EIIB domain)	Mannitol	MtlR	154
Modulation of activity of a DNA-binding protein				
E. coli	PtsG, glucose permease (EIIB subunit)	Glucose	Mlc	155
E. coli *Salmonella enterica*	LysP, lysine permease	Lysine/ pH	CadC	156
E. coli	DhaL, DhaK, dihydroxyacetone kinase (nucleotide and substrate binding subunits)	dihydroxyacetone	DhaR DhaK antagonizes DhaL	157
E. coli	MalY, bC-S lyase MalK, ATP-binding subunit of the maltose transporterAes, esterase	Maltodextrins, maltotriose	MalT	158
E. coli	YdaM, a diguaylate cyclase	GTP	MlrA	125
B. subtilis	CysK, O-acetyl-serine-thiol-lyase	O-acetyl-serine	CymR	121
S. aureus	CysM, O-acetyl-serine-thiol-lyase	O-acetyl-serine	CymR	122
B. subtilis	GlnA, glutamine synthetase	Glutamine	TnrA, GlnR	124, 133
Bacillus cereus	GlnA, glutamine synthetase	Glutamine	GlnR	159
Streptococcus mutans	GlnA, glutamine synthetase	pH	GlnR	160
Streptococcus pneumoniae	GlnA, glutamine synthetase	Glutamine	GlnR	116
B. subtilis	GudB1, glutamate dehydrogenase	Glutamate?	GltC	9
B. subtilis	RocG, glutamate dehydrogenase	Glutamate?	GltC	132
B. subtilis	SipW, signal peptidase I	?	SinR	161
S. pyogenes	LacD.1	Dihydroxyacetone phosphate, glyceraldehyde 3-phosphate (?)	RopB	143
Unknown mechanism/regulators that evolved from enzymes				
Bradyrhizobium japonicum	BzdR	Benzoyl-CoA	*bzd* operon for anaerobic benzoate utilization	137
Sphingomonas macrogolitabida	ThnY, ferredoxin-like Fe-S flavoprotein	?	*thn* genes for tetralin utilization	162
T. thermophilus	Mlc	Glucose	*glc* and *mal* operons	141
B. subtilis	RibR	FMN (?)	*rib* operon mRNA	91

transcription factors GltC and TnrA, respectively, by direct protein-protein interaction (132, 133). In both cases the control of the transcription factors depends on metabolites that are sensed by the enzymes (Fig. 4B and C). Recently, it has been shown that the DNA-binding activity of the transcription activator GltC, which regulates glutamate biosynthesis in *B. subtilis*, is modulated by the same metabolites (glutamate and

2-oxoglutarate) that are sensed by the gluta-mate dehydrogenase RocG (Fig. 4B; 134). Thus, the regulation of glutamate biosyn-thesis occurs at multiple levels in this organism. Interestingly, a single amino ex-change in the RocG protein is sufficient to turn the bifunctional enzyme into a mono-functional variant that has retained enzy-matic activity but has lost its regulatory activity (135). Consequently, in a *B. subtilis* strain expressing the monofunctional RocG variant, the DNA-binding activity of GltC is only regulated by intermediates from glutamate metabolism (Fig. 4A). It would be interesting to identify the selective pres-sure that was acting on the *rocG* glutamate dehydrogenase gene and that resulted in the emergence of this trigger enzyme in *B. subtilis*.

Many other trigger enzymes seem to be evolutionary intermediates as they are re-lated to enzymes, as well as to regulatory proteins. Two evolutionary patterns seem to be of special importance: (i) the acquisition of DNA-binding domains by enzymes, and (ii) the functional separation of enzymatic and regulatory functions by gene-duplication events. Examples for trigger enzymes that have acquired DNA-binding domains are the PutA and NadR proteins from many proteo-bacteria, including *E. coli* and other entero-bacteria (Table 1). The DNA-binding activity of these proteins is modulated by metabolites, which also serve as their substrates (Fig. 1; Fig. 4D). In contrast to the *E. coli* PutA en-zyme, in *Bradyrhizobium japonicum* and many other α-proteobacteria, PutA lacks a DNA-binding domain and the enzymes are only active in metabolism, i.e., they do not act as trigger enzymes (136). Recently, a plausible pathway for the evolutionary origin of a DNA-binding trigger enzyme has been reconstructed in the laboratory (137). The BzdR repressor protein from the β-proteobacterium *Azoarcus* sp CIB consists of an N-terminal helix-turn-helix (HTH) motif, which is connected through a linker to a C-terminal domain. The C-terminal domain shares sequence homology to the shikimate kinase I from *E. coli*. Interestingly, the *E. coli* shikimate kinase can be converted into a trigger enzyme by fusing the DNA-binding domain of BzdR to the N-terminus of the enzyme (137). The functionality of the chimeric regulator provides strong support for the evolution of trigger enzymes by gene-fusion events. Although the BzdR protein has lost its enzymatic activity, the DNA-binding activity of the protein is still modulated by an effector molecule (138). In several other cases the effector-binding pockets of DNA-binding transcriptional regulators are derived from enzymes. For instance, the MocR-like tran-scriptional factors from bacteria consist of an N-terminal HTH motif and a C-terminal domain that belongs to the superfamily of the pyridoxal-5′-phosphate dependent enzymes of fold type I (139). As yet, the MocR-like transcriptional factors have lost enzymatic activities but have retained the ability to bind to cofactors. It has been suggested that the loss of the enzymatic activity was a critical event to improve the ability of ancestral trigger enzymes to bind to DNA. Indeed, an enzymatically inactive variant of the chimeric trigger enzyme that was constructed by Durante-Rodríguez et al. was still able to bind to ligands and to elicit conformational changes in the HTH motif (137). However, DNA-binding transcription factors that have evolved from enzymes appear to under-go a strong selective pressure to maintain the residues essential for cofactor binding, suggesting that this function is crucial for the physiological role of these regulators (139).

Another fascinating example for the evolution of an enzyme to a DNA-binding regulator is provided by the ROK family of proteins comprising repressors of genes involved in sugar metabolism (140). This family includes the Mlc repressors from *E. coli* and *T. thermophilus*. Mlc is a very good example for the ongoing evolution. In *T. thermophilus*, glucose acts as the inducer by binding to a motif that is conserved in the glucose kinases (141). In contrast, *E. coli* Mlc has lost its ability to bind glucose (141), and the DNA-binding activity of the protein is controlled by interaction with the trigger enzyme PtsG. The loss of Mlc to respond to glucose was probably driven by the invention of the PTS for sugar transport with concomitant phosphorylation of the incoming sugars (92). Thus, free glucose is not available to control the *E. coli* Mlc protein.

The emergence of a trigger enzyme by the functional separation of enzymatic and regulatory functions through gene duplication events has been observed in the human pathogen *Streptococcus pyogenes*. In this bacterium, a unique form of carbon-catabolite repression is exerted by the trigger enzyme LacD.1. LacD.1 and its paralogue, LacD.2, have tagatose-1,6-bisphosphate aldolase activity. The LacD.1 protein is involved in the global control of virulence gene expression by regulating the activity of the virulence transcription factor RopB in response to the presence of lactose and galactose (142). Under conditions of high glycolytic activity, binding of dihydroxyacetone phosphate to LacD.1 (but not LacD.2) was shown to induce a conformational change in the enzyme, and in this conformational state LacD.1 binds and inhibits the transcriptional activator RopB (143). At low glycolytic activity, LacD.1 does not bind to RopB, which can then activate transcription of the *speB* gene encoding a protease that may serve to degrade host proteins as an alternative carbon source. This regulatory mechanism provides a link between carbon source availability in the host and the expression of virulence factors, among them the protease SpeB. Although the two LacD.1 and LacD.2 enzymes share 71% identical amino acids, LacD.2 is involved in sugar catabolism rather than in gene regulation (143, 144). Recent structural analyses of the LacD proteins showed that several surface patches show noticeable differences between LacD.1 and LacD.2 (145). Moreover, a biochemical characterization of the LacD.2

and LacD.1 proteins revealed that the latter enzyme has a lower catalytic activity and a reduced substrate affinity. Interestingly, an engineered mutant LacD.1 variant that has kinetic properties similar to that of LacD.2, has lost the ability to function as a regulator of the transcription activator RopB (146). Thus, the alteration of the catalytic efficiency and the structure of LacD.1 might be important determinants for RopB to distinguish between LacD.1 and LacD.2 (145, 146). However, despite the fact that the molecular details of the regulatory protein-protein interaction between LacD.1 and RopB remain to be uncovered, this example strikingly illustrates that the duplication of an enzyme might result in the acquisition of regulatory activity for only distantly related functions.

PERSPECTIVES

Given the fact that trigger enzymes have been discovered in a variety of pathogenic bacteria, it can be expected that several novel trigger enzymes await their discovery. In the future it will be interesting to study the molecular details of how the signal that is perceived by the trigger enzyme or its enzymatic domain is transmitted to the regulatory domain, which generates the regulatory output.

ACKNOWLEDGMENTS

We are grateful to all members of our labs for their enthusiasm and support. Work in the authors' labs is supported by grants from the Deutsche Forschungsgemeinschaft (CO 1139/1 to FMC and SFB 860 to JS).

CITATION

Commichau FM, Stülke J. 2015. Trigger enzymes: coordination of metabolism and virulence gene expression. Microbiol Spectrum 3(2):MBP-0010-2014.

REFERENCES

1. **Sonenshein AL.** 2007. Control of key metabolic intersections in *Bacillus subtilis*. *Nat Rev Microbiol* **5:**917–927.
2. **Halbedel S, Hames C, Stülke J.** 2007. Regulation of carbon metabolism in the mollicutes and its relation to virulence. *J Mol Microbiol Biotechnol* **12:**147–154.
3. **Poncet S, Milohanic E, Maze A, Nait-Abdallah J, Ake F, Larribe M, Deghmane AE, Taha MK, Dozot M, De Bolle X, Letesson JJ, Deutscher J.** 2009. Correlations between carbon metabolism and virulence in bacteria. *Contrib Microbiol* **16:**88–102.
4. **Commichau FM, Stülke J.** 2008. Trigger enzymes: bifunctional proteins active in metabolism and in controlling gene expression. *Mol Microbiol* **67:**692–702.
5. **Greenberg EP.** 2000. Bacterial genomics. Pump up the versatility. *Nature* **406:**947–948.
6. **Arraiano CM, Mauxion F, Viegas SC, Matos RG, Seraphin B.** 2013. Intracellular ribonucleases involved in transcript processing and decay: precision tools for RNA. *Biochim Biophys Acta* **1829:**491–513.
7. **Gao R, Stock AM.** 2010. Molecular strategies for phosphorylation-mediated regulation of response regulator activity. *Curr Opin Microbiol* **13:**160–167.
8. **Joyet P, Bouraoui H, Aké FM, Derkaoui M, Zébré AC, Cao TN, Ventroux M, Nessler S, Noirot-Gros MF, Deutscher J, Milohanic E.** 2013. Transcription regulators controlled by interaction with enzyme IIB components of the phosphoenolpyruvate:sugar phosphotransferase system. *Biochim Biophys Acta* **1834:**1415–1424.
9. **Gunka K, Commichau FM.** 2012. Control of glutamate homeostasis in *Bacillus subtilis*: a complex interplay between ammonium assimilation, glutamate biosynthesis and degradation. *Mol Microbiol* **85:**213–224.
10. **Österberg S, del Peso-Santos T, Shingler V.** 2011. Regulation of alternative sigma factor use. *Annu Rev Microbiol* **65:**37–55.
11. **Narberhaus F.** 2010. Translational control of bacterial heat shock and virulence genes by temperature-sensing mRNAs. *RNA Biol* **7:**84–89.
12. **Johansson J, Mandin P, Renzoni A, Chiaruttini C, Springer M, Cossart P.** 2002. An RNA thermosensor controls expression of virulence genes in *Listeria monocytogenes*. *Cell* **110:**551–561.
13. **Böhme K, Steinmann R, Kortmann J, Seekircher S, Heroven AK, Berger E, Pisano F, Thiermann T, Wolf-Watz H, Narberhaus F, Dersch P.** 2012. Concerted actions of a thermo-labile regulator and a unique intergenic RNA thermosensor control *Yersinia* virulence. *PLoS Pathog* **8:**e1002518. doi:10.1371/journal.ppat.1002518
14. **Loh E, Kugelberg E, Tracy A, Zhang Q, Gollan B, Ewles H, Chalmers R, Pelicic V, Tang CM.** 2013. Temperature triggers immune evasion by *Neisseria meningitidis*. *Nature* **502:**237–240.
15. **Kamp HD, Higgins DE.** 2011. A protein thermometer controls temperature-dependent transcription of flagellar motility genes in *Listeria monocytogenes*. *PLoS Pathog* **7:**e1002153. doi:10.1371/journal.ppat.1002153
16. **Quade N, Mendonca C, Herbst K, Heroven AK, Ritter C, Heinz DW, Dersch P.** 2012. Structural basis for intrinsic thermosensing by the master virulence regulator RovA of *Yersinia*. *J Biol Chem* **287:**35796–35803.
17. **Freitag NE, Port GC, Miner MD.** 2009. *Listeria monocytogenes* – from saprophyte to intracellular pathogen. *Nat Rev Microbiol* **7:**623–628.
18. **Fouet A.** 2010. AtxA, a *Bacillus anthracis* global virulence regulator. *Res Microbiol* **161:**735–742.
19. **Jeffery CJ.** 2009. Moonlighting proteins – an update. *Mol Biosyst* **5:**345–350.
20. **Henderson B, Martin A.** 2011. Bacterial virulence in the moonlight: multitasking bacterial moonlighting proteins are virulence determinants in infectious disease. *Infect Immunol* **79:**3476–3491.
21. **Copley SD.** 2012. Moonlighting is mainstream: paradigm adjustment required. *Bioessays* **34:**578–588.
22. **Chien AC, Zareh SK, Wang YM, Levin PA.** 2012. Changes in the oligomerization potential of the division inhibitor UgtP co-ordinate *Bacillus subtilis* cell size with nutrient availability. *Mol Microbiol* **86:**594–610.
23. **Weart RB, Lee AH, Chien AC, Haeusser DP, Hill NS, Levin PA.** 2007. A metabolic sensor governing cell size in bacteria. *Cell* **130:**335–347.
24. **Hill NS, Buske PJ, Shi Y, Levin PA.** 2013. A moonlighting enzyme links *Escherichia coli* cell size with central metabolism. *PLoS Genet* **9:**e1003663. doi:10.1371./journal.pgen.1003663
25. **Beckham KS, Connolly JP, Ritchie JM, Wang D, Gawthorne JA, Tahoun A, Gally DL, Burgess K, Burchmore RJ, Smith BO, Beatson SA, Byron O, Wolfe AJ, Douce GR, Roe AJ.** 2014. The metabolic enzyme AdhE controls the virulence of *Escherichia coli* O157:H7. *Mol Microbiol* **93:**199–211.

26. **Gu D, Zhou Y, Kallhoff V, Baban B, Tanner JJ, Becker DF.** 2004. Identification and characterization of the DNA-binding domain of the multifunctional PutA flavoenzyme. *J Biol Chem* **279:**31171–31176.

27. **Singh RK, Larson JD, Rambo RP, Hura GL, Becker DF, Tanner JJ.** 2011. Small-angle X-ray scattering studies of the oligomeric state and quarternary structure of the trifunctional proline utilization A (PutA) flavoprotein from *Escherichia coli. J Biol Chem* **286:**43144–43153.

28. **Singh RK, Tanner JJ.** 2012. Unique structural features and sequence motifs of proline utilization A (PutA). *Front Biosci* **17:**556–568.

29. **Muro-Pastor AM, Maloy S.** 1995. Proline dehydrogenase activity of the transcriptional repressor PutA is required for induction of the *put* operon by proline. *J Biol Chem* **270:**9819–9827.

30. **Ostrovsky de Spicer P, Maloy S.** 1993. PutA protein, a membrane-associated flavin dehydrogenase, acts as a redox-dependent transcriptional regulator. *Proc Natl Acad Sci U S A* **90:**4295–4298.

31. **Zhu W, Becker DF.** 2005. Exploring the proline-dependent conformational change in the multifunctional PutA flavoprotein by tryptophan fluorescence spectroscopy. *Biochemistry* **44:**12297–12306.

32. **Lin S, Cronan JE.** 2011. Closing in on complete pathways of biotin biosynthesis. *Mol Biosyst* **7:**1811–1821.

33. **Rodionov DA, Mironov AA, Gelfand AS.** 2002. Conservation of the biotin regulon and the BirA regulatory signal in Eubacteria and Archaea. *Genome Res* **12:**1507–1516.

34. **Wilson KP, Shewchuk LM, Brennan RG, Otsuka AJ, Matthews BW.** 1992. *Escherichia coli* biotin holoenzyme synthetase/bio repressor crystal structure delineates the biotin- and DNA-binding domains. *Proc Natl Acad Sci U S A* **89:**9257–9261.

35. **Solbiati J, Cronan JE.** 2010. The switch regulating transcription of the *Escherichia coli* biotin operon does not require extensive protein-protein contacts. *Chem Biol* **17:**11–17.

36. **Adikaram PR, Beckett D.** 2013. Protein:protein interactions in control of a transcriptional switch. *J Mol Biol* **425:**4584–4594.

37. **Chakravartty V, Cronan JE.** 2013. The wing of a winged helix-turn-helix transcription factor organizes the active site of BirA, a bifunctional repressor/ligase. *J Biol Chem* **288:**36029–36039.

38. **Henke SK, Cronan JE.** 2014. Successful conversion of the *Bacillus subtilis* BirA group

II biotin protein ligase into a group I ligase. *PLoS ONE* **9:**e96757. doi:10.1371/journal.pone.0096757

39. **Raffaelli N, Lorenzi T, Mariani PL, Emanuelli M, Amici A, Ruggieri S, Magni G.** 1999. The *Escherichia coli* NadR regulator is endowed with nicotinamide mononucleotide adenylyltransferase activity. *J Bacteriol* **181:**5509–5511.

40. **Grose JH, Bergthorsson U, Roth JR.** 2005. Regulation of NAD synthesis by the trifunctional NadR protein of *Salmonella enterica. J Bacteriol* **187:**2774–2782.

41. **Morohoshi F, Hayashi K, Munakata N.** 1990. *Bacillus subtilis ada* operon encodes two DNA alkyltransferases. *Nucleic Acids Res* **18:**5473–5480.

42. **Landini P, Volkert MR.** 2000. Regulatory responses of the adaptive response to alkylation damage: a simple regulon with complex regulatory features. *J Bacteriol* **182:**6543–6549.

43. **Takinowaki H, Matsuda Y, Yoshida T, Kobayashi Y, Ohkubo T.** 2006. The solution structure of the methylated form of the N-terminal 16-kDa domain of *Escherichia coli* Ada protein. *Protein Sci* **15:**487–497.

44. **Kholti A, Charlier D, Gigot D, Huysveld N, Roovers M, Glansdorff N.** 1998. *pyrH-*encoded UMP-kinase directly participates in pyrimidine-specific modulation of promoter activity in *Escherichia coli. J Mol Biol* **280:**571–582.

45. **Rostirolla DC, Breda A, Rosado LA, Palma MS, Basso LA, Santos DS.** 2011. UMP kinase from *Mycobacterium tuberculosis*: mode of action and allosteric interactions, and their likely role in pyrimidine metabolism regulation. *Arch Biochem Biophys* **505:**202–212.

46. **Charlier D, Hassanzadeh G, Kholti A, Gigot D, Pierard A, Glansdorff N.** 1995. *carP*, involved in pyrimidine regulation of the *Escherichia coli* carbamoylphosphate synthetase operon encodes a sequence-specific DNA-binding protein identical to XerB and PepA, also required for resolution of ColEI multimers. *J Mol Biol* **250:**392–406.

47. **Minh PN, Devroede N, Massant J, Maes D, Charlier D.** 2009. Insights into the architecture and stoichiometry of *Escherichia coli* PepA*DNA complexes involved in transcriptional control and site-specific DNA recombination by atomic force microscopy. *Nucleic Acids Res* **37:**1463–1476.

48. **Sinha SC, Krahn J, Shin BS, Tomchick DR, Zalkin H, Smith JL.** 2003. The purine repressor of *Bacillus subtilis*: a novel combination of

domains adapted for transcription regulation. *J Bacteriol* **185**:4087–4098.

49. **Putney SD, Schimmel P.** 1981. An aminoacyl tRNA synthetase binds to a specific DNA sequence and regulates its gene transcription. *Nature* **291**:632–635.

50. **Lin TH, Hu YN, Shaw GC.** 2014. Two enzymes, TilS and HprT, can form a complex to function as a transcriptional activator for the cell division protease gene *ftsH* in *Bacillus subtilis*. *J Biochem* **155**:5–16.

51. **Ellington AD, Szostack JW.** 1990. *In vitro* selection of RNA molecules that bind specific ligands. *Nature* **346**:818–822.

52. **Stülke J.** 2002. Control of transcription termination in bacteria by RNA-binding proteins that modulate RNA structures. *Arch Microbiol* **177**:433–440.

53. **Roth A, Breaker RR.** 2009. The structural and functional diversity of metabolite-binding riboswitches. *Annu Rev Biochem* **78**:305–334.

54. **Klass DM, Scheibe M, Butte F, Hogan GJ, Mann M, Brown PO.** 2013. Quantitative proteomic analysis reveals concurrent RNA-protein interactions and identifies new RNA-binding proteins in *Saccharomyces cerevisiae*. *Genome Res* **23**:1028–1038.

55. **Scherrer T, Mittal N, Janga SC, Gerber AP.** 2010. A screen for RNA-binding proteins in yeast indicates dual functions for many enzymes. *PLoS ONE* **5**:e15499. doi:10.1371/journal.pone.0015499

56. **Tsvetanova NG, Klass DM, Salzman J, Brown PO.** 2010. Proteome-wide search reveals unexpected RNA-binding proteins in *Saccharomyces cerevisiae*. *PLoS ONE* **5**:e12671. doi:10.1371/journal.pone.0012671

57. **Volz K.** 2008. The functional duality of iron regulatory protein 1. *Curr Opin Struct Biol* **18**:106–111.

58. **Leipuviene R, Theil EC.** 2007. The family of iron responsive RNA structures regulated by changes in cellular iron and oxygen. *Cell Mol Life Sci* **64**:2945–2955.

59. **Beinert H, Kennedy MC, Stout CD.** 1996. Aconitase as iron-sulfur protein, enzyme, and iron-regulatory protein. *Chem Rev* **96**:2335–2374.

60. **Rouault TA, Klausner RD.** 1996. Iron-sulfur clusters as biosensors for oxygen and iron. *Trends Biochem Sci* **21**:174–177.

61. **Weinstock GM, Hardham JM, McLeod MP, Sodergren EJ, Norris SJ.** 1998. The genome of *Treponema pallidum*: new light on the agent of syphilis. *FEMS Microbiol Rev* **22**:323–332.

62. **Vardhan H, Bhengraj AR, Jha R, Singh Mittal A.** 2009. *Chlamydia trachomatis* alters iron-regulatory protein-1 binding capacity and modulates cellular iron homeostasis in HeLa-229 cells. *J Biomed Biotechnol* **2009**:342032.

63. **Kaptain S, Downey WE, Tang C, Philpott C, Haile D, Orloff DG, Harford JB, Rouault TA, Klausner RD.** 1991. A regulated RNA binding protein also possesses aconitase activity. *Proc Natl Acad Sci U S A* **88**:10109–10113.

64. **Tang Y, Guest JR.** 1999. Direct evidence for mRNA binding and post-transcriptional regulation by *Escherichia coli* aconitases. *Microbiology* **145**:3069–3079.

65. **Alén C, Sonenshein AL.** 1999. *Bacillus subtilis* aconitase in an RNA-binding protein. *Proc Natl Acad Sci U S A* **96**:10412–10417.

66. **Austin CM, Maier RJ.** 2013. Aconitase-mediated post-transcriptional regulation of *Helicobacter pylori* peptidoglycan deacetylase. *J Bacteriol* **195**:5316–5322.

67. **Banerjee S, Nandyala AK, Raviprasad P, Ahmed N, Hasnain SE.** 2007. Iron-dependent RNA-binding activity of *Mycobacterium tuberculosis* aconitase. *J Bacteriol* **189**:4046–4052.

68. **Baothman OA, Rolfe MD, Green J.** 2013. Characterization of *Salmonella enterica* serovar typhimurium aconitase A. *Microbiology* **159**:1209–1216.

69. **Mengaud JM, Horwitz MA.** 1993. The major iron-containing protein of *Legionella pneumophila* is an aconitase homologous with the human iron-responsive element-binding protein. *J Bacteriol* **175**:5666–5676.

70. **Sadykov MR, Olson ME, Halouska S, Zhu Y, Fey PD, Powers R, Somerville GA.** 2008. Tricarboxylic acid cycle-dependent regulation of *Staphylococcus epidermidis* polysaccharide intercellular adhesion synthesis. *J Bacteriol* **190**:7621–7632.

71. **Zhu Y, Xiong YQ, Sadykov MR, Fey PD, Lei MG, Lee CY, Bayer AS, Somerville GA.** 2009. Tricarboxylic acid cycle-dependent attenuation of *Staphylococcus aureus in vivo* virulence by selective inhibition of amino acid transport. *Infect Immun* **77**:4256–4264.

72. **Somerville GA, Mikoryak CA, Reitzer L.** 1999. Physiological characterization of *Pseudomonas aeruginosa* during exotoxin A synthesis: glutamate, iron limitation, and aconitase activity. *J Bacteriol* **181**:1072–1078.

73. **Wilson TJ, Bertrand N, Tang JL, Feng JX, Pan MQ, Barber CE, Dow JM, Daniels MJ.** 1998. The *rpfA* gene of *Xanthomonas campestris* pathovar *campestris*, which is involved in the regulation of pathogenicity factor production, encodes an aconitase. *Mol Microbiol* **28**:961–970.

74. **Robbins AH, Stout CD.** 1989. Structure of activated aconitase: formation of the [4Fe-4S]

cluster in the crystal. *Proc Natl Acad Sci U S A* **86:**3639–3643.

75. **Williams CH, Stillman TJ, Barynin VV, Sedelnikova SE, Tang Y, Green J, Guest JR, Artymiuk PJ.** 2002. *E. coli* aconitase B structure reveals a HEAT-like domain with implications for protein-protein recognition. *Nat Struct Biol* **9:**447–452.

76. **Walden WE, Selezneva AI, Dupuy J, Volbeda A, Fontecilla-Camps JC, Theil C, Volz K.** 2006. Structure of dual function iron regulatory protein 1 complexed with ferritin IRE-RNA. *Science* **314:**1903–1908.

77. **Goforth JB, Anderson SA, Nizzi CP, Eisenstein RS.** 2010. Multiple determinants within iron-responsive elements dictate iron regulatory protein binding and regulatory hierarchy. *RNA* **16:**154–169.

78. **Selezneva AI, Walden WE, Volz KW.** 2013. Nucleotide-specific recognition of iron-responsive elements by iron regulatory protein 1. *J Mol Biol* **425:**3301–3310.

79. **Serio AW, Pechter KB, Sonenshein AL.** 2006. *Bacillus subtilis* aconitase is required for efficient late-sporulation gene expression. *J Bacteriol* **188:**6396–6405.

80. **Craig JE, Ford MJ, Blaydon DC, Sonenshein AL.** 1997. A null mutation in the *Bacillus subtilis* aconitase gene causes a block in Spo0A-phosphate-dependent gene expression. *J Bacteriol* **179:**7351–7359.

81. **Pechter KB, Meyer FM, Serio AW, Stülke J, Sonenshein AL.** 2013. Two roles for aconitase in the regulation of tricarboxylic acid branch gene expression in *Bacillus subtilis*. *J Bacteriol* **195:**1525–1537.

82. **Baumgart M, Mustafi N, Krug A, Bott M.** 2011. Deletion of the aconitase gene in *Corynebacterium glutamicum* causes strong selection pressure for secondary mutations inactivating citrate synthase. *J Bacteriol* **193:**6864–6873.

83. **Viollier PH, Nguyen KT, Minas W, Folcher M, Dale GE, Thompson CJ.** 2001. Roles of aconitase in growth, metabolism, and morphological differentiation of *Streptomyces coelicolor*. *J Bacteriol* **183:**3193–3203.

84. **Tang Y, Guest JR, Artymiuk PJ, Green J.** 2005. Switching aconitase B between catalytic and regulatory modes involves iron-dependent dimer formation. *Mol Microbiol* **56:**1149–1158.

85. **Tang Y, Guest JR, Artymiuk PJ, Read RC, Green J.** 2004. Post-transcriptional regulation of bacterial motility by aconitase proteins. *Mol Microbiol* **51:**1817–1826.

86. **Michta E, Schad K, Blin K, Ort-Winklbauer R, Röttig M, Kohlbacher O, Wohlleben W,** Schinko E, Mast Y. 2012. The bifunctional role of aconitase in *Streptomyces viridochromogenes* Tü494. *Environ Microbiol* **14:**3203–3219.

87. **Springer M, Plumbridge JA, Butler JS, Graffe M, Dondon J, Mayaux JF, Fayat G, Lestienne P, Blanquet S, Grunberg-Manago M.** 1985. Autogenous control of *Escherichia coli* threonyl-tRNA synthetase expression *in vivo*. *J Mol Biol* **185:**93–104.

88. **Higashitsuji Y, Angerer A, Berghaus S, Hobl B, Mack M.** 2007. RibR, a possible regulator of the *Bacillus subtilis* riboflavin biosynthetic operon, *in vivo* interacts with the 5′-untranslated leader of *rib* mRNA. *FEMS Microbiol Lett* **274:**48–54.

89. **Grabner GK, Switzer RL.** 2003. Kinetic studies of the uracil phosphoribosyltransferase reaction catalyzed by the *Bacillus subtilis* pyrimidine attenuation regulatory protein PyrR. *J Biol Chem* **278:**6921–6927.

90. **Chander P, Halbig KM, Miller JK, Fields CJ, Bonner HK, Grabner GK, Switzer RL, Smith JL.** 2005. Structure of the nucleotide complex of PyrR, the *pyr* attenuation protein from *Bacillus caldolyticus*, suggests dual regulation by pyrimidine and purine nucleotides. *J Bacteriol* **187:**1773–1782.

91. **Hobl B, Mack M.** 2007. The regulator protein PyrR of *Bacillus subtilis* specifically interacts *in vivo* with three untranslated regions within *pyr* mRNA of pyrimidine biosynthesis. *Microbiology* **153:**693–700.

92. **Deutscher J, Francke C, Postma PW.** 2006. How phosphotransferase system-related protein phosphorylation regulates carbohydrate metabolism in bacteria. *Microbiol Mol Biol Rev* **70:**939–1031.

93. **Stülke J, Hillen W.** 1998. Coupling physiology and gene regulation in bacteria:the phosphotransferase sugar uptake system delivers the signals. *Naturwissenschaften* **85:**583–592.

94. **Deutscher J, Aké FM, Derkaoui M, Zébré AC, Cao TN, Bouraoui H, Kentache T, Mokhtari A, Milohanic E, Joyet P.** 2014. The bacterial phosphoenolpyruvate:carbohydrate phosphotransferase system: regulation by protein phosphorylation and phosphorylation-dependent protein-protein interactions. *Microbiol Mol Biol Rev* **78:**231–256.

95. **Greenberg DB, Stülke J, Saier MH Jr.** 2002. Domain analysis of transcriptional regulators bearing PTS-regulatory domains. *Res Microbiol* **153:**519–526.

96. **Brehm K, Ripio MT, Kreft J, Vázquez-Boland JA.** 1999. The *bvr* locus of *Listeria monocytogenes* mediates virulence gene re-

pression by bet-glucosides. *J Bacteriol* **181**: 5024–5032.

97. **Gray MJ, Freitag NE, ad Boor KJ.** 2006. How the bacterial pathogen *Listeria monocytogenes* mediates the switch from environmental Dr. Jekyll to pathogenic Mr. Hyde. *Infect Immun* **74**:2505–2512.

98. **Schnetz K, Rak B.** 1990. Beta-glucoside permease represses the *bgl* operon of *Escherichia coli* by phosphorylation of the antiterminator protein and also interacts with glucose-specific enzyme III, the key element in catabolite control. *Proc Natl Acad Sci U S A* **87**:5074–5078.

99. **Amster-Choder O, Wright A.** 1992. Modulation of the dimerization of a transcriptional antiterminator protein by phosphorylation. *Science* **257**:1395–1398.

100. **Chen Q, Arents JC, Bader R, Postma PW, Amster-Choder O.** 1997. BglF, the sensor of the *E. coli bgl* system, uses the same site to phosphorylate both a sugar and a regulatory protein. *EMBO J* **16**:4617–4627.

101. **Rothe FM, Bahr T, Stülke J, Rak B, Görke B.** 2012. Activation of *Escherichia coli* antiterminator BglG requires its phosphorylation. *Proc Natl Acad Sci U S A* **109**:15906–15911.

102. **Himmel S, Zschiedrich CP, Becker S, Hsiao HH, Wolff S, Diethmaier C, Urlaub H, Lee D, Griesinger C, Stülke J.** 2012. Determinants of interaction specificity of the *Bacillus subtilis* GlcT antitermination protein: functionality and phosphorylation specificity depend on the arrangement of the regulatory domains. *J Biol Chem* **287**:27731–27742.

103. **Schilling O, Langbein I, Müller M, Schmalisch M, Stülke J.** 2004. A protein-dependent riboswitch controlling *ptsGHI* operon expression in *Bacillus subtilis*: RNA structure rather than sequence provides interaction specificity. *Nucleic Acids Res* **32**:2853–2864.

104. **Schilling O, Herzberg C, Hertrich T, Vörsmann H, Jessen D, Hübner S, Titgemeyer F, Stülke J.** 2006. Keeping signals straight in transcription regulation: specificity determinants for the interaction of a family of conserved bacterial RNA-protein couples. *Nucleic Acids Res* **34**:6102–6115.

105. **Hübner S, Declerck N, Diethmaier C, Le Coq D, Aymerich S, Stülke J.** 2011. Prevention of cross-talk in conserved regulatory systems: Identification of specificity determinants in RNA-binding anti-termination proteins of the BglG family. *Nucleic Acids Res* **39**:4360–4372.

106. **Lopian L, Elisha Y, Nussbaum-Schochat A, Amster-Choder O.** 2010. Spatial and temporal organization of the *E. coli* PTS components. *EMBO J* **29**:3630–3645.

107. **Rothe FM, Wrede C, Lehnik-Habrink M, Görke B, Stülke J.** 2013. Dynamic localization of a transcription factor in *Bacillus subtilis*: the LicT antiterminator relocalizes in response to inducer availability. *J Bacteriol* **195**:2146–2154.

108. **Lopian L, Nussbaum-Schochat A, O'Day-Kerstein K, Wright A, Amster-Choder O.** 2003. The BglF sensor recruits the BglG transcription regulator to the membrane and releases it on stimulation. *Proc Natl Acad Sci U S A* **100**:7099–7104.

109. **Bouraoui H, Ventroux M, Noirot-Gros MF, Deutscher J, Joyet P.** 2013. Membrane sequestration by the EIIB domain of the mannitol permease MtlA activates the *Bacillus subtilis mtl* operon regulator MtlR. *Mol Microbiol* **87**:789–801.

110. **Heravi KM, Altenbuchner J.** 2014. Regulation of the *Bacillus subtilis* mannitol utilization genes: promoter structure and transcriptional activation by the wild-type regulator (MtlR) and its mutants. *Microbiology* **160**:91–101.

111. **Tsvetanova B, Wilson AC, Bongiorni C, Chiang C, Hoch JA, Perego M.** 2007. Opposing effects of histidine phosphorylation regulate the AtxA virulence transcription factor in Bacillus anthracis. *Mol Microbiol* **63**:644–655.

112. **Tetsch L, Jung K.** 2009. How are signals transduced across the cytoplasmic membrane? Transport proteins as transmitter of information. *Amino Acids* **37**:467–477.

113. **Dintner S, Staron A, Berchtold E, Petri T, Mascher T, Gebhard S.** 2011. Coevolution of ABC transporters and two-component regulatory systems as resistance modules against antimicrobial peptides in Firmicutes bacteria. *J Bacteriol* **193**:3851–3962.

114. **Murray DS, Chinnam N, Tonthat NK, Whitfill T, Wray LV Jr, Fisher SH, Schumacher MA.** 2013. Structures of the *Bacillus subtilis* glutamine synthetase dodecamer reveal large intersubunit catalytic conformational changes linked to a unique feedback inhibition mechanism. *J Biol Chem* **288**:35801–35811.

115. **Fedorova K, Kayumov A, Woyda K, Ilinskaja O, Forchhammer K.** 2013. Transcription factor TnrA inhibits the biosynthetic activity of glutamine synthetase in *Bacillus subtilis*. *FEBS Lett* **587**:1293–1298.

116. **Kloosterman TG, Hendriksen WT, Bijlsma JJ, Bootsma HJ, van Hijum SA, Kok J, Hermans PW, Kuipers OP.** 2006. Regulation of glutamine and glutamate metabolism by GlnR and GlnA in *Streptococcus pneumoniae*. *J Biol Chem* **281**:25097–25109.

117. Hendriksen WT, Kloosterman TG, Bootsma HJ, Estevao S, de Groot R, Kuipers OP, Hermans PW. 2008. Site-specific contributions of glutamine-dependent regulator GlnR and GlnR-regulated genes to virulence of *Streptococcus pneumoniae*. *Infect Immun* **76**:1230–1238.

118. Groot Kormelink T, Koenders E, Hagemeijer Y, Overmars L, Siezen RJ, de Vos WM, Francke C. 2012. Comparative genome analysis of central nitrogen metabolism and its control by GlnR in the class Bacilli. *BMC Genomics* **13**:191.

119. Even S, Burguière P, Auger S, Soutourina O, Danchin A, Martin-Verstraete I. 2006. Global control of cysteine metabolism by CymR in *Bacillus subtilis*. *J Bacteriol* **188**:2184–2197.

120. Hullo MF, Martin-Verstraete I, Soutourina O. 2010. Complex phenotypes of a mutant inactivated for CymR, the global regulator of cysteine metabolism in *Bacillus subtilis*. *FEMS Microbiol Lett* **309**:201–207.

121. Tanous C, Soutourina O, Raynal B, Hullo MF, Mervelet P, Gilles AM, Noirot P, Danchin A, England P, Martin-Verstraete I. 2008. The CymR regulator in complex with the enzyme CysK controls cysteine metabolism in *Bacillus subtilis*. *J Biol Chem* **283**:35551–35560.

122. Soutourina O, Poupal O, Coppée JY, Danchin A, Msadek T, Martin-Verstraete I. 2009. CymR, the master regulator of cysteine metabolism in *Staphylococcus aureus*, controls host sulphur source utilization and plays a role in biofilm formation. *Mol Microbiol* **73**:194–211.

123. Zhao C, Moriga Y, Feng B, Kumada Y, Imanaka H, Imamura K, Nakanishi K. 2006. On the interaction site of serine acetyltransferase in the cysteine synthase complex in *Escherichia coli*. *Biochem Biophys Res Commun* **341**:911–916.

124. Fisher SH, Wray LV. 2008. *Bacillus subtilis* glutamine synthetase regulates its own synthesis by acting as a chaperone to stabilize GlnR-DNA complexes. *Proc Natl Acad Sci U S A* **105**:1014–1019.

125. Lindenberg S, Klauck G, Pesavento C, Klauck E, Hengge R. 2013. The EAL domain protein YciR acts as a trigger enzyme in a c-di-GMP signalling cascade in *E. coli* biofilm control. *EMBO J* **32**:2001–2014.

126. Gebhard S. 2012. ABC transporters of antimicrobial peptides in Firmicutes bacteria - phylogeny, function, and regulation. *Mol Microbiol* **86**:1295–1317.

127. Revilla-Guarinos A, Gebhard S, Mascher T, Zuñiga M. 2014. Defence against antimicrobial peptides: different strategies in Firmicutes. *Environ Microbiol* **16**:1225–1237.

128. Hiron A, Falord M, Valle J, Débarbouillé M, Msadek T. 2011. Bacitracin and nisin resistance in *Staphylococcus aureus*: a novel pathway involving the BraS/BraR two-component system (SA2417/SA2418) and both the BraD/BraE and VraD/VraE ABC transporters. *Mol Microbiol* **81**:602–622.

129. Kallenberg F, Dintner S, Schmitz R, Gebhard S. 2013. Identification of regions important for resistance and signaling within the antimicrobial peptide transporter BceAB of *Bacillus subtilis*. *J Bacteriol* **195**:3287–3297.

130. Gebhard S, Fang C, Shaaly A, Leslie DJ, Weimar MR, Kalamorz F, Carne A, Cook GM. 2014. Identification and characterization of a bacitracin resistance network in *Enterococcus faecalis*. *Antimicrob Agents Chemother* **58**:1425–1433.

131. Falord M, Karimova G, Hiron A, Msadek T. 2012. GraXSR proteins interact with the VraFG ABC transporter to form a five-component system required for cationic antimicrobial peptide sensing and resistance in *Staphylococcus aureus*. *Antimicrob Agents Chemother* **56**:1047–1058.

132. Commichau FM, Herzberg C, Tripal P, Valerius O, Stülke J. 2007. A regulatory protein-protein interaction governs glutamate biosynthesis in *Bacillus subtilis*: the glutamate dehydrogenase RocG moonlights in controlling the transcription factor GltC. *Mol Microbiol* **65**:642–654.

133. Wray LV Jr, Zalieckas JM, Fisher SH. 2001. *Bacillus subtilis* glutamine synthetase controls gene expression through a protein-protein interaction with transcription factor TnrA. *Cell* **107**:427–435.

134. Picossi S, Belitsky BR, Sonenshein AL. 2007. Molecular mechanism of the regulation of *Bacillus subtilis* gltAB expression by GltC. *J Mol Biol* **365**:1298–1313.

135. Gunka K, Newman JA, Commichau FM, Herzberg C, Rodrigues C, Hewitt L, Lewis RJ, Stülke J. 2010. Functional dissection of a trigger enzyme: mutations of the *Bacillus subtilis* glutamate dehydrogenase RocG that affect differentially its catalytic activity and regulatory properties. *J Mol Biol* **400**:815–827.

136. Krishnan N, Becker DF. 2005. Characterization of a bifunctional PutA homologue from *Bradyrhizobium japonicum* and identification of an active site residue that modulates proline reduction of the flavin adenine dinucleotide cofactor. *Biochemistry* **44**:9130–9139.

137. **Durante-Rodríguez G, Mancheño JM, Rivas G, Alfonso C, García JL, Díaz E, Carmona M.** 2013. Identification of a missing link in the evolution of an enzyme into a transcriptional regulator. *PLoS ONE* **8:**e57518. doi:10.1371/journal.pone.0057518

138. **Barragán MJ, Blázquez B, Zamarro MT, Mancheño JM, García JL, Díaz E, Carmona M.** 2005. BzdR, a repressor that controls the anaerobic catabolism of benzoate in *Azoarcus* sp. CIB, is the first member of a new subfamily of transcriptional regulators. *J Biol Chem* **280:**10683–10694.

139. **Bramucci E, Milano T, Pascarella S.** 2011. Genomic distribution and heterogeneity of MocR-like transcriptional factors containing a domain belonging to the superfamily of the pyridoxal-5′-phosphate dependent enzymes of fold type I. *Biochem Biophys Res Commun* **415:**88–93.

140. **Titgemeyer F, Reizer J, Reizer A, Saier MH Jr.** 1994. Evolutionary relationships between sugar kinases and transcriptional repressors in bacteria. *Microbiology* **140:**2349–2354.

141. **Chevance FF, Erhardt M, Lengsfeld C, Lee SJ, Boos W.** 2006. Mlc of *Thermus thermophilus*: a glucose-specific regulator for a glucose/mannose ABC transporter in the absence of the phosphotranferase system. *J Bacteriol* **188:**6561–6571.

142. **Kietzman CC, Caparon MG.** 2010. CcpA and LacD.1 affect temporal regulation of *Streptococcus pyogenes* virulence genes. *Infect Immun* **78:**241–252.

143. **Loughman JA, Caparon MG.** 2006. A novel adaptation of aldolase regulates virulence in *Streptococcus pyogenes*. *EMBO J* **25:**5414–5422.

144. **Loughman JA, Caparon MG.** 2007. Comparative functional analysis of the *lac* operons in *Streptococcus pyogenes*. *Mol Microbiol* **64:**269–280.

145. **Lee SJ, Kim HS, Kim do J, Yoon HJ, Kim KH, Yoon JY, Suh SW.** 2011. Crystal structures of LacD from *Staphylococcus aureus* and LacD.1 from *Streptococcus pyogenes*: insights into substrate specificity and virulence gene regulation. *FEBS Lett* **585:**307–312.

146. **Cusumano Z, Caparon M.** 2013. Adaptive evolution of the *Streptococcus pyogenes* regulatory aldolase LacD.1. *J Bacteriol* **195:**1294–1304.

147. **Shevell DE, Friedman BM, Walker GC.** 1990. Resistance to alkylation damage in *Escherichia coli*: role of the Ada protein in induction of the adaptive response. *Mutat Res* **233:**53–72.

148. **Kleefeld A, Ackermann B, Bauer J, Krämer J, Unden G.** 2009. The fumarate/succinate antiporter DcuB of *Escherichia coli* is a bifunctional protein with sites for regulation of DcuS-dependent gene expression. *J Biol Chem* **284:**265–275.

149. **Tomchick DR, Turner RJ, Switzer RL, Smith JL.** 1998. Adaptation of an enzyme to regulatory function: structure of *Bacillus subtilis* PyrR, a *pyr* RNA-binding attenuation protein and uracil phosphoribosyltransferase. *Structure* **6:**337–350.

150. **Bachem S, Stülke J.** 1998. Regulation of the *Bacillus subtilis* GlcT antiterminator protein by components of the phosphotransferase system. *J Bacteriol* **180:**5319–5326.

151. **Martin-Verstraete I, Charrier V, Stülke J, Galinier A, Erni B, Rapoport G, Deutscher J.** 1998. Antagonistic effects of dual PTS catalysed phosphorylation on the *Bacillus subtilis* transcriptional activator LevR. *Mol Microbiol* **28:**293–303.

152. **Tobisch S, Stülke J, Hecker M.** 1999. Regulation of the *lic* operon of *Bacillus subtilis* and characterization of potential phosphorylation sites of the LicR regulator protein by site-directed mutagenesis. *J Bacteriol* **181:**4995–5003.

153. **Wenzel M, Altenbuchner J.** 2013. The *Bacillus subtilis* mannose regulator, ManR, a DNA-binding protein regulated by HPr and its cognate PTS transporter, ManP. *Mol Microbiol* **88:**562–576.

154. **Joyet P, Derkaoui M, Poncet S, Deutscher J.** 2010. Control of *Bacillus subtilis mtl* operon expression by complex phosphorylation-dependent regulation of the transcriptional activator MtlR. *Mol Microbiol* **76:**1279–1294.

155. **Tanaka Y, Kimata K, Aiba H.** 2000. A novel regulatory role of glucose transporter of *Escherichia coli*: membrane sequestration of a global repressor Mlc. *EMBO J* **19:**5344–5352.

156. **Tetsch L, Koller C, Haneburger I, Jung K.** 2008. The membrane-integrated transcriptional activator CadC of *Escherichia coli* senses lysine indirectly via the interaction with the lysine permease LysP. *Mol Microbiol* **67:**570–583.

157. **Bächler C, Schneider P, Bähler P, Lustig A, Erni B.** 2005. *Escherichia coli* dihydroxyacetone kinase controls gene expression by binding to transcription factor DhaR. *EMBO J* **24:**283–293.

158. **Joly N, Böhm A, Boos W, Richet E.** 2004. MalK, the ATP-binding cassette component of the *Escherichia coli* maltodextrin transporter, inhibits the transcriptional activator MalT by antagonizing inducer binding. *J Biol Chem* **279:**33123–33130.

159. **Nakano Y, Kimura K.** 1991. Purification and characterization of a repressor for the *Bacillus cereus glnRA* operon. *J Biochem* **109:**223–228.

160. **Chen PM, Chen YY, Yu SL, Sher S, Lai CH, Chia JS.** 2010. Role of GlnR in acid-mediated repression of genes encoding proteins involved glutamine and glutamate metabolism in *Streptococcus mutans. Appl Environ Microbiol* **76:**2478–2486.

161. **Terra R, Stanley-Wall NR, Cao G, Lazazzera BA.** 2012. Identification of *Bacillus subtilis* SipW as a bifunctional signal peptidase that controls surface-adhered biofilm formation. *J Bacteriol* **194:**2781–2790.

162. **Garcia LL, Rivas-Marín E, Floriano B, Bernhardt R, Ewen KM, Reyes-Ramírez F, Santero E.** 2011. ThnY is a ferredoxin reductase-like iron-sulfur flavoprotein that has evolved to function as a regulator of tetralin biodegradation gene expression. *J Biol Chem* **286:**1709–1718.

Regulating the Intersection of Metabolism and Pathogenesis in Gram-positive Bacteria

7

ANTHONY R. RICHARDSON,[1]† GREG A. SOMERVILLE,[2]† and
ABRAHAM L. SONENSHEIN[3]†

For prototrophic bacteria, central metabolism (i.e., glycolysis, the pentose phosphate pathway, and the Krebs cycle) supplies the 13 biosynthetic intermediates necessary to synthesize *all* biomolecules (Fig. 1). Gram-positive bacteria (i.e., Actinobacteria and Firmicutes) exhibit a diverse collection of central metabolic capabilities that have been shaped by reductive evolution. Some Gram-positive bacteria (e.g., *Bacillus anthracis* and *Staphylococcus aureus*) have complete central metabolic pathways, but others (e.g., *Streptococcus pyogenes* and *Enterococcus faecium*) have Krebs cycle deficiencies, and some have multiple central metabolism deficiencies (e.g., *Mycoplasma genitalium* and *Ureaplasma parvum*). These differences in central metabolic capabilities are also reflected in the bacteria's ability to persist away from a host organism; specifically, the more metabolically impaired the bacterium, the more dependent it is on its host. In essence, hosts serve as a reservoir for metabolites that overcome deficiencies in central and intermediary metabolism. Metabolic deficiencies are not created by only reductive evolution; they are also created when bacteria encounter stressful environments (e.g., iron limitation or a host immune response) that alter carbon flux (1, 2). These

[1]Department of Microbiology and Immunology, University of North Carolina, Chapel Hill, NC; [2]School of Veterinary Medicine and Biomedical Sciences, University of Nebraska-Lincoln, Lincoln, NE; [3]Department of Molecular Biology and Microbiology, Tufts University School of Medicine, Boston, MA; †Equal contributors and co-corresponding authors.

Metabolism and Bacterial Pathogenesis
Edited by Tyrrell Conway and Paul Cohen
© 2015 American Society for Microbiology, Washington, DC
doi:10.1128/microbiolspec.MBP-0004-2014

Metabolism

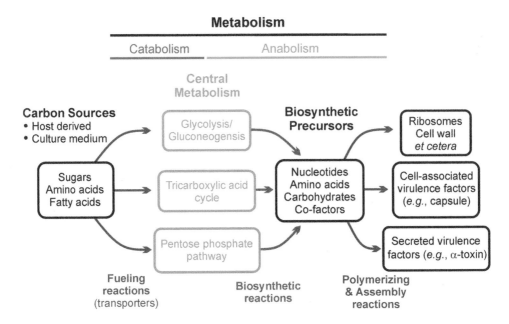

FIGURE 1 A simplified view of bacterial physiology. The 13 biosynthetic intermediates discussed in this chapter are all derived from the three metabolic pathways of central metabolism. Alterations in the availability of these biosynthetic intermediates always affect virulence factor synthesis. doi:10.1128/microbiolspec.MBP-0004-2014.f1

changes in flux alter the metabolome, which can modulate the activity of metabolite-responsive global regulators such as CodY, CcpA, Rex, and RpiR. In the first portion of this chapter, we discuss how genetic, environmental, and nutritional conditions alter the metabolome, primarily central metabolism, and in the second part, how these metabolic changes influence the activity of metabolite-responsive regulators. Finally, we discuss how metabolism and metabolite-responsive global regulators influence the outcomes of host-pathogen interactions. This review references only those manuscripts published through December 2013.

GRAM-POSITIVE METABOLISM

Glycolysis (Embden–Meyerhof–Parnas Pathway)

Glycolysis is the catabolic pathway that generates two molecules of pyruvate per

molecule of glucose and, in the process, reduces two molecules of NAD^+ and produces a net gain of two ATP molecules. In addition to providing a modest amount of ATP by substrate-level phosphorylation, glycolysis also generates seven (i.e., glucose-6-phosphate, fructose-6-phosphate, dihydroxyacetone phosphate, glyceraldehyde-3-phosphate, phosphoenolpyruvate [PEP], pyruvate, and acetyl-Coenzyme A) of the 13 biosynthetic intermediates that are used to synthesize all other organic molecules in a bacterium. Glycolysis is largely fueled by the PEP-dependent phosphotransferase system (PTS), a transport system that bacteria use to bring sugars into the cytoplasm using energy transferred from PEP (3, 4). In its simplest form, the PTS uses three enzymes to transfer the high-energy phosphate from PEP to the sugar being translocated, enzyme I (EI), enzyme II (EII, the permease), and a histidine-containing phosphocarrier protein (HPr) (3). The transfer of the high-energy phosphate from PEP to a carbohydrate initiates with transfer of phosphate from PEP to a histidine residue on EI, which is then transferred to a histidine residue on HPr. The phosphorylated HPr transfers the high-energy phosphate to EII, which transfers it to the carbohydrate. Transfer of a phosphate to a carbohydrate has two primary consequences: first, the phosphorylated carbohydrate is no longer recognized by the EII permease, preventing efflux of the carbohydrate from the cytoplasm; and second, the phosphorylated carbohydrate is activated for glycolysis. Glucose and fructose can also be activated for glycolysis intracellularly by transfer of a phosphate from ATP by hexokinase (5).

In glycolysis, activated glucose (i.e., glucose-6-phosphate) is processed in a series of key enzymatic reactions by phosphoglucose isomerase, phosphofructokinase, and aldolase to generate glyceraldehyde-3-phosphate. The importance of these enzymatic reactions is evident from the complex feedback and feedforward allosteric regulation that controls their activity. For example, PEP, ATP, and citrate are inhibitors of phosphofructokinase (6), while ADP and GDP enhance its enzymatic activity (7). Because the activity of enzymes like phosphofructokinase is modulated by allosteric effectors, changes in the intracellular concentrations of these effectors will alter carbon flow through glycolysis. Alterations in carbon flow through glycolysis will change the intracellular concentrations of glycolytic intermediates such as fructose-1,6-bisphosphate (FBP). In other words, anything that can alter the concentration of an allosteric effector will alter glycolysis.

Free iron in a eukaryotic cell is present at a concentration of 10^{-18} M, meaning that free iron is unavailable to invading pathogens. Whereas the activity of phosphofructokinase is independent of iron, phosphofructokinase activity and carbon flux through glycolysis are nonetheless affected by growth in iron-limited conditions. Two possible explanations can be considered: First, variations in the availability of iron alter transcription of the phosphofructokinase gene in *Mycobacterium smegmatis*, *Enterococcus faecalis*, and *S. aureus* (8–10). Second, some bacteria, when cultivated in an iron-limiting medium, accumulate citric acid in the cytosol and the culture medium due to a metabolic block in the Krebs cycle at aconitase (2, 11–13). Because citrate is an allosteric inhibitor of phosphofructokinase, the accumulation of citrate should lead to an increased concentration of fructose-6-phosphate or metabolites derived from fructose-6-phosphate. When the Krebs cycle in *Staphylococcus epidermidis* is genetically inactivated or the bacteria are cultivated in iron-limited medium, glucose-6-phosphate and amino sugars accumulate, which is indicative of reduced phosphofructokinase activity (1, 2, 12). Decreased phosphofructokinase activity limits the availability of downstream biosynthetic intermediates and precursors, which decreases the bacterium's ability to assemble macromolecules (Fig. 1). The allosteric and genetic regulation of phosphofructokinase provides an excellent example of the interconnection between metabolism and the bacterial environment, but these connections also rely on metabolite-responsive regulators to control the adaptive response to environmental changes (discussed section 2).

Pentose Phosphate Pathway (Warburg-Lipmann-Dickens-Horecker Shunt)

The processing of activated glucose through the pentose phosphate pathway (PPP) produces three of the 13 biosynthetic intermediates; specifically, ribose-5-phosphate, sedoheptulose-7-phosphate, and erythrose-4-phosphate (14, 15). Two of these biosynthetic intermediates, ribose-5-phosphate and erythrose-4-phosphate, are essential for the synthesis of purines, histidine, and aromatic amino acids. The third intermediate, sedoheptulose-7-phosphate, in conjunction with glyceraldehyde-3-phosphate, can be used by tranketolase to generate ribose-5-phosphate or by transaldolase to generate fructose-6-phosphate and erythrose-4-phosphate (16). In addition to providing biosynthetic intermediates, the PPP also generates two molecules of NADPH per molecule of glucose-6-phosphate, which can be used as electron donors in biosynthetic reactions such as fatty acid and glutamate biosynthesis. The enzymatic reactions that reduce $NADP^+$ to $NADPH/H^+$ occur in the oxidative portion of the PPP that produces ribulose-5-phosphate from activated glucose (15, 17). This process starts with the oxidation of glucose-6-phosphate to 6-phosphogluconolactone catalyzed by glucose-6-phosphate dehydrogenase. In Gram-positive bacteria, reductive evolution has caused the loss of glucose-6-phosphate dehydrogenase (*zwf*) in the oxidative portion of the PPP in *Mycoplasma sp.*, *Streptococcus pyogenes*, *S. mutans*, *S. agalactiae*, and *Clostridium difficile* [(18, 19), http://biocyc.org]. While these bacteria lack part of the oxidative portion of the PPP, most possess the nonoxidative portion. One notable exception is *Mycoplasma suis*, which lacks the PPP (20), and other *Mycoplasma sp.* that lack transaldolase (Somerville, unpublished observations). The metabolic consequences of the loss of glucose-6-phosphate dehydrogenase are a decreased ability to generate pentose sugars and reducing potential, while the loss of transaldolase prevents regeneration of fructose-6-phosphate from sedoheptulose-7-phosphate. Despite the fact that reductive evolution has resulted in PPP variation, it is interesting to note that pentose phosphate metabolism is frequently increased in Gram-positive pathogens in response to environmental stresses and in infection models (21–24).

Increased carbon flow through the oxidative portion of the PPP generates NADPH, while the nonoxidative branch produces fructose-6-phosphate. In addition to biosynthetic reactions, NADPH is required for the enzymatic reduction of oxidized glutathione, thioredoxin, bacillithiol, mycothiol, and coenzyme A (25–28). As an example, thioredoxin reductase catalyzes the transfer of electrons from NADPH to the active site of

thioredoxin via flavin adenine dinucleotide (29, 30). Reduced thioredoxin, in concert with other low-molecular-weight thiols, is critical for reducing protein disulfides and providing electrons to ribonucleotide reductase, methionine sulfoxide reductase, mycothiol disulfide reductase, and other enzymes (31). While NADPH is necessary for the function of reductases, transaldolase in the PPP is also important for redox homeostasis because it produces fructose-6-phosphate from sedoheptulose-7-phosphate and glyceraldehyde-3-phosphate. Fructose-6-phosphate is a precursor for *N*-acetylglucosamine, which is required for bacillithiol and mycothiol biosynthesis (26, 32). These connections provide a rationale for the observation that increased carbon flow through the PPP is often associated with stressful conditions, environments, or infection (21–23, 33).

While induction of the PPP during an infection supports redox homeostasis, it also is important for intracellular pathogens like *L. monocytogenes* to overcome one of the deleterious effects of interferon-γ: namely, the indoleamine 2,3-dioxygenase-induced depletion of tryptophan (34). Interferon-γ-activated macrophages increase synthesis of indoleamine 2,3-dioxygenase, which cleaves the 2,3-double bond in the indole ring of tryptophan, effectively depleting the cell of tryptophan and depriving bacteria of an important amino acid (35). To counter the host-mediated depletion of tryptophan, intracellular bacteria like *L. monocytogenes* are able to synthesize tryptophan from the PPP intermediate erythrose-4-phosphate (36). Synthesis proceeds via the chorismate pathway to anthranilate and, subsequently, to tryptophan by enzymes coded within the *trp* operon (*trpEGDCFBA*). While the ability to synthesize tryptophan can rescue some intracellular bacteria, bacteria that synthesize tryptophan but live predominantly extracellularly, such as *S. aureus* or *Streptococcus pneumoniae*, remain sensitive to host-mediated depletion of tryptophan (37, 38). It is unclear why bacteria that can synthesize tryptophan remain sensitive to its depletion; however, addition of exogenous tryptophan to the tissue culture media relieves this bacteriostatic condition. In summary, the PPP is important for maintaining bacterial redox homeostasis and biosynthesis during a host-pathogen interaction.

Tricarboxylic acid cycle (Krebs cycle) – As mentioned earlier, reductive evolution has strongly influenced central metabolism in Gram-positive bacteria, and the Krebs cycle is the most prominent example of this evolution (39–41). Many Gram-positive pathogens lack all or most of the Krebs cycle. Most *Streptococcus* spp. (*S. mutans* being an exception), *Enterococcus* spp., *Erysipelothrix rhusiopathiae*, *Mycoplasma* spp., and *Ureaplasma* spp. lack the Krebs cycle, which prevents the pyruvate-derived synthesis of three of the 13 biosynthetic intermediates [i.e., oxaloacetate, α-ketoglutarate (aka 2-oxoglutarate), and succinate/succinyl-CoA; (41) and Somerville, unpublished observations]. These three biosynthetic intermediates are critical for the *de novo* synthesis of many amino acids and porphyrins; for example, oxaloacetate is a precursor for biosynthesis of aspartate, asparagine, lysine, cysteine, threonine, isoleucine, and methionine; α-ketoglutarate is a precursor of glutamate, glutamine, arginine, and proline; and succinate is used in porphyrin biosynthesis. Thus, the evolutionary loss of Krebs cycle genes is reflected in the complex amino acid and vitamin requirements necessary for cultivation of these bacteria (42–44).

It is hypothesized that the Krebs cycle evolved from two amino acid biosynthetic pathways: one oxidative pathway and one reductive pathway (45). This metabolic arrangement allows the formation of a bifurcated pathway starting at pyruvate, with branches leading to succinate/succinyl-CoA and α-ketoglutarate. This bifurcated configuration is found in several Gram-positive pathogens; for example, *L. monocytogenes*, *Clostridium difficile*, and *Peptostreptococcus anaerobius* lack the genes for α-ketoglutarate dehydro-

genase, succinyl-CoA synthetase, and succinate dehydrogenase (46) and http://biocyc. org, while *Corynebacterium diphtheriae* lacks succinyl-CoA synthetase (this may be compensated for by a putative succinyl-CoA: coenzyme A transferase) (47). In these examples, bacteria have maintained the Krebs cycle in an incomplete format but one that still allows the generation of oxaloacetate, α-ketoglutarate, and succinate. In addition, these bacteria use anaerobic respiration to oxidize NADH and FADH by running part of the Krebs cycle backwards (i.e., oxaloacetate to succinate). The use of anaerobic respiration also underscores the fact that most Gram-positive pathogens using this bifurcated Krebs cycle are anaerobes, *L. monocytogenes* being the exception. That said, some facultative anaerobes (e.g., *B. subtilis*) with a complete Krebs cycle also bifurcate the pathway when grown anaerobically (48).

Though an incomplete Krebs cycle is common in Gram-positive bacteria, two of the most prevalent Gram-positive pathogens worldwide have complete Krebs cycles: namely, *M. tuberculosis* and *S. aureus*. In both of these bacteria, the Krebs cycle is important for pathogenesis (49–51); however, having complete Krebs cycles and being important for pathogenesis is where the similarities end. One major difference between *M. tuberculosis* and *S. aureus* is that *M. tuberculosis* has a glyoxylate cycle, which allows the conservation of carbon by bypassing the Krebs cycle decarboxylation reactions (52). In Gram-positive bacteria, the glyoxylate cycle is primarily restricted to Actinobacteria (i.e., *Mycobacterium* sp., *Nocardia* sp., and *Rhodococcus* sp.); however, it is also found in the Firmicute *B. anthracis*. The presence of the glyoxylate cycle in Actinobacteria is likely a reflection of the poor nutrient environment they encounter when residing inside of a phagocytic cell or when walled-off in a granuloma (53–55). A second major difference between *M. tuberculosis* and *S. aureus* is that staphylococci exhibit carbon catabolite repression of the

Krebs cycle when cultivated in media containing glucose (56, 57). This does not appear to be the case with *M. tuberculosis* (58), which grows best on non-glucose carbon sources like glycerol, acetate, and fatty acids that are degraded to acetyl CoA (59). The utilization of acetate explains why *M. tuberculosis* uses the glyoxylate shunt. Doing so prevents the formation of a futile cycle in which two carbons enter the Krebs cycle and two carbons are lost through decarboxylation reactions. For the remainder of this chapter, discussion of the Krebs cycle will be kept to the Firmicutes.

In general, Gram-positive bacteria repress transcription of Krebs cycle genes when cultivated in media containing a readily catabolizable carbohydrate and glutamate or glutamine (46, 60–63). This catabolite repression leads to the accumulation of incompletely oxidized metabolites/fermentation products in the culture media, most commonly acetic acid and lactic acid (61, 64). Once carbohydrates are depleted from the medium, these metabolites can be re-imported and used to fuel the Krebs cycle and generate the three biosynthetic intermediates. Catabolism of acetate begins with the ATP-dependent formation of a thioester bond between acetate and coenzyme A catalyzed by acetyl-CoA synthetase/acetyl-CoA ligase. At this point, acetyl-CoA can enter into the Krebs cycle via a condensation reaction with oxaloacetate that is catalyzed by citrate synthase, a process using the energy of thioester hydrolysis to drive carbon-carbon bond formation to form citric acid. As stated above, most Gram-positive pathogens lack the glyoxylate shunt; hence, two carbons are lost as CO_2 for every two carbons (i.e., acetyl-CoA) that enter the Krebs cycle. For this reason, when biosynthetic intermediates are withdrawn from the Krebs cycle for biosynthesis, anaplerotic reactions are required to maintain carbon flow through the Krebs cycle. The most commonly used substrates for the anaplerotic reactions are amino acids (50). For instance, conversion of aspartate to oxaloacetate can start a new round of the Krebs cycle,

allowing continued drawing off of intermediates. In total, catabolism of incompletely oxidized metabolites through the Krebs cycle provides biosynthetic intermediates (i.e., α-ketoglutarate, succinyl-CoA, and oxaloacetate), ATP, and reducing potential but consumes amino acids in the process.

Not only do genetic variation and catabolite repression of the Krebs cycle affect the availability of biosynthetic intermediates and ATP in Gram-positive bacteria, but Krebs cycle activity can also be altered by environmental changes (11, 60, 65–67). Like glycolysis or the PPP, altering carbon flow through the Krebs cycle will affect the intracellular concentrations of biosynthetic intermediates and precursors, ATP, and redox homeostasis. Of importance, the activity of metabolite-responsive regulators is controlled by intracellular concentrations of biosynthetic intermediates (68), amino acids (69), nucleic acids (70), and cofactors (e.g., iron) (71). In other words, altering metabolism provides a means to transduce external environmental changes into internal metabolic signals that alter the activity of metabolite-responsive regulators, which facilitate adaptation to the altered environment (72). The function of metabolite-responsive regulators will be discussed in the second part of this chapter.

Amino acid biosynthesis – As previously discussed for tryptophan, amino acids can be important factors in the host-pathogen interaction, and two of the more important amino acids for bacteria are glutamate and glutamine. These amino acids are important because they serve as the nitrogen donors in most biosynthetic processes (73). Synthesis of glutamate is dependent on the nutritional environment (74) and usually involves one of two enzymes: glutamate dehydrogenase or glutamine oxoglutarate aminotransferase (aka GOGAT or glutamate synthase). Glutamate dehydrogenase catalyzes the reductive amination of the Krebs cycle intermediate α-ketoglutarate by using the oxidation of NADH to drive the assimilation of ammonia. Glutamate synthase converts glutamine and α-ketoglutarate into two molecules of gluta-mate by a transamidation reaction using NADPH/H[+] as a reductant (75). While gluta-mate is the nitrogen donor for most amino acids (exceptions being asparagine and tryp-tophan and histidine, which uses both gluta-mate and glutamine), glutamine also has an essential function as a nitrogen donor for the synthesis of tryptophan, amino sugars, and nucleic acids (73). Glutamine is synthesized by the condensation of glutamate and NH_3 by glutamine synthetase, using the energy of ATP hydrolysis to catalyze the reaction. In all, these three enzymes and the reactions that they catalyze constitute the primary means of nitrogen assimilation in bacteria.

The assimilation of nitrogen into gluta-mate, and to a lesser extent, glutamine, allows transamination reactions that transfer amino groups from glutamate to an amino acid pre-cursor. While all amino acid biosynthesis is important, we briefly describe branched-chain amino acid (BCAA) biosynthesis be-cause the valine pathway also leads to synthesis of pantothenate and because of the importance of BCAA in modulating the activity of the Gram-positive metabolite-responsive regulator CodY (76). Like central metabolism, BCAA biosynthesis has been strongly influenced by reductive evolution. For example, *E. rhusiopathiae*, *Mycoplasma* spp., *Ureaplasma* spp., *P. anaerobius*, *S. pyo-genes*, and *S. agalactiae*, lack all or most, of the genes necessary for the *de novo* synthesis of isoleucine, leucine, and valine, specifically, the *ilv-leu* operon (http://biocyc.org). Obvi-ously, the loss of BCAA biosynthetic genes creates auxotrophies for isoleucine, leucine, and valine, but this also means that panto-thenate and coenzyme A biosynthesis are dependent on exogenous sources of valine. For Gram-positive bacteria that have BCAA biosynthetic pathways (e.g., *Staphylococcus* sp., *Bacillus* sp.), synthesis of valine and leucine begins with one of the 13 biosynthetic intermediates of central metabolism; namely, pyruvate. Acetolactate synthase (*ilvB*) cata-lyzes the thiamine pyrophosphate-dependent

conversion of two pyruvate molecules into acetolactate, which is the substrate for the ketol-acid reductoisomerase encoded by *ilvC*. The products of ketol-acid reductoisomerase are oxidized NADP[+] and 2,3-dihydroxy-isovalerate, the substrate for dihydroxy-acid dehydratase (*ilvD*) (77). Dihydroxy-acid dehydratase produces 2-oxoisovalerate, a key intermediate that sits at a branch point between valine and leucine biosynthesis (78). From this branch point, leucine is synthe-sized by the enzymes coded within the *leuABCD* cluster, and 2-oxoisovalerate is converted to valine by a branched-chain amino acid aminotransferase (e.g., YbgE or YwaA in *B. subtilis*) using the amino group of glutamate as the nitrogen donor.

In contrast to leucine and valine biosyn-thesis, isoleucine synthesis begins by conden-sation of pyruvate and 2-oxobutyrate (79). 2-oxobutyrate is not itself one of the canonical 13 precursors but is made from threonine by threonine dehydratase (*ilvA*)-catalyzed deam-ination; threonine is made from aspartate, a product of transamination of oxaloacetate, thereby connecting isoleucine biosynthesis to central metabolism. From 2-oxobutyrate, the enzymes that catalyze the synthesis of isoleucine are the same ones that catalyze the synthesis of valine; namely, acetolactate synthase (*ilvB*), ketol-acid reductoisomerase (*ilvC*), dihydroxy-acid dehydratase (*ilvD*), and branched-chain amino acid aminotransferases (*ybgE* and *ywaA*).

All biosynthetic reactions in a bacterium are dependent on the biosynthetic inter-mediates produced by central metabolism, and BCAA biosynthesis is no different (Fig. 1). Any perturbation of central metabolism has the potential to disrupt biosynthetic reactions like BCAA biosynthesis (21, 80, 81), and these disruptions affect the polymerizing and as-sembly reactions. As mentioned earlier, iron limitation creates metabolic blocks in the Krebs cycle that limit the availability of Krebs cycle intermediates (i.e., α-ketoglutarate, suc-cinate and oxaloacetate). This decreased availability of intermediates alters the synthesis of amino acids, such as aspartate, which reduces the synthesis of threonine and BCAA synthesis. Not only is BCAA synthesis limited by the availability of intermediates during iron-limited growth, but the dihydroxy-acid dehydratase (IlvD) contains a [4Fe-4S] iron-sulfur cluster, which is susceptible to inactivation by iron-limitation or oxidative inactivation. When IlvD is inactive, the metabolic block in BCAA biosynthesis induces BCAA auxotrophy. In this example, the common cofactor requirements and the interconnections of metabolism create a "ripple effect" that cause metabolic changes seemingly unrelated to the nature of the perturbation. The severity of this ripple effect is determined by the extent of the perturbation and the availability of exogenous metabolites that can compensate for the loss of biosynthetic intermediates and precursors. To overcome these perturbations, bacteria have evolved/acquired metabolite-responsive regulators that facilitate adaptation and survival.

METABOLITE-RESPONSIVE GLOBAL REGULATORS THAT INFLUENCE VIRULENCE

Regulatory proteins that coordinately control metabolic and virulence genes provide compelling evidence that, from the bacterium's point of view, virulence is one of many interrelated responses to specific environmental conditions. Such conditions may reflect nutrient availability, temperature, pH, oxidative stress, osmotic stress, or other stresses that bacteria may encounter within and outside of host environments. In Gram-positive bacteria, several such global regulatory proteins have been studied in detail. All of these regulators are found predominantly, if not exclusively, in Gram-positive species.

CcpA

Acting as a global regulator of carbon metabolism genes in response to the availability of certain preferred carbon sources, CcpA regulates dozens of metabolism and virulence genes in *Bacillus* (82, 83), *Clostridium* (84, 85), *Staphylococcus* (86, 87), *Streptococcus* (88–91), *Lactococcus* (92), and *Enterococcus* (93, 94). (It is important to note that not all pathways for metabolism of sugars or other carbon sources are under CcpA control, even though some of these pathways are affected by the availability of glucose and other rapidly metabolized sugars (95). Moreover, inducer exclusion is a major component of carbon catabolite repression in most bacteria (96).) CcpA proteins are members of the LacI family and bind to a DNA sequence (*cre* site) that is conserved among the various species. *B. subtilis* CcpA, the group member studied in the greatest detail, is activated as a DNA-binding protein by interaction with a phosphorylated form of HPr (97). HPr has two potential sites of phosphorylation. When phosphorylated on histidine-15 by EI of the PTS system, HPr is specifically involved in transferring phosphate to PTS EII. Serine-46, however, is phosphorylated by HPr kinase/phosphorylase, whose kinase activity is activated by binding of FBP. It is the HPr-Ser$_{46}$~P form that binds to CcpA and increases its activity as a DNA-binding protein (98). Moreover, HPr-His$_{15}$~P does not activate CcpA, and HPr-Ser$_{46}$~P is inactive in the PTS. Since accumulation of FBP reflects the availability of glucose and other rapidly metabolized sugars, such sugars activate CcpA indirectly via HPr. In most other Gram-positive bacteria, the same complex of CcpA and HPr-Ser$_{46}$~P is the active form of the regulator. For instance, in *S. mutans*, increased phosphorylation of HPr at serine-46 inhibits sugar uptake, presumably by preventing phosphorylation of HPr at histidine-15, thereby interrupting the PTS signaling pathway (99). In contrast, *C. difficile* CcpA appears to interact directly with FBP, bypassing the need for HPr (100).

CcpA activity responds to other signals as well. Phosphorylation of CcpA at two threonine residues by the *S. aureus* kinase

PknB leads to loss of CcpA binding to *cre* sites and disruption of the CcpA regulon, as well as overexpression of the *ccpA* gene (101). The signals activating PknB are not known.

Some clinical isolates of *S. aureus* appear to be proline auxotrophs; mutations in the major proline transporter cause loss of virulence in several models of pathogenesis (102, 103). Mutations in *ccpA* or *ptsH* relieve the auxotrophy, suggesting that the auxotrophy is due to severe repression of the proline biosynthetic pathway by CcpA (104). Because *S. aureus* does not encode the conventional glutamate-to-proline pathway, proline can be made in these bacteria only as a by-product of arginine degradation (104). Arginine degradation is strongly repressed by CcpA/HPr when cells are grown with glucose (104), explaining the apparent proline auxotrophy of glucose-grown cells.

The role of CcpA in virulence gene expression

Streptococcus

Multiple effects of CcpA on virulence gene expression have been observed in *Streptococcus* spp. A *S. pneumoniae ccpA* mutant shows reduced expression of the capsular polysaccharide locus (105) and is highly attenuated for mouse nasopharyngeal and lung infection (106). Indirect evidence that *S. suis* CcpA controls capsule biosynthesis came from analysis of the effect of a mutation in a virulence-associated surface protein on HPr phosphorylation at serine-46 (107). In *S. mutans*, the major cause of dental caries, a *ccpA* mutant produces more acid than its parent, grows better at low pH, and excretes acid more rapidly, indicating that CcpA normally holds the cariogenic potential of *S. mutans* in check when rapidly metabolizable carbon sources are in excess (108). Two other oral streptococci, *S. gordonii* and *S. sanguis*, help prevent caries by producing H_2O_2, an antagonist of other oral bacteria, including *S. mutans*, in response to carbon

limitation and changes in oxygen tension. This H_2O_2 is produced by pyruvate oxidase, and the gene coding for pyruvate oxidase (*spx* or *pox*) is repressed by CcpA (109, 110). Thus, when rapidly metabolizable carbon sources become limiting, repression by CcpA of both caries-promoting genes and *S. mutans*-antagonizing factors is relieved. Autolysis of *S. mutans*, a factor in survival of biofilms, is also regulated by CcpA in response to the availability of glucose (111). In Group A *Streptococcus*, CcpA appears to activate transcription of *mga*, the gene that encodes a positive regulator of genes whose products mediate adhesion and invasion of host cells, as well as resistance to host defenses (112).

Clostridium

The major *C. difficile* toxins, TcdA and TcdB, are synthesized during the stationary phase in rich media, as long as the media do not contain glucose or other rapidly metabolizable carbon sources (113, 114). CcpA mediates the glucose-dependent repression of the major toxin locus indirectly by binding to the promoter region of the *tcdR* gene, which encodes the alternative sigma factor necessary for high-level toxin gene transcription (Fig. 2) (84). CcpA also represses many metabolic genes whose products may be important for growth in the GI tract (discussed later). The CcpA protein

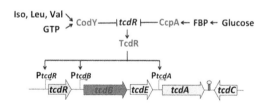

FIGURE 2 Synergistic repression of *C. difficile* toxin synthesis by CodY and CcpA. Responding independently to different nutritional signals, CcpA and CodY both bind to the regulatory region of the *tcdR* gene, repressing production of the alternative sigma factor necessary for high-level toxin gene (*tcdA* and *tcdB*) expression. doi:10.1128/microbiolspec.MBP-0004-2014.f2

of *C. perfringens* is responsible for the glucose-dependent repression of two major toxins, alpha-toxin (phospholipase C) and theta-toxin (perfringolysin), which initiate gas gangrene (115). The inhibitory effect of glucose on type IV pilus-dependent gliding motility is also mediated by CcpA (116). Gliding motility is required for efficient biofilm formation, a process that increases the bacterium's resistance to antibiotics and other environmental stresses (117). In contrast, production of the enterotoxin responsible for *C. perfringens* food poisoning is positively regulated by CcpA (118). Whether that effect is direct or indirect is uncertain.

Staphylococcus

Interestingly, an *S. aureus ccpA* mutant shows decreased replication in the liver in a murine abscess model (104). This effect may be due in part to indirect positive regulation of immunodominant antigen B, α-hemolysin, and biofilm formation in the presence of glucose (119–121). On the other hand, synthesis of toxic shock syndrome toxin-1 and capsule is repressed by glucose via CcpA (121, 122).

Bacillus

Toxin gene expression in *B. anthracis* is subject to complex regulation involving an activator of toxin gene expression, AtxA, and multiple indirect regulators. CcpA, for instance, is needed for glucose-dependent induction of *atxA* transcription (123). Thus, a *ccpA* mutant is attenuated in virulence.

Other genera

In *E. faecalis*, CcpA activates expression of a collagen-binding colonization factor (124). In *Listeria monocytogenes*, effects of carbon availability on virulence gene expression are mediated by HPr independently of CcpA. Hyperphosphorylation of serine-46 leads to reduced activity of the global virulence gene regulator PrfA and, as a consequence, reduced expression of PrfA-dependent genes (125). This effect is not mediated by CcpA; instead, hyperphosphorylation of serine-46

interferes with phosphorylation of histidine-15, thereby inhibiting PTS-dependent transport, implying that sugar transporters or imported sugars are the direct modulators of PrfA activity (125, 126).

CodY

A global regulator of metabolism found in nearly all low G+C Gram-positive bacteria, CodY controls at least 200 genes in *Bacillus* (127), *Clostridium* (128), *Staphylococcus* (129, 130), *Streptococcus* (131, 132), *Lactococcus* (133), and *Listeria* (134). *Lactobacillus* spp. typically lack CodY, as do all high G+C Gram-positive bacteria. CodY activity is controlled by two types of ligands. The branched-chain amino acids (BCAAs) isoleucine, leucine, and valine are universal effectors of CodY activity (69, 133, 135), while GTP also activates CodY (136–138) in genera other than *Streptococcus* (139) and *Lactococcus* (140). The metabolic genes regulated by CodY, directly or indirectly, include those for amino acid biosynthetic pathways (typically isoleucine, leucine, valine, threonine, arginine, glutamate, and histidine), purine biosynthesis (particularly the steps from IMP to GMP), the Krebs cycle, energy metabolism, sugar and amino acid transport, carbon overflow metabolism, chemotaxis and motility, genetic competence, and sporulation. In all cases studied to date, most metabolism genes under CodY control are repressed by CodY; the major exceptions are carbon overflow pathways, which tend to be activated by CodY. In general, the genes repressed by CodY are expressed at low level during growth in rich medium until the cells reach the stationary phase but are expressed at high level during growth in a poor medium; conversely, genes whose transcription is activated by CodY are expressed during rapid growth.

The CodY ligands interact with the protein in different ways that are only partially understood. The BCAAs bind to an N-terminal GAF domain that is unstructured in the unliganded protein but folds around the

BCAA side-chain, forming a hydrophobic pocket (141). Binding of the BCAA causes a conformational change in which the C-terminal winged helix-turn-helix motifs of the CodY dimer separate in a way that facilitates DNA binding. GTP also activates CodY as a DNA-binding protein but does so by an unknown mechanism. Whereas the BCAAs and GTP each activate CodY independently, the combination of ligands has a synergistic effect.

The bacterium's rationale for using two unrelated types of effectors to modulate the activity of a single global regulator can only be surmised. One possibility is that the GTP pool is not only an indicator of ability to synthesize purine nucleotides, but is also a measure of protein synthetic capacity. That is, in addition to being a substrate for RNA synthesis, GTP is utilized as an energy source during protein synthesis. Perhaps most important, the GTP pool in Gram-positive bacteria is strongly affected by amino acid limitation. Limitation of any amino acid leads to accumulation of uncharged tRNA, a molecule sensed by the ribosome-bound stringency factor known as RelA or RSH. The RelA-catalyzed conversion of GTP to pppGpp reduces the GTP pool, but probably not enough to have a large impact on CodY activity. Instead, the major effect of pppGpp in *B. subtilis* and, presumably, other Gram-positive bacteria is to inhibit enzymes of GTP synthesis (142, 143), thereby linking CodY activity to availability of all amino acids. In fact, inducing the stringent response in *B. subtilis* or *S. aureus* leads to altered expression of most genes of the CodY regulon (143, 144). BCAAs are also abundant amino acids in proteins, but their special nature is that they are readily interconvertible with the branched-chain keto acids, which serve as precursors of pantothenate and coenzyme A, as described previously, and branched-chain fatty acids. In many Gram-positive bacteria, the branched-chain fatty acids are the predominant membrane fatty acids under normal growth conditions. Thus, by sensing the availability of BCAAs, CodY is monitoring the cell's capacity to carry out essential reactions, including membrane biosynthesis.

In all species studied to date, the sites of CodY binding share a common 15-bp palindromic sequence, AATTTTCNGAAAATT (the "CodY box") (139, 145); specific nucleotides within this sequence are essential for CodY binding (146), but the exact relationship between the 15-bp palindrome and affinity of CodY for DNA is not fully clear. It is very rare for any bacterium to have within its genome even one CodY box that matches the consensus sequence perfectly. When an imperfect *B. subtilis* CodY box was converted by mutation to a perfect box, the gene controlled by that site was barely expressed at all, even under conditions of very low CodY activity (147). Thus, it is not in the cell's interest to have CodY boxes without mismatches. In fact, many sites of moderate-to-high affinity have 3-5 mismatches. In many of the latter cases, two overlapping copies of the palindromic sequence, each with multiple mismatches, serve as the binding site, suggesting that the protein can interact with DNA in multiple ways (148, 149).

CodY affects gene expression by multiple molecular mechanisms. As a positive regulator, CodY binds upstream of the promoter and, presumably, stabilizes the binding of RNA polymerase to promoters that are intrinsically weak (150). Like traditional repressors, CodY can inhibit transcription initiation by binding within or just downstream of a target promoter region (147), but CodY can also block transcription elongation by binding within a coding sequence, in some cases as far as 600 bp downstream of the start codon (149, 151). In at least one case, CodY acts as a negative regulator by inhibiting the binding of a positive regulator (152).

The Role of CodY in Virulence Gene Expression

CodY acts as a negative regulator of virulence in *C. difficile* and *S. aureus* but as a positive regulator of virulence in *B. anthracis, B.*

cereus, and *L. monocytogenes*. In other species, CodY seems to act as an activator of some types of virulence genes and a repressor of others. The implication is that Gram-positive bacteria use CodY to regulate metabolism in a well-conserved manner but manipulate CodY to regulate virulence according to their niches within hosts and in accordance with whether they are virulent when growing rapidly or slowly. Moreover, depending on the environmental conditions in which a bacterium finds itself, it may be advantageous to activate some types of virulence genes while repressing others.

Clostridium

Proven virulence genes of *C. difficile* are restricted to the two major toxins that cause the inflammatory response and diarrheal disease characteristic of infection. It must be the case, however, that as-yet-unidentified colonization factors and defense mechanisms play critical roles in the infection process. CodY regulates many genes that may fit these latter categories, but proof of function for the target genes is lacking. Toxin synthesis is responsive to many nutrients, including glucose (noted previously) and BCAAs (114, 153). The latter effect raised the possibility that CodY functions in toxin gene regulation. In fact, CodY, like CcpA, represses the toxin genes indirectly by binding to sites upstream of *tcdR*, the gene that encodes the major sigma factor for toxin gene transcription (Figure 2). Consistent with CodY-dependent regulation, toxin gene transcription in bacteria growing in rich medium increases dramatically during the post-exponential phase, when CodY repression would be relieved (113). In a *codY* null mutant, expression of the toxin locus during the exponential growth phase is 50-fold higher than in wild-type cells (154), and toxin protein released into the culture fluid is 100-fold higher in the mutant strain after 24 hours of growth (Bouillaut, Sun, Tzipori, and Sonenshein, unpublished). Given these results, it is not surprising that a *codY* null

mutant is hypervirulent compared with its parent strain (Bouillaut et al, unpublished).

The effect of CodY in *C. perfringens* is more subtle. A *codY* mutant of a type D strain shows reduced expression of enterotoxin (ETX), reduced sialidase (i.e., NanJ) in the culture fluid, and a reduced ability to attach to Caco-2 cells (155). By contrast, inactivation of CodY has no effect on production of phospholipase C or perfringolysin; however, it led to overexpression of NanH and increased efficiency of sporulation in a rich medium. CodY appears to bind to the *etx* promoter region, but other direct targets of CodY have not yet been defined (155).

Staphylococcus

A *codY* null mutant of *S. aureus* USA300 (the CA-MRSA prototype) is hypervirulent, correlating with increased transcription and protein accumulation for hemolysins, leukocidins, and proteases (156, 157). The situation in *S. aureus* is complex, however, because dozens of virulence-associated genes are known, and CodY directly or indirectly regulates most of them (129, 130, 157, 158). As reported by Novick (159), *S. aureus* virulence genes can be separated into two groups depending on whether their products are involved in colonization or serve as toxins, lipases, proteases, capsule-forming, biofilm-forming, or tissue-damaging proteins. Colonization genes are expressed in the laboratory during the exponential phase in rich medium, whereas the latter group of genes is expressed during the postexponential phase. CodY controls genes from both groups, activating the transcription of some growth and colonization genes (e.g., *fnbA*) and repressing the transcription of toxins (i.e., *hla, hlb, hld, lukSF, tst-1, sak*), capsule, biofilm (*ica*), lipase, nuclease, protease, and ROS detoxifying genes (129, 130, 156, 157). In some of these cases, CodY binds directly to the target gene region, but in other cases, it works indirectly through its repressive effect on expression of the quorum-sensing Agr system and the regulatory RNA, RNAIII

(129). CodY also controls the *agr* locus indirectly by a mechanism that is not yet known (252). Moreover, the regulation may be reciprocal, as an *S. epidermidis agr* mutant showed reduced expression of *codY* and derepression of CodY-repressed genes (160).

The expression of many Staphylococcal virulence genes (e.g., fibronectin-binding protein, epidermin immunity proteins, enterotoxin B) is altered when the stringent response is induced by limitation of amino acids (leucine or valine) (144) or treatment with mupirocin, an inhibitor of isoleucyl-tRNA synthetase (161). Most of the affected genes are regulated by either CodY or the alternative sigma factor, σ^B. In addition, the expression of several virulence gene regulators is induced by stringency, including ArlRS, SarA, SarR, SarS, and σ^B (161). Like many other bacteria, *S. aureus* has several pppGpp-synthesizing enzymes but only one synthetase (RelA/RSH), which also has pppGpp hydrolase activity. A *relA/rsh* null mutant of *S. aureus* is not viable, presumably because it is unable to prevent blockage of GTP synthesis, but a mutant defective only in the synthetase function can be studied. Such a mutant is less virulent than its parent strain, but the defect can be suppressed by introducing a *codY* mutation, implying that CodY is hyperactive as a repressor of virulence in the *relA/rsh* mutant (144).

It is important to note that the first reported phenotype caused by an *S. aureus codY* null mutation was reduced biofilm formation (162); all subsequent reports based on mutations in other strains showed substantial increases in biofilm gene expression and biofilm formation (158). Thus, one must be open to the possibility that global regulators can be used by different species and different clones within species in different ways that allow adaptation to particular environmental conditions.

Bacillus

B. anthracis CodY promotes virulence by increasing the abundance of AtxA, the positive regulator of toxin genes; a *codY* null mutant shows greatly decreased virulence (163, 164). Again, this effect is indirect and appears to reflect repression by CodY of a gene that encodes a protease that degrades AtxA. A *codY* mutant of *B. cereus*, a close relative of *B. anthracis*, is also attenuated for virulence (165). In strain ATCC 14579, the reduced virulence phenotype could be explained by reduced expression of cytotoxin K, non-hemolytic enterotoxin and hemolysin BL (166). In the emetic strain F4810/72, CodY also appeared to activate expression of non-hemolytic enterotoxin as well as phospholipases and immune-inhibitor metalloprotease 1 (165). In the same strain, CodY represses indirectly the regulon controlled by the PlcR-PapR quorum-sensing system (165). CodY also appears to be an activator of biofilm formation in strain UW101C (167) but seems to repress biofilm formation in strain ATCC 14579 (166).

Listeria

In *L. monocytogenes*, CodY was initially postulated to be a negative regulator of virulence because a *codY* mutation partially restored virulence to a *relA* mutant (134). This conclusion was consistent with the idea that a *relA* mutant would be expected to have a higher-than-normal GTP pool and, as a result, accumulate more active CodY, leading to hyperrepression of CodY target genes. In subsequent work, CodY was shown to be a positive regulator of *prfA*, the gene that encodes a positive regulator of key virulence genes, such as those for listeriolysin and actin-polymerizing protein (168). When a global regulator can be a positive regulator of some genes and a negative regulator of others, it is sometimes inadvisable to draw too many conclusions about a complex phenomenon such as virulence based on the phenotype of a null mutant.

Streptococcus

In *S. pyogenes*, the first pathogen in which a *codY* mutant was isolated (169), the role of

CodY is complex. By repressing some virulence genes and activating others, CodY appears to be used to express different classes of virulence genes under different environmental conditions. In part, this complexity may be accounted for by the effect of CodY on expression of the CovRS two-component regulatory system. That is, CovR can be a positive or negative regulator of some virulence-associated genes, depending on environmental conditions. By repressing the *covRS* operon, CodY seems to work at cross-purposes with CovRS and can therefore appear to be a reciprocal activator or inhibitor of virulence (132). How this complicated arrangement plays out *in vivo* is uncertain, but during growth of *S. pyogenes* in human blood *in vitro*, many genes are subject to regulation by CodY but not necessarily in the same way as in laboratory growth media (131).

The situation in *S. pneumoniae* is also complicated, in this case because of conflicting reports on the effects of a *codY* mutation. As originally reported, a *codY* mutant was defective in colonization (139). A later report challenged these findings, claiming that a *codY* null mutation in *S. pneumoniae* is lethal; the mutants previously isolated were likely to have had compensatory mutations that allowed the *codY* mutant to survive (170). Any apparent effects of a *codY* mutation on virulence, therefore, may be due to the compensatory mutations rather than the *codY* mutation itself.

PrdR

Proline is among the nutrients that severely inhibit *C. difficile* toxin synthesis (153); this inhibition is mediated via proline reductase, a "Stickland reaction" enzyme (171). *C. difficile*, like several other *Clostridium* spp., has the ability to cometabolize proline or glycine and certain other amino acids (172). That is, oxidative metabolism of isoleucine, leucine, valine, or alanine generates NADH and ATP; the NADH can then be used to reduce

proline, producing 5-aminovalerate, or glycine, generating acetate and more ATP (Fig. 3). The proline reductase and glycine reductase selenoenzyme complexes are encoded in multigene operons (173). Transcription of the proline reductase operon, which has a σ^{54}-type promoter, and inhibition of toxin synthesis by proline depend on PrdR, a σ^{54}-activating regulatory protein encoded just upstream of the proline reductase genes (171). Addition of proline to rich medium has broad effects on metabolism in *C. difficile*; these effects are mostly related to reductive pathways and are dependent on PrdR, implying that PrdR is a global regulator (Bouillaut, Dubois, Monot, Dupuy, and Sonenshein, manuscript in preparation).

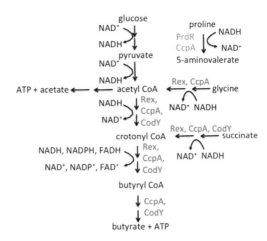

FIGURE 3 **Examples of oxidative and reductive metabolic pathways in *C. difficile*. NADH produced during glycolysis and other oxidative pathways is converted back to NAD⁺ by a series of reductive pathways. The proline pathway, catalyzed by proline reductase, appears to be the favored pathway. When proline is available, the other pathways shown are repressed by Rex. Repression by Rex is relieved when the ratio of NAD⁺ to NADH indicates the need for increased regeneration of NAD⁺. Additional repression by CcpA and CodY restricts maximal expression of the alternative pathways to conditions in which CcpA and CodY are relatively inactive. The bottom three pathways (effectively, acetyl-CoA to butyrate) are encoded in a single eight-gene operon. doi:10.1128/microbiolspec.MBP-0004-2014.f3**

This unexpected observation raises important questions, however: Is PrdR a global regulator or a specific regulator of the proline reductase operon? If the latter, why and how would *C. difficile* use proline availability and a proline-sensing regulator to control so many diverse pathways, including toxin production? The answer seems to be that proline has a favored role in NAD^+ regeneration (see next section on Rex).

Rex

First discovered in *Streptomyces coelicolor* (174), Rex is a global regulator of genes whose products interconvert NADH and NAD^+. Rex is found in virtually all low G+C Gram-positive bacteria. Ironically, *Streptomyces* is the only high G+C Gram-positive genus that encodes Rex. By monitoring the ratio of reduced to oxidized forms of nicotinamide nucleotides, Rex helps to regulate pathways that regenerate NAD^+ in cells that are actively oxidizing carbon-energy sources. For example, glycolysis, the Krebs cycle and the pentose phosphate pathway convert NAD^+ and $NADP^+$ to NADH and NADPH. Without adequate regeneration of oxidized co-factors, these metabolic pathways would come to a halt. Bacteria have three major ways of regenerating $NAD^+/NADP^+$: (i) NADH/NADPH-dependent biosynthesis; (ii) carbon overflow pathways (e.g., lactate dehydrogenase); and (iii) respiration. In some bacteria, respiration-independent NADH oxidase can also serve. Rex typically represses carbon overflow pathways and reductive pathways that are a form of respiration when the pool of NAD^+ is adequate. Examples of Rex targets in *S. coelicolor, S. aureus, C. acetobutylicum, B. subtilis,* or *E. faecalis* include lactate dehydrogenase (LDH), lactate permease, alcohol dehydrogenase, butyryl-CoA synthesis, crotonase, acetyl-CoA acetyltransferase, pyruvate formate lyase, pyruvate oxidase, NADH dehydrogenase, and cytochrome bd (175–180).

Binding of Rex to its target sites (the well-conserved sequence (TTGTGAAa/t$_6$TTCACAA) is inhibited by NADH; thus, Rex is active as a repressor only when the $NAD^+/NADH$ ratio indicates adequate NAD^+ (174). Binding of *S. aureus* and *C. difficile* Rex proteins to their target sites is stimulated by NAD^+; NADH competes with NAD^+ for interaction with Rex, thereby inhibiting DNA binding [(178) and L. Bouillaut and A. L. Sonenshein, unpublished]. *Thermus aquaticus* Rex has served as the model for structural studies of Rex-$NAD^+/NADH$ interactions. When NADH binds, a conformational change ensues that causes Rex to lose its ability to interact with DNA (181).

In *E. faecalis*, deletion of the *ldh* gene causes the carbon overflow metabolism to be redirected toward acetoin, ethanol, and formate and, at the same time, causes changes in expression of many genes that have apparent Rex binding sites upstream (179). The implication is that LDH is the major regenerator of NAD^+; in the absence of LDH, NADH accumulates, leading to derepression of Rex-repressed genes. A similar case of prioritization of NAD^+-regenerating pathways is seen in *C. difficile* (discussed later).

The role of Rex in virulence gene regulation

Staphylococcus

The relationship between pathogenesis and the maintenance of an appropriate balance of reduced and oxidized NAD/NADP is not yet clear, but in some bacteria, metabolic pathways under Rex control are implicated in virulence. For instance, in *S. aureus*, Rex is intimately involved in survival of cells exposed to nitric oxide (NO). NO inhibits the activities of terminal respiratory oxidases, nitrate reductase, and pyruvate formate lyase, leaving lactate dehydrogenase, a target of Rex, as the major means of regenerating NAD^+ (178). The known connection between LDH activity and resistance to host innate immunity means that modulating Rex function appropriately is a critical factor in pathogenesis (182).

Clostridium

In *C. difficile*, proline reductase appears to be the preferred pathway for NAD^+ regeneration. If proline is limiting in the growth medium or if proline reductase or PrdR is inactive, alternative pathways for NAD^+ regeneration [e.g., glycine reductase, alcohol dehydrogenase, succinate to crotonyl CoA, crotonyl CoA to butyrate (183, 184)] are induced (Bouillaut et al, manuscript in preparation). These alternative pathways (Fig. 3) are repressed by Rex in wild-type cells when proline is available but not in proline reductase or *prdR* mutants (Bouillaut and Sonenshein, manuscript in preparation). Since these pathways produce butyrate, a stimulator of toxin synthesis, one can imagine that *C. difficile* withholds toxin synthesis when proline is in excess, because the cells have an adequate supply of NAD^+. If butyrate is being produced, the implication is that NAD^+ is limiting. Alternatively, by sensing butyrate, the cells may be primed to detect the presence of other butyrate-producing bacteria in the intestinal tract. How the cell communicates its NAD^+/NADH status to the toxin locus is unknown.

RpiR

The RpiR family of regulatory proteins controls carbohydrate metabolism in both Gram-negative and Gram-positive bacteria. First described as a regulator of ribose metabolism in *Escherichia coli* (185), members of the family have been shown to regulate *N*-acetylmuramic acid catabolism in *E. coli* (68), the pentose phosphate pathway in *Pseudomonas aeruginosa* (186), inositol catabolism in *Sinorhizobium meliloti* (187), maltose transport and metabolism in *B. subtilis* (188), and the pentose phosphate pathway in *S. aureus* (189). Family members typically have an apparent DNA binding domain near the N-terminus and a sugar isomerase binding domain near the C-terminus, suggesting that intermediates in pentose metabolism modulate the activity of at least some RpiR

proteins. *S. aureus* encodes three RpiR homologs, of which two, RpiRb and RpiRc, are involved in pentose regulation (189). An *rpiRc* mutant shows increased levels of the regulatory RNA, RNAIII, during exponential growth, and an *rpiRb* mutant demonstrates a similar phenotype during the postexponential phase (189). This phenomenon may be attributable to increased synthesis of the regulatory protein SarA. Moreover, an *rpiRc* mutation causes increased capsule formation and increased resistance to peroxide (189). The coregulation of pentose metabolism and virulence factors by RpiR proteins suggests strongly that RpiR mediates one of several mechanisms by which virulence and metabolism are coupled.

Interactions Among the Global Regulators

Although each of these global regulators monitors a different intracellular metabolite pool, they interact with each other and with operon-specific regulators in multiple ways. For instance, in *B. subtilis, ilvBHC leuABCD*, the major operon for biosynthesis of BCAAs, is repressed by CodY and TnrA (the global regulator of nitrogen metabolism) and activated by CcpA (190–192). Since TnrA is active only under conditions of glutamine (nitrogen) limitation, and CodY is activated when BCAAs are in excess; these two regulators combine to reduce *ilvB* operon transcription when the cell has enough BCAAs or is so depleted of nitrogen that it is best to limit its consumption. When leucine is specifically in excess, most transcription that initiates is aborted before the beginning of the first coding sequence by a T-box–mediated termination/antitermination system that responds to uncharged $tRNA^{leu}$ (193). When cells are in carbon excess, CcpA stimulates transcription of the *ilvB* operon, thereby making CodY more active as a repressor of many pathways that draw intermediates away from glycolysis. On the other hand, since both CcpA and CodY activate carbon overflow pathways when glucose is in excess, the increased CodY activity generated by CcpA accentuates this cooperation. Finally, the decrease in the GTP pool caused by the stringent response also affects *ilvB* operon expression in two ways (i.e., by reducing CodY activity and the pool of GTP, the initiating nucleotide for *ilvB* transcription (143, 194)). Thus, expression of the *ilvB* operon responds positively to the pool of FBP and negatively to the pools of the three BCAAs, GTP, and glutamine.

Regulation of the Krebs cycle can also be mediated cooperatively by CcpA and CodY. In *B. subtilis*, CcpA represses the genes for citrate synthase and isocitrate dehydrogenase, whereas CodY represses the gene for aconitase. A third regulator, CcpC, is Krebs cycle-specific and represses the genes for all three enzymes (195). None of these repressors is fully effective by itself; therefore, complete repression requires high intracellular pools of FBP, BCAAs, and GTP and low citrate, the effector of CcpC. (When citrate accumulates to high level, CcpC is inactivated as a repressor of citrate synthase and isocitrate dehydrogenase but switches from a repressor to an activator of the aconitase gene (196)). A similar regulatory scheme is found in *L. monocytogenes* (196). *S. aureus* also uses CcpA and CcpE (a homolog of CcpC), as well as RpiRc, to regulate the tricarboxylic acid branch of the Krebs cycle. In this case, CcpE acts as a positive regulator of aconitase but not citrate synthase expression and binds to the aconitase promoter region, but its binding is not affected by citrate (197). CcpA represses the synthesis of citrate synthase, aconitase, and isocitrate dehydrogenase in glucose-containing medium (87). Unlike the situation in *B. subtilis*, an *S. aureus codY* mutation does not cause derepression of any TCA branch enzymes, at least not under the growth conditions tested (129, 130); of the Krebs cycle enzymes, only α-ketoglutarate dehydrogenase is derepressed in a *codY* mutant (129).

As mentioned previously, *B. anthracis* CodY and CcpA are both positive regulators of AtxA accumulation and toxin synthesis, but in *C. difficile*, both CcpA and CodY repress *tcdR* and thereby block toxin synthesis. In addition, *C. difficile* toxin synthesis is inhibited when proline or cysteine is in excess. The direct or indirect mediator of the proline effect on toxin gene expression is PrdR. Thus, *C. difficile* does not become fully virulent unless deprived of a rapidly metabolizable carbon source and sources of BCAAs, proline, and cysteine. Nonetheless, deprivation of any one of these nutrient sources leads to significant toxin production.

NAD$^+$ regeneration is also subject to multiple overlapping controls. In *S. aureus*, a mutation in *ccpA* decreases expression of the LDH-1 gene, implying that CcpA is a positive regulator. But CcpA does not bind to the *ldh1* regulatory region, and its effect on *ldh1* expression depends on Rex. In contrast, Rex does bind to the *ldh1* promoter region and represses transcription (198).

The CodY, CcpA, and Rex proteins of *C. difficile* all control reductive metabolism of acetyl-CoA, succinate, and glycine to butyryl-CoA. (In addition, PrdR appears to be an indirect regulator of these pathways through its direct effect on proline-dependent NAD$^+$ regeneration, which was previously noted.) The three operons encoding these pathways appear to be direct targets of CodY and CcpA, as determined by *in vivo* and *in vitro* binding assays (84, 128). They also have apparent Rex binding sites (199). Moreover, the proline reductase operon is positively regulated by CcpA as well as by PrdR (84). One can speculate that these pathways play at least three roles: (i) They are designed to regenerate NAD$^+$, hence their regulation by Rex; (ii) they are ATP-generating pathways, justifying their repression by CcpA when glucose is in excess; and, (iii) as producers of butyrate, an activator of toxin gene expression, they work at cross-purposes with repression of toxin synthesis by CcpA and CodY. Proline reductase is a special case;

assuming it is the preferred pathway for NAD$^+$ regeneration, it makes sense that it is positively regulated by CcpA as a means of balancing the redox state during growth on glucose. In summary, it is apparently in the interest of *C. difficile* to minimize toxin synthesis unless multiple checkpoints have been reached (i.e., lack of rapidly metaboliz-able carbon sources, limitation of BCAAs, and an unfavorable ratio of NAD$^+$ to NADH). When all those conditions are met, butyrate can be made and can signal the cells to turn on toxin synthesis.

CodY and PrdR also appear to repress in partnership the utilization by *C. difficile* of ethanolamine as a carbon and nitrogen source (Bouillaut et al, manuscript in prepa-ration). Ethanolamine is a relatively abun-dant nutrient in the GI tract by virtue of its abundance in food and its release from host and bacterial membrane phosphatidyletha-nolamine by phosphodiesterase activity pro-duced by members of the normal microflora (200). The enzyme complex that deaminates ethanolamine and then converts the carbon skeleton to acetate while producing ATP by substrate-level phosphorylation is located in a microcompartment that appears to be necessary to avoid diffusion of the volatile intermediate acetaldehyde (201). The ability to metabolize ethanolamine may contribute to virulence by serving as a nutrient source or by acting as a signal to pathogens that they are in an appropriate environment for expressing virulence functions (202). It is well established that the *eut* genes of several bacterial species are induced during growth in the intestinal tract (202), but the evidence that ethanolamine metabolism is critical for pathogenesis is limited. A *eut* mutant strain of *L. monocytogenes* is defective in intracellular growth *in vitro* in Caco-2 cells (203), and an *E. faecalis eut* mutant strain is attenuated for virulence in a worm model of infection (204). No study has yet shown that the ability to metabolize ethanolamine is critical for successful infection of an animal.

METABOLISM AND HOST: PATHOGEN INTERACTIONS

It is becoming increasingly appreciated that host:pathogen interactions are not merely limited to physical contact between bacterial adhesins, toxins, and surface molecules with their cognate host cell receptors. Rather, the interconnected metabolic network encom-passing both the host and the microbe can also have measurable effects on disease outcomes. Activated phagocytes responding to invading microbes must acquire nutrients and generate energy in order to produce an array of inflammatory mediators (reactive oxygen/nitrogen species, and antimicrobial peptides) as well as signaling molecules (e.g., cytokines, chemokines, and lipid mediators). The concerted effects of all of these immune functions are essential for clearing infections and restoring tissue homeostasis. In fact, activated phagocytes have energy demands similar to those of muscle cells and neurons, making them some of the most energetically active cells in the body (205). The metabolic pathways used by immune cells to meet the energy demands of activation depend on the immune cell type, the nature of the pathogen, and features of the infected tissue. On the other hand, for an invading microbial pathogen to succeed, it too must be able to acquire nutrients and generate energy for the purpose of persisting long enough to be transmitted to a new host. While the Gram-positive pathogens highlighted in this chapter have extensive interconnected webs of met-abolic pathways available to them (reviewed in the first section), which ones are active and functional during various stages of infection are only recently becoming under-stood. The metabolic pathways used by a pathogen during infection are dictated not only by complex regulatory networks that sense both extra- and intra-cellular cues (reviewed in second section), but also by the compatibility of metabolic pathways with available nutrients and tissue environments, as well as the status of the host immune response. In this section, we will highlight some of the recent findings regarding the role of metabolism in the host:pathogen interaction, focusing both on the metabolic pathways used by the host as well as the diversity of pathways employed by various Gram-positive pathogens.

Host Immune Cell Metabolism During Infections

One of the first immune cells to infiltrate infected tissue is the polymorphonuclear monocyte (PMN or neutrophil). PMNs extravasate from the blood stream in response to a cytokine gradient generated from stromal and epithelial cells at sites of tissue damage/infection. PMNs are short-lived cells that become highly activated, engulf and kill infecting bacteria, and then undergo one of many forms of programmed cell death. They are by far the most abundant effector cell at sites of infection and are critical for rapid control of invading microbes. Another effector cell commonly found within infectious foci is the macrophage. Macrophages mature from circulating monocytes in the blood as they extravasate into infected tissue. Like PMNs, macrophages also engulf and kill invading bacteria, however these cells are generally much more long-lived than PMNs. This is due to the fact that macrophages serve additional roles beyond merely destroying microbes. First, macrophages (as well as dendritic cells) are efficient antigen-presenting cells (APCs), presenting processed antigen in the context of MHC to T-cells, thereby linking the innate and adaptive immune compartments. Furthermore, macrophages also play key roles in restoring tissue homeostasis as the infection resolves. Macrophages engulf and dispose of necrotic and apoptotic cell debris (such as dead neutrophils) and secrete proresolving cytokines and lipid mediators that signal to surrounding stromal and epithelial cells to restore tissue integrity (206). Depending on the state of the infection, the cytokine milieu, and the needs of

the surrounding tissue, macrophages adopt distinct phenotypes that allow them to fulfill their wide array of roles within infected tissue. These phenotypes have been broadly categorized into classically or alternatively activated macrophages, the former being associated with inflammation and bacterial clearance (207).

Classically activated macrophages (aka M1 macrophages) resemble activated PMNs in many ways. Both cell types respond to proinflammatory cytokines such as TNF-α, IL-1β, and INF-γ, both are exquisitely sensitive to pathogen associated molecular patterns (PAMPs) sensed through toll-like receptors (TLRs), and both cell-types produce an array of antimicrobial effectors such as antimicrobial cationic peptides, reactive oxygen species (superoxide anion, $\cdot O_2^-$, hydrogen peroxide, H_2O_2, and hypochlorite HOCl), as well as nitric oxide (NO·). Together with PMNs, M1 macrophages produce most of the antimicrobial effectors that ensure the quick and efficient elimination of invading microbes. However, these effectors also damage host tissue, and prolonged inflammation can lead to the accumulation of damage associated molecular patterns (DAMPs). Within the site of infection, immune, stromal, and epithelial cells sense the accumulating DAMPs and, concomitant with diminishing levels of PAMPs, alter the cytokine and lipid mediator milieu towards a proresolving profile, including IL-4, IL-10, IL-13, and TGF-β (208). These signals alter the macrophage phenotypes to that of alternatively activated or M2-macrophages. The phenotype of an M2-macrophage differs from M1-macrophages or PMNs. M2-macrophages make no immune radicals and synthesize no antimicrobial peptides but are highly phagocytic to facilitate the elimination of cellular debris at the conclusion of an infection (209). They also produce numerous signals that promote cell proliferation and tissue regeneration. Thus, M2 macrophages are critical for the resolution or wound healing phase that follows an inflammatory response to invading microbes.

Despite the fact that PMNs, M1-, and M2-macrophages are all highly energetically active cells, the pathways they use to meet their energy demands differ greatly. For instance, PMNs and M1-macrophages rely heavily on rapid glucose consumption for energy (210). Activation of these phagocytes by inflammatory stimuli induces a form of metabolism reminiscent of that observed in cancer cells. Warburg metabolism was first described in cultured cancer cells and is a phenomenon whereby glucose is oxidized primarily to lactate with little flux into the Krebs cycle or mitochondrial oxidative phosphorylation (OxPhos) despite the abundance of oxygen (211). Indeed, activated

PMNs and M1-macrophages show similar tendencies, consuming large quantities of glucose and acidifying the surrounding environment through the excretion of lactate (205, 212). This form of metabolism (also known as aerobic glycolysis) involves the import of glucose, primarily through the GLUT-1 transporter, after which it is phosphorylated to glucose-6-phosphate (G6P) via hexokinase-1 (HK–1) (205, 213). G6P is then oxidized to pyruvate, which is reduced to lactate and excreted (Fig. 4). This metabolism allows rapid ATP production through substrate-level phosphorylation, albeit less efficiently than through OxPhos, and maintains redox balance via lactate production.

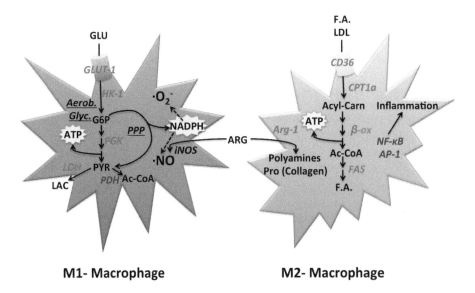

M1- Macrophage   **M2- Macrophage**

FIGURE 4 Differences in M1- versus M2-macrophage fueling reactions. In response to inflammatory stimuli, M1-macrophages upregulate a pathway known as aerobic glycolysis. This involves the import of glucose through GLUT-1 and its phosphorylation by Hexokinase-1 (HK-1). The resulting glucose-6-phosphate (G6P) can be shuttled through the pentose phosphate pathway (PPP) for NADPH generation, which fuels immune radical production, including nitric oxide (NO). At the same time, G6P is also oxidized to pyruvate (PYR) for ATP synthesis, and this PYR is primarily reduced to lactate (LAC) to conserve redox balance. Very little PYR enters the Krebs cycle as acetyl-CoA (Ac-CoA) due to the phosphorylation and inactivation of pyruvate dehydrogenase (PDH). Genes activated/repressed by HIF-1α are depicted as green/red. Upon stimulation with anti-inflammatory stimuli, M2-macrophages adopt an oxidative metabolism involving the import of free fatty acids and low-density lipoprotein (LDL)-associated lipids (fatty acids and LDL) by CD36. These fatty acids are linked to carnitine and shuttled to the mitochondria for β-oxidation, yielding ATP. In addition, some of the Ac-CoA is reused to synthesize new fatty acids. Rather than using tissue arginine for NO-production, these cells use the amino acid for proline and polyamine production, the former of which is critical for collagen synthesis. Features activated/repressed by PPAR-γ are depicted in green/red. doi:10.1128/microbiolspec.MBP-0004-2014.f4

G6P is also a substrate for the first enzyme in the PPP. This pathway is important for the *de novo* synthesis of ribonucleotides, but, more importantly for activated phagocytes, this pathway provides most cellular reducing power in the form of NADPH (214). Electrons from NADPH are used to generate immune radicals such as $\cdot O_2^-$ and $NO\cdot$, thereby necessitating significant flux through PPP for an effective immune response (Fig. 4). Thus, rapid consumption of glucose by activated PMNs and M1-macrophages allows rapid ATP production, allowing for chemotaxis and protein synthesis, as well as reducing power for the production of immune radicals. In contrast, M2-macrophages exhibit a markedly different metabolic profile upon stimulation. These cells exhibit drastic increases in fatty acid uptake and catabolism via β-oxidation, pathways not prevalent among activated PMNs or M1-macrophages (215). Curiously, significant increases in the expression of genes involved in fatty acid synthesis are also apparent in active M2-macrophages. The precise role for the balanced expression of fatty acid breakdown and synthesis programs in M2-macrophages is still unclear, but the fact remains that much more of the energy demand in M2-macrophages is met by β-oxidation rather than aerobic glycolysis (215) (Fig. 4).

Another metabolic feature that distinguishes M2- from M1-macrophages (and PMNs) is the fate of tissue arginine. L-arginine is considered a semi-essential amino acid in that it can be synthesized by mammalian cells, but the total body demand for arginine often outpaces the rate of *de novo* production, requiring dietary intake to maintain optimal levels (216). Arginine serves as a precursor to two important pathways during infection in addition to general protein synthesis. First, M1-macrophages and PMNs primarily use arginine for $NO\cdot$ synthesis via inducible $NO\cdot$-synthase (iNOS) (208). This enzyme uses electrons from NADPH to convert arginine to $NO\cdot$ and citrulline. Consequently, iNOS expression is a hallmark of M1-macrophage

activation and is required for the efficient clearance of a variety of microbial pathogens (217). Alternatively, M2-macrophages do not express iNOS; rather, arginine is metabolized to the amino acid ornithine by Arginase-1 (Arg-1) (208). Ornithine is a precursor to proline synthesis, the most abundant amino acid in collagen. Efficient collagen synthesis is critical to reestablishing basement membrane integrity for tissue regeneration. In addition, ornithine can be converted to a series of compounds known as polyamines, which promote cell proliferation. By rerouting arginine flux away from $NO\cdot$ and towards proline and polyamine production, M2-macrophages shift the tissue from an inflammatory, destructive environment to a resolving, profibrotic, and proliferative niche. Consequently, Arg-1 expression is a hallmark of M2 macrophages and is not abundantly produced by M1-macrophages or PMNs.

Just as distinct metabolic pathways fuel macrophages exhibiting M1 versus M2 phenotypes, the transcriptional regulators that drive these fueling reactions also differ between M1- and M2-macrophages. While the complex regulatory networks that differentiate M2- from M1-macrophages and/or PMNs are by no means completely understood, a few key regulators essential to each phenotype have been described. For instance, the aerobic glycolysis that fuels M1-macrophages requires the activity of the hypoxia-inducible factor 1α (HIF-1α) (213). Under normoxic or homeostatic conditions, HIF-1α is highly unstable stemming from its oxygen-dependent hydroxylation at proline residues 405 and 531 by prolyl-hydroxylases. This hydroxylation targets HIF-1α for proteosomal degradation (218). However, as oxygen tension drops, HIF-1α hydroxylation is inhibited, and the protein is therefore stabilized and efficiently translocates to the nucleus. There, it dimerizes with the constitutively expressed HIF-1β (aka aryl hydrocarbon nuclear translocator [ARNT]) to bind to HIF-responsive elements (HREs) and modulate gene expression. During inflammation, the massive consumption of oxygen for the production of reactive oxygen species results in hypoxic conditions sensed by infiltrating phagocytes (210). Thus, HIF-1α is highly stable within inflamed tissue. HIF-1α activates expression of genes involved in aerobic glycolysis, including the glucose importer GLUT-1, hexokinase-1 (HK-1), phosphoglycerate kinase (PGK), and lactate dehydrogenase (LDHA) (219) (Fig. 4). HIF-1α activity also limits the flux of pyruvate into the Krebs cycle by activating pyruvate dehydrogenase kinase (PDK) (219). PDK-mediated phosphorylation of pyruvate dehydrogenase (PDH) limits the conversion of pyruvate to acetyl-CoA, ensuring that glycolytic pyruvate is reduced to lactate via LDHA for maintaining redox balance. Thus, HIF-1α is essential for the metabolic priming of M1-macrophages, as well as PMNs. Mutant phagocytes lacking HIF-1α show diminished ATP levels, reduced chemotaxis, and limited immune effector production upon stimulation, demonstrating the importance of proper metabolic fueling to an effective immune response (213). Like M1-macrohpages and PMNs, M2-macrophages also express a regulator central to their metabolic strategies. Expression of Arg-1, as well as genes involved in β-oxidation and fatty acid synthesis, are all dependent on the activity of a nuclear receptor known as the peroxisome proliferator-activated receptor γ (PPAR-γ) (220). PPAR-γ is a member of a family of related nuclear receptors (PPAR-α and PPARβ/δ), all of which generally affect lipid homeostasis (221). These receptors enter the nucleus upon binding of their activator ligands and dimerize with the retinoid-x receptor (RXR) to bind PPAR-response elements (PPREs) and affect gene expression. In response to anti-inflammatory cytokines such as IL-4, macrophages stimulate PPAR-γ transcription in a STAT-6 dependent fashion (207). Elevated cytosolic PPAR-γ sensitizes M2-macrophages to the presence of activating ligands. The activating ligand for PPAR-γ is still unknown, but certain prostaglandins and

oxidized fatty acids are known to bind and activate the receptor (222). These lipid-based signals accumulate in tissue following prolonged inflammation and oxidative/nitrosative radical production. Thus, PPAR-γ-activation serves as a negative feedback loop to limit the duration of inflammation and reprogram tissue to a resolution phenotype. Indeed, in addition to activating the metabolic fueling reactions of M2-macrophages, PPAR-γ is known to directly limit the activity of pro-inflammatory regulators, including NF-κβ and AP-1 (223). The central role of PPAR-γ in the M2-macrophage phenotype is exemplified by the fact that macrophages deficient in PPAR-γ cannot express Arg-1 and do not exhibit robust β-oxidation or fatty acid synthesis (220). Thus, PPAR-γ activity is essential for limiting inflammation and activating the metabolic fueling pathways that maintain the M2 macrophage phenotype. These cells are essential for wound healing and restoration of tissue homeostasis following an inflammatory response to invading microbes.

Pathogen metabolism during infections

Gram-positive pathogens represent a diverse group of bacteria that, in general, span two phyla, the Firmicutes and the Actinobacteria. Just as marked diversity exists among the different species of Gram-positive pathogens with respect to disease presentation and virulence factor production, the metabolic pathways used by these pathogens to establish infection and persist are also quite varied.

Some of the most in-depth knowledge of bacterial metabolism critical to disease progression has been gathered by studying the intracellular pathogen *M. tuberculosis*, the causative agent of tuberculosis. *M. tuberculosis* thrives within the phagosome of infected macrophages, which traffic the bacterium to multicellular caseous granulomas in the lung. Here, *M. tuberculosis* resides in a quiescent state in which it awaits activation for the elaboration of disease symptoms and transmission. In general, *M. tuberculosis* relies primarily on β-oxidation of host lipids, fatty acids, and cholesterol for energy and carbon sources (224). This generalization is supported by the presence of more than 36 acyl-CoA ligase and 35 acyl-CoA dehydrogenase genes in the *M. tuberculosis* genome (compared with two acyl-CoA ligase and one acyl-CoA dehydrogenase genes encoded by *E. coli*) (http://biocyc.org). Moreover, these genes are highly expressed within macrophages, and many are required for normal *M. tuberculosis* growth during infection (225–227). Fatty acids are oxidized to acetyl-CoA (and proprionyl-CoA in the case of cholesterol and odd-chained fatty acids) and therefore require both gluconeogenesis and the Krebs cycle to generate all precursor metabolites. This is supported by the requirement of a key gluconeogenic enzyme, pyruvate carboxykinase (PckA), for full virulence (227). *M. tuberculosis* does not flux lipid-derived carbon through the canonical Krebs pathway because it is incompatible with two-carbon nutrient sources (fatty acids are mainly oxidized to acetyl-CoA and can be thought of as polymers of two-carbon nutrients). As discussed in the first section, the two tandem decarboxylation steps in the Krebs cycle (isocitrate dehydrogenase followed by α-ketoglutarate dehydrogenase) would essentially oxidize both carbons of acetyl-CoA to CO_2, precluding the synthesis of precursor metabolites. Instead, *M. tuberculosis* relies on the glyoxylate shunt, a pathway that involves isocitrate lyase and malate synthase; bypasses the two decarboxylation steps; and essentially combines two molecules of two-carbon acetyl-CoA into one molecule of four-carbon malate. The latter can be oxidized to oxaloacetate and serve as a gluconeogenic substrate via PckA. The importance of the glyoxylate shunt to *M. tuberculosis* pathogenesis is demonstrated by the requirement of both isoforms of isocitrate lyase (ICL1 and ICL2) for full virulence (228). However, ICL1 provides an additional role in *M. tuberculosis* metabolism, being critical in the methylcitrate cycle (MCC) as well as the glyoxylate shunt (229). Utilization of odd-chained fatty

acids via β-oxidation results in the generation of a single proprionyl-CoA molecule, and oxidation of cholesterol results in the production of three proprionyl-CoA molecules. Propionyl-CoA can be toxic to cells in that it can inhibit pyruvate dehydrogenase (230). Accordingly, propionyl-CoA is shuttled through the MCC, a series of reactions analogous to the Krebs cycle, by ligation to oxaloacetate via methylcitrate synthase, generating methylcitrate. Methylcitrate accumulation, however, can also be toxic to cells relying on gluconeogenesis in that it can inhibit fructose bisphosphate phosphatase, an essential enzyme for complete gluconeogenesis (231). The fact that ICL1 uses methylcitrate as well as isocitrate as substrates allows *M. tuberculosis* to use ICL1 to reduce the toxic pool of methylcitrate. In addition, when proprionyl-CoA production outpaces ICL1 and the MCC, two other pathways assist in reducing propionyl CoA levels in *M. tuberculosis*: the vitamin B_{12}-dependent methylmalonyl-CoA pathway and the direct incorporation of proprionyl-CoA into cell wall lipids (232, 233). The importance of the glyoxylate shunt, gluconeogenesis, the MCC, and cholesterol and fatty acid import to *M. tuberculosis* underscores the importance of host lipids and membrane cholesterol as fueling sources for this pathogen. However, *M. tuberculosis* does not rely on mere chance encounters with membrane lipids as nutrient sources. *M. tuberculosis* residing within granulomas comes in contact with macrophages that have accumulated high levels of lipids and cholesterol within large lipid droplets. These "foam cells" are thought to be major sources of nutrients for the infecting bacterium. Recent studies have suggested that direct interactions between macrophages and cell wall lipids of *M. tuberculosis* (particularly lipoarabanomannan, ManLAM) alter host cell gene expression, favoring the development of the foam cell phenotype (234). This interaction relies on PPAR-γ activation, which results in the robust fatty acid synthesis necessary for lipid droplet and foam cell formation (234, 235). PPAR-$\gamma^{-/-}$ macrophages (or macrophages treated with chemical PPAR-γ inhibitors) are far more bactericidal to intracellular *M. tuberculosis*. It is likely that some of this phenotype is due to the "starvation" of *M. tuberculosis* within macrophages defective in PPAR-γ activity and unable to perform robust lipid synthesis. However, ManLAM stimulation of PPAR-γ could have other important implications for *M. tuberculosis* infection outcomes. The PPAR-γ-dependent induction of Arg-1 competes for available arginine used by iNOS to generate NO·. This radical is a potent antituberculosis factor, and diminished NO· production, as well as inhibition of iNOS expression mediated by PPAR-γ, likely contribute to enhanced *M. tuberculosis* survival within macrophages (236). In total, modulation of macrophage PPAR-γ activity by *M. tuberculosis* serves many functions in the pathogenesis of this organism, including the dampened production of inflammatory mediators as well as the activation of fueling pathways that "feed" the pathogen during infection.

Another Gram-positive pathogen that relies heavily on gluconeogenic carbon sources is the intracellular bacillus, *L. monocytogenes*. Unlike *M. tuberculosis*, however, *L. monocytogenes* does not reside within a phagosome, but rather escapes to replicate within the host cell cytosol. Here, the bacterium appears to primarily use host glycerol as a preferred source of carbon and energy. Because glycerol is a gluconeogenic carbon source, *L. monocytogenes* downregulates glycolytic enzyme expression and activates the synthesis of gluconeogenic genes during infection (237, 238). Moreover, the expression of genes involved in glycerol uptake and utilization are also induced intracellularly in *L. monocytogenes*, and mutations in these genes limit intracellular growth of the bacterium (237, 238). In addition, G6P can serve as a nutrient source for intracellular *L. monocytogenes*, albeit to a lesser extent than glycerol. Interestingly, the G6P uptake system is upregulated intracellularly, but the glucose transporter is not, suggesting that only carbohydrate phosphorylated by hexokinase is used by *L. monocytogenes* (239, 240). The reliance on glycerol and G6P rather than glucose intracellularly likely reflects the metabolic adaptation of infected host cells to *L. monocytogenes*. A marked increase in glycolysis and glutaminolysis accompanies infection of primary macrophages with the bacterium, which could lead to the intracellular accumulation of glycerol and G6P (241). While the regulatory pathways that control this host response are still unknown, HIF-1α induction of glycolysis may play an as-of-yet uncharacterized role.

In contrast to the obligate intracellular pathogens discussed previously that require gluconeogenesis for full virulence, *S. aureus* appears to require both glycolysis and gluconeogenesis to be fully pathogenic (242). This has been documented in an invertebrate infection model, and we have observed similar findings in murine infection models (253). Recent findings have shed light on this curious observation. While *S. aureus* can survive within phagocytes for prolonged periods, it often lyses host phagocytes and escapes. Within these phagocytes, however, it must endure an impressive onslaught of innate immune effectors, including immune radicals ($\cdot O_2^-$ and NO·). Interestingly, *S. aureus* is remarkably resistant to the effects of these immune radicals, particularly NO· (182, 243). Most bacteria are unable to replicate in the presence of NO· due to its propensity to attack redox centers of key metabolic enzymes (244). *S. aureus* elicits an incompletely defined metabolic state that allows the bacterium to circumvent the metabolic constraints imposed by host NO· (245). Of all the nutrients used by *S. aureus*, only certain glycolytic carbon sources support NO·-resistance (246). Therefore, glycolysis is essential for *S. aureus* to resist high-level NO· encountered within PMNs and M1-macrophages. Indeed, glycolytic mutants exhibit severe attenuation within mouse macrophages, but this defect is completely reversed upon inhibition of iNOS activity (253). Thus, because of host immune response, intracellular *S. aureus* relies on glycolysis to persist within the phagocyte until the bacterium can lyse the host cell and escape into the extracellular space. Once there, it then replicates to form bacterial aggregates at the center of tissue abscesses. Here, *S. aureus* resides within lysed host cell debris surrounded by newly infiltrating phagocytes. Inside these highly inflamed hypoxic abscesses, HIF-1α activity is high, thereby driving excessive glucose consumption by host cells (210). Therefore, glucose is likely scarce, and the bacterium must rely on abundant gluconeogenic substrates such as lactate resulting from host cell aerobic glycolysis. Accordingly, *S. aureus* encodes three independent enzymes that can use lactate as a carbon source (246). Furthermore, the low oxygen tension in this abscess environment likely limits the robust production of inflammatory radicals, including NO·, thereby relieving the strict dependence on glycolysis for *S. aureus* to thrive. Thus, lactate and host peptide-derived amino acids could serve as the major energy/carbon source to *S. aureus* at the center of an abscess, necessitating gluconeogenesis. In essence, *S. aureus* relies on different nutrient sources throughout the course of animal infections, explaining the reliance on both glycolysis and gluconeogenesis for full virulence. However, more research is needed to solidify our understanding of the metabolic adaptions of *S. aureus* to the dynamic host immune environment.

The spore-forming Gram-positive pathogens, including *Bacillus anthracis* and *Clostridium difficile*, must germinate once inside the host in order to cause disease. Metabolite cues within host tissue interact with specific germinant receptors in the spore membrane and mechanically initiate the germination program. *B. anthracis* spores germinate once the nucleoside inosine and an accompanying cogerminant (usually an amino acid) are encountered (247). Ingested *C. difficile* spores germinate once they encounter bile salts such as taurocholate combined with glycine in the gut (248). Once germinated, little is known about the metabolism of either species. Curiously, as described above, glucose has opposing effects on the production of the major virulence toxins in the two species. Glucose stimulates the production of anthrax toxin, whereas it suppresses production of the *C. difficile* toxin. Both effects are mediated through CcpA (see section 2). It is tempting to hypothesize that *C. difficile* primarily relies on gluconeogenic substrates during infection; however, direct experimental confirmation of this is lacking. *C. difficile* replicates in the large intestine and, because most digestible carbohydrates are absorbed in the small intestine, the bacterium likely encounters low levels of simple sugars. Moreover, the organism encodes amino acid reductases (see second section) that allow *C. difficile* to thrive on peptide nutrient sources in anaerobic environments such as the gut. One of these reductases, the proline reductase, was induced upon inoculation into germ-free mice compared with laboratory-cultured bacteria (249). However, whether these systems are required for virulence has not been directly tested. In addition, it has long been appreciated that antibiotic treatment drastically predisposes patients to *C. difficile* infection. Thus, the normal gut microflora must provide an effective competitive barrier to *C. difficile* colonization. As mentioned earlier, the lumen of the large intestine is largely devoid of dietary carbohydrate, but it is rich in protein-associated sugars found in mucin and on enterocyte cell surfaces. These carbohydrates can be liberated by glycosidases expressed by various microflora and subsequently taken up by these symbiotic bacteria. Consequently, the steady-state levels of carbohydrate in the lumen of the large intestine are low. However, antibiotic treatment can interfere with the homeostasis of the gut flora and disrupt the balance between glycosidase-mediated carbohydrate liberation and consumption (250).

This generally leads to a transient increase in luminal carbohydrate content in the large intestine, which may be exploited by *C. difficile*. Indeed, microflora-liberated sialic acid was shown to promote *C. difficile* expansion in a murine model of infection. Moreover, antibiotic treatment of mice resulted in elevated luminal sialic acid, and *C. difficile* mutants unable to metabolize sialic acid were defective in colonization of these antibiotic-treated animals (251). Thus, antibiotic treatment of the host may affect the balance of microflora-dependent carbohydrate metabolism in the large intestine, providing *C. difficile* with a glycolytic nutrient source during infection.

While we have gained significant insights into the roles of metabolism in the host: pathogen interaction, there is much still to be learned about what pathways fuel both host immune cells and invading pathogens. This information might yield possibilities for new therapeutics aimed at inhibiting key bacterial metabolic pathways as well as modulating host metabolism to boost immune production.

ACKNOWLEDGMENTS

G.A.S. was supported by funds provided through the Hatch Act to the University of Nebraska Institute of Agriculture and Natural Resources and by funds provided through the National Institutes of Health (R01AI087668). Unpublished work from the A.L.S. laboratory was supported by a grant (R01GM042219) from the National Institutes of Health. A.R.R. was supported by funds provided through the National Institutes of Health (R01AI093613).

CITATION

Richardson AR, Somerville GA, Sonenshein AL. 2015. Regulating the intersection of metabolism and pathogenesis in Gram-positive bacteria. Microbiol Spectrum 3(3): MBP-0004-2014.

REFERENCES

1. **Sadykov MR, Olson ME, Halouska S, Zhu Y, Fey PD, Powers R, Somerville GA.** 2008. Tricarboxylic acid cycle-dependent regulation of *Staphylococcus epidermidis* polysaccharide intercellular adhesin synthesis. *J Bacteriol* **190:**7621–7632.

2. **Sadykov MR, Zhang B, Halouska S, Nelson JL, Kreimer LW, Zhu Y, Powers R, Somerville GA.** 2010. Using NMR metabolomics to investigate tricarboxylic acid cycle dependent signal transduction in *Staphylococcus epidermidis*. *J Biol Chem* **285:**36616–36624.

3. **Kundig W, Ghosh S, Roseman S.** 1964. Phosphate Bound to Histidine in a Protein as an Intermediate in a Novel Phospho-Transferase System. *Proc Natl Acad Sci U S A* **52:**1067–1074.

4. **Saier MH Jr.** 1989. Protein phosphorylation and allosteric control of inducer exclusion and catabolite repression by the bacterial phosphoenolpyruvate: sugar phosphotransferase system. *Microbiol Rev* **53:**109–120.

5. **Klein HP, Doudoroff M.** 1950. The mutation of *Pseudomonas putrefaciens* to glucose utilization and its enzymatic basis. *J Bacteriol* **59:** 739–750.

6. **Sanwal BD.** 1970. Allosteric controls of amphibolic pathways in bacteria. *Bacteriological Reviews* **34:**20–39.

7. **Blangy D, Buc H, Monod J.** 1968. Kinetics of the allosteric interactions of phosphofructokinase from *Escherichia coli*. *J Mol Biol* **31:**13–35.

8. **Lopez G, Latorre M, Reyes-Jara A, Cambiazo V, Gonzalez M.** 2012. Transcriptomic response of *Enterococcus faecalis* to iron excess. *Biometals* **25:**737–747.

9. **Ojha A, Hatfull GF.** 2007. The role of iron in *Mycobacterium smegmatis* biofilm formation: the exochelin siderophore is essential in limiting iron conditions for biofilm formation but not for planktonic growth. *Mol Micro* **66:** 468–483.

10. **Friedman DB, Stauff DL, Pishchany G, Whitwell CW, Torres VJ, Skaar EP.** 2006. *Staphylococcus aureus* redirects central metabolism to increase iron availability. *PLoS Pathog* **2:**e87.

11. **Somerville G, Mikoryak CA, Reitzer L.** 1999. Physiological characterization of *Pseudomonas aeruginosa* during exotoxin A synthesis: glutamate, iron limitation, and aconitase activity. *J Bacteriol* **181:**1072–1078.

12. **Zhang B, Halouska S, Schiaffo CE, Sadykov MR, Somerville GA, Powers R.** 2011. NMR analysis of a stress response metabolic signaling network. *J Proteome Res* **10:**3743–3754.

13. Craig JE, Ford MJ, Blaydon DC, Sonenshein AL. 1997. A null mutation in the *Bacillus subtilis* aconitase gene causes a block in Spo0A-phosphate-dependent gene expression. *J Bacteriol* **179:**7351–7359.

14. Dickens F. 1938. Oxidation of phosphohexonate and pentose phosphoric acids by yeast enzymes: Oxidation of phosphohexonate. II. Oxidation of pentose phosphoric acids. *Biochem J* **32:**1626–1644.

15. Warburg O, Christian W, Griese A. 1935. Hydrogen-transferring coenzyme, its composition and mode of action. *Biochem Z* **282:**157–205.

16. Horecker BL, Smyrniotis PZ. 1955. Purification and properties of yeast transaldolase. *J Biol Chem* **212:**811–825.

17. Scott DB, Cohen SS. 1953. The oxidative pathway of carbohydrate metabolism in *Escherichia coli*. 1. The isolation and properties of glucose 6-phosphate dehydrogenase and 6-phosphogluconate dehydrogenase. *Biochem J* **55:**23–33.

18. Dutow P, Schmidl SR, Ridderbusch M, Stulke J. 2010. Interactions between glycolytic enzymes of *Mycoplasma pneumoniae*. *J Mol Microbiol Biotechnol* **19:**134–139.

19. Tettelin H, Nelson KE, Paulsen IT, Eisen JA, Read TD, Peterson S, Heidelberg J, DeBoy RT, Haft DH, Dodson RJ, Durkin AS, Gwinn M, Kolonay JF, Nelson WC, Peterson JD, Umayam LA, White O, Salzberg SL, Lewis MR, Radune D, Holtzapple E, Khouri H, Wolf AM, Utterback TR, Hansen CL, McDonald LA, Feldblyum TV, Angiuoli S, Dickinson T, Hickey EK, Holt IE, Loftus BJ, Yang F, Smith HO, Venter JC, Dougherty BA, Morrison DA, Hollingshead SK, Fraser CM. 2001. Complete genome sequence of a virulent isolate of *Streptococcus pneumoniae*. *Science* **293:**498–506.

20. Guimaraes AM, Santos AP, SanMiguel P, Walter T, Timenetsky J, Messick JB. 2011. Complete genome sequence of *Mycoplasma suis* and insights into its biology and adaption to an erythrocyte niche. *PloS One* **6:**e19574.

21. Bergman NH, Anderson EC, Swenson EE, Janes BK, Fisher N, Niemeyer MM, Miyoshi AD, Hanna PC. 2007. Transcriptional profiling of *Bacillus anthracis* during infection of host macrophages. *Infect Immun* **75:**3434–3444.

22. Chatterjee SS, Hossain H, Otten S, Kuenne C, Kuchmina K, Machata S, Domann E, Chakraborty T, Hain T. 2006. Intracellular gene expression profile of *Listeria monocytogenes*. *Infect Immun* **74:**1323–1338.

23. Fleury B, Kelley WL, Lew D, Gotz F, Proctor RA, Vaudaux P. 2009. Transcriptomic and metabolic responses of *Staphylococcus aureus* exposed to supra-physiological temperatures. *BMC Microbiol* **9:**76.

24. Weigoldt M, Meens J, Bange FC, Pich A, Gerlach GF, Goethe R. 2013. Metabolic adaptation of *Mycobacterium avium* subsp. *paratuberculosis* to the gut environment. *Microbiol* **159:**380–391.

25. delCardayre SB, Stock KP, Newton GL, Fahey RC, Davies JE. 1998. Coenzyme A disulfide reductase, the primary low molecular weight disulfide reductase from *Staphylococcus aureus*. Purification and characterization of the native enzyme. *J Biol Chem* **273:**5744–5751.

26. Helmann JD. 2011. Bacillithiol, a new player in bacterial redox homeostasis. *Antioxid Redox Signal* **15:**123–133.

27. Newton GL, Fahey RC, Rawat M. 2012. Detoxification of toxins by bacillithiol in *Staphylococcus aureus*. *Microbiol* **158:**1117–1126.

28. Spies HS, Steenkamp DJ. 1994. Thiols of intracellular pathogens. Identification of ovothiol A in *Leishmania donovani* and structural analysis of a novel thiol from *Mycobacterium bovis*. *Eur J Biochem* **224:**203–213.

29. Holmgren A. 1985. Thioredoxin. *Annu Rev Biochem* **54:**237–271.

30. Moore EC, Reichard P, Thelander L. 1964. Enzymatic synthesis of deoxyribonucleotides. V. Purification and properties of thioredoxin reductase from *Escherichia coli* B. *J Biol Chem* **239:**3445–3452.

31. Gustafsson TN, Sahlin M, Lu J, Sjoberg BM, Holmgren A. 2012. *Bacillus anthracis* thioredoxin systems, characterization and role as electron donors for ribonucleotide reductase. *J Biol Chem* **287:**39686–39697.

32. Newton GL, Buchmeier N, Fahey RC. 2008. Biosynthesis and functions of mycothiol, the unique protective thiol of Actinobacteria. *Microbiol Mol Biol Rev* **72:**471–494.

33. Eisenreich W, Slaghuis J, Laupitz R, Bussemer J, Stritzker J, Schwarz C, Schwarz R, Dandekar T, Goebel W, Bacher A. 2006. ^{13}C isotopologue perturbation studies of *Listeria monocytogenes* carbon metabolism and its modulation by the virulence regulator PrfA. *Proc Natl Acad Sci U S A* **103:**2040–2045.

34. Pfefferkorn ER, Rebhun S, Eckel M. 1986. Characterization of an indoleamine 2,3-dioxygenase induced by gamma-interferon in cultured human fibroblasts. *J Interferon Res* **6:**267–279.

35. **Daubener W, MacKenzie CR.** 1999. IFN-gamma activated indoleamine 2,3-dioxygenase activity in human cells is an antiparasitic and an antibacterial effector mechanism. *Adv Exp Med Biol* **467:**517–524.

36. **Mraheil MA, Billion A, Mohamed W, Rawool D, Hain T, Chakraborty T.** 2011. Adaptation of Listeria monocytogenes to oxidative and nitrosative stress in IFN-gamma-activated macrophages. *Int J Med Microbiol* **301:**547–555.

37. **Hucke C, MacKenzie CR, Adjogble KD, Takikawa O, Daubener W.** 2004. Nitric oxide-mediated regulation of gamma interferon-induced bacteriostasis: inhibition and degradation of human indoleamine 2,3-dioxygenase. *Infect Immun* **72:**2723–2730.

38. **MacKenzie CR, Hadding U, Daubener W.** 1998. Interferon-gamma-induced activation of indoleamine 2,3-dioxygenase in cord blood monocyte-derived macrophages inhibits the growth of group B streptococci. *J Infect Dis* **178:**875–878.

39. **Huynen MA, Dandekar T, Bork P.** 1999. Variation and evolution of the citric-acid cycle: a genomic perspective. *Trends Microbiol* **7:**281–291.

40. **Srinivasan V, Morowitz HJ.** 2006. Ancient genes in contemporary persistent microbial pathogens. *Biol Bull* **210:**1–9.

41. **Sonenshein AL.** 2002. The Krebs citric acid cycle, p 151–162. *In* Sonenshein AL, Hoch JA, Losick R (ed), Bacillus subtilis *and its closest relatives: From genes to cells.* ASM Press, Washington, D.C.

42. **Bondi A, Kornblum J, De St Phalle M.** 1954. The amino acid requirements of penicillin resistant and penicillin sensitive strains of *Micrococcus pyogenes. J Bacteriol* **68:**617–621.

43. **Murray BE, Singh KV, Ross RP, Heath JD, Dunny GM, Weinstock GM.** 1993. Generation of restriction map of *Enterococcus faecalis* OG1 and investigation of growth requirements and regions encoding biosynthetic function. *J Bacteriol* **175:**5216–5223.

44. **Tourtellotte ME, Morowitz HJ, Kasimer P.** 1964. Defined medium for *Mycoplasma laidlawii. J Bacteriol* **88:**11–15.

45. **Melendez-Hevia E, Waddell TG, Cascante M.** 1996. The puzzle of the Krebs citric acid cycle: assembling the pieces of chemically feasible reactions, and opportunism in the design of metabolic pathways during evolution. *J Mol Evol* **43:**293–303.

46. **Kim HJ, Mittal M, Sonenshein AL.** 2006. CcpC-dependent regulation of *citB* and lmo0847 in *Listeria monocytogenes. J Bacteriol* **188:**179–190.

47. **Cerdeno-Tarraga AM, Efstratiou A, Dover LG, Holden MT, Pallen M, Bentley SD, Besra GS, Churcher C, James KD, De Zoysa A, Chillingworth T, Cronin A, Dowd L, Feltwell T, Hamlin N, Holroyd S, Jagels K, Moule S, Quail MA, Rabbinowitsch E, Rutherford KM, Thomson NR, Unwin L, Whitehead S, Barrell BG, Parkhill J.** 2003. The complete genome sequence and analysis of Corynebacterium diphtheriae NCTC13129. *Nucleic Acids Res* **31:**6516–6523.

48. **Nakano MM, Zuber P, Sonenshein AL.** 1998. Anaerobic regulation of *Bacillus subtilis* Krebs cycle genes. *J Bacteriol* **180:**3304–3311.

49. **Eoh H, Rhee KY.** 2013. Multifunctional essentiality of succinate metabolism in adaptation to hypoxia in *Mycobacterium tuberculosis. Proc Natl Acad Sci U S A* **110:**6554–6559.

50. **Somerville GA, Chaussee MS, Morgan CI, Fitzgerald JR, Dorward DW, Reitzer LJ, Musser JM.** 2002. *Staphylococcus aureus* aconitase inactivation unexpectedly inhibits post-exponential-phase growth and enhances stationary-phase survival. *Infect Immun* **70:**6373–6382.

51. **Mei JM, Nourbakhsh F, Ford CW, Holden DW.** 1997. Identification of *Staphylococcus aureus* virulence genes in a murine model of bacteraemia using signature-tagged mutagenesis. *Mol Microbiol* **26:**399–407.

52. **Kornberg HL, Krebs HA.** 1957. Synthesis of cell constituents from C_2-units by a modified tricarboxylic acid cycle. *Nature* **179:**988–991.

53. **Beaman BL, Beaman L.** 1994. *Nocardia* species: host-parasite relationships. *Clin Microbiol Rev* **7:**213–264.

54. **Muscatello G.** 2012. Rhodococcus equi pneumonia in the foal–part 1: pathogenesis and epidemiology. *Vet J* **192:**20–26.

55. **Lorenz MC, Fink GR.** 2002. Life and death in a macrophage: role of the glyoxylate cycle in virulence. Eukaryot. *Cell* **1:**657–662.

56. **Titgemeyer F, Hillen W.** 2002. Global control of sugar metabolism: a gram-positive solution. *Antonie Van Leeuwenhoek* **82:**59–71.

57. **Warner JB, Lolkema JS.** 2003. CcpA-dependent carbon catabolite repression in bacteria. *Microbiol Mol Biol Rev* **67:**475–490.

58. **Pimentel-Schmitt EF, Thomae AW, Amon J, Klieber MA, Roth HM, Muller YA, Jahreis K, Burkovski A, Titgemeyer F.** 2007. A glucose kinase from *Mycobacterium smegmatis. J Mol Microbiol Biotechnol* **12:**75–81.

59. **Bowles JA, Segal W.** 1965. Kinetics of utilization of organic compounds in the growth of *Mycobacterium tuberculosis. J Bacteriol* **90:**157–163.

60. **Collins FM, Lascelles J**. 1962. The effect of growth conditions on oxidative and dehydrogenase activity in *Staphylococcus aureus*. *J Gen Microbiol* **29:**531–535.

61. **Hanson RS, Srinivasan VR, Halvorson HO**. 1963. Biochemistry of sporulation. I. Metabolism of acetate by vegetative and sporulating cells. *J Bacteriol* **85:**451–460.

62. **Hanson RS, Blicharska J, Arnaud M, Szulmajster J**. 1964. Observation on the regulation of the synthesis of the tricarboxylic acid cycle enzymes in *Bacillus subtilis*, Marburg. *Biochem Biophys Res Commun* **17:**690–695.

63. **Holmgren NB, Millman I, Youmans GP**. 1954. Studies on the metabolism of *Mycobacterium tuberculosis*. VI. The effect of Krebs' tricarboxylic acid cycle intermediates and precursors on the growth and respiration of *Mycobacterium tuberculosis*. *J Bacteriol* **68:**405–410.

64. **Somerville GA, Saïd-Salim B, Wickman JM, Raffel SJ, Kreiswirth BN, Musser JM**. 2003. Correlation of acetate catabolism and growth yield in *Staphylococcus aureus*: Implications for host-pathogen interactions. *Infect Immun* **71:**4724–4732.

65. **Somerville GA, Cockayne A, Dürr M, Peschel A, Otto M, Musser JM**. 2003. Synthesis and deformylation of *Staphylococcus aureus* delta-toxin are linked to tricarboxylic acid cycle activity. *J Bacteriol* **185:**6686–6694.

66. **Vuong C, Kidder JB, Jacobson ER, Otto M, Proctor RA, Somerville GA**. 2005. *Staphylococcus epidermidis* polysaccharide intercellular adhesin production significantly increases during tricarboxylic acid cycle stress. *J Bacteriol* **187:**2967–2973.

67. **Varghese S, Tang Y, Imlay JA**. 2003. Contrasting sensitivities of *Escherichia coli* aconitases A and B to oxidation and iron depletion. *J Bacteriol* **185:**221–230.

68. **Jaeger T, Mayer C**. 2008. The transcriptional factors MurR and catabolite activator protein regulate N-acetylmuramic acid catabolism in *Escherichia coli*. *J Bacteriol* **190:**6598–6608.

69. **Shivers RP, Sonenshein AL**. 2004. Activation of the *Bacillus subtilis* global regulator CodY by direct interaction with branched-chain amino acids. *Mol Microbiol* **53:**599–611.

70. **Romling U, Galperin MY, Gomelsky M**. 2013. Cyclic di-GMP: the first 25 years of a universal bacterial second messenger. *Microbiol Mol Biol Rev* **77:**1–52.

71. **Ernst JF, Bennett RL, Rothfield LI**. 1978. Constitutive expression of the iron-enterochelin and ferrichrome uptake systems in a mutant strain of *Salmonella typhimurium*. *J Bacteriol* **135:**928–934.

72. **Somerville GA, Proctor RA**. 2009. At the crossroads of bacterial metabolism and virulence factor synthesis in staphylococci. *Microbiol Mol Biol Rev* **73:**233–248.

73. **Reitzer L**. 2003. Nitrogen assimilation and global regulation in *Escherichia coli*. *Annu Rev Microbiol* **57:**155–176.

74. **Fisher SH**. 1999. Regulation of nitrogen metabolism in *Bacillus subtilis*: *vive la* difference! *Mol Microbiol* **32:**223–232.

75. **Tempest DW, Meers JL, Brown CM**. 1970. Synthesis of glutamate in *Aerobacter aerogenes* by a hitherto unknown route. *Biochem J* **117:** 405–407.

76. **Slack FJ, Serror P, Joyce E, Sonenshein AL**. 1995. A gene required for nutritional repression of the *Bacillus subtilis* dipeptide permease operon. *Mol Microbiol* **15:**689–702.

77. **Armstrong FB, Wagner RP**. 1961. Biosynthesis of valine and isoleucine. IV. alpha-hydroxy-beta-keto acid reductoisomerase of *Salmonella*. *J Biol Chem* **236:**2027–2032.

78. **Myers JW**. 1961. Dihydroxy acid dehydrase: an enzyme involved in the biosynthesis of isoleucine and valine. *J Biol Chem* **236:**1414–1418.

79. **Leitzmann C, Bernlohr RW**. 1968. Threonine dehydratase of Bacillus licheniformis. I. Purification and properties. *Biochim Biophys Acta* **151:**449–460.

80. **Nobre LS, Saraiva LM**. 2013. Effect of combined oxidative and nitrosative stresses on *Staphylococcus aureus* transcriptome. *Appl Microbiol Biotechnol* **97:**2563–2573.

81. **Thorsing M, Klitgaard JK, Atilano ML, Skov MN, Kolmos HJ, Filipe SR, Kallipolitis BH**. 2013. Thioridazine induces major changes in global gene expression and cell wall composition in methicillin-resistant *Staphylococcus aureus* USA300. *PLoS One* **8:**e64518.

82. **Moreno MS, Schneider BL, Maile RR, Weyler W, Saier MH Jr**. 2001. Catabolite repression mediated by the CcpA protein in *Bacillus subtilis*: novel modes of regulation revealed by whole-genome analyses. *Mol Microbiol* **39:**1366–1381.

83. **Yoshida K, Kobayashi K, Miwa Y, Kang CM, Matsunaga M, Yamaguchi H, Tojo S, Yamamoto M, Nishi R, Ogasawara N, Nakayama T, Fujita Y**. 2001. Combined transcriptome and proteome analysis as a powerful approach to study genes under glucose repression in *Bacillus subtilis*. *Nucleic Acids Res* **29:**683–692.

84. **Antunes A, Camiade E, Monot M, Courtois E, Barbut F, Sernova NV, Rodionov DA,**

Martin-Verstraete I, Dupuy B. 2012. Global transcriptional control by glucose and carbon regulator CcpA in *Clostridium difficile*. *Nucleic Acids Res* **40**:10701–10718.

85. Ren C, Gu Y, Wu Y, Zhang W, Yang C, Yang S, Jiang W. 2012. Pleiotropic functions of catabolite control protein CcpA in Butanol-producing *Clostridium acetobutylicum*. *BMC Genomics* **13**:349.

86. Jankovic I, Egeter O, Bruckner R. 2001. Analysis of catabolite control protein A-dependent repression in *Staphylococcus xylosus* by a genomic reporter gene system. *J Bacteriol* **183**:580–586.

87. Seidl K, Muller S, Francois P, Kriebitzsch C, Schrenzel J, Engelmann S, Bischoff M, Berger-Bachi B. 2009. Effect of a glucose impulse on the CcpA regulon in *Staphylococcus aureus*. *BMC Microbiol* **9**:95.

88. Zeng L, Choi SC, Danko CG, Siepel A, Stanhope MJ, Burne RA. 2013. Gene regulation by CcpA and catabolite repression explored by RNA-Seq in *Streptococcus mutans*. *PloS One* **8**:e60465.

89. Willenborg J, Fulde M, de Greeff A, Rohde M, Smith HE, Valentin-Weigand P, Goethe R. 2011. Role of glucose and CcpA in capsule expression and virulence of *Streptococcus suis*. *Microbiology* **157**:1823–1833.

90. Carvalho SM, Kloosterman TG, Kuipers OP, Neves AR. 2011. CcpA ensures optimal metabolic fitness of *Streptococcus pneumoniae*. *PloS One* **6**:e26707.

91. Kinkel TL, McIver KS. 2008. CcpA-mediated repression of streptolysin S expression and virulence in the group A *streptococcus*. *Infect Immun* **76**:3451–3463.

92. Zomer AL, Buist G, Larsen R, Kok J, Kuipers OP. 2007. Time-resolved determination of the CcpA regulon of *Lactococcus lactis* subsp. *cremoris* MG1363. *J Bacteriol* **189**:1366–1381.

93. Leboeuf C, Leblanc L, Auffray Y, Hartke A. 2000. Characterization of the *ccpA* gene of *Enterococcus faecalis*: identification of starvation-inducible proteins regulated by CcpA. *J Bacteriol* **182**:5799–5806.

94. Asanuma N, Yoshii T, Hino T. 2004. Molecular characterization of CcpA and involvement of this protein in transcriptional regulation of lactate dehydrogenase and pyruvate formate-lyase in the ruminal bacterium *Streptococcus bovis*. *Appl Environ Microbiol* **70**:5244–5251.

95. Fujita Y. 2009. Carbon catabolite control of the metabolic network in *Bacillus subtilis*. *Biosci Biotechnol Biochem* **73**:245–259.

96. Dahl MK. 2002. CcpA-independent carbon catabolite repression in *Bacillus subtilis*. *J Mol Microbiol Biotechnol* **4**:315–321.

97. Deutscher J, Herro R, Bourand A, Mijakovic I, Poncet S. 2005. P-Ser-HPr--a link between carbon metabolism and the virulence of some pathogenic bacteria. *Biochim Biophys Acta* **1754**:118–125.

98. Luesink EJ, Beumer CM, Kuipers OP, De Vos WM. 1999. Molecular characterization of the *Lactococcus lactis ptsHI* operon and analysis of the regulatory role of HPr. *J Bacteriol* **181**:764–771.

99. Zeng L, Burne RA. 2010. Seryl-phosphorylated HPr regulates CcpA-independent carbon catabolite repression in conjunction with PTS permeases in *Streptococcus mutans*. *Mol Microbiol* **75**:1145–1158.

100. Antunes A, Martin-Verstraete I, Dupuy B. 2011. CcpA-mediated repression of *Clostridium difficile* toxin gene expression. *Mol Microbiol* **79**:882–899.

101. Leiba J, Hartmann T, Cluzel ME, Cohen-Gonsaud M, Delolme F, Bischoff M, Molle V. 2012. A novel mode of regulation of the *Staphylococcus aureus* catabolite control protein A (CcpA) mediated by Stk1 protein phosphorylation. *J Biol Chem* **287**:43607–43619.

102. Bayer AS, Coulter SN, Stover CK, Schwan WR. 1999. Impact of the high-affinity proline permease gene (*putP*) on the virulence of *Staphylococcus aureus* in experimental endocarditis. *Infect Immun* **67**:740–744.

103. Schwan WR, Wetzel KJ, Gomez TS, Stiles MA, Beitlich BD, Grunwald S. 2004. Low-proline environments impair growth, proline transport and in vivo survival of *Staphylococcus aureus* strain-specific *putP* mutants. *Microbiology* **150**:1055–1061.

104. Li C, Sun F, Cho H, Yelavarthi V, Sohn C, He C, Schneewind O, Bae T. 2010. CcpA mediates proline auxotrophy and is required for *Staphylococcus aureus* pathogenesis. *J Bacteriol* **192**:3883–3892.

105. Giammarinaro P, Paton JC. 2002. Role of RegM, a homologue of the catabolite repressor protein CcpA, in the virulence of *Streptococcus pneumoniae*. *Infect Immun* **70**:5454–5461.

106. Iyer R, Baliga NS, Camilli A. 2005. Catabolite control protein A (CcpA) contributes to virulence and regulation of sugar metabolism in *Streptococcus pneumoniae*. *J Bacteriol* **187**:8340–8349.

107. Zhang A, Chen B, Yuan Z, Li R, Liu C, Zhou H, Chen H, Jin M. 2012. HP0197 contributes to CPS synthesis and the virulence of *Streptococcus suis* via CcpA. *PloS One* **7**:e50987.

108. Abranches J, Nascimento MM, Zeng L, Browngardt CM, Wen ZT, Rivera MF, Burne RA. 2008. CcpA regulates central

metabolism and virulence gene expression in *Streptococcus mutans*. *J Bacteriol* **190**:2340–2349.

109. **Zheng L, Itzek A, Chen Z, Kreth J.** 2011. Environmental influences on competitive hydrogen peroxide production in *Streptococcus gordonii*. *Appl Environ Microbiol* **77**:4318–4328.

110. **Zheng L, Chen Z, Itzek A, Ashby M, Kreth J.** 2011. Catabolite control protein A controls hydrogen peroxide production and cell death in *Streptococcus sanguinis*. *J Bacteriol* **193**:516–526.

111. **Ahn SJ, Rice KC, Oleas J, Bayles KW, Burne RA.** 2010. The *Streptococcus mutans* Cid and Lrg systems modulate virulence traits in response to multiple environmental signals. *Microbiology* **156**:3136–3147.

112. **Almengor AC, Kinkel TL, Day SJ, McIver KS.** 2007. The catabolite control protein CcpA binds to P$_{mga}$ and influences expression of the virulence regulator Mga in the Group A *Streptococcus*. *J Bacteriol* **189**:8405–8416.

113. **Dupuy B, Sonenshein AL.** 1998. Regulated transcription of *Clostridium difficile* toxin genes. *Mol Microbiol* **27**:107–120.

114. **Karlsson S, Burman LG, Akerlund T.** 1999. Suppression of toxin production in *Clostridium difficile* VPI 10463 by amino acids. *Microbiology* **145**:1683–1693.

115. **Mendez MB, Goni A, Ramirez W, Grau RR.** 2012. Sugar inhibits the production of the toxins that trigger clostridial gas gangrene. *Microb Pathog* **52**:85–91.

116. **Mendez M, Huang IH, Ohtani K, Grau R, Shimizu T, Sarker MR.** 2008. Carbon catabolite repression of type IV pilus-dependent gliding motility in the anaerobic pathogen *Clostridium perfringens*. *J Bacteriol* **190**:48–60.

117. **Varga JJ, Therit B, Melville SB.** 2008. Type IV pili and the CcpA protein are needed for maximal biofilm formation by the gram-positive anaerobic pathogen *Clostridium perfringens*. *Infect Immun* **76**:4944–4951.

118. **Varga J, Stirewalt VL, Melville SB.** 2004. The CcpA protein is necessary for efficient sporulation and enterotoxin gene (cpe) regulation in *Clostridium perfringens*. *J Bacteriol* **186**:5221–5229.

119. **Mackey-Lawrence NM, Jefferson KK.** 2013. Regulation of *Staphylococcus aureus* immunodominant antigen B (IsaB). *Microbiol Res* **168**:113–118.

120. **Seidl K, Goerke C, Wolz C, Mack D, Berger-Bachi B, Bischoff M.** 2008. *Staphylococcus aureus* CcpA affects biofilm formation. *Infect Immun* **76**:2044–2050.

121. **Seidl K, Stucki M, Ruegg M, Goerke C, Wolz C, Harris L, Berger-Bachi B, Bischoff M.** 2006. *Staphylococcus aureus* CcpA affects virulence determinant production and antibiotic resistance. *Antimicrob Agents Chemother* **50**:1183–1194.

122. **Seidl K, Bischoff M, Berger-Bachi B.** 2008. CcpA mediates the catabolite repression of *tst* in *Staphylococcus aureus*. *Infect Immun* **76**:5093–5099.

123. **Chiang C, Bongiorni C, Perego M.** 2011. Glucose-dependent activation of *Bacillus anthracis* toxin gene expression and virulence requires the carbon catabolite protein CcpA. *J Bacteriol* **193**:52–62.

124. **Gao P, Pinkston KL, Bourgogne A, Cruz MR, Garsin DA, Murray BE, Harvey BR.** 2013. Library screen identifies *Enterococcus faecalis* CcpA, the catabolite control protein A, as an effector of Ace, a collagen adhesion protein linked to virulence. *J Bacteriol* **195**:4761–4768.

125. **Herro R, Poncet S, Cossart P, Buchrieser C, Gouin E, Glaser P, Deutscher J.** 2005. How seryl-phosphorylated HPr inhibits PrfA, a transcription activator of *Listeria monocytogenes* virulence genes. *J Mol Microbiol Biotechnol* **9**:224–234.

126. **Mertins S, Joseph B, Goetz M, Ecke R, Seidel G, Sprehe M, Hillen W, Goebel W, Muller-Altrock S.** 2007. Interference of components of the phosphoenolpyruvate phosphotransferase system with the central virulence gene regulator PrfA of *Listeria monocytogenes*. *J Bacteriol* **189**:473–490.

127. **Molle V, Nakaura Y, Shivers RP, Yamaguchi H, Losick R, Fujita Y, Sonenshein AL.** 2003. Additional targets of the *Bacillus subtilis* global regulator CodY identified by chromatin immunoprecipitation and genome-wide transcript analysis. *J Bacteriol* **185**:1911–1922.

128. **Dineen SS, McBride SM, Sonenshein AL.** 2010. Integration of metabolism and virulence by *Clostridium difficile* CodY. *J Bacteriol* **192**:5350–5362.

129. **Majerczyk CD, Dunman PM, Luong TT, Lee CY, Sadykov MR, Somerville GA, Bodi K, Sonenshein AL.** 2010. Direct targets of CodY in *Staphylococcus aureus*. *J Bacteriol* **192**:2861–2877.

130. **Pohl K, Francois P, Stenz L, Schlink F, Geiger T, Herbert S, Goerke C, Schrenzel J, Wolz C.** 2009. CodY in *Staphylococcus aureus*: a regulatory link between metabolism and virulence gene expression. *J Bacteriol* **191**:2953–2963.

131. **Malke H, Ferretti JJ.** 2007. CodY-affected transcriptional gene expression of *Streptococcus*

pyogenes during growth in human blood. *J Med Microbiol* **56:**707–714.

132. **Kreth J, Chen Z, Ferretti J, Malke H.** 2011. Counteractive balancing of transcriptome expression involving CodY and CovRS in *Streptococcus pyogenes*. *J Bacteriol* **193:**4153–4165.

133. **Guedon E, Serror P, Ehrlich SD, Renault P, Delorme C.** 2001. Pleiotropic transcriptional repressor CodY senses the intracellular pool of branched-chain amino acids in *Lactococcus lactis*. *Mol Microbiol* **40:**1227–1239.

134. **Bennett HJ, Pearce DM, Glenn S, Taylor CM, Kuhn M, Sonenshein AL, Andrew PW, Roberts IS.** 2007. Characterization of relA and codY mutants of *Listeria monocytogenes*: identification of the CodY regulon and its role in virulence. *Mol Microbiol* **63:**1453–1467.

135. **Brinsmade SR, Kleijn RJ, Sauer U, Sonenshein AL.** 2010. Regulation of CodY activity through modulation of intracellular branched-chain amino acid pools. *J Bacteriol* **192:**6357–6368.

136. **Ratnayake-Lecamwasam M, Serror P, Wong KW, Sonenshein AL.** 2001. *Bacillus subtilis* CodY represses early-stationary-phase genes by sensing GTP levels. *Genes Dev* **15:**1093–1103.

137. **Handke LD, Shivers RP, Sonenshein AL.** 2008. Interaction of *Bacillus subtilis* CodY with GTP. *J Bacteriol* **190:**798–806.

138. **Brinsmade SR, Sonenshein AL.** 2011. Dissecting complex metabolic integration provides direct genetic evidence for CodY activation by guanine nucleotides. *J Bacteriol* **193:**5637–5648.

139. **Hendriksen WT, Bootsma HJ, Estevao S, Hoogenboezem T, de Jong A, de Groot R, Kuipers OP, Hermans PW.** 2008. CodY of *Streptococcus pneumoniae*: link between nutritional gene regulation and colonization. *J Bacteriol* **190:**590–601.

140. **Petranovic D, Guedon E, Sperandio B, Delorme C, Ehrlich D, Renault P.** 2004. Intracellular effectors regulating the activity of the *Lactococcus lactis* CodY pleiotropic transcription regulator. *Mol Microbiol* **53:**613–621.

141. **Levdikov VM, Blagova E, Colledge VL, Lebedev AA, Williamson DC, Sonenshein AL, Wilkinson AJ.** 2009. Structural rearrangement accompanying ligand binding in the GAF domain of CodY from *Bacillus subtilis*. *J Mol Biol* **390:**1007–1018.

142. **Kriel A, Bittner AN, Kim SH, Liu K, Tehranchi AK, Zou WY, Rendon S, Chen R, Tu BP, Wang JD.** 2012. Direct regulation of GTP homeostasis by (p)ppGpp: a critical component of viability and stress resistance. *Mol Cell* **48:**231–241.

143. **Kriel A, Brinsmade SR, Tse JL, Tehranchi A, Bittner A, Sonenshein AL, Wang JD.** 2014. GTP dysregulation in *Bacillus subtilis* cells lacking (p)ppGpp results in phenotypic amino acid auxotrophy and failure to adapt to nutrient downshift and regulate biosynthesis genes. *J Bacteriol* **196:**189–201.

144. **Geiger T, Goerke C, Fritz M, Schafer T, Ohlsen K, Liebeke M, Lalk M, Wolz C.** 2010. Role of the (p)ppGpp synthase RSH, a RelA/SpoT homolog, in stringent response and virulence of *Staphylococcus aureus*. *Infect Immun* **78:**1873–1883.

145. **Guedon E, Sperandio B, Pons N, Ehrlich SD, Renault P.** 2005. Overall control of nitrogen metabolism in *Lactococcus lactis* by CodY, and possible models for CodY regulation in Firmicutes. *Microbiology* **151:**3895–3909.

146. **Belitsky BR, Sonenshein AL.** 2008. Genetic and biochemical analysis of CodY-binding sites in *Bacillus subtilis*. *J Bacteriol* **190:**1224–1236.

147. **Belitsky BR, Sonenshein AL.** 2011. Contributions of multiple binding sites and effector-independent binding to CodY-mediated regulation in *Bacillus subtilis*. *J Bacteriol* **193:**473–484.

148. **Wray LV Jr, Fisher SH.** 2011. Bacillus subtilis CodY operators contain overlapping CodY binding sites. *J Bacteriol* **193:**4841–4848.

149. **Belitsky BR, Sonenshein AL.** 2013. Genome-wide identification of *Bacillus subtilis* CodY-binding sites at single-nucleotide resolution. *Proc Natl Acad Sci U S A* **110:**7026–7031.

150. **Shivers RP, Dineen SS, Sonenshein AL.** 2006. Positive regulation of *Bacillus subtilis* ackA by CodY and CcpA: establishing a potential hierarchy in carbon flow. *Mol Microbiol* **62:**811–822.

151. **Belitsky BR, Sonenshein AL.** 2011. Roadblock repression of transcription by *Bacillus subtilis* CodY. *J Mol Biol* **411:**729–743.

152. **Belitsky BR.** 2011. Indirect repression by *Bacillus subtilis* CodY via displacement of the activator of the proline utilization operon. *J Mol Biol* **413:**321–336.

153. **Karlsson S, Lindberg A, Norin E, Burman LG, Akerlund T.** 2000. Toxins, butyric acid, and other short-chain fatty acids are coordinately expressed and down-regulated by cysteine in *Clostridium difficile*. *Infect Immun* **68:**5881–5888.

154. **Dineen SS, Villapakkam AC, Nordman JT, Sonenshein AL.** 2007. Repression of *Clostridium difficile* toxin gene expression by CodY. *Mol Microbiol* **66:**206–219.

155. **Li J, Ma M, Sarker MR, McClane BA.** 2013. CodY is a global regulator of virulence-

associated properties for *Clostridium perfringens* type D strain CN3718. *mBio* **4**:e00770-00713.

156. **Montgomery CP, Boyle-Vavra S, Roux A, Ebine K, Sonenshein AL, Daum RS.** 2012. CodY deletion enhances in vivo virulence of community-associated methicillin-resistant *Staphylococcus aureus* clone USA300. *Infect Immun* **80**:2382–2389.

157. **Rivera FE, Miller HK, Kolar SL, Stevens SM Jr, Shaw LN.** 2012. The impact of CodY on virulence determinant production in community-associated methicillin-resistant *Staphylococcus aureus*. *Proteomics* **12**:263–268.

158. **Majerczyk CD, Sadykov MR, Luong TT, Lee C, Somerville GA, Sonenshein AL.** 2008. *Staphylococcus aureus* CodY negatively regulates virulence gene expression. *J Bacteriol* **190**:2257–2265.

159. **Novick RP.** 2003. Autoinduction and signal transduction in the regulation of staphylococcal virulence. *Mol Microbiol* **48**:1429–1449.

160. **Batzilla CF, Rachid S, Engelmann S, Hecker M, Hacker J, Ziebuhr W.** 2006. Impact of the accessory gene regulatory system (Agr) on extracellular proteins, *codY* expression and amino acid metabolism in *Staphylococcus epidermidis*. *Proteomics* **6**:3602–3613.

161. **Reiss S, Pane-Farre J, Fuchs S, Francois P, Liebeke M, Schrenzel J, Lindequist U, Lalk M, Wolz C, Hecker M, Engelmann S.** 2012. Global analysis of the *Staphylococcus aureus* response to mupirocin. *Antimicrob Agents Chemother* **56**:787–804.

162. **Tu Quoc PH, Genevaux P, Pajunen M, Savilahti H, Georgopoulos C, Schrenzel J, Kelley WL.** 2007. Isolation and characterization of biofilm formation-defective mutants of *Staphylococcus aureus*. *Infect Immun* **75**:1079–1088.

163. **van Schaik W, Chateau A, Dillies MA, Coppee JY, Sonenshein AL, Fouet A.** 2009. The global regulator CodY regulates toxin gene expression in *Bacillus anthracis* and is required for full virulence. *Infect Immun* **77**:4437–4445.

164. **Chateau A, van Schaik W, Six A, Aucher W, Fouet A.** 2011. CodY regulation is required for full virulence and heme iron acquisition in *Bacillus anthracis*. *FASEB J* **25**:4445–4456.

165. **Frenzel E, Doll V, Pauthner M, Lucking G, Scherer S, Ehling-Schulz M.** 2012. CodY orchestrates the expression of virulence determinants in emetic *Bacillus cereus* by impacting key regulatory circuits. *Mol Microbiol* **85**:67–88.

166. **Lindback T, Mols M, Basset C, Granum PE, Kuipers OP, Kovacs AT.** 2012. CodY, a pleiotropic regulator, influences multicellular behaviour and efficient production of virulence factors in *Bacillus cereus*. *Environ Microbiol* **14**:2233–2246.

167. **Hsueh YH, Somers EB, Wong AC.** 2008. Characterization of the codY gene and its influence on biofilm formation in *Bacillus cereus*. *Arch Microbiol* **189**:557–568.

168. **Lobel L, Sigal N, Borovok I, Ruppin E, Herskovits AA.** 2012. Integrative genomic analysis identifies isoleucine and CodY as regulators of *Listeria monocytogenes* virulence. *PLoS Genet* **8**:e1002887.

169. **Malke H, Steiner K, McShan WM, Ferretti JJ.** 2006. Linking the nutritional status of *Streptococcus pyogenes* to alteration of transcriptional gene expression: the action of CodY and RelA. *Int J Med Microbiol* **296**:259–275.

170. **Caymaris S, Bootsma HJ, Martin B, Hermans PW, Prudhomme M, Claverys JP.** 2010. The global nutritional regulator CodY is an essential protein in the human pathogen *Streptococcus pneumoniae*. *Mol Microbiol* **78**:344–360.

171. **Bouillaut L, Self WT, Sonenshein AL.** 2013. Proline-dependent regulation of *Clostridium difficile* Stickland metabolism. *J Bacteriol* **195**:844–854.

172. **Stickland LH.** 1935. Studies in the metabolism of the strict anaerobes (Genus *Clostridium*): The reduction of proline by Cl. sporogenes. *Biochem J* **29**:288–290.

173. **Jackson S, Calos M, Myers A, Self WT.** 2006. Analysis of proline reduction in the nosocomial pathogen *Clostridium difficile*. *J Bacteriol* **188**:8487–8495.

174. **Brekasis D, Paget MS.** 2003. A novel sensor of NADH/NAD+ redox poise in *Streptomyces coelicolor* A3(2). *EMBO J* **22**:4856–4865.

175. **Schau M, Chen Y, Hulett FM.** 2004. *Bacillus subtilis* YdiH is a direct negative regulator of the *cydABCD* operon. *J Bacteriol* **186**:4585–4595.

176. **Larsson JT, Rogstam A, von Wachenfeldt C.** 2005. Coordinated patterns of cytochrome bd and lactate dehydrogenase expression in *Bacillus subtilis*. *Microbiology* **151**:3323–3335.

177. **Gyan S, Shiohira Y, Sato I, Takeuchi M, Sato T.** 2006. Regulatory loop between redox sensing of the NADH/NAD(+) ratio by Rex (YdiH) and oxidation of NADH by NADH dehydrogenase Ndh in *Bacillus subtilis*. *J Bacteriol* **188**:7062–7071.

178. **Pagels M, Fuchs S, Pane-Farre J, Kohler C, Menschner L, Hecker M, McNamarra PJ, Bauer MC, von Wachenfeldt C, Liebeke M, Lalk M, Sander G, von Eiff C, Proctor RA, Engelmann S.** 2010. Redox sensing by a Rex-

family repressor is involved in the regulation of anaerobic gene expression in *Staphylococcus aureus*. *Mol Microbiol* **76:**1142–1161.

179. **Mehmeti I, Jonsson M, Fergestad EM, Mathiesen G, Nes IF, Holo H.** 2011. Transcriptome, proteome, and metabolite analyses of a lactate dehydrogenase-negative mutant of *Enterococcus faecalis* V583. *Appl Environ Microbiol* **77:**2406–2413.

180. **Wietzke M, Bahl H.** 2012. The redox-sensing protein Rex, a transcriptional regulator of solventogenesis in *Clostridium acetobutylicum*. *Appl Microbiol Biotechnol* **96:**749–761.

181. **Sickmier EA, Brekasis D, Paranawithana S, Bonanno JB, Paget MS, Burley SK, Kielkopf CL.** 2005. X-ray structure of a Rex-family repressor/NADH complex insights into the mechanism of redox sensing. *Structure* **13:**43–54.

182. **Richardson AR, Libby SJ, Fang FC.** 2008. A nitric oxide-inducible lactate dehydrogenase enables *Staphylococcus aureus* to resist innate immunity. *Science* **319:**1672–1676.

183. **Aboulnaga el H, Pinkenburg O, Schiffels J, El-Refai A, Buckel W, Selmer T.** 2013. Effect of an oxygen-tolerant bifurcating butyryl coenzyme A dehydrogenase/electron-transferring flavoprotein complex from *Clostridium difficile* on butyrate production in *Escherichia coli*. *J Bacteriol* **195:**3704–3713.

184. **Sohling B, Gottschalk G.** 1996. Molecular analysis of the anaerobic succinate degradation pathway in *Clostridium kluyveri*. *J Bacteriol* **178:**871–880.

185. **Sorensen KI, Hove-Jensen B.** 1996. Ribose catabolism of *Escherichia coli*: characterization of the *rpiB* gene encoding ribose phosphate isomerase B and of the rpiR gene, which is involved in regulation of *rpiB* expression. *J Bacteriol* **178:**1003–1011.

186. **Daddaoua A, Krell T, Ramos JL.** 2009. Regulation of glucose metabolism in *Pseudomonas*: the phosphorylative branch and Entner-Doudoroff enzymes are regulated by a repressor containing a sugar isomerase domain. *J Biol Chem* **284:**21360–21368.

187. **Kohler PR, Choong EL, Rossbach S.** 2011. The RpiR-like repressor IolR regulates inositol catabolism in *Sinorhizobium meliloti*. *J Bacteriol* **193:**5155–5163.

188. **Yamamoto H, Serizawa M, Thompson J, Sekiguchi J.** 2001. Regulation of the *glv* operon in *Bacillus subtilis*: YfiA (GlvR) is a positive regulator of the operon that is repressed through CcpA and cre. *J Bacteriol* **183:**5110–5121.

189. **Zhu Y, Nandakumar R, Sadykov MR, Madayiputhiya N, Luong TT, Gaupp R, Lee CY, Somerville GA.** 2011. RpiR homologues may link *Staphylococcus aureus* RNAIII synthesis and pentose phosphate pathway regulation. *J Bacteriol* **193:**6187–6196.

190. **Tojo S, Satomura T, Morisaki K, Yoshida K, Hirooka K, Fujita Y.** 2004. Negative transcriptional regulation of the *ilv-leu* operon for biosynthesis of branched-chain amino acids through the *Bacillus subtilis* global regulator TnrA. *J Bacteriol* **186:**7971–7979.

191. **Shivers RP, Sonenshein AL.** 2005. *Bacillus subtilis ilvB* operon: an intersection of global regulons. *Mol Microbiol* **56:**1549–1559.

192. **Tojo S, Satomura T, Morisaki K, Deutscher J, Hirooka K, Fujita Y.** 2005. Elaborate transcription regulation of the *Bacillus subtilis ilv-leu* operon involved in the biosynthesis of branched-chain amino acids through global regulators of CcpA, CodY and TnrA. *Mol Microbiol* **56:**1560–1573.

193. **Grandoni JA, Fulmer SB, Brizzio V, Zahler SA, Calvo JM.** 1993. Regions of the *Bacillus subtilis ilv-leu* operon involved in regulation by leucine. *J Bacteriol* **175:**7581–7593.

194. **Tojo S, Kumamoto K, Hirooka K, Fujita Y.** 2010. Heavy involvement of stringent transcription control depending on the adenine or guanine species of the transcription initiation site in glucose and pyruvate metabolism in *Bacillus subtilis*. *J Bacteriol* **192:**1573–1585.

195. **Jourlin-Castelli C, Mani N, Nakano MM, Sonenshein AL.** 2000. CcpC, a novel regulator of the LysR family required for glucose repression of the *citB* gene in *Bacillus subtilis*. *J Mol Biol* **295:**865–878.

196. **Mittal M, Pechter KB, Picossi S, Kim HJ, Kerstein KO, Sonenshein AL.** 2013. Dual role of CcpC protein in regulation of aconitase gene expression in *Listeria monocytogenes* and *Bacillus subtilis*. *Microbiology* **159:**68–76.

197. **Hartmann T, Zhang B, Baronian G, Schulthess B, Homerova D, Grubmuller S, Kutzner E, Gaupp R, Bertram R, Powers R, Eisenreich W, Kormanec J, Herrmann M, Molle V, Somerville GA, Bischoff M.** 2013. Catabolite control protein E (CcpE) is a LysR-type transcriptional regulator of TCA cycle activity in *Staphylococcus aureus*. *J Biol Chem* **288:**36116–361128.

198. **Crooke AK, Fuller JR, Obrist MW, Tomkovich SE, Vitko NP, Richardson AR.** 2013. CcpA-independent glucose regulation of lactate dehydrogenase 1 in *Staphylococcus aureus*. *PloS One* **8:**e54293.

199. **Ravcheev DA, Li X, Latif H, Zengler K, Leyn SA, Korostelev YD, Kazakov AE, Novichkov PS, Osterman AL, Rodionov DA.** 2012. Transcriptional regulation of central carbon and

energy metabolism in bacteria by redox-responsive repressor Rex. *J Bacteriol* **194:**1145–1157.

200. **Larson TJ, Ehrmann M, Boos W.** 1983. Periplasmic glycerophosphodiester phosphodiesterase of *Escherichia coli*, a new enzyme of the *glp* regulon. *J Biol Chem* **258:**5428–5432.

201. **Pitts AC, Tuck LR, Faulds-Pain A, Lewis RJ, Marles-Wright J.** 2012. Structural insight into the *Clostridium difficile* ethanolamine utilisation microcompartment. *PloS One* **7:**e48360.

202. **Garsin DA.** 2010. Ethanolamine utilization in bacterial pathogens: roles and regulation. *Nat Rev Microbiol* **8:**290–295.

203. **Joseph B, Przybilla K, Stuhler C, Schauer K, Slaghuis J, Fuchs TM, Goebel W.** 2006. Identification of *Listeria monocytogenes* genes contributing to intracellular replication by expression profiling and mutant screening. *J Bacteriol* **188:**556–568.

204. **Maadani A, Fox KA, Mylonakis E, Garsin DA.** 2007. *Enterococcus faecalis* mutations affecting virulence in the *Caenorhabditis elegans* model host. *Infect Immun* **75:**2634–2637.

205. **Newsholme P, Curi R, Gordon S, Newsholme EA.** 1986. Metabolism of glucose, glutamine, long-chain fatty acids and ketone bodies by murine macrophages. *Biochem J* **239:**121–125.

206. **Gordon S, Martinez FO.** 2010. Alternative activation of macrophages: mechanism and functions. *Immunity* **32:**593–604.

207. **Tugal D, Liao X, Jain MK.** 2013. Transcriptional control of macrophage polarization. *Arterioscler Thromb Vasc Biol* **33:**1135–1144.

208. **Mantovani A, Biswas SK, Galdiero MR, Sica A, Locati M.** 2013. Macrophage plasticity and polarization in tissue repair and remodelling. *J Pathol* **229:**176–185.

209. **Mahdavian Delavary B, van der Veer WM, van Egmond M, Niessen FB, Beelen RH.** 2011. Macrophages in skin injury and repair. *Immunobiology* **216:**753–762.

210. **Nizet V, Johnson RS.** 2009. Interdependence of hypoxic and innate immune responses. *Nat Rev Immunol* **9:**609–617.

211. **Palsson-McDermott EM, O'Neill LA.** 2013. The Warburg effect then and now: from cancer to inflammatory diseases. *Bioessays* **35:**965–973.

212. **Newsholme P, Gordon S, Newsholme EA.** 1987. Rates of utilization and fates of glucose, glutamine, pyruvate, fatty acids and ketone bodies by mouse macrophages. *Biochem J* **242:**631–636.

213. **Cramer T, Yamanishi Y, Clausen BE, Forster I, Pawlinski R, Mackman N, Haase VH, Jaenisch R, Corr M, Nizet V, Firestein GS, Gerber HP, Ferrara N, Johnson RS.** 2003.

214. **Hothersall JS, Gordge M, Noronha-Dutra AA.** 1998. Inhibition of NADPH supply by 6-aminonicotinamide: effect on glutathione, nitric oxide and superoxide in J774 cells. *FEBS Lett* **434:**97–100.

215. **Vats D, Mukundan L, Odegaard JI, Zhang L, Smith KL, Morel CR, Wagner RA, Greaves DR, Murray PJ, Chawla A.** 2006. Oxidative metabolism and PGC-1beta attenuate macrophage-mediated inflammation. *Cell Metab* **4:**13–24.

216. **Wu G, Bazer FW, Davis TA, Kim SW, Li P, Marc Rhoads J, Carey Satterfield M, Smith SB, Spencer TE, Yin Y.** 2009. Arginine metabolism and nutrition in growth, health and disease. *Amino Acids* **37:**153–168.

217. **De Groote MA, Fang FC.** 1995. NO inhibitions: antimicrobial properties of nitric oxide. *Clin Infect Dis* **21**(Suppl 2):**S162–S165.

218. **Shay JE, Celeste Simon M.** 2012. Hypoxia-inducible factors: crosstalk between inflammation and metabolism. *Semin Cell Dev Biol* **23:**389–394.

219. **Semenza GL.** 2010. HIF-1: upstream and downstream of cancer metabolism. *Curr Opin Genet Dev* **20:**51–56.

220. **Odegaard JI, Ricardo-Gonzalez RR, Goforth MH, Morel CR, Subramanian V, Mukundan L, Red Eagle A, Vats D, Brombacher F, Ferrante AW, Chawla A.** 2007. Macrophage-specific PPARgamma controls alternative activation and improves insulin resistance. *Nature* **447:**1116–1120.

221. **Lehrke M, Lazar MA.** 2005. The many faces of PPARgamma. *Cell* **123:**993–999.

222. **Kiss M, Czimmerer Z, Nagy L.** 2013. The role of lipid-activated nuclear receptors in shaping macrophage and dendritic cell function: From physiology to pathology. *J Allergy Clin Immunol* **132:**264–286.

223. **Mandard S, Patsouris D.** 2013. Nuclear control of the inflammatory response in mammals by peroxisome proliferator-activated receptors. *PPAR Res* **2013:**613864.

224. **Bloch H, Segal W.** 1956. Biochemical differentiation of *Mycobacterium tuberculosis* grown in vivo and in vitro. *J Bacteriol* **72:**132–141.

225. **Schnappinger D, Ehrt S, Voskuil MI, Liu Y, Mangan JA, Monahan IM, Dolganov G, Efron B, Butcher PD, Nathan C, Schoolnik GK.** 2003. Transcriptional adaptation of *Mycobacterium tuberculosis* within macrophages: insights into the phagosomal environment. *J Exp Med* **198:**693–704.

226. **de Carvalho LP, Fischer SM, Marrero J, Nathan C, Ehrt S, Rhee KY.** 2010. Metabolomics

of *Mycobacterium tuberculosis* reveals compartmentalized co-catabolism of carbon substrates. *Chem Biol* **17:**1122–1131.

227. **Marrero J, Rhee KY, Schnappinger D, Pethe K, Ehrt S.** 2010. Gluconeogenic carbon flow of tricarboxylic acid cycle intermediates is critical for *Mycobacterium tuberculosis* to establish and maintain infection. *Proc Natl Acad Sci U S A* **107:**9819–9824.

228. **Munoz-Elias EJ, McKinney JD.** 2005. *Mycobacterium tuberculosis* isocitrate lyases 1 and 2 are jointly required for in vivo growth and virulence. *Nat Med* **11:**638–644.

229. **Gould TA, van de Langemheen H, Munoz-Elias EJ, McKinney JD, Sacchettini JC.** 2006. Dual role of isocitrate lyase 1 in the glyoxylate and methylcitrate cycles in *Mycobacterium tuberculosis*. *Mol Microbiol* **61:**940–947.

230. **Brock M, Buckel W.** 2004. On the mechanism of action of the antifungal agent propionate. *Eur J Biochem* **271:**3227–3241.

231. **Rocco CJ, Escalante-Semerena JC.** 2010. In *Salmonella enterica*, 2-methylcitrate blocks gluconeogenesis. *J Bacteriol* **192:**771–778.

232. **Lee W, VanderVen BC, Fahey RJ, Russell DG.** 2013. Intracellular *Mycobacterium tuberculosis* exploits host-derived fatty acids to limit metabolic stress. *J Biol Chem* **288:**6788–6800.

233. **Savvi S, Warner DF, Kana BD, McKinney JD, Mizrahi V, Dawes SS.** 2008. Functional characterization of a vitamin B12-dependent methylmalonyl pathway in *Mycobacterium tuberculosis*: implications for propionate metabolism during growth on fatty acids. *J Bacteriol* **190:**3886–3895.

234. **Rajaram MV, Brooks MN, Morris JD, Torrelles JB, Azad AK, Schlesinger LS.** 2010. *Mycobacterium tuberculosis* activates human macrophage peroxisome proliferator-activated receptor gamma linking mannose receptor recognition to regulation of immune responses. *J Immunol* **185:**929–942.

235. **Mahajan S, Dkhar HK, Chandra V, Dave S, Nanduri R, Janmeja AK, Agrewala JN, Gupta P.** 2012. *Mycobacterium tuberculosis* modulates macrophage lipid-sensing nuclear receptors PPARgamma and TR4 for survival. *J Immunol* **188:**5593–5603.

236. **Almeida PE, Carneiro AB, Silva AR, Bozza PT.** 2012. PPARgamma expression and function in mycobacterial infection: roles in lipid metabolism, immunity, and bacterial killing. *PPAR Res* **2012:**383829.

237. **Chatterjee SS, Hossain H, Otten S, Kuenne C, Kuchmina K, Machata S, Domann E,**

Chakraborty T, Hain T. 2006. Intracellular gene expression profile of *Listeria monocytogenes*. *Infect Immun* **74:**1323–1338.

238. **Joseph B, Przybilla K, Stuhler C, Schauer K, Slaghuis J, Fuchs TM, Goebel W.** 2006. Identification of *Listeria monocytogenes* genes contributing to intracellular replication by expression profiling and mutant screening. *J Bacteriol* **188:**556–568.

239. **Chico-Calero I, Suarez M, Gonzalez-Zorn B, Scortti M, Slaghuis J, Goebel W, Vazquez-Boland JA.** 2002. Hpt, a bacterial homolog of the microsomal glucose- 6-phosphate translocase, mediates rapid intracellular proliferation in *Listeria*. *Proc Natl Acad Sci U S A* **99:**431–436.

240. **Stoll R, Goebel W.** 2010. The major PEP-phosphotransferase systems (PTSs) for glucose, mannose and cellobiose of *Listeria monocytogenes*, and their significance for extra- and intracellular growth. *Microbiology* **156:**1069–1083.

241. **Gillmaier N, Gotz A, Schulz A, Eisenreich W, Goebel W.** 2012. Metabolic responses of primary and transformed cells to intracellular *Listeria monocytogenes*. *PloS One* **7:**e52378.

242. **Purves J, Cockayne A, Moody PC, Morrissey JA.** 2010. Comparison of the regulation, metabolic functions, and roles in virulence of the glyceraldehyde-3-phosphate dehydrogenase homologues gapA and gapB in *Staphylococcus aureus*. *Infect Immun* **78:**5223–5232.

243. **Hochgrafe F, Wolf C, Fuchs S, Liebeke M, Lalk M, Engelmann S, Hecker M.** 2008. Nitric oxide stress induces different responses but mediates comparable protein thiol protection in *Bacillus subtilis* and *Staphylococcus aureus*. *J Bacteriol* **190:**4997–5008.

244. **Richardson AR, Payne EC, Younger N, Karlinsey JE, Thomas VC, Becker LA, Navarre WW, Castor ME, Libby SJ, Fang FC.** 2011. Multiple targets of nitric oxide in the tricarboxylic acid cycle of *Salmonella enterica* serovar *typhimurium*. *Cell Host Microbe* **10:**33–43.

245. **Richardson AR, Dunman PM, Fang FC.** 2006. The nitrosative stress response of *Staphylococcus aureus* is required for resistance to innate immunity. *Mol Microbiol* **61:**927–939.

246. **Fuller JR, Vitko NP, Perkowski EF, Scott E, Khatri D, Spontak JS, Thurlow LR, Richardson AR.** 2011. Identification of a lactate-quinone oxidoreductase in *Staphylococcus aureus* that is essential for virulence. *Front Cell Infect Microbiol* **1:**19.

247. **Weiner MA, Read TD, Hanna PC.** 2003. Identification and characterization of the *gerH* operon of *Bacillus anthracis* endospores:

a differential role for purine nucleosides in germination. *J Bacteriol* **185:**1462–1464.

248. **Sorg JA, Sonenshein AL.** 2008. Bile salts and glycine as cogerminants for *Clostridium difficile* spores. *J Bacteriol* **190:**2505–2512.

249. **Janoir C, Deneve C, Bouttier S, Barbut F, Hoys S, Caleechum L, Chapeton-Montes D, Pereira FC, Henriques AO, Collignon A, Monot M, Dupuy B.** 2013. Adaptive strategies and pathogenesis of *Clostridium difficile* from in vivo transcriptomics. *Infect Immun* **81:**3757–3769.

250. **Antunes LC, Han J, Ferreira RB, Lolic P, Borchers CH, Finlay BB.** 2011. Effect of antibiotic treatment on the intestinal metabolome. *Antimicrob Agents Chemother* **55:**1494–1503.

251. **Ng KM, Ferreyra JA, Higginbottom SK, Lynch JB, Kashyap PC, Gopinath S, Naidu N, Choudhury B, Weimer BC, Monack DM, Sonnenburg JL.** 2013. Microbiota-liberated host sugars facilitate post-antibiotic expansion of enteric pathogens. *Nature* **502:**96–99.

252. **Roux A, Todd DA, Velázquez JV, Cech NB, Sonenshein AL.** 2014. CodY-mediated regulation of the *Staphylococcus aureus* Agr system integrates nutritional and population density signals. *J Bacteriol* **196:**1184–1196.

253. **Vitko NP, Spahich NA, Richardson AR.** 2015. Glycolytic dependency of high-level nitric oxide resistance and virulence in *Staphylococcus aureus*. *MBio.* **6**(2):0045–15. doi: 10.1128/mBio.00045-15.

Borrelia burgdorferi: Carbon Metabolism and the Tick-Mammal Enzootic Cycle

8

ARIANNA CORONA[1] and IRA SCHWARTZ[1]

THE ENZOOTIC CYCLE

Borrelia burgdorferi is the spirochetal agent of Lyme disease, the most commonly reported arthropod-borne disease in the United States (1–3). *B. burgdorferi* is a zoonotic pathogen that is maintained in a natural cycle involving mammalian reservoir hosts such as field mice, squirrels, and birds and an arthropod vector of the *Ixodes* species (4–6) (Fig. 1). In the United States, the principal vector is *Ixodes scapularis*, the common deer tick (5, 6). Because there is no transovarial transmission of *B. burgdorferi*, newly hatched larvae acquire the spirochete during their first blood meal on an infected mammalian host reservoir (7, 8). The spirochete is maintained in the midgut of the tick during molting to the nymphal stage. At this point, the spirochete is in a nonmotile state until the nymph begins to feed on the next mammalian host (9). The spirochete then begins rapidly replicating in the feeding nymphal midgut, leaves the midgut and enters the hemolymph, from which the bacteria migrate to the salivary glands and are transmitted to the next mammalian host (9–12) (Fig. 1).

During each stage of the enzootic cycle, *B. burgdorferi* is exposed to different environments. Each milieu varies by temperature, pH, small molecules, and most important, nutrient sources. The drastic changes in environmental

[1]Department of Microbiology and Immunology, New York Medical College, Valhalla, NY.
Metabolism and Bacterial Pathogenesis
Edited by Tyrrell Conway and Paul Cohen
© 2015 American Society for Microbiology, Washington, DC
doi:10.1128/microbiolspec.MBP-0011-2014

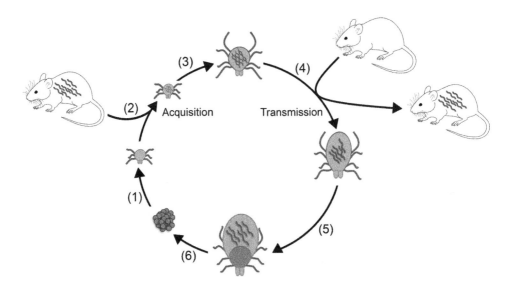

FIGURE 1 *Borrelia burgdorferi* **enzootic cycle. (1) Uninfected larva emerges from eggs. (2) Larval acquisition of** *B. burgdorferi* **during a blood meal on an infected reservoir host. (3) Infected fed larva molts to an unfed nymph. (4) Transmission of** *B. burgdorferi* **from a feeding nymph to an uninfected reservoir host during the nymphal blood meal. (5) Infected fed nymph molts to an adult. (6) Female and male adults mate on a large mammal (typically deer). The female adult feeds on the large mammal and lays eggs. doi:10.1128/microbiolspec.MBP-0011-2014.f1**

attributes require the spirochete to not only modulate the expression of colonization factors necessary for persistence in either the vector or the mammalian host, but also to adjust its metabolic state to adapt to the changing nutrient profile. While in the mammalian host, *B. burgdorferi* can be found in the skin at the tick bite site, in the circulatory system, and at distal tissue sites such as joints and heart (13). During larval acquisition, spirochetes enter the midgut concurrent with the blood meal and its associated nutrients and host factors. After molting to the nymphal stage, *B. burgdorferi* is confined to the midgut lumen of the unfed nymph. The lumen is a nutrient-poor environment consistent with the metabolically dormant state of the spirochetes at this stage (9, 14). When the nymph begins to feed, the incoming blood meal is surrounded by a peritrophic membrane that sequesters the blood meal and its accompanying nutrients away from the spirochetes within the midgut lumen (15). The ability to use a variety of

available carbohydrates during the enzootic cycle is likely essential for the survival of *B. burgdorferi* during the tick phases and pathogenesis in mammalian hosts, as has been observed for other pathogenic bacteria (16).

B. BURGDORFERI GENOMICS

B. burgdorferi has a complex genome. The segmented genome of the B31 type strain consists of one large linear chromosome (about 910 Kb) and 21 circular (cp) and linear plasmids (lp) (17, 18). A substantial portion of the predicted open reading frames (ORFs) is annotated as hypothetical or as having homology to other spirochete genes without any functional annotation (17, 18). ORFs located on the linear chromosome encode primarily housekeeping functions, including most genes associated with metabolism (17, 19). Plasmid content in individual *B. burgdorferi* isolates is variable, although lp54 and cp26 are uniformly present in most characterized isolates (20–22). Plasmids can be lost during *in vitro* cultivation, which typically has no effect on *in vitro* growth; however, several plasmids play an essential role *in vivo* (22–30). cp26 is absolutely required for survival of the bacterium (25, 31); it encodes several proteins involved in purine biosynthesis, transport, telomere resolution, and outer surface protein C (OspC), an essential virulence factor (17, 25, 32–34). lp25 is required for mammalian infection (35). It has been shown that the critical gene on lp25 is *pncA*, which encodes a nicotinamidase and is likely involved in NAD$^+$ salvage (24, 32, 36). In addition, lp28-1 is required for persistence during mammalian infection, likely as the result of antigenic variation in *vlsE* (23, 35, 37–39).

REGULATION OF GENE EXPRESSION DURING THE ENZOOTIC CYCLE

RpoS

Bacteria typically regulate gene expression in response to environmental cues through processes that are often mediated by alternative sigma factors or two-component signaling systems (TCS) (40, 41). The *B. burgdorferi* genome encodes for only two alternative sigma factors and two TCS (17). Thus, to adapt to the different environments encountered during the enzootic cycle, *B. burgdorferi* must rely on a limited repertoire of such components. Seminal studies by Norgard and colleagues demonstrated a link between one TCS, specifically Hk2-Rrp2, and the alternative sigma factors RpoN and RpoS (42, 43). The expression of several virulence genes is dependent on RpoS (42, 44). RpoS is also essential for repression of genes whose expression is required in the tick but not in the mammalian host (45). As a result, RpoS is absolutely essential for mammalian infection as well as migration through the tick during transmission (44, 46). Global transcriptome analyses of wild type and RpoS mutant

strains under mammalian-like conditions defined the RpoS regulon and included both genes that are induced and repressed by RpoS. On this basis, RpoS has been referred to as a "gatekeeper" that controls the reciprocal expression of genes required for mammalian infection or maintenance in ticks (47).

In *B. burgdorferi*, RpoS expression is controlled by multiple layers of transcriptional and posttranscriptional regulation (48). The Rrp2-RpoN-RpoS pathway is required for transcription of *rpoS* (49–51). RpoN binds directly to a canonical -24/-12 sequence in the *rpoS* promoter to induce transcription (52). This interaction requires the activation of Rrp2, the encoded response regulator of the Hk2-Rrp2 two-component system (43, 50). It was assumed that Rrp2 would be phosphorylated and activated by Hk2; however, it was later demonstrated that Rrp2 could induce virulence gene expression independent of Hk2 (53). Xu et al subsequently showed that acetyl phosphate can function as the phosphate group donor for Rrp2 (54). RpoS expression is also modulated by several other effectors, including BosR (discussed later), DsrA (a small RNA), Hfq (an RNA chaperone), and CsrA (BB184) (55–58), although the role of CsrA has been recently challenged (59). It has also been suggested that BadR (BB693), a protein with sequence homology to the ROK family of proteins, represses the transcription of *rpoS* (60).

BosR

B. burgdorferi encodes a Fur/Per homolog, BosR (*Borrelia* oxidative stress regulator) (17, 61, 62). Its role in mediating the oxidative stress response is not clear, as *B. burgdorferi* BosR mutants are only slightly more sensitive to oxidative stress than wild-type spirochetes (63, 64). Several studies have demonstrated that BosR is a transcriptional activator of *rpoS* (63, 65, 66). Ouyang et al identified a "BosR box" to which the protein binds and demonstrated that the *rpoS* promoter contains three such binding sites; interestingly, bioinformatic analysis revealed the presence of 60 additional *cis*-acting "BosR boxes" in the *B. burgdorferi* genome, suggesting that BosR likely regulates the expression of additional genes in addition to its role as an activator of RpoS (67, 68). More recently, BosR was reported to be directly involved in repression of lipoprotein gene expression (69). The environmental signal or signals that control *bosR* expression have not been fully elucidated. It has been suggested that transition metals may be involved, because *bosR* expression is Zn^{2+}-dependent and is posttranscriptionally inhibited by Mn^{2+} (61, 70).

Hk1-Rrp1

The second TCS in *B. burgdorferi* has been designated as Hk1-Rrp1. The response regulator Rrp1 contains a GGDEF motif characteristic of diguanylate cyclases that convert two GTP molecules to a molecule of c-di-GMP and is the sole protein in the genome containing this motif (17, 71). Ryjenkov et al demonstrated that Rrp1 functions as a diguanylate cyclase and its activity is dependent on phosphorylation of its receiver domain (72). The secondary messenger c-di-GMP has gained attention as a global regulator in bacteria that is associated with virulence, motility, and central metabolism (73, 74). Rrp1 receives its signal from the membrane-bound sensor histidine kinase Hk1 (75, 76). *B. burgdorferi* also encodes the other components of the c-di-GMP signaling system, including two phosphodiesterases (PdeA [BB363] and PdeB [BB374]) and a cyclic-di-GMP binding protein PlzA (BB733) (17, 77). The *hk1-rrp1* operon appears to be constitutively expressed (76), although *rrp1* and *plzA* expression may be elevated during tick feeding (71, 77). Both Rrp1 and Hk1 deletion mutants are infectious in mice but are unable to survive in the tick vector (76, 78, 79); Rrp1 mutants also have defective motility (76, 79). Global transcriptome analyses of wild-type and Rrp1 mutant spirochetes revealed that c-di-GMP regulates the expression of a substantial number of genes (71, 78). Taken together, the data suggest that c-di-GMP signaling plays a critical role in tick colonization (80).

CARBOHYDRATE METABOLISM

B. burgdorferi has a very restricted metabolic capacity. Genes encoding functions related to carbohydrate transport and metabolism are listed in Table 1. The genome encodes enzymes of the glycolytic pathway but not of the tricarboxylic acid cycle or oxidative phosphorylation (17). The spirochete does encode the oxidative branch of the pentose phosphate pathway (17), but because ribose cannot support *in vitro* growth when supplied as the principal carbon source (81), this pathway is not likely to play a role in energy production. Thus, *B. burgdorferi* relies solely on glycolysis for ATP generation. The bacterium does not encode any complete pathways for *de novo* biosynthesis of fatty acids, amino acids, or nucleotides (17). For this reason, *B. burgdorferi* is completely dependent on the transport of nutrients and cofactors from extracellular sources and several salvage pathways (17, 25, 32, 33, 82–84). Consequently, *B. burgdorferi* has more than 50 genes encoding transporters for carbohydrates, oligopeptides, and amino acids (17, 32).

Typically, glycolysis yields a net of only two or three ATP molecules per glucose molecule; the reliance on only glycolysis for production of ATP could account for the spirochete's slow growth, even under optimal *in vitro* cultivation conditions. *B. burgdorferi* is not only restricted to the carbohydrate sources available in the distinct environments it encounters during the enzootic cycle but also by the limited number of encoded carbohydrate uptake systems and catabolic pathways (17). von Lackum and Stevenson reported that *B. burgdorferi* can use six carbohydrates as the principal carbon source during *in vitro* growth in BSK medium—glucose, glycerol, maltose, mannose, N-acetylglucosamine (GlcNAc), and chitobiose

(81). More recently, Hoon-Hanks et al demonstrated that trehalose can also support *B. burgdorferi in vitro* growth (85).

Glycolysis

B. burgdorferi encodes all enzymes of the glycolytic pathway (17). Glucose enters as either glucose (via a putative ABC transporter [dicussed later]) or as glucose 6-phosphate (via a phosphoenolpyruvate-phosphotransferase system [PEP-PTS]) (Fig. 2). Acquired glucose would be converted to glucose 6-phosphate by the action of a putative glucokinase (BB831) and then to fructose 6-phosphate by glucose 6-phosphate isomerase (Pgi, BB730) (Fig. 3). *B. burgdorferi* encodes two enzymes with putative phosphofructokinase activity (BB020, BB727). Both genes have been expressed and characterized *in vitro*; BB020 is an active pyrophosphate-dependent phosphofructokinase that functions as a dimer (86, 87), whereas BB727 exists as a multimer in solution, with no measurable enzymatic activity *in vitro* (86). The fact that phosphofructokinase uses pyrophosphate as the phosphate donor rather than ATP is of particular interest because this would conserve intracellular ATP and increase the ATP yield per molecule of glucose. Cleavage of fructose 1,6-bisphosphate to dihydroxyacetone phosphate and glyceraldehyde-3-phosphate and their ultimate conversion to pyruvate proceeds by the action of the glycolytic pathway, as expected. *tpi*, *pgk*, and *gapdh* comprise a single operon in *B. burgdorferi* (*bb055-057*) and are transcribed from a single promoter; this type of genetic organization has also been observed in other bacteria (88). *B. burgdorferi* does not encode for a pyruvate dehydrogenase or pyruvate oxidase (17); as a result, the only disposition of pyruvate is conversion to lactate by lactate dehydrogenase. This results in regeneration of NAD^+ that is required for continued glycolysis (Fig. 3). The potential role of $NADH/NAD^+$ ratio in regulating metabolite flux through the glycolytic pathway is discussed later in this chapter.

Other Carbohydrate Utilization Pathways

Glucose is the preferred carbohydrate for ATP generation, as in most bacteria. In addition to glycolysis, glucose-6-P can also enter the oxidative branch of the pentose phosphate pathway through conversion to 6-phosphogluconolactone through the action of glucose 6-phosphate dehydrogenase. As noted, several additional carbohydrates can support *B. burgdorferi* growth *in vitro*. Mannose can be taken up via a mannose-specific PTS, and the resultant mannose-6-phosphate is converted to fructose-6-phosphate by the action of ManA (BB407) and can be directed into glycolysis (17) (Table 1).

B. burgdorferi requires GlcNAc to reach high density during *in vitro* cultivation, making its import and use an absolute requirement for growth (81, 89, 90). It is assumed that GlcNAc is taken up through a glucose-specific PTS. It is used for peptidoglycan biosynthesis, but it can also be converted to fructose-6-phosphate through the combined actions of NagA (BB151) and NagB (BB152) and thereby used as a substrate for glycolysis. Chitobiose, a dimer of GlcNAc, is a constituent of the tick cuticle and peritrophic membrane (15, 91). It has been demonstrated that chitobiose can substitute for the GlcNAc requirement during *in vitro* growth (90, 92). Studies have shown that a chitobiose-specific PTS is encoded on plasmid cp26 (*bbb04*, *bbb05*, *bbb06*) (Fig. 2) and that ChbC mutants cannot use chitobiose to support *in vitro* growth (90). After uptake via the chitobiose-specific PTS, chitobiose-6-P is cleaved into GlcNAc and GlcNAc-6-P by chitobiase (BB002), and the monomers enter the glycolytic pathway as fructose-6-P. The disaccharides maltose and trehalose can also be used after hydrolysis to glucose monomers by either

TABLE 1 *Borrelia burgdorferi* genes encoding proteins involved in carbohydrate metabolism

Gene Locus	Annotated Name[a]
Transporters	
bb448	PTS; phosphocarrier protein Hpr
bb557	PTS; phosphocarrier protein Hpr, *ptsH*
bb558	PTS; phosphoenolpyruvate protein phosphocarrier EI, *pstP*
bb559	PTS; glucose-specific EIIA, *crr*
bb645	PTS; glucose-specific EIIBC, *ptsG*
bb116	PTS; glucose-specific EIIABC, *malX1*
bbb29	PTS; glucose-specific EIIABC, *malX2*
bb408	PTS; fructose, mannose-specific EIIABC, *fruA1*
bb629	PTS; fructose, mannose-specific EIIABC, *fruA2*
bbb04	PTS; chitobiose-specific EIIC, *chbC*
bbb05	PTS; chitobiose-specific EIIA, *chbA*
bbb06	PTS; chitobiose-specific EIIB, *chbB*
bb240	glycerol facilitator, *glpF*
bb677	ABC transporter (glucose, ribose, galactose), ATP-binding protein, *mglA*
bb678	ABC transporter (glucose, ribose, galactose), permease protein, $mglC_1$
bb679	ABC transporter (glucose, ribose, galactose), permease protein, $mglC_2$
bb604	lactate permease, *lctP*
Glycolysis	
bb730	glucose-6-phosphate isomerase, *pgi*
bb727	phosphofructokinase, *pfk*
bb020	diphosphate-fructose-6-phosphate-1-phosphotransferase, *pfpB*
bb445	fructose-bisphosphate aldolase, class II, *fbaA*
bb055	triose-phosphate isomerase, *tpiA*
bb056	phosphoglycerate kinase, *pgk*
bb658	phosphoglycerate mutase
bb337	enolase, *eno*
bb348	pyruvate kinase, *pyk*
bb087	L-lactate dehydrogenase, *ldh*
Pentose Phosphate Pathway	
bb636	glucose-6-phosphate-1-dehydrogenase, *zwf*
bb222	6-phosphoglyconolactonase, *pgl*
bb561	6-phophogluconate dehydrogenase, *gnd*
bb657	ribose-5-phosphate isomerase, *rpi*
Other Carbohydrate Utilization Pathways	
Mannose	
bb407	mannose-6-phosphate isomerase, class I, *manA*
bb630	1-phosphofructokinase, *pfkB*
bb835	phosphomannomutase
bb644	N-acetylmannosamine-6-phosphate epimerase
GlcNAc	
bb004	phosphoglucomutase
bb151	N-acetylglucosamine-6-phosphate isomerase, *nagA*
bb152	glucosamine-6-phosphate isomerase, *nagB*
Chitobiose	
bb002	chitobiase
bb620	beta-glucosidase
bb831	glucokinase
Maltose	
bb116	4-alpha-glucanotransferase, *malQ*
Trehalose	
bb381	trehalase, *treA*

(Continued on next page)

TABLE 1 *Borrelia burgdorferi* **genes encoding proteins involved in carbohydrate metabolism** *(Continued)*

Gene Locus	Annotated Name[a]
Glycerol	
bb241	glycerol kinase, *glpK*
bb243	glycerol-3-phosphate dehydrogenase, *glpD*

[a]Annotations based on Fraser et al (17) and NCBI *Borrelia burgdorferi* B31 genome (NC_001318.1 and NC_001903.1).

amylomaltase (MalQ; BB166) or trehalase (TreA; BB381) (Fig. 2).

In addition to monosaccharides and disaccharides, *B. burgdorferi* also contains a pathway for uptake and use of glycerol. The genes comprise an operon (*bb240-bb243*), which encodes an uptake facilitator (GlpF), glycerol kinase (GlpK), and glycerol-3-phosphate dehydrogenase (GlpD) (17). Once glycerol has entered into the cytosol, it is phosphorylated by GlpK to yield glycerol-3-phosphate; GlpD converts glycerol-3-phosphate to dihydroxy-acetone phosphate, which can enter the glycolytic pathway after conversion to glyceraldehyde 3-phosphate (81) (Fig. 2).

Carbohydrate Transporters

Phosphotransferase systems (PTS or PEP-PTS) simultaneously import and phosphorylate a sugar substrate by coupling phosphorelay of a phosphoryl group from phosphoenolpyruvate (PEP) to carbohydrate transport (93). PTS are composed of a number of proteins referred to

as enzyme I (EI), enzyme II (EII), and histidine phosphocarrier protein (HPr). EII is composed of two cytoplasmic domains (EIIA, EIIB) and a transmembrane domain (EIIC). These domains can be located on a single polypeptide or separate protein molecules (94). PEP transfers a phosphoryl group to EI, which in turn transfers it to HPr and, ultimately, the phosphate is transferred to the sugar concomitant with its EIIC-mediated uptake (94, 95). Specificity of these systems is defined by the EII components downstream of the initial phosphorelay among PEP, EI, and HPr (93, 94).

Based on genome sequence, *B. burgdorferi* contains the complete PTS machinery (17) (Table 1). EI is encoded by *bb558* and, interestingly, two genes are annotated as encoding HPr (*bb557*, *bb448*). It is important to note, however, that very few of the PTS components have been definitively shown to function in their putative roles by direct biochemical or genetic studies. A schematic diagram showing the specific carbohydrate transporters is

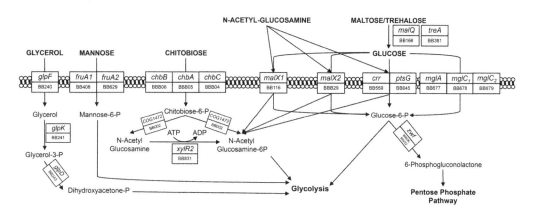

FIGURE 2 *Borrelia burgdorferi* **carbohydrate transporters. Schematic diagram indicates predicted or experimentally verified transport systems.** *B. burgdorferi* **numbers indicate gene locus in** *B. burgdorferi* **strain B31 (17). Based on von Lackum and Stevenson (81). doi:10.1128/microbiolspec.MBP-0011-2014.f2**

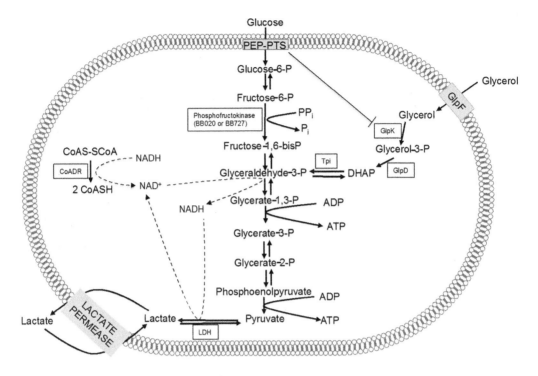

FIGURE 3 The glycolytic pathway and control of glycolytic flux during the enzootic cycle. doi:10.1128/microbiolspec.MBP-0011-2014.f3

presented in Fig. 2. Glucose import is mediated by the glucose-specific EII proteins (BB559, BB645, BB116, and BBB29); by homology, BB559 (EIIA) and BB645 (EIIBC) are annotated as glucose-specific, whereas BB116 and BBB29 (EIIABC) are annotated as maltose- and glucose-specific (17). GlcNAc can also be transported into the cell via these EII components. It was thought that the disaccharides, maltose (α1-4-glucose-glucose) and trehalose (α1-1-glucose-glucose), are cleaved into two molecules of glucose by MalQ or TreA before transport into the cytosol (17, 96). However, it was recently demonstrated that *malQ* mutants are able to grow normally when maltose or trehalose are provided as the principal carbon source *in vitro*, and these mutants successfully complete the experimental mouse-tick-mouse enzootic cycle (85). It is possible that a simultaneous disruption of both enzymes might prevent use of both disaccharides, but this has not been tested.

B. burgdorferi encodes two mannose-specific EIIABC components (BB408, BB629). They were initially annotated as fructose-specific (17), but fructose does not support *in vitro* growth (81). As a result, they are now presumed to be mannose-specific transporters (32, 81). In addition, the spirochete also possesses a dedicated chitobiose PTS (17). *chbC* (*bbb04*) is required for chitobiose transport and is induced during the tick phase of the enzootic cycle (12, 90, 92). Although the genes for the three EII components are adjacently located on cp26, the gene encoding the EIIC component, *chbC*, is divergently transcribed from those for *chbA* (*bbb05*) and *chbB* (*bbb06*); these latter genes are apparently not differentially expressed in ticks (92).

Glucose could be transported into the cytosol by a putative ABC transporter MglAC$_1$C$_2$ (BB677-79) (17, 32, 97). This operon was initially annotated to encode a ribose/galactose transporter. It is unlikely that either sugar is used for ATP production, because neither can support spirochetal growth *in vitro* (81). *B. burgdorferi* membranes contain mongalactosyl diacylglycerol (98), and its synthesis would require a source of galactose. If galactose were imported through the MglAC$_1$C$_2$ transporter, a uridylyltransferase would be required for synthesis of UDP-galactose; such an activity has not been identified in the *B. burgdorferi* genome. Alternatively, UDP-galactose could be produced from UDP-glucose by an epimerase (BB444) (32). If ribose were imported via this ABC transporter, it could be used for nucleotide biosynthesis. The route from ribose to ribose-5-phosphate is unclear, but *bb545* is annotated as a xylulokinase, and it has been suggested that this enzyme could convert ribose to ribulose-5-phosphate and then ribose 5-phosphate through the action of ribose phosphate isomerase (BB657) (32). Ribose-5-phosphate could also be produced from glucose-6-phosphate via the oxidative branch of the pentose-phosphate pathway. Taken together, it seems unlikely that the MglAC$_1$C$_2$ transporter is used for galactose or ribose uptake; rather, it could serve as an additional route for glucose transport.

B. burgdorferi encodes one carbon-specific major facilitator super family protein, GlpF (17). GlpF is a member of a family of conserved aquaglyceroporins that mediate diffusion of glycerol into the cytosol (99, 100). Interestingly, expression of the *glp* operon (*bb240-243*) is significantly induced during the tick phase of the spirochete enzootic cycle, and glycerol use is vital for maximal fitness of the spirochetes in the tick vector (12).

NAD⁺/NADH BALANCE

The balance between NAD⁺ and NADH is an indicator of the intracellular redox state for both eukaryotic and prokaryotic cells, and NAD⁺ is a required cofactor for many cellular enzymes (101, 102). During glycolysis, NAD⁺ is reduced to NADH by GAPDH, and the cell must regenerate NAD⁺ to maintain a balanced redox state and allow glycolysis to continue (Fig. 3). NAD⁺ pools can be replenished by biosynthesis or oxidative metabolism of pyruvate. Many bacteria encode biosynthetic pathways that use tryptophan or aspartic acid to generate NAD⁺ (101, 102). *B. burgdorferi* does not encode this biosynthetic capacity and cannot metabolize pyruvate oxidatively (17). Instead, it must depend on alternative strategies to replenish NAD⁺. As noted, the only metabolic fate for pyruvate is its conversion to lactate by pyruvate dehydrogenase. During lactogenesis, NADH is oxidized, producing NAD⁺ that can be recycled back into the glycolytic pathway (103), which directly couples NAD⁺ regeneration to glycolysis.

In an oxidative environment, organisms must deal with the presence of reactive oxygen species. Among the consequences is formation of the disulphide form of coenzyme A (CoA). To deal with this, *B. burgdorferi* produces a CoA disulphide reductase (CoADR) (encoded by *bb728*) that uses NADH exclusively as a cofactor, producing reduced CoA and regenerating NAD⁺ (104, 105). A *B. burgdorferi* CoADR mutant was avirulent in mice and had reduced survival in feeding nymphs, suggesting an important role in maintaining an optimal redox state in the spirochete (105).

Microbes contain pathways for nicotinamide salvage that can be employed for NAD⁺ production (102, 106). *B. burgdorferi pncA* (*bbe22*) encodes a functional nicotinamidase that can complement *E. coli* and *S. typhimurium pncA* deletion strains (17, 36). *B. burgdorferi pncA* is absolutely required for both mammalian infection and persistence within the tick vector, suggesting the importance of this pathway for maintenance of the intracellular NAD⁺ pool in *B. burgdorferi* (24, 36). Although a nicotinamide uptake system has not yet been identified in *B. burgdorferi*, enzymes that would mediate the stepwise conversion of nicotinamide to NAD⁺ (*pncB* [*bb635*], *nadD* [*bb782*], *nadE* [*bb311*]) are apparently encoded in the *B. burgdorferi* genome (17).

REGULATION OF GLYCEROL AND CHITOBIOSE UTILIZATION

Numerous global analyses of the wild-type *B. burgdorferi* transcriptome during *in vitro* growth in a variety of environmental conditions (e.g., temperature, pH, redox state) or in a mammalian host-adapted state have been reported. Similar studies have been performed to identify members of the RpoS, Rrp1, Rrp2, and BosR regulons (reviewed in 48, 107). A comprehensive survey of these studies revealed no definitive transcriptional control of carbohydrate uptake and utilization genes/pathways except for those involving glycerol and chitobiose (Fig. 4). Perhaps this can best be understood with the reasonable assumption that glucose is the preferred carbohydrate throughout the enzootic cycle

and therefore constitutively used except for certain tick stages during which glucose availability would be limited.

Glycerol Uptake and Utilization

Glycerol is a readily available carbohydrate in the tick vector and is produced by *Ixodes spp.* to serve as an antifreeze during overwintering (108, 109). The *B. burgdorferi glp* operon encodes the capacity for uptake of glycerol and its conversion to dihydroxyacetone phosphate (see above). Early global transcriptome studies showed that the *glp* operon is expressed at higher levels in cells grown at 23°C than at 35°C, suggesting a potential role in the vector (110). This was definitively confirmed by Pappas et al, who reported that *glpF* and *glpD* transcripts are substantially elevated during all tick stages compared with mouse joints (12). The *glp* operon is subject to RpoS-mediated repression (47), and its expression is induced by

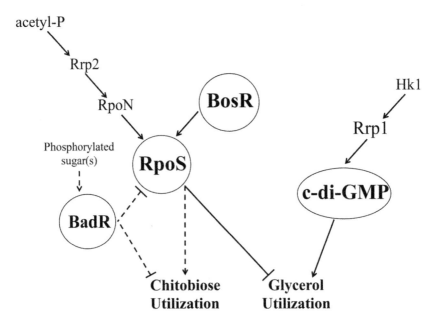

FIGURE 4 Schematic diagram depicting reported regulatory circuits controlling glycerol and chitobiose utilization. Solid lines indicate interactions confirmed by *in vivo* studies; dashed lines indicate interactions observed *in vitro* only. Diagram is a summary of data from references 43, 47, 50, 54, 60, 71, 72, 78, 111, and 112. doi:10.1128/microbiolspec.MBP-0011-2014.f4

Rrp1 (71, 78). The inability of Rrp1 mutants to survive in ticks was partially restored by complementation with the *glp* operon (78). Although a GlpD mutant could complete the experimental mouse-tick-mouse infection cycle, it had a fitness defect in ticks that was manifested in impaired growth and lower spirochete numbers in fed nymphs (12). Taken together, the findings demonstrate that the ability to use glycerol as a nutrient during the tick phase is critical and that modulation of *glp* operon expression is dependent, at least in part, on reciprocal regulation by RpoS and Rrp1.

Chitobiose Uptake and Utilization

Chitin, a polymer composed of chitobiose (glucosamine or GlcNAc dimer) units, is the primary component of the tick exoskeleton and the peritrophic membrane (15, 91). Chitobiose can substitute for GlcNAc, an essential requirement for *in vitro* cultivation, and can serve as the principal carbon source *in vitro* (81, 90). As described earlier, chitobiose can be taken up via a dedicated PTS, ChbABC, cleaved to GlcNAc and GlcNAc-6-P, and enter glycolysis after conversion to fructose-6-P. *chbC* is highly expressed during all tick stages and has little expression in mouse joints (12), an expected finding given that chitobiose is likely available as a nutrient only in the vector. However, the transcriptional regulators responsible for this differential expression are a matter of controversy. Rhodes et al suggested an RpoS involvement in *chbC* expression (111), a finding at odds with the global transcriptome analysis of an RpoS mutant that found no effect on *chbC* expression (47).

Recently, Sze et al reported that *chbC* transcript levels were significantly repressed in an Rrp1 mutant (112), but in this case as well, previous global transcriptome analyses of Rrp1 mutants did not reveal any differences in *chbC* expression (71, 78). Curiously, Sze et al suggested that *chbC* regulation by Rrp1 is mediated through BosR and RpoS,

because expression of these transcriptional regulators was also repressed in their Rrp1 mutant. However, it must be noted that *chbC* expression is not altered in a BosR mutant, and a BosR binding site has not been identified upstream of *chbC* (65, 67, 113). BadR has also been implicated as a repressor of *chbC* expression. Based on current understanding of the roles for RpoS and Rrp1 (the former being responsible for the mammalian phase regulon and the latter a tick phase regulon) and the likelihood that chitobiose would only be available to the spirochete in the vector, it seems reasonable to conclude that *chbC* expression is regulated primarily by Rrp1 (and c-di-GMP), which functions during the vector phase of the enzootic cycle. Further studies to more clearly define the regulation of chitobiose uptake are warranted.

CONTROL OF CARBOHYDRATE UPTAKE AND GLYCOLYTIC FLUX DURING THE ENZOOTIC CYCLE

When multiple nutrient sources are available, bacteria initially use the preferred carbon source, typically glucose, and in an effort to conserve energy, also repress genes associated with the use of alternate nutrient sources; this process is referred to as carbon catabolite repression (95, 114). Unlike free-living bacteria, *B. burgdorferi* is restricted during its enzootic cycle to using only nutrients found in the tick vector or the mammalian host. Concentrations of glucose and glycerol in mouse plasma are about 150 and 2.8 mg/100 mL, respectively (115), and glycerol is abundantly present during all tick stages (109). It may be assumed that *B. burgdorferi* uses glucose while in the mammal and at early stages of tick feeding. At later feeding stages or after cessation of vector feeding, glucose is depleted due to its uptake by tick midgut epithelial cells. This triggers a switch to glycerol utilization that is mediated by elevated levels of glycerol pathway proteins.

Pappas et al proposed that this switch in carbohydrate utilization may represent the *B. burgdorferi* version of carbon catabolite repression but did not provide a mechanism for this process (12).

B. burgdorferi carbohydrate metabolism has several unusual features (Fig. 3). Among them are the sole reliance on glycolysis for ATP generation, a phosphofructokinase (PFK) that uses pyrophosphate as the donor for conversion of fructose-6-P to fructose-1,6-bisphosphate, the inability to use pyruvate as an energy source, and the presence of lactate permease (LctP; BB604). The rate-determining step of glycolysis in most eubacteria and eukaryotes is catalyzed by PFK, an allosteric enzyme whose activity is typically controlled by the ADP/ATP ratio (116). *B. burgdorferi* contains genes for two PFK enzymes (BB020, BB727). Interestingly, purified BB020 exists as a dimer in solution and, in comparison with other pyrophosphate-dependent PFKs (e.g., *Treponema pallidum* TP0542), is likely to be nonallosteric (86, 87, 117). If PFK does not sense the ADP/ATP ratio, how is the flux of glycolytic intermediates through the pathway regulated in *B. burgdorferi*?

The end product of glycolysis is pyruvate, and continued function of the pathway requires regeneration of NAD⁺. In *B. burgdorferi*, this can only be accomplished by conversion of pyruvate to lactate, and this reaction is the only metabolic fate for pyruvate, because the spirochete lacks pyruvate dehydrogenase (and a TCA cycle), pyruvate-formate lyase, or lactate oxidase (17). Extracellular lactate could be re-acquired through the action of lactate permease, likely functioning as a lactate: H^+ symporter (118) (Fig. 3); interestingly, *lctP* transcription is significantly induced in feeding ticks (Schwartz et al., unpublished observations). Typically, the acquired lactate would be used as an energy source by stepwise conversion to pyruvate (by lactate dehydrogenase [LDH]) and acetyl-CoA (via pyruvate dehydrogenase) and entry into the TCA cycle. However, the acquired lactate cannot be used in this manner in *B. burgdorferi*, because pyruvate cannot be converted to any other metabolic intermediate. In addition, conversion of lactate to pyruvate by lactate dehydrogenase would generate NADH, altering the NAD^+/NADH ratio in a manner unfavorable for continued glycolysis.

Taken together, these findings suggest a model for control of the glycolytic pathway in *B. burgdorferi*, particularly during the tick phases of the enzootic cycle, by the intracellular pyruvate/lactate ratio (Fig. 3). When *B. burgdorferi* grows in an environment with abundant glucose (e.g., a mammal), the sugar would enter the glycolytic pathway by uptake through its PTS and would ultimately be converted to lactate and excreted. In an open system (mammalian tissue), excreted lactate would diffuse away, and glycolysis would continue to function normally as long as sufficient glucose is available. During tick feeding, glucose would also be available to *B. burgdorferi*, and glycolysis would proceed initially as in the mammal. Furthermore, this metabolic state enables a process referred to as inducer exclusion, whereby phosphorylated EIIA^Glc interacts directly with GlpK and prevents conversion of available glycerol to glycerol 3-phosphate (119) (Fig. 3). At later stages of feeding, glucose becomes limiting (as a result of uptake by tick midgut epithelial cells). In addition, excreted lactate would accumulate in the tick midgut because this is essentially a physically closed environment, and LctP would be induced. As a result, *B. burgdorferi* would take up lactate, which would be converted to pyruvate by LDH. Elevated pyruvate levels would alter the PEP/pyruvate ratio, a key factor in controlling glucose uptake through its cognate PTS.

In *E. coli*, when the PEP/pyruvate ratio is low (i.e., when pyruvate is high), the EIIA component is dephosphorylated and the import of glucose as glucose-6-P by action of the PTS is blocked; indeed, the PEP/pyruvate ratio is directly correlated with the phosphorylation state of EIIA^Glc (119–121).

Furthermore, GlpK-mediated conversion of glycerol to glycerol-3-phosphate would resume. Under these circumstances, glycerol would become an important carbohydrate source for continued energy generation via glycolysis. The significant increase in *glp* operon expression during the tick feeding stages is consistent with this model. A second consequence of lactate uptake and conversion to pyruvate is the generation of an unfavorable NAD^+/NADH ratio. This problem could be solved by the action of CoADR (see above), which would regenerate NAD^+ and facilitate continued glycolysis (104, 105).

The precise details of the mechanism by which *B. burgdorferi* senses the levels of extracellular lactate, glucose, and glycerol and transmits signals to the transcription machinery remain elusive. The Rrp2/RpoN/RpoS and Hk1/Rrp1 (c-di-GMP) circuits regulate transcription of genes during the mammalian and vector phases of the enzootic cycle. The involvement of these transcriptional regulators in modulating the expression of genes whose products facilitate transport and utilization of the alternative carbon sources glycerol and chitobiose argues that they play an important role in these processes. Elucidation of the interactions between these two regulatory systems would help clarify how *B. burgdorferi* can alternate among different metabolic states in order to survive the drastically different environments encountered during its lifecycle.

ACKNOWLEDGMENTS

We thank Justin Radolf for many helpful discussions. Studies conducted in the authors' laboratory were supported by National Institutes of Health grants AR41511 and AI45801.

CITATION

Corona A, Schwartz I. 2015. *Borrelia burgdorferi*: carbon metabolism and the tick-mammal enzootic cycle. Microbiol Spectrum 3(3):MBP-0011-2014

REFERENCES

1. **Bacon RM, Kugeler KJ, Mead PS.** 2008. Surveillance for Lyme disease—United States, 1992–2006. *MMWR Surveill Summ* **57:**1–9.
2. **Steere AC, Grodzicki RL, Kornblatt AN, Craft JE, Barbour AG, Burgdorfer W, Schmid GP, Johnson E, Malawista SE.** 1983. The spirochetal etiology of Lyme disease. *N Engl J Med* **308:**733–740.
3. **Benach JL, Bosler EM, Hanrahan JP, Coleman JL, Habicht GS, Bast TF, Cameron DJ, Ziegler JL, Barbour AG, Burgdorfer W, Edelman R, Kaslow RA.** 1983. Spirochetes isolated from the blood of two patients with Lyme disease. *N Engl J Med* **308:**740–742.
4. **Levine JF, Wilson ML, Spielman A.** 1985. Mice as reservoirs of the Lyme disease spirochete. *Am J Trop Med Hyg* **34:**355–360.
5. **Tsao JI.** 2009. Reviewing molecular adaptations of Lyme borreliosis spirochetes in the context of reproductive fitness in natural transmission cycles. *Vet Res* **40:**36.
6. **Radolf JD, Caimano MJ, Stevenson B, Hu LT.** 2012. Of ticks, mice and men: understanding the dual-host lifestyle of Lyme disease spirochaetes. *Nat Rev Microbiol* **10:**87–99.
7. **Magnarelli LA, Anderson JF, Fish D.** 1987. Transovarial transmission of *Borrelia burgdorferi* in *Ixodes dammini* (Acari:Ixodidae). *J Infect Dis* **156:**234–236.
8. **Rollend L, Fish D, Childs JE.** 2013. Transovarial transmission of *Borrelia* spirochetes by *Ixodes scapularis*: a summary of the literature and recent observations. *Ticks Tick-borne Dis* **4:**46–51.
9. **Dunham-Ems SM, Caimano MJ, Pal U, Wolgemuth CW, Eggers CH, Balic A, Radolf JD.** 2009. Live imaging reveals a biphasic mode of dissemination of *Borrelia burgdorferi* within ticks. *J Clin Invest* **119:**3652–3665.
10. **De Silva AM, Fikrig E.** 1995. Growth and migration of *Borrelia burgdorferi* in *Ixodes* ticks during blood feeding. *Am J Trop Med Hyg* **53:**397–404.
11. **Piesman J, Schneider BS, Zeidner NS.** 2001. Use of quantitative PCR to measure density of *Borrelia burgdorferi* in the midgut and salivary glands of feeding tick vectors. *J Clin Microbiol* **39:**4145–4148.
12. **Pappas CJ, Iyer R, Petzke MM, Caimano MJ, Radolf JD, Schwartz I.** 2011. *Borrelia burgdorferi* requires glycerol for maximum fitness during the tick phase of the enzootic cycle. *PLoS Pathog* **7:**e1002102.
13. **Steere AC, Coburn J, Glickstein L.** 2004. The emergence of Lyme disease. *J Clin Invest* **113:**1093–1101.

14. **Pal U, Fikrig E.** 2010. Tick Interactions, p 279–298. *In* Samuels DS, Radolf JD (ed.), *Borrelia: Molecular biology, host interaction and pathogenesis.* Caister Academic Press, Norfolk, UK.

15. **Shao L, Devenport M, Jacobs-Lorena M.** 2001. The peritrophic matrix of hematophagous insects. *Arch Insect Biochem Physiol* **47:**119–125.

16. **Rohmer L, Hocquet D, Miller SI.** 2011. Are pathogenic bacteria just looking for food? Metabolism and microbial pathogenesis. *Trends Microbiol* **19:**341–348.

17. **Fraser CM, Casjens S, Huang WM, Sutton GG, Clayton R, Lathigra R, White O, Ketchum KA, Dodson R, Hickey EK, Gwinn M, Dougherty B, Tomb JF, Fleischmann RD, Richardson D, Peterson J, Kerlavage AR, Quackenbush J, Salzberg S, Hanson M, van VR, Palmer N, Adams MD, Gocayne J, Weidman J, Utterback T, Watthey L, McDonald L, Artiach P, Bowman C, Garland S, Fuji C, Cotton MD, Horst K, Roberts K, Hatch B, Smith HO, Venter JC.** 1997. Genomic sequence of a Lyme disease spirochaete, *Borrelia burgdorferi. Nature* **390:**580–586.

18. **Casjens S, Palmer N, van VR, Huang WM, Stevenson B, Rosa P, Lathigra R, Sutton G, Peterson J, Dodson RJ, Haft D, Hickey E, Gwinn M, White O, Fraser CM.** 2000. A bacterial genome in flux: the twelve linear and nine circular extrachromosomal DNAs in an infectious isolate of the Lyme disease spirochete *Borrelia burgdorferi. Mol Microbiol* **35:** 490–516.

19. **Brisson D, Drecktrah D, Eggers CH, Samuels DS.** 2012. Genetics of *Borrelia burgdorferi. Annu Rev Genet* **46:**515–536.

20. **Iyer R, Kalu O, Purser J, Norris S, Stevenson B, Schwartz I.** 2003. Linear and circular plasmid content in *Borrelia burgdorferi* clinical isolates. *Infect Immun* **71:**3699–3706.

21. **Terekhova D, Iyer R, Wormser GP, Schwartz I.** 2006. Comparative genome hybridization reveals substantial variation among clinical isolates of *Borrelia burgdorferi* sensu stricto with different pathogenic properties. *J Bacteriol* **188:**6124–6134.

22. **Casjens SR, Mongodin EF, Qiu WG, Luft BJ, Schutzer SE, Gilcrease EB, Huang WM, Vujadinovic M, Aron JK, Vargas LC, Freeman S, Radune D, Weidman JF, Dimitrov GI, Khouri HM, Sosa JE, Halpin RA, Dunn JJ, Fraser CM.** 2012. Genome stability of Lyme disease spirochetes: comparative genomics of *Borrelia burgdorferi* plasmids. *PLoS ONE* **7:** e33280.

23. **Purser JE, Norris SJ.** 2000. Correlation between plasmid content and infectivity in *Borrelia burgdorferi. Proc Natl Acad Sci U S A* **97:**13865–13870.

24. **Purser JE, Lawrenz MB, Caimano MJ, Howell JK, Radolf JD, Norris SJ.** 2003. A plasmid-encoded nicotinamidase (PncA) is essential for infectivity of *Borrelia burgdorferi* in a mammalian host. *Mol Microbiol* **48:**753–764.

25. **Byram R, Stewart PE, Rosa P.** 2004. The essential nature of the ubiquitous 26-kilobase circular replicon of *Borrelia burgdorferi. J Bacteriol* **186:**3561–3569.

26. **Stewart PE, Byram R, Grimm D, Tilly K, Rosa PA.** 2005. The plasmids of *Borrelia burgdorferi*: essential genetic elements of a pathogen. *Plasmid* **53:**1–13.

27. **Grimm D, Tilly K, Bueschel DM, Fisher MA, Policastro PF, Gherardini FC, Schwan TG, Rosa PA.** 2005. Defining plasmids required by *Borrelia burgdorferi* for colonization of tick vector *Ixodes scapularis* (Acari: Ixodidae). *J Med Entomol* **42:**676–684.

28. **Chaconas G, Kobryn K.** 2010. Structure, function, and evolution of linear replicons in *Borrelia. Annu Rev Microbiol* **64:**185–202.

29. **Chaconas G, Norris SJ.** 2013. Peaceful coexistence amongst Borrelia plasmids: getting by with a little help from their friends? *Plasmid* **70:**161–167.

30. **Dulebohn DP, Bestor A, Rosa PA.** 2013. *Borrelia burgdorferi* linear plasmid 28-3 confers a selective advantage in an experimental mouse-tick infection model. *Infect Immun* **81:**2986–2996.

31. **Jewett MW, Byram R, Bestor A, Tilly K, Lawrence K, Burtnick MN, Gherardini F, Rosa PA.** 2007. Genetic basis for retention of a critical virulence plasmid of *Borrelia burgdorferi. Mol Microbiol* **66:**975–990.

32. **Gherardini F, Boylan J, Lawrence K, Skare J.** 2010. Metabolism and Physiology of *Borrelia*, p 103–138. *In* Samuels DS, Radolf JD (ed.), *Borrelia: Molecular biology, host interaction and pathogenesis.* Caister Academic Press, Norfolk, UK.

33. **Jewett MW, Lawrence KA, Bestor A, Byram R, Gherardini F, Rosa PA.** 2009. GuaA and GuaB are essential for *Borrelia burgdorferi* survival in the tick-mouse infection cycle. *J Bacteriol* **191:**6231–6241.

34. **Tilly K, Krum JG, Bestor A, Jewett MW, Grimm D, Bueschel D, Byram R, Dorward D, Vanraden MJ, Stewart P, Rosa P.** 2006. *Borrelia burgdorferi* OspC protein required exclusively in a crucial early stage of mammalian infection. *Infect Immun* **74:**3554–3564.

35. **Labandeira-Rey M, Skare JT.** 2001. Decreased infectivity in *Borrelia burgdorferi* strain B31 is associated with loss of linear plasmid 25 or 28-1. *Infect Immun* **69:**446–455.

36. **Jewett MW, Jain S, Linowski AK, Sarkar A, Rosa PA.** 2011. Molecular characterization of the *Borrelia burgdorferi* in vivo-essential protein PncA. *Microbiology* **157:**2831–2840.

37. **Zhang JR, Hardham JM, Barbour AG, Norris SJ.** 1997. Antigenic variation in Lyme disease borreliae by promiscuous recombination of VMP-like sequence cassettes. *Cell* **89:**275–285.

38. **Zhang JR, Norris SJ.** 1998. Genetic variation of the *Borrelia burgdorferi* gene vlsE involves cassette-specific, segmental gene conversion. *Infect Immun* **66:**3698–3704.

39. **Norris SJ.** 2006. Antigenic variation with a twist–the *Borrelia* story. *Mol Microbiol* **60:** 1319–1322.

40. **Kazmierczak MJ, Wiedmann M, Boor KJ.** 2005. Alternative sigma factors and their roles in bacterial virulence. *Microbiol Mol Biol Rev* **69:**527–543.

41. **Beier D, Gross R.** 2006. Regulation of bacterial virulence by two-component systems. *Curr Opin Microbiol* **9:**143–152.

42. **Hubner A, Yang X, Nolen DM, Popova TG, Cabello FC, Norgard MV.** 2001. Expression of *Borrelia burgdorferi* OspC and DbpA is controlled by a RpoN-RpoS regulatory pathway. *Proc Natl Acad Sci U S A* **98:**12724–12729.

43. **Yang XF, Alani SM, Norgard MV.** 2003. The response regulator Rrp2 is essential for the expression of major membrane lipoproteins in *Borrelia burgdorferi. Proc Natl Acad Sci U S A* **100:**11001–11006.

44. **Caimano MJ, Eggers CH, Hazlett KR, Radolf JD.** 2004. RpoS is not central to the general stress response in *Borrelia burgdorferi* but does control expression of one or more essential virulence determinants. *Infect Immun* **72:**6433–6445.

45. **Caimano MJ, Eggers CH, Gonzalez CA, Radolf JD.** 2005. Alternate sigma factor RpoS is required for the in vivo-specific repression of *Borrelia burgdorferi* plasmid lp54-borne *ospA* and *lp6.6* genes. *J Bacteriol* **187:**7845–7852.

46. **Dunham-Ems SM, Caimano MJ, Eggers CH, Radolf JD.** 2012. *Borrelia burgdorferi* requires the alternative sigma factor RpoS for dissemination within the vector during tick-to-mammal transmission. *PLoS Pathog* **8:**e1002532.

47. **Caimano MJ, Iyer R, Eggers CH, Gonzalez C, Morton EA, Gilbert MA, Schwartz I, Radolf JD.** 2007. Analysis of the RpoS regulon in *Borrelia burgdorferi* in response to mammalian host signals provides insight into RpoS function during the enzootic cycle. *Mol Microbiol* **65:**1193–1217.

48. **Samuels DS.** 2011. Gene regulation in *Borrelia burgdorferi. Annu Rev Microbiol* **65:**479–499.

49. **Ouyang Z, Blevins JS, Norgard MV.** 2008. Transcriptional interplay among the regulators Rrp2, RpoN and RpoS in *Borrelia burgdorferi. Microbiology* **154:**2641–2658.

50. **Blevins JS, Xu H, He M, Norgard MV, Reitzer L, Yang XF.** 2009. Rrp2, a sigma54-dependent transcriptional activator of *Borrelia burgdorferi*, activates *rpoS* in an enhancer-independent manner. *J Bacteriol* **191:**2902–2905.

51. **Ouyang Z, Narasimhan S, Neelakanta G, Kumar M, Pal U, Fikrig E, Norgard MV.** 2012. Activation of the RpoN-RpoS regulatory pathway during the enzootic life cycle of *Borrelia burgdorferi. BMC Microbiol* **12:**44.

52. **Smith AH, Blevins JS, Bachlani GN, Yang XF, Norgard MV.** 2007. Evidence that RpoS (sigmaS) in *Borrelia burgdorferi* is controlled directly by RpoN (sigma54/sigmaN). *J Bacteriol* **189:**2139–2144.

53. **Burtnick MN, Downey JS, Brett PJ, Boylan JA, Frye JG, Hoover TR, Gherardini FC.** 2007. Insights into the complex regulation of *rpoS* in *Borrelia burgdorferi. Mol Microbiol* **65:** 277–293.

54. **Xu H, Caimano MJ, Lin T, He M, Radolf JD, Norris SJ, Gherardini F, Wolfe AJ, Yang XF.** 2010. Role of acetyl-phosphate in activation of the Rrp2-RpoN-RpoS pathway in *Borrelia burgdorferi. PLoS Pathog* **6:**e1001104.

55. **Lybecker MC, Samuels DS.** 2007. Temperature-induced regulation of RpoS by a small RNA in *Borrelia burgdorferi. Mol Microbiol* **64:**1075–1089.

56. **Lybecker MC, Abel CA, Feig AL, Samuels DS.** 2010. Identification and function of the RNA chaperone Hfq in the Lyme disease spirochete *Borrelia burgdorferi. Mol Microbiol* **78:**622–635.

57. **Karna SL, Sanjuan E, Esteve-Gassent MD, Miller CL, Maruskova M, Seshu J.** 2011. CsrA modulates levels of lipoproteins and key regulators of gene expression critical for pathogenic mechanisms of *Borrelia burgdorferi. Infect Immun* **79:**732–744.

58. **Sze CW, Li C.** 2011. Inactivation of *bb0184*, which encodes carbon storage regulator A, represses the infectivity of *Borrelia burgdorferi. Infect Immun* **79:**1270–1279.

59. **Ouyang Z, Zhou J, Norgard MV.** 2014. CsrA (BB0184) is not involved in activation of the RpoN-RpoS regulatory pathway in *Borrelia burgdorferi. Infect Immun* **82:**1511–1522.

60. **Miller CL, Karna SL, Seshu J.** 2013. Borrelia host adaptation Regulator (BadR) regulates *rpoS* to modulate host adaptation and virulence factors in *Borrelia burgdorferi. Mol Microbiol* **88:**105–124.

61. **Boylan JA, Posey JE, Gherardini FC.** 2003. *Borrelia* oxidative stress response regulator, BosR: a distinctive Zn-dependent transcriptional activator. *Proc Natl Acad Sci U S A* **100:**11684–11689.

62. **Katona LI, Tokarz R, Kuhlow CJ, Benach J, Benach JL.** 2004. The *fur* homologue in *Borrelia burgdorferi. J Bacteriol* **186:**6443–6456.

63. **Hyde JA, Shaw DK, Smith R III, Trzeciakowski JP, Skare JT.** 2010. Characterization of a conditional *bosR* mutant in *Borrelia burgdorferi. Infect Immun* **78:**265–274.

64. **Samuels DS, Radolf JD.** 2009. Who is the BosR around here anyway? *Mol Microbiol* **74:**1295–1299.

65. **Ouyang Z, Kumar M, Kariu T, Haq S, Goldberg M, Pal U, Norgard MV.** 2009. BosR (BB0647) governs virulence expression in *Borrelia burgdorferi. Mol Microbiol* **74:**1331–1343.

66. **Hyde JA, Shaw DK, Smith IR, Trzeciakowski JP, Skare JT.** 2009. The BosR regulatory protein of *Borrelia burgdorferi* interfaces with the RpoS regulatory pathway and modulates both the oxidative stress response and pathogenic properties of the Lyme disease spirochete. *Mol Microbiol* **74:**1344–1355.

67. **Ouyang Z, Deka RK, Norgard MV.** 2011. BosR (BB0647) controls the RpoN-RpoS regulatory pathway and virulence expression in *Borrelia burgdorferi* by a novel DNA-binding mechanism. *PLoS Pathog* **7:**e1001272.

68. **Ouyang Z, Zhou J, Brautigam CA, Deka R, Norgard MV.** 2014. Identification of a core sequence for the binding of BosR to the *rpoS* promoter region in *Borrelia burgdorferi. Microbiology* **160:**851–862.

69. **Wang P, Dadhwal P, Cheng Z, Zianni MR, Rikihisa Y, Liang FT, Li X.** 2013. *Borrelia burgdorferi* oxidative stress regulator BosR directly represses lipoproteins primarily expressed in the tick during mammalian infection. *Mol Microbiol* **89:**1140–1153.

70. **Troxell B, Ye M, Yang Y, Carrasco SE, Lou Y, Yang XF.** 2013. Manganese and zinc regulate virulence determinants in *Borrelia burgdorferi. Infect Immun* **81:**2743–2752.

71. **Rogers EA, Terekhova D, Zhang HM, Hovis KM, Schwartz I, Marconi RT.** 2009. Rrp1, a cyclic-di-GMP-producing response regulator, is an important regulator of *Borrelia burgdorferi* core cellular functions. *Mol Microbiol* **71:**1551–1573.

72. **Ryjenkov DA, Tarutina M, Moskvin OV, Gomelsky M.** 2005. Cyclic diguanylate is a ubiquitous signaling molecule in bacteria: insights into biochemistry of the GGDEF protein domain. *J Bacteriol* **187:**1792–1798.

73. **Hengge R.** 2009. Principles of c-di-GMP signalling in bacteria. *Nat Rev Microbiol* **7:**263–273.

74. **Sondermann H, Shikuma NJ, Yildiz FH.** 2012. You've come a long way: c-di-GMP signaling. *Curr Opin Microbiol* **15:**140–146.

75. **Novak EA, Sultan SZ, Motaleb MA.** 2014. The cyclic-di-GMP signaling pathway in the Lyme disease spirochete, *Borrelia burgdorferi.* Front. *Cell Infect Microbiol* **4:**56.

76. **Caimano MJ, Kenedy MR, Kairu T, Desrosiers DC, Harman M, Dunham-Ems S, Akins DR, Pal U, Radolf JD.** 2011. The hybrid histidine kinase Hk1 is part of a two-component system that is essential for survival of *Borrelia burgdorferi* in feeding Ixodes scapularis ticks. *Infect Immun* **79:**3117–3130.

77. **Freedman JC, Rogers EA, Kostick JL, Zhang H, Iyer R, Schwartz I, Marconi RT.** 2010. Identification and molecular characterization of a cyclic-di-GMP effector protein, PlzA (BB0733): additional evidence for the existence of a functional cyclic-di-GMP regulatory network in the Lyme disease spirochete, *Borrelia burgdorferi. FEMS Immunol Med Microbiol* **58:**285–294.

78. **He M, Ouyang Z, Troxell B, Xu H, Moh A, Piesman J, Norgard MV, Gomelsky M, Yang XF.** 2011. Cyclic di-GMP is essential for the survival of the Lyme disease spirochete in ticks. *PLoS Pathog* **7:**e1002133.

79. **Kostick JL, Szkotnicki LT, Rogers EA, Bocci P, Raffaelli N, Marconi RT.** 2011. The diguanylate cyclase, Rrp1, regulates critical steps in the enzootic cycle of the Lyme disease spirochetes. *Mol Microbiol* **81:**219–231.

80. **Groshong AM, Blevins JS.** 2014. Insights into the biology of *Borrelia burgdorferi* gained through the application of molecular genetics. *Adv Appl Microbiol* **86:**41–143.

81. **von Lackum K, Stevenson B.** 2005. Carbohydrate utilization by the Lyme borreliosis spirochete, *Borrelia burgdorferi. FEMS Microbiol Lett* **243:**173–179.

82. **Das R, Hegyi H, Gerstein M.** 2000. Genome analyses of spirochetes: a study of the protein structures, functions and metabolic pathways in *Treponema pallidum* and *Borrelia burgdorferi. J Mol Microbiol Biotechnol* **2:**387–392.

83. **Jain S, Sutchu S, Rosa PA, Byram R, Jewett MW.** 2012. *Borrelia burgdorferi* harbors a transport system essential for purine salvage and mammalian infection. *Infect Immun* **80:**3086–3093.

84. **Ouyang Z, He M, Oman T, Yang XF, Norgard MV.** 2009. A manganese transporter, BB0219 (BmtA), is required for virulence by the Lyme

disease spirochete, *Borrelia burgdorferi*. *Proc Natl Acad Sci U S A* **106**:3449–3454.

85. **Hoon-Hanks LL, Morton EA, Lybecker MC, Battisti JM, Samuels DS, Drecktrah D.** 2012. *Borrelia burgdorferi malQ* mutants utilize disaccharides and traverse the enzootic cycle. *FEMS Immunol Med Microbiol* **66**:157–165.

86. **Deng Z, Roberts D, Wang X, Kemp RG.** 1999. Expression, characterization, and crystallization of the pyrophosphate-dependent phosphofructo-1-kinase of *Borrelia burgdorferi*. *Arch Biochem Biophys* **371**:326–331.

87. **Moore SA, Ronimus RS, Roberson RS, Morgan HW.** 2002. The structure of a pyrophosphate-dependent phosphofructokinase from the Lyme disease spirochete *Borrelia burgdorferi*. *Structure* **10**:659–671.

88. **Gebbia JA, Backenson PB, Coleman JL, Anda P, Benach JL.** 1997. Glycolytic enzyme operon of *Borrelia burgdorferi*: characterization and evolutionary implications. *Gene* **188**:221–228.

89. **Barbour AG.** 1984. Isolation and cultivation of Lyme disease spirochetes. *Yale J Biol Med* **57**:521–525.

90. **Tilly K, Elias AF, Errett J, Fischer E, Iyer R, Schwartz I, Bono JL, Rosa P.** 2001. Genetics and regulation of chitobiose utilization in *Borrelia burgdorferi*. *J Bacteriol* **183**:5544–5553.

91. **Hackman RH.** 1982. Structure and function in tick cuticle. *Annu Rev Entomol* **27**:75–95.

92. **Rhodes RG, Atoyan JA, Nelson DR.** 2010. The chitobiose transporter, chbC, is required for chitin utilization in *Borrelia burgdorferi*. *BMC Microbiol* **10**:21.

93. **Barabote RD, Saier MH Jr.** 2005. Comparative genomic analyses of the bacterial phosphotransferase system. *Microbiol Mol Biol Rev* **69**:608–634.

94. **Clore GM, Venditti V.** 2013. Structure, dynamics and biophysics of the cytoplasmic protein-protein complexes of the bacterial phosphoenolpyruvate: sugar phosphotransferase system. *Trends Biochem Sci* **38**:515–530.

95. **Gorke B, Stulke J.** 2008. Carbon catabolite repression in bacteria: many ways to make the most out of nutrients. *Nat Rev Microbiol* **6**:613–624.

96. **Godany A, Vidova B, Janecek S.** 2008. The unique glycoside hydrolase family 77 amylomaltase from *Borrelia burgdorferi* with only catalytic triad conserved. *FEMS Microbiol Lett* **284**:84–91.

97. **Jahreis K, Pimentel-Schmitt EF, Bruckner R, Titgemeyer F.** 2008. Ins and outs of glucose transport systems in eubacteria. *FEMS Microbiol Rev* **32**:891–907.

98. **Ostberg Y, Berg S, Comstedt P, Wieslander A, Bergstrom S.** 2007. Functional analysis of a lipid galactosyltransferase synthesizing the major envelope lipid in the Lyme disease spirochete *Borrelia burgdorferi*. *FEMS Microbiol Lett* **272**:22–29.

99. **Jensen MO, Park S, Tajkhorshid E, Schulten K.** 2002. Energetics of glycerol conduction through aquaglyceroporin GlpF. *Proc Natl Acad Sci U S A* **99**:6731–6736.

100. **Chen LY.** 2013. Glycerol modulates water permeation through *Escherichia coli* aquaglyceroporin GlpF. *Biochim Biophy Acta* **1828**:1786–1793.

101. **Foster JW, Moat AG.** 1980. Nicotinamide adenine dinucleotide biosynthesis and pyridine nucleotide cycle metabolism in microbial systems. *Microbiol Rev* **44**:83–105.

102. **Gazzaniga F, Stebbins R, Chang SZ, McPeek MA, Brenner C.** 2009. Microbial NAD metabolism: lessons from comparative genomics. *Microbiol Mol Biol Rev* **73**:529–541.

103. **Wolfe AJ.** 2005. The acetate switch. *Microbiol Mol Biol Rev* **69**:12–50.

104. **Boylan JA, Hummel CS, Benoit S, Garcia-Lara J, Treglown-Downey J, Crane EJ III, Gherardini FC.** 2006. *Borrelia burgdorferi bb0728* encodes a coenzyme A disulphide reductase whose function suggests a role in intracellular redox and the oxidative stress response. *Mol Microbiol* **59**:475–486.

105. **Eggers CH, Caimano MJ, Malizia RA, Kariu T, Cusack B, Desrosiers DC, Hazlett KR, Claiborne A, Pal U, Radolf JD.** 2011. The coenzyme A disulphide reductase of *Borrelia burgdorferi* is important for rapid growth throughout the enzootic cycle and essential for infection of the mammalian host. *Mol Microbiol* **82**:679–697.

106. **Gazanion E, Garcia D, Silvestre R, Gerard C, Guichou JF, Labesse G, Seveno M, Cordeiro-Da-Silva A, Ouaissi A, Sereno D, Vergnes B.** 2011. The Leishmania nicotinamidase is essential for NAD$^+$ production and parasite proliferation. *Mol Microbiol* **82**:21–38.

107. **Skare JT, Carroll JA, Yang XF, Samuels DS, Akins DR.** 2010. Gene Regulation, Transcriptomics and Proteomics, p 67–101. *In* Samuels DS, Radolf JD (ed.), *Borrelia: Molecular Biology, Host Interaction and Pathogenesis*. Caister Academic Press, Norfolk, UK.

108. **Lee RJ, Baust JG.** 1987. Cold-hardiness in the Antartic tick, *Ixodes uriae* Physiol. *Zool* **60**:499–506.

109. **Vandyk JK, Bartholomew DM, Rowley WA, Platt KB.** 1996. Survival of *Ixodes scapularis* (Acari: Ixodidae) exposed to cold. *J Med Entomol* **33**:6–10.

110. **Ojaimi C, Brooks C, Casjens S, Rosa P, Elias A, Barbour A, Jasinskas A, Benach J, Katona L, Radolf J, Caimano M, Skare J, Swingle K, Akins D, Schwartz I.** 2003. Profiling of temperature-induced changes in *Borrelia burgdorferi* gene expression by using whole genome arrays. *Infect Immun* **71:**1689–1705.

111. **Rhodes RG, Coy W, Nelson DR.** 2009. Chitobiose utilization in *Borrelia burgdorferi* is dually regulated by RpoD and RpoS. *BMC Microbiol* **9:**108.

112. **Sze CW, Smith A, Choi YH, Yang X, Pal U, Yu A, Li C.** 2013. Study of the response regulator Rrp1 reveals its regulatory role in chitobiose utilization and virulence of *Borrelia burgdorferi*. *Infect Immun* **81:**1775–1787.

113. **Hyde JA, Seshu J, Skare JT.** 2006. Transcriptional profiling of *Borrelia burgdorferi* containing a unique *bosR* allele identifies a putative oxidative stress regulon. *Microbiology* **152:**2599–2609.

114. **Deutscher J.** 2008. The mechanisms of carbon catabolite repression in bacteria. *Curr Opin Microbiol* **11:**87–93.

115. **Maeda N, Funahashi T, Hibuse T, Nagasawa A, Kishida K, Kuriyama H, Nakamura T, Kihara S, Shimomura I, Matsuzawa Y.** 2004. Adaptation to fasting by glycerol transport through aquaporin 7 in adipose tissue. *Proc Natl Acad Sci U S A* **101:**17801–17806.

116. **Moat AG, Foster JW, Spector MP.** 2002. *Microbial Physiology*, 4th ed. Wiley-Liss, New York.

117. **Roberson RS, Ronimus RS, Gephard S, Morgan HW.** 2001. Biochemical characterization of an active pyrophosphate-dependent phosphofructokinase from *Treponema pallidum*. *FEMS Microbiol Lett* **194:**257–260.

118. **Saier MH Jr, Eng BH, Fard S, Garg J, Haggerty DA, Hutchinson WJ, Jack DL, Lai EC, Liu HJ, Nusinew DP, Omar AM, Pao SS, Paulsen IT, Quan JA, Sliwinski M, Tseng TT, Wachi S, Young GB.** 1999. Phylogenetic characterization of novel transport protein families revealed by genome analyses. *Biochim Biophys Acta* **1422:**1–56.

119. **Deutscher J, Ake FM, Derkaoui M, Zebre AC, Cao TN, Bouraoui H, Kentache T, Mokhtari A, Milohanic E, Joyet P.** 2014. The bacterial phosphoenolpyruvate:carbohydrate phosphotransferase system: regulation by protein phosphorylation and phosphorylation-dependent protein-protein interactions. *Microbiol Mol Biol Rev* **78:**231–256.

120. **Lengeler JW, Jahreis K.** 2009. Bacterial PEP-dependent carbohydrate: phosphotransferase systems couple sensing and global control mechanisms. *Contrib Microbiol* **16:**65–87.

121. **Hogema BM, Arents JC, Bader R, Eijkemans K, Yoshida H, Takahashi H, Aiba H, Postma PW.** 1998. Inducer exclusion in *Escherichia coli* by non-PTS substrates: the role of the PEP to pyruvate ratio in determining the phosphorylation state of enzyme IIAGlc. *Mol Microbiol* **30:**487–498.

Metabolism and Pathogenicity of *Pseudomonas aeruginosa* Infections in the Lungs of Individuals with Cystic Fibrosis

9

GREGORY C. PALMER[1] and MARVIN WHITELEY[2]

INTRODUCTION

The lung is typically considered a sterile environment; however, this is rarely true in practice. Healthy individuals constantly inhale various microbes, and the innate immune system is responsible for their removal through a system of mucociliary clearance. Some individuals experience altered mucociliary clearance due to any of a number of diseases, and one of the best-studied examples of this occurs in individuals with the genetic disease cystic fibrosis (CF). A hallmark phenotype of CF is dehydration of the mucosal layer across all epithelial surfaces inside the body, resulting in a thickening of the mucus that cannot be adequately cleared despite frequent expectoration. This mucus is an excellent growth substrate for a range of bacteria, and the lung disease associated with these infections further alters the mucus, resulting in a complex mixture of host and microbe derived macromolecules called sputum that is not present in healthy lungs. The opportunistic pathogen *Pseudomonas aeruginosa* is generally the most difficult of these invaders to eradicate, and infections with this organism are the leading cause of morbidity and mortality for individuals with CF (1, 2).

[1]Institute for Cell and Molecular Biology, Freshman Research Initiative, College of Natural Sciences, The University of Texas at Austin, Austin, TX 78712; [2]Institute for Cell and Molecular Biology, Department of Molecular Biosciences, The University of Texas at Austin, Austin, TX 78712

Metabolism and Bacterial Pathogenesis
Edited by Tyrrell Conway and Paul Cohen
© 2015 American Society for Microbiology, Washington, DC
doi:10.1128/microbiolspec.MBP-0003-2014

The goal of this chapter is to summarize our current understanding of the role CF sputum plays in establishment and maintenance of *P. aeruginosa* infections in the CF lung. The introduction will summarize the causes of CF, compare available nutrients in healthy and CF lungs, and introduce *P. aeruginosa* as an opportunistic pathogen. The following section will describe methods for studying *P. aeruginosa* infections both *in vivo* and *in vitro*, and special attention is paid to whether these models accurately reflect CF lung disease in humans. The third section will describe how specific nutritional cues, like mucins, DNA, lipids, and amino acids, that are present in CF sputum impact *P. aeruginosa* pathogenicity, while the final section will discuss relevant future areas of research in this field.

CF DISEASE CAUSES AND PHENOTYPES

In 1989, the genetic locus responsible for CF was identified on the seventh chromosome in a gene encoding a cAMP-dependent chloride ion transporter called the **C**ystic **F**ibrosis **T**ransmembrane Conductance **R**egulator (CFTR) (3, 4). The CFTR protein broadly regulates ion flux across the epithelial surface through interactions with other ion transporters like the sodium ion (Na^+) transporter ENaC (5, 6). CFTR has been localized to ciliated cells in airway epithelia, and importantly, no CFTR protein was detected in ciliated cells from individuals carrying the most common CF mutation, ΔF508 CFTR (7, 8). This is attributed to post-translational degradation of CFTR in CF cells (7, 8). Further, a study by Tucker and colleagues determined that expression of ΔF508 CFTR can spoil processing of WT CFTR in the same cells (9). This observation is significant to both heterozygous CFTR carriers as well as scientists' attempts to cure CF using gene therapy or other strategies (10). CFTR is also an important component of the innate immune response; a study

by Pier and colleagues demonstrated that the absence of CFTR diminished the ability of airway epithelial cells to internalize *P. aeruginosa*, which is a significant determinant of the host's ability to clear this pathogen (11). Interestingly, epithelial cells lacking CFTR also possess greater levels of AMP-activated Kinase (AMPK) activity, which may be an adaptive response, because it is associated with decreased production of inflammatory effector molecules (12). Thus, the absence of CFTR both generates the thick mucus required for CF lung disease and renders the lung particularly susceptible to infection.

NUTRIENTS PRESENT IN THE LUNG: HEALTHY VERSUS DISEASED STATE

Healthy airway surfaces are coated in a mucus layer that is largely composed of water, salts, mucins, and surfactants. Among the ions present in sputum, iron is generally viewed as a critical nutrient for any pathogenic bacterium, and it has been observed to be present at 0.9 micromolar (μM) to 63 μM in CF sputum compared to largely undetectable levels in healthy airway secretions (13–18). Like iron, many nutrients present in CF sputum are not found in healthy lungs, and the goal of this section is to delineate these differences. We will begin by discussing differences in mucins and surfactants between healthy individuals and those with CF, while the remainder of the section will be dedicated to nutrients generally not present in healthy lungs.

Of the several gel-forming, glycoprotein mucins produced by mammals, MUC5AC, MUC5B, and to a lesser extent MUC2 have been detected in airways (19–21). Mucin-like glycoprotein was quantified from healthy individuals and asthmatic individuals with a history of sputum production, and 3- to 6-fold more glycoprotein was quantified from the asthmatic individuals (0.8 to 1.6 mg/mL versus 4.3 to 4.8 mg/mL) (19, 22). A similar study compared mucin levels in CF sputum and healthy sputum, and found that CF sputum contained significantly less MUC5AC and MUC5B (93% and 70%, respectively) (23). However, during times of heightened *P. aeruginosa* infection called exacerbations, MUC5AC and MUC5B levels increased 908% and 59%, respectively (24). Enhanced production of mucins during exacerbations may be due to the presence of *P. aeruginosa*, as several studies have indicated that the bacterium induces mucin secretion (25–29). Whether enhanced mucin production during exacerbations is a cause or effect of the exacerbations remains unclear; however, the increased presence of mucins can have a profound effect on the infecting bacteria as will become clear in future sections of this chapter. The chemical composition of mucin in individuals with CF may also be different compared to healthy mucin, as a study by Carnoy and colleagues found greater levels of glycoproteins in CF salivary mucin and those glycoproteins displayed greater sulfate and fucose content (30).

Healthy airway epithelial cells also secrete surfactants, which provide protection and facilitate gas exchange by reducing surface tension of the mucosal layer that coats alveoli. The composition of pulmonary surfactant is largely lipids (90%), of which phosphatidylcholine (PC) and its derivatives are the most common (62%) followed by phosphatidylglycerol, phosphatidylinositol, phoshpatidylserine, phosphatidylethanolamine, sphingomyelin, and other lipids (28%) (31–33). The remainder of surfactant (10%) is comprised of four surfactant proteins: SP-A and SP-D are large proteins that are involved in protecting the surface from invasion, and SP-B and SP-C are smaller, hydrophobic proteins that are associated with lipids and interact with the surface of the epithelium (31–33). Several studies have examined the extent to which surfactant composition changes during CF disease. Significant conclusions include the fact that surfactant composition is not remarkably altered under treatment with recombinant human DNase (a common CF therapeutic which degrades

sputum), and surfactant composition is largely stable until chronic infection is established, which occurs early in the lives of most individuals with CF (34, 35). Decreased PC levels and an increase in typically minor lipids in healthy individuals or those with chronic obstructive pulmonary diseases (COPD) have also been observed (34–39).

There are many nutrients present in CF sputum that are essentially absent in healthy lung secretions, and many of these are derived from lysed host and microbial cells. DNA is one such nutrient, and it appears that most of the DNA in CF sputum is derived from host cells that were lysed as a consequence of CF lung disease (40). Many studies have measured DNA concentrations in sputum samples that had been either expectorated or acquired using bronchoalveolar lavage (BAL), and reported concentrations range widely from 5.4 µg/mL to 17.6 mg/mL, with most reports falling into a range of 400 µg/mL to 700 µg/mL (24, 34, 41–46). The use of recombinant human DNase (rhDNase) as an inhaled therapeutic has been widespread since the 1990s, because charged DNA molecules are known to crosslink sputum, making it more difficult to clear. Many studies have examined the effect of rhDNase on DNA levels in sputum. While these studies varied in their conclusions as to whether rhDNase treatment decreased total DNA in sputum, there was a consensus that treatment changed the rheological properties of sputum, allowing for easier expectoration (24, 34, 41–46).

Several small organic acids and sugars are also present in CF sputum. Lactate appears to be an indicator of lung inflammation, and several groups have measured its concentration in CF sputum resulting in a range of 3.0 mM to 14.1 mM (15, 47, 48). Bensel and colleagues also measured lactate levels in individuals with chronic obstructive pulmonary disease and acute lung inflammation finding 1.6 mM in the former and nondetectable levels in the latter (47). Our own group has measured glucose levels in expectorated sputum and found an average concentration of 2.9 mM (15). In a separate study, we also measured levels of N-acetyl-glucosamine (GlcNAc), which is a component of macromolecules present in sputum, and we found it to be present at hundred micro-molar levels (49). *P. aeruginosa* is known to rapidly consume small organic acids like lactate, pyruvate, and succinate, and can grow on glucose or GlcNAc as sole sources of carbon and energy.

Proteins and free amino acids other than the previously discussed mucins and surfactant proteins are final examples of CF sputum nutrients not typically found at high levels in the mucus of healthy individuals. For example, the sputum of individuals with CF contains high levels of proteins from immune cells and microbes, and these protein levels are correlated to exacerbations and the accompanied enhanced immune response (50–52). Both the immune and pathogenic cells present in sputum release extracellular proteases capable of cleaving proteins into free amino acids. Increased levels of free amino acids have been detected in CF sputum (5.70 mg/mL), when compared to non-CF sputum samples (2.52 mg/mL) (53). Our own measurements of sputum have indicated 4.4 to 24.7 mM free amino acids present in sputum (15). These two studies also quantified levels of individual amino acids in CF sputum, and the former found all 22 standard amino acids except tryptophan, proline, glutamate, glutamine, asparagine, cysteine, and aspartate, though some amino acids present at low levels may not have been detectable (53). In the latter study, 19 of the 22 standard amino acids were detectable, with the exceptions being asparagine and glutamine, while hydroxyproline was not measured (15).

It is clear that the lungs of individuals with CF contain a remarkably different set of nutrients compared to those of healthy individuals. The inability to clear secreted mucins and surfactants combined with degradation products from lysed host and bacterial cells results in a complex, dynamic nutritional environment. As we will see in later sections, nutrients, like mucins, lipids, DNA, and amino acids, serve as sources of carbon and energy to promote *P. aeruginosa* growth as well as potent cues for enhanced virulence in CF sputum.

THE OPPORTUNISTIC PATHOGEN *PSEUDOMONAS AERUGINOSA*

Pseudomonas aeruginosa is a Gram-negative opportunistic pathogen. It is commonly found in a range of diverse environments across the world including soil, fresh, and marine water. As an opportunistic pathogen, it lives a dual lifestyle as a commensal and environmental organism, but can occasionally switch to a pathogenic state, causing costly and difficult to treat infections. The most notable of these is in the lungs of individuals with the genetic disease cystic fibrosis, as *P. aeruginosa* colonization is associated with declining clinical outcomes, and is the leading cause of morbidity and mortality in these individuals (1, 2). It is believed that *P. aeruginosa* infections are acquired from the environment, and recent studies demonstrating a correlation between the sinus and lung microbiota suggest initial colonization of the sinus proceeds to the lower respiratory system (54–57). *P. aeruginosa* utilizes a plethora of virulence factors to colonize and maintain infections in immune-compromised hosts, and production of many of these virulence factors is regulated in a cell density-dependent manner known as quorum sensing (QS, reviewed below). *P. aeruginosa* infections are notoriously difficult to treat, because the organism is intrinsically resistant to most conventional antibiotics. This is in part due to the fact that it produces antibiotic-resistant, surface-attached communities surrounded in a polymeric matrix known as biofilms or microcolonies.

In order to understand how CF sputum nutrients can affect the pathogenicity of *P. aeruginosa*, it is first critical to establish the infection-causing toolkit available to this organism. *P. aeruginosa* produces an arsenal of virulence factors, which are generally defined as pathogen-derived molecules that help initiate or exacerbate an infection. In the following paragraphs, we will discuss these tools that *P. aeruginosa* uses to cause infections with special emphasis on how they may affect CF sputum and the nutrients available to this organism.

Quorum Sensing

Quorum sensing (QS) describes a process whereby bacteria in groups coordinately regulate gene expression (58, 59). In QS, growing cells secrete small molecule signals called autoinducers, and as the number of cells increases, the concentration of the autoinducer also increases to a point where it can interact with transcriptional regulators inside the cell to alter gene expression (58, 59). *P. aeruginosa* possesses three known QS systems: the LasI/LasR system for which 3-oxododecanoyl-homoserine lactone (3OC12-HSL) is the autoinducer (60–62); the RhlI/RhlR system for which butyryl-homoserine lactone (C4-HSL) is the autoinducer (63–65); and a quinolone-based system for which 2-heptyl-3-hydroxy-4-quinolone (*Pseudomonas* quinolone signal; PQS) is the most potent autoinducer (66–68). Collectively these QS systems represent an integrated regulatory network that affects transcription of up to 10% of the *P. aeruginosa* genome (69), including genes for production of the many virulence factors that will be discussed in the following paragraphs. Additionally, several studies have indicated that CF sputum enhances *P. aeruginosa* QS, emphasizing its importance for this work (15, 70, 71).

Secretion of Virulence Determinants and Degradative Enzymes

In the 1960s, it was determined that *P. aeruginosa* produces several secreted factors that impact its virulence toward host cells (72–74), though some secreted factors were identified much earlier (75). The organism possesses five of the six known classes of secretion systems in Gram-negative bacteria and multiple copies of some systems exist with differential regulation and/or secreted substrates (76). The two type I secretion systems of *P. aeruginosa* are responsible for secreting an alkaline protease and a heme-binding protein that both function as virulence determinants (76–78). *P. aeruginosa* utilizes a type II secretion system to release several virulence factors beginning with Exotoxin A, which inhibits host protein synthesis in a manner similar to diphtheria toxin (79). Other protein virulence factors secreted through the type II system include several phospholipases, alkaline phosphatases, and proteases that both attack host cells and degrade macromolecules in sputum (76). *P. aeruginosa* possesses a type III secretion system in which toxins are injected directly into host cells via a needle like structure made of proteins related to those of the flagellum (80–82). The four known effector proteins that are injected to host cells (ExoS, ExoT, ExoU, ExoY) disrupt cytoskeleton formation, signal transduction, and have phospholipase activity (80–83). *P. aeruginosa* lacks a type IV secretion system, and its type V systems are known to secrete an esterase, another protease, and a hemagglutinin-like protein (76). It also possesses three type VI secretion systems that form phage-derived nanotube structures similar to the flagellar-derived type III secretion system (76). Among the few type VI secreted effectors identified in *P. aeruginosa* are a peptidoglycanase and a phospholipase that are both likely involved in degrading neighboring bacterial cells (84, 85). Evidence that the type VI secretion system apparatus is important for virulence in the CF lung includes the fact that antibodies targeting these structures have been identified in CF sputum (76, 86). Finally, *P. aeruginosa* toxins and competitive factors including β-lactamase, alkaline phosphatase, hemolytic phospholipase C, and CFTR inhibitory factor (Cif) are also packaged into outer membrane vesicles (OMVs)

(87, 88). Such vesicles are the result of blebs of the outer membrane and are ubiquitous among Gram-negative bacteria (89). In *P. aeruginosa*, OMVs are produced in a QS-dependent manner as our group has shown that the quinolone signal PQS is trafficked in and promotes formation of OMVs (90).

The degradative enzymes secreted by *P. aeruginosa* are capable of both attacking host cells and degrading sputum macromolecules. In many cases, the breakdown products of these macromolecules serve as both nutrients and mediators of *P. aeruginosa* pathogenicity. *P. aeruginosa* secretes many extracellular proteases, such as LasA, LasB, and Alkaline protease, whose production are regulated in a QS-dependent manner. These factors work alongside copious host-secreted proteases (91, 92) to profoundly alter the nature of CF sputum and contribute to the inflammatory lung disease phenotype common to nearly all individuals with CF (93, 94). Additionally, a proteomic study by Scott and colleagues examining the transmissible *P. aeruginosa* strain AES-1R found enhanced production of secreted proteases, particularly in an artificial sputum medium described in a later section (95). In addition to altering sputum by cleaving substrates, secreted proteases may also worsen the disease state as Butterworth and colleagues have shown that the alkaline protease of *P. aeruginosa* can activate ENaC channels in the epithelium (96). *P. aeruginosa* also produces several lipases that attack the host (97, 98), and degradation of lipids may be critical for growth and heightened virulence in CF sputum (84, 99, 100). *P. aeruginosa* secretes another toxin that has been shown to inhibit CFTR (CFTR inhibiting factor, Cif), which may play a role in colonization of CF lungs (101). Finally, proteomic studies by Folders and colleagues have identified a staphylolytic/chitin-binding protein and chitinase as QS-induced degradative molecules that act as virulence factors and also likely release free GlcNAc, which potentiates *P. aeruginosa* virulence (49, 94, 102–104). While chitin may not be a common nutrient in CF sputum, the activities of these enzymes may still impact virulence in the CF lung by affecting *P. aeruginosa* interactions with co-infecting yeast and host cells (102, 105). It is clear that *P. aeruginosa* contains a myriad of degradative enzymes that both attack the host and provide nutrients for growth and pathogenicity. More information on how these nutrients affect *P. aeruginosa* virulence can be found in the section "Nutritional cues in CF sputum that affect *P. aeruginosa* metabolism and pathogenicity."

Phenazines and Other Redox Active Toxic Factors

Cultures of *P. aeruginosa* appear a characteristically striking blue/green color, and this is largely due to the liberal secretion of a redox-active phenazine compound called pyocyanin. Pyocyanin production is QS-regulated via the PQS system (106), and it is capable of destroying host cells by altering critical cellular processes (107), generating reactive oxygen species (108), altering cell signaling (109), and inducing neutrophil apoptosis (110). Recently a study by Hunter and colleagues measured levels of pyocyanin and its biosynthetic precursor, phenazine-1-carboxylic acid in expectorated CF sputum samples and found pyocyanin levels negatively correlated to lung function (111). Pyocyanin also appears to act as a signaling molecule, as a study by Dietrich and colleagues demonstrated that the phenazine up-regulates genes for transport and redox control and suppresses genes involved in iron acquisition (112).

Other redox-active factors produced by *P. aeruginosa* include a battery of 4-quinolone molecules with a range of activities (113). Lepine and colleagues used electrospray/mass spectrometry to survey the 4-quinolones produced by *P. aeruginosa* and found 56 unique molecules, which differed by the presence of 3-position substituents, n-oxide groups in place of the quinolone nitrogen, and the alkyl chains at the 2 position (114). The

Pseudomonas quinolone signal (PQS) is one of the most well studied of these quinolones because of its ability to regulate virulence factor production in a cell-density dependent manner (66–68). In terms of its role in redox balance, PQS is both helpful and harmful to *P. aeruginosa* as Haussler and Becker demonstrated it can induce an anti-oxidative stress response and it also sensitizes cells to oxidative and other stresses (115). Another redox-active toxin produced by *P. aeruginosa* is the cytotoxic poison, hydrogen cyanide, which has been demonstrated to be critical for pathogenicity in a *Caenorhabditis elegans* infection model (116, 117). Regulation of its production is complex and partially dependent on QS (118, 119).

P. aeruginosa is a deadly opportunistic pathogen, and this stems from the extensive array of virulence factors it is capable of producing. We have seen that many of these virulence factors are QS-regulated, which facilitates colonization of host environments like the CF lung. It is also clear that the nutritional environment of the CF lung is radically different from that of healthy individuals, and we will see how the presence of these nutrients impacts *P. aeruginosa* virulence. CF lung infections are unique in that they are chronic, and generally persist throughout the individual's lifetime. Thus, it is critical to study *P. aeruginosa* in the context of the CF lung over extended periods of time, and the next section will examine methods to recapitulate these chronic CF lung infections in the laboratory setting.

APPROACHES AND CHALLENGES TO DEVELOPING CF SPUTUM PATHOGENESIS MODELS

P. aeruginosa causes decades-long, chronic infections in the lungs of individuals who have the genetic disease cystic fibrosis. These infections are the leading cause of morbidity and mortality for individuals with CF (1, 2, 120). The complexity of CF sputum and the

lung environment in which these infections occur has already been addressed in this chapter; however, we have yet to describe the physiology of these infections. An obvious place to begin this study is to isolate bacteria from genuine CF infections, and much information has been gained from studying these clinical isolates (121–127). Additionally, studying sequential isolates over time has yielded information on the adaptive responses of *P. aeruginosa* to the CF lung environment (122, 128–130). While these studies have been instrumental in understanding critical aspects of the physiology of *P. aeruginosa* infections, they are entirely retrospective and could potentially miss heterogeneity within the population causing the infection. An ideal alternative would involve use of an animal model system that is capable of mimicking the CF lung as well as the chronic infections observed there. Several model systems have been used to study *P. aeruginosa* infections, and somewhat surprisingly, these studies indicate there is conservation of the virulence factors required for colonization of a broad range of eukaryotic hosts (131–137). As the focus of this chapter is *P. aeruginosa* infections in the lungs of humans with cystic fibrosis, the following discussion of model systems will focus on mammalian CF model organisms. This section will detail different approaches to solving this problem, and particular emphasis will be placed on overcoming the challenges of accurately replicating the environment/physiology of human CF-related lung infections.

Mouse Models

Mammalian animal models are a reasonable first approach to recreating the chronic *P. aeruginosa* infections of the CF lung, and the mouse has been a workhorse model organism for a range of CF-related studies reviewed in the following reference (138). In 1992, several groups generated mice with a homozygous disruption of the murine CFTR gene, and phenotypic characterization of these mutants demonstrated similarities to the human CF disease, including aberrant ion transport across epithelial surfaces, meconium ileus, failure to thrive, and histological abnormalities (138–142). Subsequent mouse mutants have been made that more closely replicate the human mutations responsible for CF disease, and the ability to over-express CFTR in different tissues has been employed to mitigate these mutations (143). While these models have been able to replicate CF lung disease and colonization with *P. aeruginosa* to varying degrees of success, they generally require artificial enhancements like repeated aerosolized exposure of *P. aeruginosa* to mice missing a second Cl⁻ ion channel (144, 145), or inoculating with *P. aeruginosa* coated agar/agarose beads to establish infections (146, 147). Importantly, these infections were largely transient and failed to model the chronic infections that are hallmarks of CF lung disease. Coleman and colleagues developed the most successful protocol for chronic infection of ΔF508-CFTR mice (143), though only ~33% of mice retained *P. aeruginosa* in their lungs 24–53 weeks after inoculation (148). Hoffmann and colleagues reported a chronic infection model that utilizes a stable mucoid strain of *P. aeruginosa*, though these experiments were not carried out beyond 2 weeks, and genuine CF infections last years to decades (149). A conditional mouse CFTR mutant has been made in which CFTR null phenotypes can be examined tissue by tissue (150, 151). These mice have been used to show that the loss of expression of CFTR in myeloid-derived cells results in mice that are more susceptible to challenge with *P. aeruginosa* embedded in agarose beads (152). This study lends credence to the hypothesis that the impaired innate immune system of the CF lung contributes to CF lung disease in addition to the physical and nutritional nature of CF sputum.

Pig and Ferret Models

The inability of mouse models to accurately replicate important CF phenotypes like chronic *P. aeruginosa* infections has led researchers to pursue alternative animal models that may possess physiologies closer to humans. One of the most intriguing alternative models to the mouse was a CF pig that was first developed in 2008 (153, 154). New insights are beginning to be gained from the CF pig model, as a study by Pezzulo and colleagues demonstrated that newborn CF pigs are less able to kill *Staphylococcus aureus*, a common pathogen present in young individuals with CF (155). The authors attached *S. aureus* to gold grids, surgically placed them into the trachea, and used live/dead staining to evaluate *S. aureus* killing. The authors determined diminished killing was due to the lower pH of CF airway surface fluids, which inhibited antimicrobial factors like lysozyme and lactoferrin (155). Additionally, a homozygous ΔF508-CFTR pig has been developed (156). Despite the fact that these ΔF508-CFTR pigs retain some CFTR activity in the airway epithelia, some bacteria (10 CFU [colony forming units]/g to 743 CFU/g of lung tissue) were able to be cultured from the lungs of these animals, while no bacteria were able to be cultured form wild-type controls (156).

In 2010, Sun and colleagues were able to generate a ferret CFTR knockout mutant that displayed several phenotypes similar to those of humans with CF, though these animals also did not suffer severe lung infections beyond their first week of life (157). A study by Fisher and colleagues compared processing of CFTR in humans and ferrets and found that the ferret CFTR protein was processed more efficiently and displayed a longer half-life in the epithelium than human CFTR, and the authors argued this could yield approaches to stabilize human CFTR (158). The same study examined processing of a ferret ΔF508-CFTR, and while maturation of the ferret mutant protein was enhanced compared to human ΔF508-CFTR, expression of neither protein resulted in cAMP-dependent chloride ion conduction in airway epithelial cells (158). Like the pig, the ferret serves as another useful model for studying initiation of CF lung infections, (157) however, neither animal model has been spontaneously or chronically colonized by *P. aeruginosa*, which bodes poorly for use of these organisms to study chronic infections (151).

CF Sputum Media

Due to the modest success of generating animal models that accurately recapitulate chronic CF lung disease, several groups have mimicked the CF lung environment by generating sputum-based growth media. An early example involved the use of sputum as the growth substrate for mucoid strains of *P. aeruginosa*, and the observed phenotypes of growth in this medium were consistent with the hypothesis that mucoidy is an adaptive aspect of chronic infection (159). Our own group used a medium featuring sputum as the sole source of carbon and energy for *P. aeruginosa* transcriptome profiling, and the results indicated that the production of several key virulence factors is enhanced in sputum (71). The same medium was also used to demonstrate that a *P. aeruginosa* membrane-bound nitrate reductase is critical for anaerobic growth in CF sputum (160). While oxygen is not a nutrient for the purposes of this work, several studies have suggested that cells grow in microaerophilic pockets within sputum, and anaerobic *P. aeruginosa* biofilms are more resistant to antibiotics, making them more difficult to eradicate from the lung (160–162).

The complexity of sputum and the difficulty in obtaining it from lavage, expectoration, or extransplanted lungs led several groups to investigate synthetic sputum media that nutritionally (and rheologically) mimic the contents of genuine CF sputum (15, 163, 164). Ghani and Soothill used such a medium to determine whether nutrients present in CF sputum would affect the ability of combinations of antibiotics to kill biofilm-grown *P. aeruginosa*, and the authors were able to identify a cocktail

of ceftazidime, gentamicin, and rifampicin as a successful combination (163). Building off this work, Sriramulu and colleagues developed a similar artificial sputum medium (ASM) that spontaneously promoted *P. aeruginosa* growth as microcolonies or the sputum component-attached biofilms believed to in part explain the difficulty of eradicating *P. aeruginosa* from the CF lung (164). Fung and colleagues generated an enhanced version of ASM that included albumin and altered the concentrations of mucin and DNA to reflect more recent definition of genuine CF sputum (165). This medium was used for *P. aeruginosa* transcriptome profiling and identified several metabolic and virulence genes that were up-regulated in the presence of CF sputum nutrients (165). Another recent study by Haley and colleagues used this medium to determine optimal concentrations of several nutrients for biofilm formation as well as to dissect the role of QS in the same process (166).

Our group has measured the concentration of several nutrients in sputum samples expectorated from individuals with CF and used these concentrations to develop an alternative synthetic CF sputum medium (SCFM) (15). The particular utility of this medium stems from the fact that its entire nutrient composition is defined, thus nutrient utilization can be quantified. Using SCFM as such a tool, we were able to delineate a hierarchy of nutrient preference for *P. aeruginosa* in sputum where proline, alanine, arginine, glutamate, aspartate, and lactate were among a set of preferentially consumed carbon sources (15). Another utility of SCFM is the ability to add and subtract nutrients from this medium in order to understand their roles in *P. aeruginosa* pathogenicity, which will be discussed in the following section.

P. aeruginosa infections in the CF lung are unique in that they are chronic and generally last for years. While this makes for an excellent system to study bacterial adaptation to infection sites, our ability to adequately replicate these infections in model systems has been a significant challenge to the field.

That said, progress has been made using various CF animal models and synthetic media for *in vivo* and *in vitro* experiments, as these tools have been invaluable in expanding our knowledge of the roles played by specific nutrients in CF sputum in cuing *P. aeruginosa* pathogenicity. The next section will summarize our current knowledge of the roles certain nutrients play in cuing *P. aeruginosa* pathogenicity in CF sputum.

NUTRITIONAL CUES IN CF SPUTUM THAT AFFECT *P. AERUGINOSA* METABOLISM AND PATHOGENICITY

The introduction to this chapter outlined the significant differences between healthy and diseased lungs, and it was made clear that there is a substantial increase in the availability of nutrients for microbial growth in diseased lungs like those of individuals with cystic fibrosis. This is likely due to the inability of these individuals to clear the dehydrated airway secretions as well as the release of charged, polymeric molecules like mucin,

DNA, and proteins from lysed host and microbial cells. We have also seen that many *P. aeruginosa* virulence factors are capable of degrading these macromolecules to generate an even more complex extracellular environment at these infection sites. The following subsections will detail what is known about how these nutrients specifically cue *P. aeruginosa* behaviors that are essential for colonization and maintenance of infections in the CF lung (for summary see Table 1).

Iron

Iron is typically sequestered within a host, and many pathogens have evolved extensive mechanisms to acquire iron while causing infections (167, 168). These mechanisms include production of energetically expensive iron scavenging molecules called siderophores, and a study by Martin and colleagues indicated that the *P. aeruginosa* siderophore pyoverdine is an important, though nonexclusive means of iron acquisition in CF sputum (169). Additionally our group has proposed that co-culture with *S. aureus*

TABLE 1 Summary of nutrients present in CF sputum and their roles in *P. aeruginosa* pathogenicity

Nutrient (concentration*)	Carbon source?	Pathogenic cue(s) and other notes
Iron (0.9–63 µM)	–	Important component for microcolony formation; iron acquisition key to infection, may explain transmissibility of Australian epidemic strain
Mucins (~7–14 mg/mL)	+	May play a role in attachment/colonization; decreases motility and induces microcolony formation
Phospholipids (281–53 µg/mL)	+	Carbon source during high cell density growth; phospholipase mutants less virulent in animal models; chemotaxis toward PE; degradation products (phosphorylcholine, choline) also cue virulence
DNA (400–700 µg/mL)	+	Crosslinks sputum, treated with rhDNase; aids microcolony formation; may sequester neutrophil proteases
Small acids Lactate (3–15 mM)	+	Preferred carbon source, levels correlated with inflammation, lactate consumption may be adaptive strategy
Sugars GlcNAc (~300 µM)	+	Induces expression of pyocyanin and other virulence factors; hyaluronic acid is antagonistic
Glucose (2.9 mM)	+	Induces heterogeneous, mushroom-shaped microcolonies; simple sugars may also inhibit biofilm formation
Amino acids (~20 mM total)	+	Several preferred carbon sources; aromatic amino acids involved in QS-mediated virulence; alanine catabolism aids in rat infection model; promote tight microcolony formation; promote swarming motility

*Concentrations reflect consensus of reported values. See "nutrients present in the lung: healthy versus diseased state" for references.

results in *P. aeruginosa* killing *S. aureus* to acquire iron (170). This may be particularly relevant in the CF lung, which is frequently co-colonized by *P. aeruginosa* and *S. aureus*. Iron levels also tend to be higher in CF airways compared to healthy airways (17, 18), and there is evidence that this helps *P. aeruginosa* infections beyond simply being a more available essential nutrient, as iron is a critical component of the extracellular matrix of biofilms (171). In a study by Moreau-Marquis and colleagues, *P. aeruginosa* biofilms grown on airway epithelial cells were 25-fold more resistant to tobramycin than those grown on an abiotic surface (14). Additionally, growth on ΔF508-CFTR mutant epithelial cells enhanced biofilm production compared to wild-type epithelial cells. The authors attributed this to greater iron levels that were observed outside these cells, because removal of excess iron abolished the biofilm phenotype (14). The picture of iron in the CF lung has been complicated with studies by Koley and Hunter and colleagues demonstrating ferrous iron is generated near *P. aeruginosa* biofilms (172) and that a proportion of ferrous relative to ferric iron increases with the progress of infection (173). The presence of ferrous iron has implications for redox active virulence factors like phenazines and biofilm formation (173). Finally, a proteomic study by Hare and colleagues comparing the Australian epidemic strain 1 (AES-1R) to a burn wound/septic strain PA14 and lab strain PAO1 grown in a synthetic CF sputum medium found an increase in enzymes for synthesis of pyocyanin and siderophores (174). The authors argued that iron acquisition could explain why AES-1R is associated with person-to-person transmission of *P. aeruginosa*, which is more commonly thought to be acquired from the environment (174).

Mucins

Mucins are substantial components of sputum, and they can serve as a source of carbon and energy for *P. aeruginosa*, as Sriramulu and colleagues observed that growth in ASM was substantially diminished in the absence of mucin (164). Another study by Henke and colleagues demonstrated that host and microbe-derived serine proteases are responsible for degrading mucins in CF sputum, and their degradation may contribute to airway inflammation by removing a barrier to the epithelium (175). Finally, a study by Aristoteli and Wilcox characterized mucin degradation by ocular infection strains, demonstrating that the ability to metabolize mucin provided a long-term growth advantage (176). Thus, mucin may be a critical nutrient for sustaining infections.

Several studies have demonstrated that *P. aeruginosa* binds mucin, and this may be the critical first step in colonizing CF airways (177–179). Interestingly, a study by Sajjan and colleagues indicated that *P. aeruginosa* interactions with mucin were not necessarily specific to mucin itself, arguing that colonization of the CF lung is not based on tropism for mucin (180). In support of the theory that pathogen-mucin interactions are nonspecific, several groups have reported that other CF-associated opportunistic pathogens bind mucin (181–185). Investigations comparing CF and non-CF mucin indicated that CF mucins are bound better by *P. aeruginosa*, though this is likely due to the fact that CF-mucins are degraded in the presence of *P. aeruginosa* into a form to which the pathogen more readily adheres (186). Mucin may also help facilitate the maintenance of *P. aeruginosa* infections, as a study by Landry and colleagues demonstrated that the presence of mucin on a surface diminished the organism's motility, resulting in large cellular aggregates with increased resistance to the antibiotic tobramycin (187). By contrast, more recent reports have indicated that mucin promotes motility across surfaces (188), and mucin biopolymers can enhance motility of planktonic cells, preventing their attachment to underlying surfaces (189). These studies collectively implicate mucin as a critical nutrient for both initiating and sustaining *P. aeruginosa* infections.

Lipids

Multiple studies have indicated that lipids may be a significant carbon source for *P. aeruginosa* in CF sputum, and this may be particularly true in high cell density (HCD) infections like those with >10^8 CFU/mL observed during exacerbations in the CF lung (1, 190, 191). It is believed that the many secreted lipases of this organism cleave off fatty acid chains in sputum, which can then be internalized and oxidized as carbon sources (190, 192, 193). Another study by Wargo and colleagues demonstrated that a mutant in hemolytic phospholipase C, PlcHR, caused less severe infections in a mouse model, and inhibition of this enzyme with the drug miltefosine also decreased the effect of PlcHR on mouse pulmonary surfactant (193). Son and colleagues were able to perform transcriptome profiling of *P. aeruginosa in vivo* undergoing HCD exacerbations, and found genes for lipid metabolism to be up-regulated (190). In a subsequent study, the same group identified several fatty acid synthetases (Fads) that are involved in metabolism of fatty acids in sputum. The ability to catabolize fatty acids seems to be important for virulence as mutants in these Fads were diminished in their ability to cause infections in a mouse model (192).

A study by Krieg and colleagues found *P. aeruginosa* increased capsule formation and adhesion during growth on phosphorylcholine, a degradation product generated by phospholipase C acting on phosphatidylcholine (194). Subsequent work by the same group demonstrated that growth on phosphorylcholine promotes conversion to mucoid variants of *P. aeruginosa*, which is believed to be an adaptation consistent with long-term colonization (195). *P. aeruginosa* possesses an enzyme capable of degrading phosphorylcholine to choline (196), and choline also appears to play a role in *P. aeruginosa* virulence, as it has been shown to induce phospholipase C activity (197, 198). Additionally, a study by Wargo determined that mutants defective in catabolism of choline to glycine betaine displayed a diminished capacity to infect mouse lungs (199). Phosphatidylethanolamine (PE) is another sputum lipid that likely promotes *P. aeruginosa* virulence. A study by Kearns and colleagues demonstrated that *P. aeruginosa* is capable of twitching motility-mediated chemotaxis up a PE gradient and that the rate of twitching motility was increased in the presence of this nutrient (200). The role this movement may play in *P. aeruginosa* virulence is not entirely clear, but the fact that the organism displayed chemotaxis toward both host- and microbe-derived PE suggests it may be both a signal for movement toward host cells, and/or an autoaggregative signal (200). Subsequent work by Barker and colleagues characterized a novel PE phospholipase that is required for this chemotaxis (100).

A final example of sputum lipids involved in *P. aeruginosa* pathogenicity is the sphingomyelin degradation product ceramide. A study by Grassme and colleagues demonstrated that the presence of *P. aeruginosa* stimulates the release of ceramide, which promotes formation of sphingomyelin rafts that are critical for internalization of *P. aeruginosa*, inducing apoptosis, and regulating the release of cytokines (201). A subsequent study by the same group found that the ceramide accumulates with age in the airways of CF mice, and this accumulation of ceramide was correlated with increased susceptibility to *P. aeruginosa* infection (202). Enhanced susceptibility of these mice to infection is likely due ceramide-induced cell death, which releases DNA on the epithelium and facilitates *P. aeruginosa* colonization (202). The complex roles of sphingolipids and ceramides in the CF lung was reviewed recently by Becker and colleagues (203).

DNA

DNA was known to be present in the polymeric extracellular matrix (ECM) that coats

P. aeruginosa biofilms, but a significant study by Whitchurch and colleagues demonstrated that DNA is actually required for effective biofilm formation (204, 205). Much of the DNA in sputum has been attributed to lysed host cells (40), though *P. aeruginosa* is also known to release DNA for biofilm formation (206). A study by Walker and colleagues demonstrated that the presence of lysed neutrophil DNA and F-actin enhances *P. aeruginosa* biofilm formation (207), and these DNA-containing biofilms appear to be more resistant to antimicrobial peptides and aminoglycosides (208, 209). DNA secreted from neutrophils as neutrophil extracellular traps (NETs) may also sequester neutrophil derived proteases, and it has been argued that DNase treatment releases these proteases (210), thus like the ECM of biofilms, the presence of DNA protects *P. aeruginosa* in the CF lung. These authors also suggest DNase treatment could be accompanied by protease and/or elastase inhibitors to more effectively combat *P. aeruginosa* CF lung infections (210).

Organic Acids and Carbohydrates

Work from our group indicated that lactate is among six preferred carbon sources that are readily consumed by *P. aeruginosa* in synthetic sputum medium, and the pathogen is known to readily utilize organic acids as sources of carbon and energy (15). This may be particularly relevant in sputum as a study by Wolak and colleagues found that an increase in lactic acid in sputum correlated with increased inflammation; though, it is not clear if this lactic acid increase is due to the inflammation, the generation of hypoxic areas in the lung, or some combination of the two (48). Another study by Bensel and colleagues found that lactate levels decreased with antibiotic treatment of exacerbations (47). Organic acids have been also implicated in *P. aeruginosa* pathogenicity as a study by Petrova and colleagues indicated that the presence of pyruvate and its

fermentation induces microcolony formation (211). The authors argued that pyruvate fermentation is an adaptive strategy used by *P. aeruginosa* in the CF lung (211), and this is consistent with work by Eschbach and colleagues, which demonstrated that pyruvate fermentation is important for long-term survival under anaerobic conditions, such as the hypoxic zones with in the CF lung (212).

Simple and modified sugars also appear to affect *P. aeruginosa* virulence and persistence as growth of *P. aeruginosa* on glucose is known to affect biofilm formation. Klausen and colleagues noted that several studies have reported that glucose promotes formation of heterogeneous, mushroom-shaped microcolonies, while growth on citrate or amino acids results in dense, flat biofilms (213–215). Other studies have shown that simple sugars like fructose, galactose, mannose, and fucose, mitigate *P. aeruginosa* infections by inhibiting attachment and diminishing lung damage (216, 217). Aminoglycosides are a common class of antibiotics used to treat *P. aeruginosa* infections in the CF lung, and several studies have indicated that subinhibitory concentrations of these drugs (and other antibiotics) may enhance *P. aeruginosa* biofilm formation and virulence (218–220).

Studies from our own group have indicated the importance of another amino-modified sugar, n-acetylglucosamine (GlcNAc), for *P. aeruginosa* pathogenicity. These studies began with the observation that genes for GlcNAc catabolism were up-regulated when *P. aeruginosa* was grown in CF sputum medium, suggesting both that GlcNAc is present in sputum and that it is a relevant carbon source for *P. aeruginosa* in the CF lung (71). Subsequent work by our group determined that GlcNAc is indeed present in CF sputum at hundred micromolar levels, and that it cues enhanced production of pyocyanin (49). There are many potential sources of GlcNAc in sputum and additional work by Korgaonkar and colleagues used colonization of *P. aeruginosa* in the fly gut to demonstrate that shedding of the GlcNAc-containing polymer peptidoglycan by Gram-positive bacteria is responsible for enhanced virulence factor production (104). The authors also identified a gene (*PA0601*) required for GlcNAc sensing, and found that mutants unable to sense GlcNAc were less competitive when co-infected with *S. aureus* in a mouse chronic wound model (104). These studies emphasize the importance of studying pathogens in the context of co-infection with other microbes since most infections in nature are polymicrobial, including the CF lung. Finally, while it appears that GlcNAc activates virulence and makes *P. aeruginosa* more competitive in the CF lung, other studies have indicated that the GlcNAc-containing polymer hyaluronic acid is antagonistic to the pathogen (221–223).

Amino Acids

The many proteases released into sputum by host and bacterial cells result in high levels (~20 mM) of free amino acids in the infection site. Work from our own group has demonstrated that aromatic amino acids are both a significant source of carbon and energy in sputum, but also cue enhanced virulence factor production in a QS-dependent manner (15, 71). While growing *P. aeruginosa* in genuine CF sputum medium, our group noticed enhanced production of the QS-signal PQS, which regulates a range of virulence factors (71). This combined with the observation by Farrow and colleagues that tryptophan enhances PQS production by serving as a source of anthranilate for PQS biosynthesis, led us to question whether aromatic amino acids could serve as metabolic cues for PQS production (71, 224). We tested this hypothesis by measuring the concentration of many nutrients in sputum samples, finding essentially no free tryptophan was present, though phenylalanine and tyrosine were present at hundred micromolar levels (15). Enhanced production of

PQS in SCFM was abolished when these aromatic amino acids were replaced with equimolar amounts of serine (15). Subsequent work in our group has determined that enhanced PQS production in the presence of aromatic amino acids is not due to coregulation of aromatic amino acid biosynthesis and PQS production (225). Thus, we favor a hypothesis in which the presence of aromatic amino acids results in altered central metabolite flux toward PQS biosynthesis (15, 225). When aromatic amino acids are present, their biosynthesis is feedback inhibited, resulting in greater flux of the central metabolite chorismate toward PQS. When aromatic amino acids are absent, more chorismate must be channeled toward synthesis of these critical amino acids, resulting in less PQS biosynthesis. Given that PQS signaling has been demonstrated to be critical for virulence in several animal models (116, 226, 227), we believe that enhanced PQS production in sputum is a powerful contributor to enhanced *P. aeruginosa* virulence in the CF lung.

In addition to aromatic amino acids, several other amino acids have been identified as cues for *P. aeruginosa* pathogenicity in sputum. Sriramulu and colleagues used artificial sputum medium (ASM) supplemented with amino acids (ASM+) resulted in "tight" microcolonies, while the standard ASM-medium resulted in "loose" microcolonies (164). A later study by Bernier and colleagues may explain this phenomenon as they used physiologically relevant concentrations of amino acids to demonstrate that arginine, ornithine, isoleucine, leucine, valine, phenylalanine and tyrosine promote *P. aeruginosa* cyclic di-GMP-mediated biofilm formation (228). Among these amino acids, arginine also inhibited *P. aeruginosa* swarming motility, suggesting that it is a potent nutritional cue for promoting the antibiotic-resistant, sessile lifestyle believed to be common in the CF lung (228).

After demonstrating that alanine was a preferred source of carbon and energy in sputum (15), our group has gone on to characterize alanine catabolism in *P. aeruginosa*, and we demonstrated that a mutant unable to catabolize DL-alanine was diminished in its capacity to cause infection in a rat lung infection model (229). A study by Rau and colleagues demonstrated that leucine catabolism genes were up-regulated in longitudinal clinical isolates of *P. aeruginosa* from individuals with CF (129). While leucine was not a preferred carbon source in SCFM (15), it was present at high levels (low millimolar), and the authors suggest optimization for growth on leucine may play a role in maintaining *P. aeruginosa* infections in the CF lung (129). Finally, Kohler and colleagues noted that growth with glutamate, aspartate, histidine, or proline as the sole source of nitrogen promoted *P. aeruginosa* swarming motility (230), which may also impact colonization and maintenance.

FUTURE AREAS OF RESEARCH

P. aeruginosa infections in the CF lung are the leading cause of morbidity and mortality for individuals with this genetic disease (1, 2), and we have seen that a great deal of research has been conducted to better understand the basic physiology of these infections. The importance of nutritional cues in sputum for *P. aeruginosa* is rapidly becoming more appreciated. As a consequence, the field is working to develop better *in vitro* and *in vivo* models that more accurately represent the nutritional environment of CF sputum. With a renewed emphasis on studying the nature of these infections *in situ* and newly available technologies for studying these infections, the future of these studies appears bright. This section will highlight some new areas of research, including experiments involving next-generation sequencing technologies, examining multispecies interactions in the CF lung, and the possibility of metabolism-based therapeutics for eradicating these infections.

Next-generation Sequencing Technology

Next-generation sequencing technologies are revolutionizing many groups' approaches to understanding *P. aeruginosa* infections. The ability to rapidly and inexpensively sequence bacterial genomes has resulted in a blossoming of available draft genome sequences (231–233). Another recently available technology, RNA-seq, involves sequencing of cDNA generated from total RNA preparations and mapping those sequence reads back to the genome (234–236). This type of transcriptome profiling has advantages over other methods like DNA microarrays in that it accounts for transcripts not present on a gene chip, can easily identify small RNAs, antisense transcripts, transcriptional start sites, and importantly, can be used on more than one species at a time (i.e. to study the host and pathogen or multispecies infections) (234–236). Some examples of RNA-seq transcriptome profiling experiments include investigation of planktonic verses biofilm growth (237), global identification of small RNAs (238), investigation of QS-regulated gene expression from different strains (239), and determining the effect of temperature on virulence (240).

Transposon insertion site sequencing (Tn–seq) is another example of a transformative technique that takes advantage of next-generation sequencing technology. Tn-seq can be used to identify fitness determinants by generating a saturated library of insertion mutants and counting the relative numbers of these mutants present after applying a selective pressure (e.g., mutant pool is grown in a mouse infection model) (241, 242). Insertions that are over-represented in the output pool represent genes that are not required for fitness, while insertions that are under-represented or absent in the output pool are important for fitness. Using this kind of approach, Skurnik and colleagues determined the contribution of every nonessential gene in the

P. aeruginosa genome toward fitness in a murine model infection (242). Additional Tn-seq work by Skurnik and colleagues identified *P. aeruginosa* mutants in *oprD* that in addition to being carbapenem-resistant, also displayed an enhanced survival in a murine model infection (241). As we will see in the next section, several groups are also taking advantage of next-generation sequencing technologies to survey the microbiome of the CF lung, and these studies are generating a greater appreciation for the role that multispecies interactions play in CF lung infections.

Polymicrobial Infections and Multispecies Interactions

Several studies have used next-generation sequencing technology to examine diversity of the CF airway microbiome (243–250). In general, these studies have reported considerable numbers of organisms, which is changing the perception that CF lung disease is dominated by handful of common opportunistic pathogens, like *P. aeruginosa, S. aureus, Haemophilus influenzae,* and *Burkholderia cepacia.* On the other hand, a recent report by Goddard and colleagues that compared 16S rRNA pyrosequencing data from sputum samples taken from transplanted lungs to those taken from throat found that much of the diversity observed in other microbiome surveys could be explained by oropharyngeal contamination (247). Regardless of how diverse the CF microbiome may be, it is clear that these infections are not monocultures, and several groups have started investigating multispecies interactions in the CF lung. A study by Hubert and colleagues found that individuals co-colonized with *P. aeruginosa* and *S. aureus* displayed greater declines in respiratory function than those colonized with *S. aureus* alone (251). Another example of these types of interactions comes from a study by Twomey and colleagues, which demonstrated that diffusible signaling factors from

Burkholderia cenocepacia and/or *Stenotrophomonas maltophila* were present in CF sputum samples, promoted *P. aeruginosa* persistence in a mouse CF model, and enhanced antibiotic resistance of *P. aeruginosa* biofilms grown on airway epithelial cells (252). Studies from our own group have shown that the presence of other pathogens can enhance *P. aeruginosa* infections (104, 170). As more of these cross-species interactions are delineated, we will have a better understanding of the entirety of the infection that causes CF lung disease, and these discoveries will likely lead to novel approaches for mitigating these infections.

Metabolism-based Therapeutics in the CF Lung

As we continue to learn more about the role of metabolism in *P. aeruginosa* infections in the CF lung, several targets for metabolism-based therapeutics have been identified. We have already seen that inhibition of phospholipase activity by miltefosine protects pulmonary surfactant from *P. aeruginosa* degradation in mouse model infections, and depriving *P. aeruginosa* of lipids as potentially significant carbon sources may contribute to the drug's positive effect in the CF lung (193). While it is obvious that the activity of degradative enzymes in sputum could be culpable for enhanced *P. aeruginosa* virulence, the use of recombinant human DNase (rhDNase) as an inhaled therapeutic capable of breaking up sputum cross-linked by DNA improves outcomes in individuals with CF (41, 44, 45). The successful use of this therapeutic suggests that other enzymes capable of degrading key nutrients in sputum may also help mitigate CF lung disease. rhDNase treatment combined with an alginate lyase that degrades the extracellular matrix of mucoid biofilms, enhanced aminoglycoside activity in CF sputum (253). Our own studies of the role phenylalanine and tyrosine play in cueing virulence factor production in the CF lung suggest that degradation of these nutrients could mitigate the capacity of *P. aeruginosa* to damage lung tissue (15). The phenylalanine- and tyrosine-degrading enzyme phenylalanine ammonia lyse from *Anabaena variabilis* is currently being prepared for use as a therapeutic to treat the buildup of phenylalanine associated with the genetic disease phenylketonuria (254–256). A therapeutic like this may also help to treat *P. aeruginosa* infections in the CF lung. The advantage of these metabolically-based therapeutics is that, unlike traditional antimicrobials which kill or inhibit the growth of a pathogen, these new therapeutics alter the environment around bacteria to diminish virulence. This removes the immediate incentive to evolve/acquire resistance. As the rise of antibiotic-resistant strains continues to threaten the global health community, it will require alternative approaches like metabolism-based therapeutics to develop the antibiotics of the future, and this makes a strong case for studying the metabolism and physiology of all infections.

SUMMARY AND CONCLUSIONS

This chapter has reviewed the current state of knowledge regarding how nutritional cues in CF sputum affect *P. aeruginosa* colonization and persistence. These devastating, life-long infections are incredibly difficult to treat, and the nutrients available in CF lung secretions have a profound influence on the infection. The nutritional contents of CF and healthy sputum were compared, and the struggles to generate both *in vitro* and *in vivo* systems for recapitulating chronic *P. aeruginosa* infections were described. Specific nutritional cues in sputum, such as mucin, lipids, DNA, and amino acids, promote *P. aeruginosa* virulence. The future of research in this field was discussed with particular emphasis on the ways in which next-generation sequencing technologies are driving our capacity to study *P. aeruginosa* infections systematically, and importantly,

examine the polymicrobial nature of these infections. The potential for new antimicrobial therapeutics based on knowledge gained from these metabolic studies is also of interest, given the rise of strains resistant to traditional antibiotics. While much is already known about *P. aeruginosa* infections of the CF lung, these infections continue to be incredibly difficult to defeat. However, continued study of *P. aeruginosa* metabolism at the site of infection has significantly enhanced understanding of the physiology of these infections and substantially increased the likelihood that this organism will cease to be a problem in the future.

CITATION

Palmer GC, Whiteley M. 2015. Metabolish and pathogenicity of pseudomonas aeruginosa infections in the lungs of individuals with cystic fibrosis. Microbiol Spectrum 3(2): MBP-0003-2014.

REFERENCES

1. **Hoiby N.** 1998. *Pseudomonas in cystic fibrosis: past, present, and future.* Cystic Fibrosis Trust, London, United Kingdom.
2. **Lyczak JB, Cannon CL, Pier GB.** 2002. Lung infections associated with cystic fibrosis. *Clin Microbiol Rev* **15:**194–222.
3. **Anderson MP, Gregory RJ, Thompson S, Souza DW, Paul S, Mulligan RC, Smith AE, Welsh MJ.** 1991. Demonstration that CFTR is a chloride channel by alteration of its anion selectivity. *Science* **253:**202–205.
4. **Riordan JR, Rommens JM, Kerem B, Alon N, Rozmahel R, Grzelczak Z, Zielenski J, Lok S, Plavsic N, Chou JL, et al.** 1989. Identification of the cystic fibrosis gene: cloning and characterization of complementary DNA. *Science* **245:**1066–1073.
5. **Reddy MM, Light MJ, Quinton PM.** 1999. Activation of the epithelial Na+ channel (ENaC) requires CFTR Cl- channel function. *Nature* **402:**301–304.
6. **Schwiebert EM, Egan ME, Hwang TH, Fulmer SB, Allen SS, Cutting GR, Guggino WB.** 1995. CFTR regulates outwardly rectifying chloride channels through an autocrine mechanism involving ATP. *Cell* **81:**1063–1073.

7. **Collawn JF, Fu L, Bebok Z.** 2010. Targets for cystic fibrosis therapy: proteomic analysis and correction of mutant cystic fibrosis transmembrane conductance regulator. *Expert Rev Proteomics* 7:495–506.

8. **Kreda SM, Mall M, Mengos A, Rochelle L, Yankaskas J, Riordan JR, Boucher RC.** 2005. Characterization of wild-type and deltaF508 cystic fibrosis transmembrane regulator in human respiratory epithelia. *Mol Biol Cell* 16:2154–2167.

9. **Tucker TA, Fortenberry JA, Zsembery A, Schwiebert LM, Schwiebert EM.** 2012. The DeltaF508-CFTR mutation inhibits wild-type CFTR processing and function when co-expressed in human airway epithelia and in mouse nasal mucosa. *BMC Physiol* 12:12.

10. **Derichs N.** 2013. Targeting a genetic defect: cystic fibrosis transmembrane conductance regulator modulators in cystic fibrosis. *Eur Respir Rev* 22:58–65.

11. **Pier GB, Grout M, Zaidi TS, Olsen JC, Johnson LG, Yankaskas JR, Goldberg JB.** 1996. Role of mutant CFTR in hypersusceptibility of cystic fibrosis patients to lung infections. *Science* 271:64–67.

12. **Hallows KR, Fitch AC, Richardson CA, Reynolds PR, Clancy JP, Dagher PC, Witters LA, Kolls JK, Pilewski JM.** 2006. Up-regulation of AMP-activated kinase by dysfunctional cystic fibrosis transmembrane conductance regulator in cystic fibrosis airway epithelial cells mitigates excessive inflammation. *J Biol Chem* 281:4231–4241.

13. **de Montalembert M, Fauchere JL, Bourdon R, Lenoir G, Rey J.** 1989. [Iron deficiency and *Pseudomonas aeruginosa* colonization in cystic fibrosis]. *Arch Fr Pediatr* 46:331–334.

14. **Moreau-Marquis S, Bomberger JM, Anderson GG, Swiatecka-Urban A, Ye S, O'Toole GA, Stanton BA.** 2008. The DeltaF508-CFTR mutation results in increased biofilm formation by *Pseudomonas aeruginosa* by increasing iron availability. *Am J Physiol Lung Cell Mol Physiol* 295:L25–L37.

15. **Palmer KL, Aye LM, Whiteley M.** 2007. Nutritional cues control *Pseudomonas aeruginosa* multicellular behavior in cystic fibrosis sputum. *J Bacteriol* 189:8079–8087.

16. **Reid DW, Withers NJ, Francis L, Wilson JW, Kotsimbos TC.** 2002. Iron deficiency in cystic fibrosis: relationship to lung disease severity and chronic *Pseudomonas aeruginosa* infection. *Chest* 121:48–54.

17. **Stites SW, Plautz MW, Bailey K, O'Brien-Ladner AR, Wesselius LJ.** 1999. Increased concentrations of iron and isoferritins in the lower respiratory tract of patients with stable cystic fibrosis. *Am J Respir Crit Care Med* 160:796–801.

18. **Stites SW, Walters B, O'Brien-Ladner AR, Bailey K, Wesselius LJ.** 1998. Increased iron and ferritin content of sputum from patients with cystic fibrosis or chronic bronchitis. *Chest* 114:814–819.

19. **Kirkham S, Sheehan JK, Knight D, Richardson PS, Thornton DJ.** 2002. Heterogeneity of airways mucus: variations in the amounts and glycoforms of the major oligomeric mucins MUC5AC and MUC5B. *Biochem J* 361:537–546.

20. **Rogers DF, Lethem MI.** 1997. *Airway mucus : basic mechanisms and clinical perspectives.* Birkhauser, Basel; Boston, MA.

21. **Voynow JA, Rubin BK.** 2009. Mucins, mucus, and sputum. *Chest* 135:505–512.

22. **Fahy JV, Steiger DJ, Liu J, Basbaum CB, Finkbeiner WE, Boushey HA.** 1993. Markers of mucus secretion and DNA levels in induced sputum from asthmatic and from healthy subjects. *Am Rev Respir Dis* 147:1132–1137.

23. **Henke MO, Renner A, Huber RM, Seeds MC, Rubin BK.** 2004. MUC5AC and MUC5B Mucins Are Decreased in Cystic Fibrosis Airway Secretions. *Am J Respir Cell Mol Biol* 31:86–91.

24. **Henke MO, John G, Germann M, Lindemann H, Rubin BK.** 2007. MUC5AC and MUC5B mucins increase in cystic fibrosis airway secretions during pulmonary exacerbation. *Am J Respir Crit Care Med* 175:816–821.

25. **Li S, Intini G, Bobek LA.** 2006. Modulation of MUC7 mucin expression by exogenous factors in airway cells in vitro and in vivo. *Am J Respir Cell Mol Biol* 35:95–102.

26. **Kohri K, Ueki IF, Shim JJ, Burgel PR, Oh YM, Tam DC, Dao-Pick T, Nadel JA.** 2002. *Pseudomonas aeruginosa* induces MUC5AC production via epidermal growth factor receptor. *Eur Respir J* 20:1263–1270.

27. **Song JS, Hyun SW, Lillihoj E, Kim BT.** 2001. Mucin secretion in the rat tracheal epithelial cells by epidermal growth factor and *Pseudomonas aeruginosa* extracts. *Korean J Intern Med* 16:167–172.

28. **Yan F, Li W, Jono H, Li Q, Zhang S, Li JD, Shen H.** 2008. Reactive oxygen species regulate *Pseudomonas aeruginosa* lipopolysaccharide-induced MUC5AC mucin expression via PKC-NADPH oxidase-ROS-TGF-alpha signaling pathways in human airway epithelial cells. *Biochem Biophys Res Commun* 366:513–519.

29. **Hao Y, Kuang Z, Walling BE, Bhatia S, Sivaguru M, Chen Y, Gaskins HR, Lau GW.** 2012. Pseudomonas aeruginosa pyocyanin causes airway goblet cell hyperplasia and

metaplasia and mucus hypersecretion by inactivating the transcriptional factor FoxA2. *Cell Microbiol* **14:**401–415.

30. **Carnoy C, Ramphal R, Scharfman A, Lo-Guidice JM, Houdret N, Klein A, Galabert C, Lamblin G, Roussel P.** 1993. Altered carbohydrate composition of salivary mucins from patients with cystic fibrosis and the adhesion of *Pseudomonas aeruginosa*. *Am J Respir Cell Mol Biol* **9:**323–334.

31. **Glasser JR, Mallampalli RK.** 2012. Surfactant and its role in the pathobiology of pulmonary infection. *Microbes Infect* **14:**17–25.

32. **Guillot L, Nathan N, Tabary O, Thouvenin G, Le Rouzic P, Corvol H, Amselem S, Clement A.** 2013. Alveolar epithelial cells: Master regulators of lung homeostasis. *Int J Biochem Cell Biol* **45:**2568–2573.

33. **Proud D.** 2008. *The pulmonary epithelium in health and disease*. John Wiley & Sons, Chichester, England ; Hoboken, NJ.

34. **Griese M, App EM, Duroux A, Burkert A, Schams A.** 1997. Recombinant human DNase (rhDNase) influences phospholipid composition, surface activity, rheology and consecutively clearance indices of cystic fibrosis sputum. *Pulm Pharmacol Ther* **10:**21–27.

35. **Hull J, South M, Phelan P, Grimwood K.** 1997. Surfactant composition in infants and young children with cystic fibrosis. *Am J Respir Crit Care Med* **156:**161–165.

36. **Gilljam H, Andersson O, Ellin A, Robertson B, Strandvik B.** 1988. Composition and surface properties of the bronchial lipids in adult patients with cystic fibrosis. *Clin Chim Acta* **176:**29–37.

37. **Girod S, Galabert C, Lecuire A, Zahm JM, Puchelle E.** 1992. Phospholipid composition and surface-active properties of tracheobronchial secretions from patients with cystic fibrosis and chronic obstructive pulmonary diseases. *Pediatr Pulmonol* **13:**22–27.

38. **Meyer KC, Sharma A, Brown R, Weatherly M, Moya FR, Lewandoski J, Zimmerman JJ.** 2000. Function and composition of pulmonary surfactant and surfactant-derived fatty acid profiles are altered in young adults with cystic fibrosis. *Chest* **118:**164–174.

39. **Griese M, Birrer P, Demirsoy A.** 1997. Pulmonary surfactant in cystic fibrosis. *Eur Respir J* **10:**1983–1988.

40. **Lethem MI, James SL, Marriott C, Burke JF.** 1990. The origin of DNA associated with mucus glycoproteins in cystic fibrosis sputum. *Eur Respir J* **3:**19–23.

41. **Brandt T, Breitenstein S, von der Hardt H, Tummler B.** 1995. DNA concentration and length in sputum of patients with cystic fibrosis during inhalation with recombinant human DNase. *Thorax* **50:**880–882.

42. **Kirchner KK, Wagener JS, Khan TZ, Copenhaver SC, Accurso FJ.** 1996. Increased DNA levels in bronchoalveolar lavage fluid obtained from infants with cystic fibrosis. *Am J Respir Crit Care Med* **154:**1426–1429.

43. **Ratjen F, Paul K, van Koningsbruggen S, Breitenstein S, Rietschel E, Nikolaizik W.** 2005. DNA concentrations in BAL fluid of cystic fibrosis patients with early lung disease: influence of treatment with dornase alpha. *Pediatr Pulmonol* **39:**1–4.

44. **Wagener JS, Rock MJ, McCubbin MM, Hamilton SD, Johnson CA, Ahrens RC.** 1998. Aerosol delivery and safety of recombinant human deoxyribonuclease in young children with cystic fibrosis: a bronchoscopic study. Pulmozyme Pediatric Broncoscopy Study Group. *J Pediatr* **133:**486–491.

45. **Riethmueller J, Vonthein R, Borth-Bruhns T, Grassme H, Eyrich M, Schilbach K, Stern M, Gulbins E.** 2008. DNA quantification and fragmentation in sputum after inhalation of recombinant human deoxyribonuclease. *Cell Physiol Biochem* **22:**347–352.

46. **Smith AL, Redding G, Doershuk C, Goldmann D, Gore E, Hilman B, Marks M, Moss R, Ramsey B, Rubio T, et al.** 1988. Sputum changes associated with therapy for endobronchial exacerbation in cystic fibrosis. *J Pediatr* **112:**547–554.

47. **Bensel T, Stotz M, Borneff-Lipp M, Wollschlager B, Wienke A, Taccetti G, Campana S, Meyer KC, Jensen PO, Lechner U, Ulrich M, Doring G, Worlitzsch D.** 2011. Lactate in cystic fibrosis sputum. *J Cyst Fibros* **10:**37–44.

48. **Wolak JE, Esther CR Jr, O'Connell TM.** 2009. Metabolomic analysis of bronchoalveolar lavage fluid from cystic fibrosis patients. *Biomarkers* **14:**55–60.

49. **Korgaonkar AK, Whiteley M.** 2011. *Pseudomonas aeruginosa* enhances production of an antimicrobial in response to N-acetylglucosamine and peptidoglycan. *J Bacteriol* **193:**909–917.

50. **Postle AD, Mander A, Reid KB, Wang JY, Wright SM, Moustaki M, Warner JO.** 1999. Deficient hydrophilic lung surfactant proteins A and D with normal surfactant phospholipid molecular species in cystic fibrosis. *Am J Respir Cell Mol Biol* **20:**90–98.

51. **Sloane AJ, Lindner RA, Prasad SS, Sebastian LT, Pedersen SK, Robinson M, Bye PT, Nielson DW, Harry JL.** 2005. Proteomic analysis of sputum from adults and children

with cystic fibrosis and from control subjects. *Am J Respir Crit Care Med* **172**:1416–1426.

52. **McMorran BJ, Patat SA, Carlin JB, Grimwood K, Jones A, Armstrong DS, Galati JC, Cooper PJ, Byrnes CA, Francis PW, Robertson CF, Hume DA, Borchers CH, Wainwright CE, Wainwright BJ.** 2007. Novel neutrophil-derived proteins in bronchoalveolar lavage fluid indicate an exaggerated inflammatory response in pediatric cystic fibrosis patients. *Clin Chem* **53**:1782–1791.

53. **Barth AL, Pitt TL.** 1996. The high amino-acid content of sputum from cystic fibrosis patients promotes growth of auxotrophic *Pseudomonas aeruginosa*. *J Med Microbiol* **45**:110–119.

54. **Dosanjh A, Lakhani S, Elashoff D, Chin C, Hsu V, Hilman B.** 2000. A comparison of microbiologic flora of the sinuses and airway among cystic fibrosis patients with maxillary antrostomies. *Pediatr Transplant* **4**:182185.

55. **Lavin J, Bhushan B, Schroeder JW Jr.** 2013. Correlation between respiratory cultures and sinus cultures in children with cystic fibrosis. *Int J Pediatr Otorhinolaryngol* **77**:686–689.

56. **Roby BB, McNamara J, Finkelstein M, Sidman J.** 2008. Sinus surgery in cystic fibrosis patients: comparison of sinus and lower airway cultures. *Int J Pediatr Otorhinolaryngol* **72**:1365–1369.

57. **Ciofu O, Hansen CR, Hoiby N.** 2013. Respiratory bacterial infections in cystic fibrosis. *Curr Opin Pulm Med* **19**:251–258.

58. **Fuqua C, Winans SC, Greenberg EP.** 1996. Census and consensus in bacterial ecosystems: the LuxR-LuxI family of quorum-sensing transcriptional regulators. *Annu Rev Microbiol* **50**:727–751.

59. **Fuqua WC, Winans SC, Greenberg EP.** 1994. Quorum sensing in bacteria: the LuxR-LuxI family of cell density-responsive transcriptional regulators. *J Bacteriol* **176**:269–275.

60. **Pearson JP, Gray KM, Passador L, Tucker KD, Eberhard A, Iglewski BH, Greenberg EP.** 1994. Structure of the autoinducer required for expression of *Pseudomonas aeruginosa* virulence genes. *Proc Natl Acad Sci USA* **91**:197–201.

61. **Passador L, Cook JM, Gambello MJ, Rust L, Iglewski BH.** 1993. Expression of *Pseudomonas aeruginosa* virulence genes requires cell-to-cell communication. *Science* **260**:1127–1130.

62. **Gambello MJ, Iglewski BH.** 1991. Cloning and characterization of the *Pseudomonas aeruginosa* lasR gene, a transcriptional activator of elastase expression. *J Bacteriol* **173**:3000–3009.

63. **Ochsner UA, Koch AK, Fiechter A, Reiser J.** 1994. Isolation and characterization of a regulatory gene affecting rhamnolipid biosurfactant synthesis in *Pseudomonas aeruginosa*. *J Bacteriol* **176**:2044–2054.

64. **Ochsner UA, Reiser J.** 1995. Autoinducer-mediated regulation of rhamnolipid biosurfactant synthesis in *Pseudomonas aeruginosa*. *Proc Natl Acad Sci USA* **92**:6424–6428.

65. **Pearson JP, Passador L, Iglewski BH, Greenberg EP.** 1995. A second N-acylhomoserine lactone signal produced by *Pseudomonas aeruginosa*. *Proc Natl Acad Sci USA* **92**:14901494.

66. **Pesci EC, Milbank JB, Pearson JP, McKnight S, Kende AS, Greenberg EP, Iglewski BH.** 1999. Quinolone signaling in the cell-to-cell communication system of *Pseudomonas aeruginosa*. *Proc Natl Acad Sci USA* **96**:11229–11234.

67. **Xiao G, Deziel E, He J, Lepine F, Lesic B, Castonguay MH, Milot S, Tampakaki AP, Stachel SE, Rahme LG.** 2006. MvfR, a key *Pseudomonas aeruginosa* pathogenicity LTTR-class regulatory protein, has dual ligands. *Mol Microbiol* **62**:1689–1699.

68. **Cao H, Krishnan G, Goumnerov B, Tsongalis J, Tompkins R, Rahme LG.** 2001. A quorum sensing-associated virulence gene of *Pseudomonas aeruginosa* encodes a LysR-like transcription regulator with a unique self-regulatory mechanism. *Proc Natl Acad Sci USA* **98**:14613–14618.

69. **Schuster M, Greenberg EP.** 2006. A network of networks: quorum-sensing gene regulation in *Pseudomonas aeruginosa*. *Int J Med Microbiol* **296**:73–81.

70. **Duan K, Surette MG.** 2007. Environmental regulation of *Pseudomonas aeruginosa* PAO1 Las and Rhl quorum-sensing systems. *J Bacteriol* **189**:4827–4836.

71. **Palmer KL, Mashburn LM, Singh PK, Whiteley M.** 2005. Cystic fibrosis sputum supports growth and cues key aspects of *Pseudomonas aeruginosa* physiology. *J Bacteriol* **187**:5267–5277.

72. **Liu PV.** 1966. The roles of various fractions of *Pseudomonas aeruginosa* in its pathogenesis. 3. Identity of the lethal toxins produced in vitro and in vivo. *J Infect Dis* **116**:481–489.

73. **Liu PV.** 1966. The roles of various fractions of *Pseudomonas aeruginosa* in its pathogenesis. II. Effects of lecithinase and protease. *J Infect Dis* **116**:112–116.

74. **Liu PV, Abe Y, Bates JL.** 1961. The roles of various fractions of *Pseudomonas aeruginosa* in its pathogenesis. *J Infect Dis* **108**:218–228.

75. **Jordan EO.** 1899. *Bacillus* Pyocyaneus and Its Pigments. *J Exp Med* **4**:627–647.

76. **Bleves S, Viarre V, Salacha R, Michel GP, Filloux A, Voulhoux R.** 2010. Protein secretion systems in *Pseudomonas aeruginosa*: A wealth

of pathogenic weapons. *Int J Med Microbiol* **300**:534–543.

77. **Duong F, Lazdunski A, Murgier M.** 1996. Protein secretion by heterologous bacterial ABC-transporters: the C-terminus secretion signal of the secreted protein confers high recognition specificity. *Mol Microbiol* **21**:459–470.

78. **Letoffe S, Redeker V, Wandersman C.** 1998. Isolation and characterization of an extracellular haem-binding protein from *Pseudomonas aeruginosa* that shares function and sequence similarities with the *Serratia marcescens* HasA haemophore. *Mol Microbiol* **28**:1223–1234.

79. **Iglewski BH, Kabat D.** 1975. NAD-dependent inhibition of protein synthesis by *Pseudomonas aeruginosa* toxin. *Proc Natl Acad Sci USA* **72**:2284–2288.

80. **Frank DW.** 1997. The exoenzyme S regulon of *Pseudomonas aeruginosa*. *Mol Microbiol* **26**:621–629.

81. **Yahr TL, Goranson J, Frank DW.** 1996. Exoenzyme S of *Pseudomonas aeruginosa* is secreted by a type III pathway. *Mol Microbiol* **22**:991–1003.

82. **Yahr TL, Mende-Mueller LM, Friese MB, Frank DW.** 1997. Identification of type III secreted products of the *Pseudomonas aeruginosa* exoenzyme S regulon. *J Bacteriol* **179**:7165–7168.

83. **Lee VT, Smith RS, Tummler B, Lory S.** 2005. Activities of *Pseudomonas aeruginosa* effectors secreted by the Type III secretion system in vitro and during infection. *Infect Immun* **73**:1695–1705.

84. **Russell AB, LeRoux M, Hathazi K, Agnello DM, Ishikawa T, Wiggins PA, Wai SN, Mougous JD.** 2013. Diverse type VI secretion phospholipases are functionally plastic antibacterial effectors. *Nature* **496**:508–512.

85. **Russell AB, Hood RD, Bui NK, LeRoux M, Vollmer W, Mougous JD.** 2011. Type VI secretion delivers bacteriolytic effectors to target cells. *Nature* **475**:343–347.

86. **Mougous JD, Cuff ME, Raunser S, Shen A, Zhou M, Gifford CA, Goodman AL, Joachimiak G, Ordonez CL, Lory S, Walz T, Joachimiak A, Mekalanos JJ.** 2006. A virulence locus of *Pseudomonas aeruginosa* encodes a protein secretion apparatus. *Science* **312**:1526–1530.

87. **Bomberger JM, Maceachran DP, Coutermarsh BA, Ye S, O'Toole GA, Stanton BA.** 2009. Long-distance delivery of bacterial virulence factors by *Pseudomonas aeruginosa* outer membrane vesicles. *PLoS Pathog* **5**:e1000382.

88. **Kadurugamuwa JL, Beveridge TJ.** 1995. Virulence factors are released from *Pseudomonas aeruginosa* in association with membrane vesicles during normal growth and exposure to gentamicin: a novel mechanism of enzyme secretion. *J Bacteriol* **177**:3998–4008.

89. **Beveridge TJ.** 1999. Structures of gram-negative cell walls and their derived membrane vesicles. *J Bacteriol* **181**:4725–4733.

90. **Mashburn LM, Whiteley M.** 2005. Membrane vesicles traffic signals and facilitate group activities in a prokaryote. *Nature* **437**:422–425.

91. **Suter S, Schaad UB, Tegner H, Ohlsson K, Desgrandchamps D, Waldvogel FA.** 1986. Levels of free granulocyte elastase in bronchial secretions from patients with cystic fibrosis: effect of antimicrobial treatment against *Pseudomonas aeruginosa*. *J Infect Dis* **153**:902–909.

92. **Voynow JA, Fischer BM, Zheng S.** 2008. Proteases and cystic fibrosis. *Int J Biochem Cell Biol* **40**:1238–1245.

93. **Smith RS, Iglewski BH.** 2003. *P. aeruginosa* quorum-sensing systems and virulence. *Curr Opin Microbiol* **6**:56–60.

94. **Nouwens AS, Beatson SA, Whitchurch CB, Walsh BJ, Schweizer HP, Mattick JS, Cordwell SJ.** 2003. Proteome analysis of extracellular proteins regulated by the las and rhl quorum sensing systems in *Pseudomonas aeruginosa* PAO1. *Microbiology* **149**:1311–1322.

95. **Scott NE, Hare NJ, White MY, Manos J, Cordwell SJ.** 2013. Secretome of Transmissible *Pseudomonas aeruginosa* AES-1R Grown in a Cystic Fibrosis Lung-Like Environment. *J Proteome Res* **12**(12):5357–5369.

96. **Butterworth MB, Zhang L, Heidrich EM, Myerburg MM, Thibodeau PH.** 2012. Activation of the epithelial sodium channel (ENaC) by the alkaline protease from *Pseudomonas aeruginosa*. *J Biol Chem* **287**:32556–32565.

97. **Terada LS, Johansen KA, Nowbar S, Vasil AI, Vasil ML.** 1999. *Pseudomonas aeruginosa* hemolytic phospholipase C suppresses neutrophil respiratory burst activity. *Infect Immun* **67**:2371–2376.

98. **Ostroff RM, Wretlind B, Vasil ML.** 1989. Mutations in the hemolytic-phospholipase C operon result in decreased virulence of *Pseudomonas aeruginosa* PAO1 grown under phosphate-limiting conditions. *Infect Immun* **57**:1369–1373.

99. **Ostroff RM, Vasil AI, Vasil ML.** 1990. Molecular comparison of a nonhemolytic and a hemolytic phospholipase C from *Pseudomonas aeruginosa*. *J Bacteriol* **172**:5915–5923.

100. **Barker AP, Vasil AI, Filloux A, Ball G, Wilderman PJ, Vasil ML.** 2004. A novel extracellular phospholipase C of *Pseudomonas aeruginosa* is required for phospholipid chemotaxis. *Mol Microbiol* **53**:1089–1098.

101. **MacEachran DP, Ye S, Bomberger JM, Hogan DA, Swiatecka-Urban A, Stanton BA, O'Toole GA.** 2007. The *Pseudomonas aeruginosa* secreted protein PA2934 decreases apical membrane expression of the cystic fibrosis transmembrane conductance regulator. *Infect Immun* **75**:3902–3912.

102. **Folders J, Tommassen J, van Loon LC, Bitter W.** 2000. Identification of a chitin-binding protein secreted by *Pseudomonas aeruginosa*. *J Bacteriol* **182**:1257–1263.

103. **Folders J, Algra J, Roelofs MS, van Loon LC, Tommassen J, Bitter W.** 2001. Characterization of *Pseudomonas aeruginosa* chitinase, a gradually secreted protein. *J Bacteriol* **183:** 7044–7052.

104. **Korgaonkar A, Trivedi U, Rumbaugh KP, Whiteley M.** 2013. Community surveillance enhances *Pseudomonas aeruginosa* virulence during polymicrobial infection. *Proc Natl Acad Sci USA* **110**:1059–1064.

105. **Ovchinnikova ES, Krom BP, Harapanahalli AK, Busscher HJ, van der Mei HC.** 2013. Surface thermodynamic and adhesion force evaluation of the role of chitin-binding protein in the physical interaction between *Pseudomonas aeruginosa* and *Candida albicans*. *Langmuir* **29**:4823–4829.

106. **Gallagher LA, McKnight SL, Kuznetsova MS, Pesci EC, Manoil C.** 2002. Functions required for extracellular quinolone signaling by *Pseudomonas aeruginosa*. *J Bacteriol* **184**:6472–6480.

107. **Ran H, Hassett DJ, Lau GW.** 2003. Human targets of *Pseudomonas aeruginosa* pyocyanin. *Proc Natl Acad Sci USA* **100**:14315–14320.

108. **Britigan BE, Roeder TL, Rasmussen GT, Shasby DM, McCormick ML, Cox CD.** 1992. Interaction of the *Pseudomonas aeruginosa* secretory products pyocyanin and pyochelin generates hydroxyl radical and causes synergistic damage to endothelial cells. Implications for *Pseudomonas*-associated tissue injury. *J Clin Invest* **90**:2187–2196.

109. **Denning GM, Railsback MA, Rasmussen GT, Cox CD, Britigan BE.** 1998. *Pseudomonas* pyocyanine alters calcium signaling in human airway epithelial cells. *Am J Physiol* **274**:L893–L900.

110. **Allen L, Dockrell DH, Pattery T, Lee DG, Cornelis P, Hellewell PG, Whyte MK.** 2005. Pyocyanin production by *Pseudomonas aeruginosa* induces neutrophil apoptosis and impairs neutrophil-mediated host defenses in vivo. *J Immunol* **174**:3643–3649.

111. **Hunter RC, Klepac-Ceraj V, Lorenzi MM, Grotzinger H, Martin TR, Newman DK.** 2012. Phenazine content in the cystic fibrosis respiratory tract negatively correlates with lung function and microbial complexity. *Am J Respir Cell Mol Biol* **47**:738–745.

112. **Dietrich LE, Price-Whelan A, Petersen A, Whiteley M, Newman DK.** 2006. The phenazine pyocyanin is a terminal signalling factor in the quorum sensing network of *Pseudomonas aeruginosa*. *Mol Microbiol* **61**:1308–1321.

113. **Huse H, Whiteley M.** 2011. 4-Quinolones: smart phones of the microbial world. *Chem Rev* **111**:152–159.

114. **Lepine F, Milot S, Deziel E, He J, Rahme LG.** 2004. Electrospray/mass spectrometric identification and analysis of 4-hydroxy-2-alkylquinolines (HAQs) produced by *Pseudomonas aeruginosa*. *J Am Soc Mass Spectrom* **15**:862–869.

115. **Haussler S, Becker T.** 2008. The pseudomonas quinolone signal (PQS) balances life and death in *Pseudomonas aeruginosa* populations. *PLoS Pathog* **4**:e1000166.

116. **Gallagher LA, Manoil C.** 2001. *Pseudomonas aeruginosa* PAO1 kills *Caenorhabditis elegans* by cyanide poisoning. *J Bacteriol* **183**:6207–6214.

117. **Goldfarb WB, Margraf H.** 1964. Cyanide Production by *Pseudomonas aeruginosa*. *Surg Forum* **15**:467–469.

118. **Anderson RD, Roddam LF, Bettiol S, Sanderson K, Reid DW.** 2010. Biosignificance of bacterial cyanogenesis in the CF lung. *J Cyst Fibros* **9**:158–164.

119. **Pessi G, Haas D.** 2000. Transcriptional control of the hydrogen cyanide biosynthetic genes hcnABC by the anaerobic regulator ANR and the quorum-sensing regulators LasR and RhlR in *Pseudomonas aeruginosa*. *J Bacteriol* **182**:6940–6949.

120. **Emerson J, Rosenfeld M, McNamara S, Ramsey B, Gibson RL.** 2002. *Pseudomonas aeruginosa* and other predictors of mortality and morbidity in young children with cystic fibrosis. *Pediatr Pulmonol* **34**:91–100.

121. **Hancock RE, Mutharia LM, Chan L, Darveau RP, Speert DP, Pier GB.** 1983. *Pseudomonas aeruginosa* isolates from patients with cystic fibrosis: a class of serum-sensitive, nontypable strains deficient in lipopolysaccharide O side chains. *Infect Immun* **42**:170–177.

122. **Huse HK, Kwon T, Zlosnik JE, Speert DP, Marcotte EM, Whiteley M.** 2010. Parallel evolution in *Pseudomonas aeruginosa* over 39,000 generations in vivo. *MBio* **1**(4):e00199–10.

123. **Luzar MA, Montie TC.** 1985. Avirulence and altered physiological properties of cystic fibrosis strains of *Pseudomonas aeruginosa*. *Infect Immun* **50:**572–576.

124. **Mahenthiralingam E, Campbell ME, Speert DP.** 1994. Nonmotility and phagocytic resistance of *Pseudomonas aeruginosa* isolates from chronically colonized patients with cystic fibrosis. *Infect Immun* **62:**596–605.

125. **Wahba AH, Darrell JH.** 1965. The Identification of Atypical Strains of *Pseudomonas aeruginosa*. *J Gen Microbiol* **38:**329–342.

126. **Wilder CN, Allada G, Schuster M.** 2009. Instantaneous within-patient diversity of *Pseudomonas aeruginosa* quorum-sensing populations from cystic fibrosis lung infections. *Infect Immun* **77:**5631–5639.

127. **Ernst RK, Yi EC, Guo L, Lim KB, Burns JL, Hackett M, Miller SI.** 1999. Specific lipopolysaccharide found in cystic fibrosis airway *Pseudomonas aeruginosa*. *Science* **286:**1561–1565.

128. **Smith EE, Buckley DG, Wu Z, Saenphimmachak C, Hoffman LR, D'Argenio DA, Miller SI, Ramsey BW, Speert DP, Moskowitz SM, Burns JL, Kaul R, Olson MV.** 2006. Genetic adaptation by *Pseudomonas aeruginosa* to the airways of cystic fibrosis patients. *Proc Natl Acad Sci USA* **103:**8487–8492.

129. **Rau MH, Hansen SK, Johansen HK, Thomsen LE, Workman CT, Nielsen KF, Jelsbak L, Hoiby N, Yang L, Molin S.** 2010. Early adaptive developments of *Pseudomonas aeruginosa* after the transition from life in the environment to persistent colonization in the airways of human cystic fibrosis hosts. *Environ Microbiol* **12:**1643–1658.

130. **Folkesson A, Jelsbak L, Yang L, Johansen HK, Ciofu O, Hoiby N, Molin S.** 2012. Adaptation of *Pseudomonas aeruginosa* to the cystic fibrosis airway: an evolutionary perspective. *Nat Rev Microbiol* **10:**841–851.

131. **Boman HG, Nilsson I, Rasmuson B.** 1972. Inducible antibacterial defence system in *Drosophila*. *Nature* **237:**232–235.

132. **Elrod RP, Braun AC.** 1942. *Pseudomonas aeruginosa*: Its Role as a Plant Pathogen. *J Bacteriol* **44:**633–645.

133. **Jander G, Rahme LG, Ausubel FM.** 2000. Positive correlation between virulence of *Pseudomonas aeruginosa* mutants in mice and insects. *J Bacteriol* **182:**3843–3845.

134. **Mahajan-Miklos S, Tan MW, Rahme LG, Ausubel FM.** 1999. Molecular mechanisms of bacterial virulence elucidated using a *Pseudomonas aeruginosa-Caenorhabditis elegans* pathogenesis model. *Cell* **96:**47–56.

135. **Phennicie RT, Sullivan MJ, Singer JT, Yoder JA, Kim CH.** 2010. Specific resistance to *Pseudomonas aeruginosa* infection in zebrafish is mediated by the cystic fibrosis transmembrane conductance regulator. *Infect Immun* **78:**4542–4550.

136. **Tan MW, Mahajan-Miklos S, Ausubel FM.** 1999. Killing of *Caenorhabditis elegans* by *Pseudomonas aeruginosa* used to model mammalian bacterial pathogenesis. *Proc Natl Acad Sci USA* **96:**715–720.

137. **Tan MW, Rahme LG, Sternberg JA, Tompkins RG, Ausubel FM.** 1999. *Pseudomonas aeruginosa* killing of *Caenorhabditis elegans* used to identify *P. aeruginosa* virulence factors. *Proc Natl Acad Sci USA* **96:** 2408–2413.

138. **Wilke M, Buijs-Offerman RM, Aarbiou J, Colledge WH, Sheppard DN, Touqui L, Bot A, Jorna H, de Jonge HR, Scholte BJ.** 2011. Mouse models of cystic fibrosis: phenotypic analysis and research applications. *J Cyst Fibros* **10**(Suppl 2):S152–S171.

139. **Clarke LL, Grubb BR, Gabriel SE, Smithies O, Koller BH, Boucher RC.** 1992. Defective epithelial chloride transport in a gene-targeted mouse model of cystic fibrosis. *Science* **257:**1125–1128.

140. **Colledge WH, Ratcliff R, Foster D, Williamson R, Evans MJ.** 1992. Cystic fibrosis mouse with intestinal obstruction. *Lancet* **340:**680.

141. **Dorin JR, Dickinson P, Alton EW, Smith SN, Geddes DM, Stevenson BJ, Kimber WL, Fleming S, Clarke AR, Hooper ML, et al.** 1992. Cystic fibrosis in the mouse by targeted insertional mutagenesis. *Nature* **359:**211–215.

142. **Snouwaert JN, Brigman KK, Latour AM, Malouf NN, Boucher RC, Smithies O, Koller BH.** 1992. An animal model for cystic fibrosis made by gene targeting. *Science* **257:**1083–1088.

143. **Colledge WH, Abella BS, Southern KW, Ratcliff R, Jiang C, Cheng SH, MacVinish LJ, Anderson JR, Cuthbert AW, Evans MJ.** 1995. Generation and characterization of a delta F508 cystic fibrosis mouse model. *Nat Genet* **10:**445–452.

144. **Kent G, Iles R, Bear CE, Huan LJ, Griesenbach U, McKerlie C, Frndova H, Ackerley C, Gosselin D, Radzioch D, O'Brodovich H, Tsui LC, Buchwald M, Tanswell AK.** 1997. Lung disease in mice with cystic fibrosis. *J Clin Invest* **100:**3060–3069.

145. **Yu H, Hanes M, Chrisp CE, Boucher JC, Deretic V.** 1998. Microbial pathogenesis in cystic fibrosis: pulmonary clearance of mucoid *Pseudomonas aeruginosa* and inflammation in

a mouse model of repeated respiratory challenge. *Infect Immun* **66**:280–288.

146. Gosselin D, Stevenson MM, Cowley EA, Griesenbach U, Eidelman DH, Boule M, Tam MF, Kent G, Skamene E, Tsui LC, Radzioch D. 1998. Impaired ability of Cftr knockout mice to control lung infection with *Pseudomonas aeruginosa*. *Am J Respir Crit Care Med* **157**:1253–1262.

147. Heeckeren A, Walenga R, Konstan MW, Bonfield T, Davis PB, Ferkol T. 1997. Excessive inflammatory response of cystic fibrosis mice to bronchopulmonary infection with *Pseudomonas aeruginosa*. *J Clin Invest* **100**:2810–2815.

148. Coleman FT, Mueschenborn S, Meluleni G, Ray C, Carey VJ, Vargas SO, Cannon CL, Ausubel FM, Pier GB. 2003. Hypersusceptibility of cystic fibrosis mice to chronic *Pseudomonas aeruginosa* oropharyngeal colonization and lung infection. *Proc Natl Acad Sci USA* **100**:1949–1954.

149. Hoffmann N, Rasmussen TB, Jensen PO, Stub C, Hentzer M, Molin S, Ciofu O, Givskov M, Johansen HK, Hoiby N. 2005. Novel mouse model of chronic *Pseudomonas aeruginosa* lung infection mimicking cystic fibrosis. *Infect Immun* **73**:2504–2514.

150. Hodges CA, Cotton CU, Palmert MR, Drumm ML. 2008. Generation of a conditional null allele for Cftr in mice. *Genesis* **46**:546–552.

151. Keiser NW, Engelhardt JF. 2011. New animal models of cystic fibrosis: what are they teaching us? *Curr Opin Pulm Med* **17**:478–483.

152. Bonfield TL, Hodges CA, Cotton CU, Drumm ML. 2012. Absence of the cystic fibrosis transmembrane regulator (Cftr) from myeloid-derived cells slows resolution of inflammation and infection. *J Leukoc Biol* **92**:1111–1122.

153. Stoltz DA, Meyerholz DK, Pezzulo AA, Ramachandran S, Rogan MP, Davis GJ, Hanfland RA, Wohlford-Lenane C, Dohrn CL, Bartlett JA, Nelson GAt, Chang EH, Taft PJ, Ludwig PS, Estin M, Hornick EE, Launspach JL, Samuel M, Rokhlina T, Karp PH, Ostedgaard LS, Uc A, Starner TD, Horswill AR, Brogden KA, Prather RS, Richter SS, Shilyansky J, McCray PB Jr, Zabner J, Welsh MJ. 2010. Cystic fibrosis pigs develop lung disease and exhibit defective bacterial eradication at birth. *Sci Transl Med* **2**:29ra31.

154. Rogers CS, Stoltz DA, Meyerholz DK, Ostedgaard LS, Rokhlina T, Taft PJ, Rogan MP, Pezzulo AA, Karp PH, Itani OA, Kabel AC, Wohlford-Lenane CL, Davis GJ, Hanfland RA, Smith TL, Samuel M, Wax D, Murphy CN, Rieke A, Whitworth K, Uc A, Starner TD, Brogden KA, Shilyansky J, McCray PB Jr, Zabner J, Prather RS, Welsh MJ. 2008. Disruption of the CFTR gene produces a model of cystic fibrosis in newborn pigs. *Science* **321**:1837–1841.

155. Pezzulo AA, Tang XX, Hoegger MJ, Alaiwa MH, Ramachandran S, Moninger TO, Karp PH, Wohlford-Lenane CL, Haagsman HP, van Eijk M, Banfi B, Horswill AR, Stoltz DA, McCray PB Jr, Welsh MJ, Zabner J. 2012. Reduced airway surface pH impairs bacterial killing in the porcine cystic fibrosis lung. *Nature* **487**:109–113.

156. Ostedgaard LS, Meyerholz DK, Chen JH, Pezzulo AA, Karp PH, Rokhlina T, Ernst SE, Hanfland RA, Reznikov LR, Ludwig PS, Rogan MP, Davis GJ, Dohrn CL, Wohlford-Lenane C, Taft PJ, Rector MV, Hornick E, Nassar BS, Samuel M, Zhang Y, Richter SS, Uc A, Shilyansky J, Prather RS, McCray PB Jr, Zabner J, Welsh MJ, Stoltz DA. 2011. The DeltaF508 mutation causes CFTR misprocessing and cystic fibrosis-like disease in pigs. *Sci Transl Med* **3**:74ra24.

157. Sun X, Sui H, Fisher JT, Yan Z, Liu X, Cho HJ, Joo NS, Zhang Y, Zhou W, Yi Y, Kinyon JM, Lei-Butters DC, Griffin MA, Naumann P, Luo M, Ascher J, Wang K, Frana T, Wine JJ, Meyerholz DK, Engelhardt JF. 2010. Disease phenotype of a ferret CFTR-knockout model of cystic fibrosis. *J Clin Invest* **120**:3149–3160.

158. Fisher JT, Liu X, Yan Z, Luo M, Zhang Y, Zhou W, Lee BJ, Song Y, Guo C, Wang Y, Lukacs GL, Engelhardt JF. 2012. Comparative processing and function of human and ferret cystic fibrosis transmembrane conductance regulator. *J Biol Chem* **287**:21673–21685.

159. Ohman DE, Chakrabarty AM. 1982. Utilization of human respiratory secretions by mucoid *Pseudomonas aeruginosa* of cystic fibrosis origin. *Infect Immun* **37**:662–669.

160. Palmer KL, Brown SA, Whiteley M. 2007. Membrane-bound nitrate reductase is required for anaerobic growth in cystic fibrosis sputum. *J Bacteriol* **189**:4449–4455.

161. Schobert M, Tielen P. 2010. Contribution of oxygen-limiting conditions to persistent infection of *Pseudomonas aeruginosa*. *Future Microbiol* **5**:603–621.

162. Su S, Hassett DJ. 2012. Anaerobic *Pseudomonas aeruginosa* and other obligately anaerobic bacterial biofilms growing in the thick airway mucus of chronically infected cystic fibrosis

patients: an emerging paradigm or "Old Hat"? *Expert Opin Ther Targets* **16:**859–873.

163. **Ghani M, Soothill JS.** 1997. Ceftazidime, gentamicin, and rifampicin, in combination, kill biofilms of mucoid *Pseudomonas aeruginosa*. *Can J Microbiol* **43:**999–1004.

164. **Sriramulu DD, Lunsdorf H, Lam JS, Romling U.** 2005. Microcolony formation: a novel biofilm model of *Pseudomonas aeruginosa* for the cystic fibrosis lung. *J Med Microbiol* **54:**667–676.

165. **Fung C, Naughton S, Turnbull L, Tingpej P, Rose B, Arthur J, Hu H, Harmer C, Harbour C, Hassett DJ, Whitchurch CB, Manos J.** 2010. Gene expression of *Pseudomonas aeruginosa* in a mucin-containing synthetic growth medium mimicking cystic fibrosis lung sputum. *J Med Microbiol* **59:**1089–1100.

166. **Haley CL, Colmer-Hamood JA, Hamood AN.** 2012. Characterization of biofilm-like structures formed by *Pseudomonas aeruginosa* in a synthetic mucus medium. *BMC Microbiol* **12:**181.

167. **Poole K, McKay GA.** 2003. Iron acquisition and its control in *Pseudomonas aeruginosa*: many roads lead to Rome. *Front Biosci* **8:**d661–d686.

168. **Vasil ML, Ochsner UA.** 1999. The response of *Pseudomonas aeruginosa* to iron: genetics, biochemistry and virulence. *Mol Microbiol* **34:**399–413.

169. **Martin LW, Reid DW, Sharples KJ, Lamont IL.** 2011. *Pseudomonas* siderophores in the sputum of patients with cystic fibrosis. *Biometals* **24:**1059–1067.

170. **Mashburn LM, Jett AM, Akins DR, Whiteley M.** 2005. *Staphylococcus aureus* serves as an iron source for *Pseudomonas aeruginosa* during in vivo coculture. *J Bacteriol* **187:**554–566.

171. **Chen X, Stewart PS.** 2002. Role of electrostatic interactions in cohesion of bacterial biofilms. *Appl Microbiol Biotechnol* **59:**718–720.

172. **Koley D, Ramsey MM, Bard AJ, Whiteley M.** 2011. Discovery of a biofilm electrocline using real-time 3D metabolite analysis. *Proc Natl Acad Sci USA* **108:**19996–20001.

173. **Hunter RC, Asfour F, Dingemans J, Osuna BL, Samad T, Malfroot A, Cornelis P, Newman DK.** 2013. Ferrous iron is a significant component of bioavailable iron in cystic fibrosis airways. *MBio* **4**(4):e00557–13.

174. **Hare NJ, Soe CZ, Rose B, Harbour C, Codd R, Manos J, Cordwell SJ.** 2012. Proteomics of Pseudomonas aeruginosa Australian epidemic strain 1 (AES-1) cultured under conditions mimicking the cystic fibrosis lung reveals

increased iron acquisition via the siderophore pyochelin. *J Proteome Res* **11:**776–795.

175. **Henke MO, John G, Rheineck C, Chillappagari S, Naehrlich L, Rubin BK.** 2011. Serine proteases degrade airway mucins in cystic fibrosis. *Infect Immun* **79:**3438–3444.

176. **Aristoteli LP, Willcox MD.** 2003. Mucin degradation mechanisms by distinct *Pseudomonas aeruginosa* isolates in vitro. *Infect Immun* **71:**5565–5575.

177. **Ramphal R, Pyle M.** 1983. Evidence for mucins and sialic acid as receptors for *Pseudomonas aeruginosa* in the lower respiratory tract. *Infect Immun* **41:**339–344.

178. **Scharfman A, Degroote S, Beau J, Lamblin G, Roussel P, Mazurier J.** 1999. *Pseudomonas aeruginosa* binds to neoglycoconjugates bearing mucin carbohydrate determinants and predominantly to sialyl-Lewis x conjugates. *Glycobiology* **9:**757–764.

179. **Vishwanath S, Ramphal R.** 1984. Adherence of *Pseudomonas aeruginosa* to human tracheobronchial mucin. *Infect Immun* **45:**197–202.

180. **Sajjan U, Reisman J, Doig P, Irvin RT, Forstner G, Forstner J.** 1992. Binding of nonmucoid *Pseudomonas aeruginosa* to normal human intestinal mucin and respiratory mucin from patients with cystic fibrosis. *J Clin Invest* **89:**657–665.

181. **Kubiet M, Ramphal R.** 1995. Adhesion of nontypeable *Haemophilus influenzae* from blood and sputum to human tracheobronchial mucins and lactoferrin. *Infect Immun* **63:**899–902.

182. **Reddy MS, Bernstein JM, Murphy TF, Faden HS.** 1996. Binding between outer membrane proteins of nontypeable *Haemophilus influenzae* and human nasopharyngeal mucin. *Infect Immun* **64:**1477–1479.

183. **Sajjan SU, Forstner JF.** 1992. Identification of the mucin-binding adhesin of *Pseudomonas cepacia* isolated from patients with cystic fibrosis. *Infect Immun* **60:**1434–1440.

184. **Sajjan US, Corey M, Karmali MA, Forstner JF.** 1992. Binding of *Pseudomonas cepacia* to normal human intestinal mucin and respiratory mucin from patients with cystic fibrosis. *J Clin Invest* **89:**648–656.

185. **Shuter J, Hatcher VB, Lowy FD.** 1996. *Staphylococcus aureus* binding to human nasal mucin. *Infect Immun* **64:**310–318.

186. **Ramphal R, Houdret N, Koo L, Lamblin G, Roussel P.** 1989. Differences in adhesion of *Pseudomonas aeruginosa* to mucin glycopeptides from sputa of patients with cystic fibrosis and chronic bronchitis. *Infect Immun* **57:**3066–3071.

187. **Landry RM, An D, Hupp JT, Singh PK, Parsek MR.** 2006. Mucin-*Pseudomonas aeruginosa* interactions promote biofilm formation and antibiotic resistance. *Mol Microbiol* **59:**142–151.

188. **Yeung AT, Parayno A, Hancock RE.** 2012. Mucin promotes rapid surface motility in *Pseudomonas aeruginosa. M Bio* **3**(3):e00073-12.

189. **Caldara M, Friedlander RS, Kavanaugh NL, Aizenberg J, Foster KR, Ribbeck K.** 2012. Mucin biopolymers prevent bacterial aggregation by retaining cells in the free-swimming state. *Curr Biol* **22:**2325–2330.

190. **Son MS, Matthews WJ Jr, Kang Y, Nguyen DT, Hoang TT.** 2007. In vivo evidence of *Pseudomonas aeruginosa* nutrient acquisition and pathogenesis in the lungs of cystic fibrosis patients. *Infect Immun* **75:**5313–5324.

191. **Storey DG, Ujack EE, Rabin HR.** 1992. Population transcript accumulation of *Pseudomonas aeruginosa* exotoxin A and elastase in sputa from patients with cystic fibrosis. *Infect Immun* **60:**4687–4694.

192. **Kang Y, Zarzycki-Siek J, Walton CB, Norris MH, Hoang TT.** 2010. Multiple FadD acyl-CoA synthetases contribute to differential fatty acid degradation and virulence in *Pseudomonas aeruginosa. PLoS One* **5:**e13557.

193. **Wargo MJ, Gross MJ, Rajamani S, Allard JL, Lundblad LK, Allen GB, Vasil ML, Leclair LW, Hogan DA.** 2011. Hemolytic phospholipase C inhibition protects lung function during *Pseudomonas aeruginosa* infection. *Am J Respir Crit Care Med* **184:**345–354.

194. **Krieg DP, Bass JA, Mattingly SJ.** 1988. Phosphorylcholine stimulates capsule formation of phosphate-limited mucoid *Pseudomonas aeruginosa. Infect Immun* **56:**864–873.

195. **Terry JM, Pina SE, Mattingly SJ.** 1991. Environmental conditions which influence mucoid conversion *Pseudomonas aeruginosa* PAO1. *Infect Immun* **59:**471–477.

196. **Domenech CE, Otero LH, Beassoni PR, Lisa AT.** 2011. Phosphorylcholine Phosphatase: A Peculiar Enzyme of *Pseudomonas aeruginosa. Enzyme Res* **2011:**561841.

197. **Lucchesi GI, Lisa TA, Domenech CE.** 1989. Choline and betaine as inducer agents of *Pseudomonas aeruginosa* phospholipase C activity in high phosphate medium. *FEMS Microbiol Lett* **48:**335–338.

198. **Shortridge VD, Lazdunski A, Vasil ML.** 1992. Osmoprotectants and phosphate regulate expression of phospholipase C in *Pseudomonas aeruginosa. Mol Microbiol* **6:**863–871.

199. **Wargo MJ.** 2013. Choline catabolism to glycine betaine contributes to *Pseudomonas aeruginosa* survival during murine lung infection. *PLoS One* **8:**e56850.

200. **Kearns DB, Robinson J, Shimkets LJ.** 2001. *Pseudomonas aeruginosa* exhibits directed twitching motility up phosphatidylethanolamine gradients. *J Bacteriol* **183:**763–767.

201. **Grassme H, Jendrossek V, Riehle A, von Kurthy G, Berger J, Schwarz H, Weller M, Kolesnick R, Gulbins E.** 2003. Host defense against *Pseudomonas aeruginosa* requires ceramide-rich membrane rafts. *Nat Med* **9:**322–330.

202. **Teichgraber V, Ulrich M, Endlich N, Riethmuller J, Wilker B, De Oliveira-Munding CC, van Heeckeren AM, Barr ML, von Kurthy G, Schmid KW, Weller M, Tummler B, Lang F, Grassme H, Doring G, Gulbins E.** 2008. Ceramide accumulation mediates inflammation, cell death and infection susceptibility in cystic fibrosis. *Nat Med* **14:**382–391.

203. **Becker KA, Riethmuller J, Zhang Y, Gulbins E.** 2010. The role of sphingolipids and ceramide in pulmonary inflammation in cystic fibrosis. *Open Respir Med J* **4:**39–47.

204. **Whitchurch CB, Tolker-Nielsen T, Ragas PC, Mattick JS.** 2002. Extracellular DNA required for bacterial biofilm formation. *Science* **295:**1487.

205. **Sutherland IW.** 2001. The biofilm matrix—an immobilized but dynamic microbial environment. *Trends Microbiol* **9:**222–227.

206. **Allesen-Holm M, Barken KB, Yang L, Klausen M, Webb JS, Kjelleberg S, Molin S, Givskov M, Tolker-Nielsen T.** 2006. A characterization of DNA release in *Pseudomonas aeruginosa* cultures and biofilms. *Mol Microbiol* **59:**1114–1128.

207. **Walker TS, Tomlin KL, Worthen GS, Poch KR, Lieber JG, Saavedra MT, Fessler MB, Malcolm KC, Vasil ML, Nick JA.** 2005. Enhanced *Pseudomonas aeruginosa* biofilm development mediated by human neutrophils. *Infect Immun* **73:**3693–3701.

208. **Chiang WC, Nilsson M, Jensen PO, Hoiby N, Nielsen TE, Givskov M, Tolker-Nielsen T.** 2013. Extracellular DNA shields against aminoglycosides in *Pseudomonas aeruginosa* biofilms. *Antimicrob Agents Chemother* **57:**2352–2361.

209. **Mulcahy H, Charron-Mazenod L, Lewenza S.** 2008. Extracellular DNA chelates cations and induces antibiotic resistance in *Pseudomonas aeruginosa* biofilms. *PLoS Pathog* **4:**e1000213.

210. **Dubois AV, Gauthier A, Brea D, Varaigne F, Diot P, Gauthier F, Attucci S.** 2012. Influence of DNA on the activities and inhibition of

neutrophil serine proteases in cystic fibrosis sputum. *Am J Respir Cell Mol Biol* **47**:80–86.

211. **Petrova OE, Schurr JR, Schurr MJ, Sauer K.** 2012. Microcolony formation by the opportunistic pathogen *Pseudomonas aeruginosa* requires pyruvate and pyruvate fermentation. *Mol Microbiol* **86**:819–835.

212. **Eschbach M, Schreiber K, Trunk K, Buer J, Jahn D, Schobert M.** 2004. Long-term anaerobic survival of the opportunistic pathogen *Pseudomonas aeruginosa* via pyruvate fermentation. *J Bacteriol* **186**:4596–4604.

213. **Klausen M, Heydorn A, Ragas P, Lambertsen L, Aaes-Jorgensen A, Molin S, Tolker-Nielsen T.** 2003. Biofilm formation by *Pseudomonas aeruginosa* wild type, flagella and type IV pili mutants. *Mol Microbiol* **48**:1511–1524.

214. **Shrout JD, Chopp DL, Just CL, Hentzer M, Givskov M, Parsek MR.** 2006. The impact of quorum sensing and swarming motility on *Pseudomonas aeruginosa* biofilm formation is nutritionally conditional. *Mol Microbiol* **62**:1264–1277.

215. **De Kievit TR, Gillis R, Marx S, Brown C, Iglewski BH.** 2001. Quorum-sensing genes in *Pseudomonas aeruginosa* biofilms: their role and expression patterns. *Appl Environ Microbiol* **67**:1865–1873.

216. **Hauber HP, Schulz M, Pforte A, Mack D, Zabel P, Schumacher U.** 2008. Inhalation with fucose and galactose for treatment of *Pseudomonas aeruginosa* in cystic fibrosis patients. *Int J Med Sci* **5**:371–376.

217. **Bucior I, Abbott J, Song Y, Matthay MA, Engel JN.** 2013. Sugar administration is an effective adjunctive therapy in the treatment of *Pseudomonas aeruginosa* pneumonia. *Am J Physiol Lung Cell Mol Physiol* **305**:L352–L363.

218. **Marr AK, Overhage J, Bains M, Hancock RE.** 2007. The Lon protease of *Pseudomonas aeruginosa* is induced by aminoglycosides and is involved in biofilm formation and motility. *Microbiology* **153**:474–482.

219. **Linares JF, Gustafsson I, Baquero F, Martinez JL.** 2006. Antibiotics as intermicrobial signaling agents instead of weapons. *Proc Natl Acad Sci USA* **103**:19484–19489.

220. **Hoffman LR, D'Argenio DA, MacCoss MJ, Zhang Z, Jones RA, Miller SI.** 2005. Aminoglycoside antibiotics induce bacterial biofilm formation. *Nature* **436**:1171–1175.

221. **Ardizzoni A, Neglia RG, Baschieri MC, Cermelli C, Caratozzolo M, Righi E, Palmieri B, Blasi E.** 2011. Influence of hyaluronic acid on bacterial and fungal species, including clinically relevant opportunistic pathogens. *J Mater Sci Mater Med* **22**:2329–2338.

222. **Carlson GA, Dragoo JL, Samimi B, Bruckner DA, Bernard GW, Hedrick M, Benhaim P.** 2004. Bacteriostatic properties of biomatrices against common orthopaedic pathogens. *Biochem Biophys Res Commun* **321**:472–478.

223. **Yadav MK, Chuck RS, Park CY.** 2013. Composition of artificial tear solution affects in vitro *Pseudomonas aeruginosa* biofilm formation on silicone hydrogel lens. *J Ocul Pharmacol Ther* **29**:591–594.

224. **Farrow JM 3rd, Pesci EC.** 2007. Two distinct pathways supply anthranilate as a precursor of the *Pseudomonas* quinolone signal. *J Bacteriol* **189**:3425–3433.

225. **Palmer GC, Palmer KL, Jorth PA, Whiteley M.** 2010. Characterization of the *Pseudomonas aeruginosa* transcriptional response to phenylalanine and tyrosine. *J Bacteriol* **192**:2722–2728.

226. **Deziel E, Gopalan S, Tampakaki AP, Lepine F, Padfield KE, Saucier M, Xiao G, Rahme LG.** 2005. The contribution of MvfR to *Pseudomonas aeruginosa* pathogenesis and quorum sensing circuitry regulation: multiple quorum-sensing-regulated genes are modulated without affecting lasRI, rhlRI or the production of N-acyl-L-homoserine lactones. *Mol Microbiol* **55**:998–1014.

227. **Zaborin A, Romanowski K, Gerdes S, Holbrook C, Lepine F, Long J, Poroyko V, Diggle SP, Wilke A, Righetti K, Morozova I, Babrowski T, Liu DC, Zaborina O, Alverdy JC.** 2009. Red death in *Caenorhabditis elegans* caused by *Pseudomonas aeruginosa* PAO1. *Proc Natl Acad Sci USA* **106**:6327–6332.

228. **Bernier SP, Ha DG, Khan W, Merritt JH, O'Toole GA.** 2011. Modulation of Pseudomonas aeruginosa surface-associated group behaviors by individual amino acids through c-di-GMP signaling. *Res Microbiol* **162**:680–688.

229. **Boulette ML, Baynham PJ, Jorth PA, Kukavica-Ibrulj I, Longoria A, Barrera K, Levesque RC, Whiteley M.** 2009. Characterization of alanine catabolism in *Pseudomonas aeruginosa* and its importance for proliferation in vivo. *J Bacteriol* **191**:6329–6334.

230. **Kohler T, Curty LK, Barja F, van Delden C, Pechere JC.** 2000. Swarming of *Pseudomonas aeruginosa* is dependent on cell-to-cell signaling and requires flagella and pili. *J Bacteriol* **182**:5990–5996.

231. **Silby MW, Winstanley C, Godfrey SA, Levy SB, Jackson RW.** 2011. *Pseudomonas* genomes: diverse and adaptable. *FEMS Microbiol Rev* **35**:652680.

232. **Snyder L, Loman N, Faraj L, Levi K, Weinstock G, Boswell T, Pallen M, Ala Aldeen D.** 2013.

Epidemiological investigation of *Pseudomonas aeruginosa* isolates from a six-year-long hospital outbreak using high-throughput whole genome sequencing. *Euro Surveill* 18(42):20611.

233. **Stewart L, Ford A, Sangal V, Jeukens J, Boyle B, Claims S, Crossman L, Hoskisson PA, Levesque R, Tucker NP.** 2013. Draft genomes of twelve host adapted and environmental isolates of *Pseudomonas aeruginosa* and their position in the core genome phylogeny. *Pathog Dis* 71(1):20–25.

234. **Sharma CM, Vogel J.** 2009. Experimental approaches for the discovery and characterization of regulatory small RNA. *Curr Opin Microbiol* 12:536–546.

235. **Westermann AJ, Gorski SA, Vogel J.** 2012. Dual RNA-seq of pathogen and host. *Nat Rev Microbiol* 10:618–630.

236. **Febrer M, McLay K, Caccamo M, Twomey KB, Ryan RP.** 2011. Advances in bacterial transcriptome and transposon insertion-site profiling using second-generation sequencing. *Trends Biotechnol* 29:586–594.

237. **Dotsch A, Eckweiler D, Schniederjans M, Zimmermann A, Jensen V, Scharfe M, Geffers R, Haussler S.** 2012. The *Pseudomonas aeruginosa* transcriptome in planktonic cultures and static biofilms using RNA sequencing. *PLoS One* 7:e31092.

238. **Gomez-Lozano M, Marvig RL, Molin S, Long KS.** 2012. Genome-wide identification of novel small RNAs in *Pseudomonas aeruginosa*. *Environ Microbiol* 14:2006–2016.

239. **Chugani S, Kim BS, Phattarasukol S, Brittnacher MJ, Choi SH, Harwood CS, Greenberg EP.** 2012. Strain-dependent diversity in the *Pseudomonas aeruginosa* quorum-sensing regulon. *Proc Natl Acad Sci USA* 109:E2823–E2831.

240. **Wurtzel O, Yoder-Himes DR, Han K, Dandekar AA, Edelheit S, Greenberg EP, Sorek R, Lory S.** 2012. The single-nucleotide resolution transcriptome of *Pseudomonas aeruginosa* grown in body temperature. *PLoS Pathog* 8:e1002945.

241. **Skurnik D, Roux D, Cattoir V, Danilchanka O, Lu X, Yoder-Himes DR, Han K, Guillard T, Jiang D, Gaultier C, Guerin F, Aschard H, Leclercq R, Mekalanos JJ, Lory S, Pier GB.** 2013. Enhanced in vivo fitness of carbapenem-resistant oprD mutants of *Pseudomonas aeruginosa* revealed through high-throughput sequencing. *Proc Natl Acad Sci USA* 110 (51):20747–20752.

242. **Skurnik D, Roux D, Aschard H, Cattoir V, Yoder-Himes D, Lory S, Pier GB.** 2013. A comprehensive analysis of in vitro and in vivo genetic fitness of *Pseudomonas aeruginosa* using high-throughput sequencing of transposon libraries. *PLoS Pathog* 9:e1003582.

243. **Armougom F, Bittar F, Stremler N, Rolain JM, Robert C, Dubus JC, Sarles J, Raoult D, La Scola B.** 2009. Microbial diversity in the sputum of a cystic fibrosis patient studied with 16S rDNA pyrosequencing. *Eur J Clin Microbiol Infect Dis* 28:1151–1154.

244. **Carmody LA, Zhao J, Schloss PD, Petrosino JF, Murray S, Young VB, Li JZ, LipPuma JJ.** 2013. Changes in cystic fibrosis airway microbiota at pulmonary exacerbation. *Ann Am Thorac Soc* 10:179–187.

245. **Delhaes L, Monchy S, Frealle E, Hubans C, Salleron J, Leroy S, Prevotat A, Wallet F, Wallaert B, Dei-Cas E, Sime-Ngando T, Chabe M, Viscogliosi E.** 2012. The airway microbiota in cystic fibrosis: a complex fungal and bacterial community—Implications for therapeutic management. *PLoS One* 7:e36313.

246. **Fodor AA, Klem ER, Gilpin DF, Elborn JS, Boucher RC, Tunney MM, Wolfgang MC.** 2012. The adult cystic fibrosis airway microbiota is stable over time and infection type, and highly resilient to antibiotic treatment of exacerbations. *PLoS One* 7:e45001.

247. **Goddard AF, Staudinger BJ, Dowd SE, Joshi-Datar A, Wolcott RD, Aitken ML, Fligner CL, Singh PK.** 2012. Direct sampling of cystic fibrosis lungs indicates that DNA-based analyses of upper-airway specimens can misrepresent lung microbiota. *Proc Natl Acad Sci USA* 109:13769–13774.

248. **Rogers GB, Hart CA, Mason JR, Hughes M, Walshaw MJ, Bruce KD.** 2003. Bacterial diversity in cases of lung infection in cystic fibrosis patients: 16S ribosomal DNA (rDNA) length heterogeneity PCR and 16S rDNA terminal restriction fragment length polymorphism profiling. *J Clin Microbiol* 41:3548–3558.

249. **Sibley CD, Grinwis ME, Field TR, Eshaghurshan CS, Faria MM, Dowd SE, Parkins MD, Rabin HR, Surette MG.** 2011. Culture enriched molecular profiling of the cystic fibrosis airway microbiome. *PLoS One* 6:e22702.

250. **Zhao J, Schloss PD, Kalikin LM, Carmody LA, Foster BK, Petrosino JF, Cavalcoli JD, VanDevanter DR, Murray S, Li JZ, Young VB, LiPuma JJ.** 2012. Decade-long bacterial community dynamics in cystic fibrosis airways. *Proc Natl Acad Sci USA* 109:5809–5814.

251. **Hubert D, Reglier-Poupet H, Sermet-Gaudelus I, Ferroni A, Le Bourgeois M, Burgel PR, Serreau R, Dusser D, Poyart C, Coste J.** 2013. Association between Staphylococcus aureus alone or combined with *Pseudomonas aeruginosa* and the clinical condition

of patients with cystic fibrosis. *J Cyst Fibros* **12:**497–503.

252. **Twomey KB, O'Connell OJ, McCarthy Y, Dow JM, O'Toole GA, Plant BJ, Ryan RP.** 2012. Bacterial cis-2-unsaturated fatty acids found in the cystic fibrosis airway modulate virulence and persistence of *Pseudomonas aeruginosa*. *Isme J* **6:**939–950.

253. **Alipour M, Suntres ZE, Omri A.** 2009. Importance of DNase and alginate lyase for enhancing free and liposome encapsulated aminoglycoside activity against *Pseudomonas aeruginosa*. *J Antimicrob Chemother* **64:**317–325.

254. **Sarkissian CN, Gamez A, Wang L, Charbonneau M, Fitzpatrick P, Lemontt JF, Zhao B, Vellard M, Bell SM, Henschell C, Lambert A, Tsuruda L, Stevens RC, Scriver CR.** 2008. Preclinical evaluation of multiple species of PEGylated recombinant phenylalanine ammonia lyase for the treatment of phenylketonuria. *Proc Natl Acad Sci USA* **105:**20894–20899.

255. **Sarkissian CN, Kang TS, Gamez A, Scriver CR, Stevens RC.** 2011. Evaluation of orally administered PEGylated phenylalanine ammonia lyase in mice for the treatment of Phenylketonuria. *Mol Genet Metab* **104:**249–254.

256. **Kang TS, Wang L, Sarkissian CN, Gamez A, Scriver CR, Stevens RC.** 2010. Converting an injectable protein therapeutic into an oral form: phenylalanine ammonia lyase for phenylketonuria. *Mol Genet Metab* **99:**4–9.

10

Metabolism and Fitness of Urinary Tract Pathogens

CHRISTOPHER J. ALTERI[1] and HARRY L. T. MOBLEY[1]

INTRODUCTION

Among common infections, urinary tract infections (UTI) are the most frequently diagnosed urologic disease. The majority of UTIs are caused by *Escherichia coli* and these uropathogenic *E. coli* (UPEC) infections place a significant financial burden on the healthcare system by generating annual costs in excess of two billion dollars (1, 2) in the United States alone.

OVERVIEW OF EXTRAINTESTINAL PATHOGENIC *E. COLI* (EXPEC) AND UROPATHOGENIC SUBSET (UPEC)

Escherichia coli, one of the most important model organisms in the laboratory, is the best studied microorganism. The primary niche occupied by *E. coli* is the lower intestinal tract of mammals, where it resides as a beneficial component of the commensal microbiota. Although it is well-known that *E. coli* resides in the human intestine as a harmless commensal, specific strains or pathotypes have the potential to cause a wide spectrum of intestinal and diarrheal diseases. For example, at least six pathotypes have been described: enterohemorrhagic (EHEC), enteropathogenic (EPEC),

[1]Department of Microbiology and Immunology, University of Michigan Medical School, Ann Arbor, MI 48109.
Metabolism and Bacterial Pathogenesis
Edited by Tyrrell Conway and Paul Cohen
© 2015 American Society for Microbiology, Washington, DC
doi:10.1128/microbiolspec.MBP-0016-2015

enterotoxigenic (ETEC), enteroaggregative (EAEC), diffuse-adherent (DAEC), and enteroinvasive *E. coli* (EIEC). On the other hand, extraintestinal diseases that include urinary tract infection (UTI), bacteremia, septicemia, and meningitis can be caused by additional pathotypes known as extraintestinal pathogenic *E. coli* (ExPEC) (3). The loss or gain of mobile genetic elements is responsible for the ability of *E. coli* to cause a broad range of human diseases (4). The core genome shared by all *E. coli* strains represents approximately 3,200 gene families, while the pan genome that represents the collective gene content for all sequenced *E. coli* strains exceeds 60,000 gene families (5). Thus, for each *E. coli* strain, which contains 4,800 genes on average, it is the specific composition of horizontally acquired genetic material that determines its ability to cause a certain disease and be defined as a specific pathotype (6).

Intestinal *E. coli* pathotypes, like the infamous EHEC O157:H7 serotype, reside in the bovine intestine as commensal bacteria and cause severe diarrheal disease only when accidentally introduced into the human intestinal tract. In contrast, extraintestinal *E. coli* pathotypes reside harmlessly in the human intestinal microenvironment but, upon access to sites outside of the intestine, become a major cause of human morbidity and mortality as a consequence of invasive UTI (pyelonephritis, bacteremia, or septicemia) (7, 8). Thus, extraintestinal pathotypes like uropathogenic *E. coli* (UPEC) possess an enhanced ability to cause infection outside of the intestinal tract and colonize the urinary tract, the bloodstream, or cerebrospinal fluid of human hosts (8, 9). It follows that ExPEC possess the unique ability to shift its behavior between harmless colonizer of the nutrient-rich human intestine and virulent pathogen of the nutritionally limited bladder (10–13) (Fig. 1). Here, we discuss the current understanding of the role for UPEC metabolism and physiology in adapting to these diverse host microenvironments.

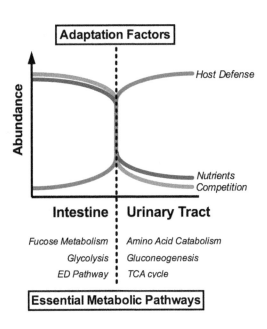

FIGURE 1 Adaptation of metabolism and basic physiology allows *E. coli* to replicate in diverse host microenvironments. ExPEC that cause urinary tract infection, bacteremia, sepsis, and meningitis, have adapted to grow as a harmless commensal in the nutrient-replete, carbon-rich human intestine but rapidly transition to pathogenic lifestyle in the nutritionally poor, nitrogen-rich urinary tract. In order to establish a commensal association within the human intestine, adaptive factors such as metabolic flexibility allow *E. coli* to successfully compete for carbon and energy sources with a large and diverse bacterial population. *E. coli* acquires nutrients from the intestinal mucus, including N-acetylglucosamine, sialic acid, glucosamine, gluconate, arabinose, fucose and simple sugars released upon breakdown of complex polysaccharides by anaerobic gut residents. When UPEC transition to the urinary tract, the bacteria encounter a drastic reduction in the abundance of nutrients and bacterial competition. Consequently, to replicate in a new host microenvironment, UPEC utilization of metabolic pathways required for growth in the dilute mixture of amino acids and peptides in the bladder signals the bacterium to elaborate virulence properties to successfully cause invasive disease and survive the onslaught of bactericidal host defenses. These adaptations are a unique and essential characteristic of ExPEC that enable a successful transition between disparate microenvironments within the same individual (13). doi:10.1128/microbiolspec.MBP-0016-2015.f1

Traditional Extraintestinal *E. coli* Virulence Factors

Studies in uropathogenesis are largely focused on pathogen-specific virulence properties including toxins, adhesins, secretion, motility, and iron-acquisition systems, and mechanisms to avoid the innate and adaptive immune response. Epidemiological studies have identified a number of specific virulence factors or genetic determinants associated with ExPEC isolates. Extraintestinal virulence genes like those encoding P-fimbriae or hemolysin are frequently clustered in genomic islands known as pathogenicity-associated islands (PAIs) (14) and encode a variety of fimbrial and non-fimbrial adhesins, toxins, and iron-acquisition systems (7, 15). Of *E. coli* pathotypes, ExPEC isolates generally have the largest number of PAIs and horizontally acquired genes; specifically, the prototype pyelonephritis UPEC strain CFT073 has 13 PAIs and is the largest genome (5,388 predicted genes) of sequenced *E. coli* strains (4). One outstanding question, however, is whether these horizontally acquired determinants are maintained in UPEC because they confer an advantage during intestinal colonization or are selected for increased extraintestinal fitness (16, 17).

An intriguing aspect of the pathogenesis of *E. coli* UTI is the lack of a single dominant virulence factor or common set of virulence determinants shared by all UPEC strains but absent from commensal *E. coli* or intestinal pathogens. Bacterial metabolism during infection has only recently been appreciated to contribute to persistence as much as their virulence properties. Due to the requirement for these *E. coli* to replicate in and colonize both the intestine and extraintestinal environments, we posit that physiology and metabolism of ExPEC strains is paramount. Indeed, we propose that the ability to survive in the urinary tract depends as much on bacterial physiology and metabolism as it does on the well-considered virulence determinants (18).

From the Intestine to the Urinary Tract

Studies of ExPEC or UPEC growth in the intestine are limited; however, we know *E. coli* strains reside and grow in the nutrient-rich mucus lining of the mammalian intestine (19, 20). Colonization studies with other *E. coli* have shown the Entner-Duodoroff pathway, and gluconate or other sugar acids, are required for intestinal growth of commensal *E. coli* (21). *E. coli* acquires nutrients from the intestinal mucus, a complex mixture of glycoconjugates, and up-regulates genes that encode enzymes involved in the catabolism of N-acetylglucosamine, sialic acid, glucosamine, gluconate, arabinose, and fucose (21, 22). EHEC O157:H7 requires the same central carbon pathways as do commensal strains, and mutations in pathways that utilize galactose, hexuronates, mannose, and ribose resulted in colonization defects (22). Further, multiple mutations in a single strain had an additive effect on colonization levels suggesting that some *E. coli* strains depend on the simultaneous metabolism of up to six sugars to support colonization of the intestine (22). These findings suggest that *E. coli* uses multiple limiting sugars for growth in the intestine and supports the assertion that *E. coli* grows in the intestine using simple sugars released upon breakdown of complex polysaccharides by anaerobic gut residents (23, 24).

Synthesis and degradation of glycogen, an endogenous glucose polymer, plays an important role for *E. coli* during colonization of the mouse intestine by functioning as an internal carbon source for the bacterium during nutrient limitation (25, 26). When faced with limiting sugars due to consumption by other colonizing bacteria, *E. coli* can also switch from glycolytic to gluconeogenic substrates to sustain growth in the intestine (27). For example, EHEC utilizes glycolytic substrates and switches to gluconeogenic substrates when present in the intestine with commensal *E. coli* that are utilizing glycolytic pathways for *in vivo* growth (27).

That competition *in vivo* can alter preferred routes of carbon flux through the central pathways, introducing the notion that metabolic flexibility and increased capacity to utilize diverse carbon sources are likely important for successful long-term intestinal colonization by extraintestinal *E. coli* (20).

Commensal *E. coli* that are resident in the intestine are three times more likely to belong to the phylogenetic group B2 (ExPEC primarily belong in B2) than transient intestinal colonizers (28, 29). If ExPEC strains are superior colonizers of the human intestine, then it is expected that horizontally acquired genomic islands and PAIs must be contributing to enhanced persistence in the intestine. Recent studies have shown that acquisition and regulation of genes that encode proteins to transport and catabolize prebiotic fructooligosaccharides provides a fitness advantage in the intestine for ExPEC strains (30, 31). The gene cluster required for ExPEC growth on fructooligosaccharides encodes two glycoside hydrolases that belong to family 32 in the carbohydrate-active enzyme database (CAZY) and a predicted cytoplasmic fructokinase (30). The latter is important because fructose must be phosphorylated to be catabolized by *E. coli*. Another horizontally acquired gene cluster proposed to provide a fitness advantage in the intestine has been discovered in ExPEC, the *frz* operon, which also encodes a sugar kinase in addition to two aldolases, and IIA, IIB, and IIC fructose family phosphotransferase system PTS-transporter subunits (32). These findings suggest that ExPEC genomic islands and PAIs may indeed provide added metabolic flexibility and promote persistence in the competitive intestinal microenvironment.

Models for Studying UTI

There are a number of models that have been successfully used to study uropathogenesis and identify virulence and fitness factors. The two primary models discussed here are the murine model of ascending infection (33) and bacterial cultivation in human urine *ex vivo*. For the latter, urine is collected and pooled from healthy donors that have not been exposed to antibiotics. The pooled urine can be filter-sterilized and used as a growth medium (34–36). Artificial urine has also been used as a growth medium (37). It is notable that growth in urine *ex vivo* is not a uropathogenesis trait *per se* as a number of non-pathogenic bacteria are capable of growth in urine (34, 38). Unlike growth in urine, the ascending model of UTI is capable of differentiating non-pathogenic *E. coli* from uropathogenic isolates. Only UPEC or ExPEC strains are able to successfully colonize the bladder and kidneys of mice in this model. It is notable that the ascending model can be used with a number of inbred mice; CBA/J, C3H, and C57BL/6 have been used to study experimental UTI.

Experimental Assessment of Gene Expression

Cultivation of uropathogenic *E. coli* in human urine has been useful to identify genes and proteins that are differentially expressed in urine as compared to during growth in lysogeny broth (LB) or defined medium. The largest group of uniformly upregulated genes and proteins are those involved in iron acquisition (34, 39, 40). Microarray experiments comparing transcription of genes between urine and LB identified numerous genes that are upregulated during growth in human urine. Functions that are highly upregulated in urine include sialic acid transport and catabolism, siderophore biosynthesis and uptake, arginine and branched-chain amino acid transport, histidine transport, serine metabolism, nitrate and formate respiration, mannonate catabolism, and galactoside transport (39). Similar experiments have also been performed with *E. coli* strain 83972 in asymptomatic bacteriuria and compared gene expression between human urine and 3-(*N*-morpholino)propanesulfonic acid (MOPS)-defined medium. In that work it was found that arginine, methionine, valine, uracil, adenine, and isoleucine are limiting nutrients in human urine and are essential for efficient growth of *E. coli* 83972 in urine (41).

Proteomic analysis of UPEC during growth in human urine has produced complementary findings. The outer membrane proteome is highly enriched for TonB-dependent receptors for siderophores and other iron-containing compounds (34). Outer membrane lipoproteins that are produced in human urine are D-methionine binding and uptake, a pectin methylesterase, and an uroporphyrin methyltransferase. This study also found increased production of BtuB, the cobalamin (B_{12}) receptor, and Tsx, which is a nucleoside-specific transport protein (34). Soluble proteins that are upregulated in human urine include many periplasmic substrate-binding proteins. These include DppA and OppA, involved in peptide transport, LivK that binds leucine, and HisJ involved in histine uptake (18). Similar to what was found by microarray, proteomic analysis also identified upregulated proteins involved with sialic acid catabolism and transport, mannonate metabolism, and serine and arginine biosynthesis (18). Overall, the findings from using human urine as a model for UPEC has revealed that this milieu is iron-limited, contains some amino acids but is limited for branched-chain amino acids, and appears to induce a profile of carbon metabolism consistent with scavenging mucosal sugars.

Description of the Urinary Tract Host Niche

In contrast to the nutritionally diverse intestine, urine in the bladder is a high-osmolarity, moderately oxygenated, iron-limited environment that contains mostly amino acids and small peptides (34, 37, 39, 42). It is therefore not surprising that defects in both branches of the pentose phosphate pathway, the Entner-Doudoroff pathway, and glycolysis have limited or no impact on *E. coli* fitness in the bladder and kidney microenviron-

ments (18, 43). Studies on UPEC metabolism during UTI have revealed that the ability to catabolize the amino acid D-serine in urine, which is both a nutrient and a signaling mechanism to trigger virulence, supports UPEC growth in the nitrogen-rich urinary tract (44). The utilization of short peptides and amino acids as a carbon source during bacterial infection of the bladder and kidneys is also supported by the observation that UPEC mutants defective in peptide import have reduced fitness during UTI while auxotrophic strains do not (18, 43). Metabolism of nucleobases is also required for *E. coli* colonization of the bladder. Signature-tagged mutagenesis screening identified a mutant in the dihydroorotate dehydrogenase gene *pyrD* (45) and, in a separate transposon screen, a gene involved in guanine biosynthesis, *guaA*, was also identified; a *guaA* mutant was found to be attenuated *in vivo* during UTI (36). Both are supported by the recent finding that *E. coli* are rapidly replicating in the bladder (46, 47).

The host urinary tract niche has also been defined by using transcriptome analysis of bacteria stabilized immediately in the urine from patients experiencing acute UTI. These studies have identified *E. coli* genes that are upregulated during human infection. It is notable that there are many upregulated genes that are also upregulated in human urine, for example, iron-acquisition genes are highly expressed in humans and in urine *ex vixo*; however, there are also many genes that are upregulated during UTI that are not upregulated in human urine, e.g., host-specific genes (46–48). Many of the host-specific UPEC genes are involved in metabolic processes. Entire gene clusters for ethanolamine and phosphonate metabolism were upregulated by UPEC isolated directly from human bladders (48). Genes encoding proteins for sulfate/thiosulfate uptake, taurine uptake, and alkane sulfonate uptake were identified as host-specific induced genes. This study also found that potassium import and nickel import are also highly upregulated processes during UTI. Consistent with increased demand for nickel was also the observed upregulation of genes that encode nickel-containing metallo-enzymes, such as formate-hydrogen lyase N and hydrogenase (47, 48). RNAseq analysis of bacteria directly from UTI also identified upregulation of genes encoding adenosine triphosphate (ATP) synthase, mannose-specific PTS components, carbamoyl-phosphate synthase, pyruvate dehydrogenase, nitrate reductase, ribonucleotide reductase, branched-chain amino acid uptake, and methionine biosynthesis (47).

Iron Acquisition

There are three classes of iron-uptake systems: siderophores, hemophores (or heme-binding and uptake systems), and direct ferrous iron (FeII)-uptake systems (such as ferric-citrate uptake) in UPEC. Siderophores are small molecules that are secreted into the environment, have very high affinity for ferric iron, and can strip the metal ion from other complexes within the host, or bind rare free ferric iron. Once a siderophore binds iron, it may be bound by specific outer-membrane receptors. Hemophores, on the other hand, are small secreted proteins that bind heme with high affinity before being imported back into the bacterium through specific importers (49). Finally, *E. coli* contains conserved ATP-binding cassette (ABC) transporters, encoded by the *feo* operon, capable of directly importing ferrous iron (50).

Both siderophore- and hemophore-mediated iron uptake depend on ferric iron. TonB is an inner-membrane protein also necessary for all ferric iron-uptake receptors in *E. coli*. TonB mutants are defective for colonizing kidneys in coinfection in a mouse model of UTI, a defect that was complementable (51). Additionally, in independent challenge the *tonB* mutant strain caused UTI at reduced colony-forming units (CFU) and reduced kidney colonization compared to the parental and complemented strains. The prototype

UPEC strain CFT073 contains at least ten ferric-uptake systems and several putative systems (52). Mutations that disrupt salmochelin, enterobactin, heme, and other siderophores are outcompeted by wild-type UPEC in coinfection (53–57)).

Iron acquisition appears to be especially important in UTI. The iron-chelating hydroxamate siderophore aerobactin is more common among UPEC strains than in fecal/commensal *E. coli* (58, 59), and the presence of at least one siderophore appears to be a common feature of UPEC strains (60). Gene-expression studies in urine from patients suffering from bacterial UTI using microarray suggests that bacterial iron-acquisition systems are up-regulated during infection (46). Isotope-dilution studies of siderophore activity, which can be more sensitive than transcript/expression analysis, confirm that siderophore activity is increased in UPEC compared to commensal strains of *E. coli* (61). Highlighting their importance, there is substantial notable redundancy in iron-acquisition systems in UPEC. Aerobactin-, enterobactin-, and heme-mediated iron uptake can each complement the activity of one another. Isogenic mutants of aerobactin or enterobactin were shown to be no different from wild-type in kidney colonization in a mouse model of UTI (51). Vaccine studies using siderophore receptors or heme-binding proteins as antigens and a mucosal route of delivery show protection from transurethral challenge in the mouse model of UTI (42), offering further support that iron systems are requisite virulence factors during the course of a UTI.

Amino Acid and Peptide Transport and Catabolism

UPEC growing in human urine induces expression of multiple isoforms of both dipeptide- and oligopeptide-binding proteins, both of which were found to be required for UPEC to effectively colonize the bladder and kidneys (18). Since the nutrient content in the kidney is expected to be very different from urine in the bladder, it is surprising that mutants lacking the ability to produce peptide-transport proteins were attenuated in both the kidneys and bladder because growth in urine mainly mimics only the bladder microenvironment. The host renal physiology might be expected to provide UPEC with several readily metabolized carbon sources during reabsorption of the kidney glomerular filtrate in the tubules; however, the ischemic damage caused by UPEC during pyelonephritis could alter nutrient availability (62). Lack of a fitness defect for UPEC amino acid auxotrophs during bladder and kidney colonization and impaired colonization of bladder and kidneys for peptide-transport-deficient mutants indicates that these bacteria actively import short peptides found in urine and suggests that peptides or amino acids represent the primary carbon source for *E. coli* during UTI (18). In fact, dissimilatory acetate metabolism coupled to the degradation of amino acids during *E. coli* colonization of the bladder and kidneys shows that ExPEC are adapted to acetogenic growth rather than acetate assimilation (63). Further, prolonged asymptomatic carriage of *E. coli* in the bladder selects for mutations that increase expression of D-serine deaminase and peptide/amino acid transport in *E. coli* (64). Gluconeogenic amino acids, like D- and L-serine, can be degraded to oxaloacetate or to pyruvate that can enter the tricarboxylic acid (TCA) cycle, which is necessary to provide substrates for gluconeogenesis when *E. coli* use amino acids as a carbon source. Consistent with peptides and amino acids being important carbon sources during UTI, only bacteria with defects in peptide transport, gluconeogenesis, or the TCA cycle demonstrate a fitness defect during colonization of the host urinary tract (Fig. 2) (18, 43, 65).

Central Carbon Pathways

Transcriptome and proteomic studies have been useful to identify iron acquisition and many nutrient-transport systems that are important for *E. coli* urinary-tract colonization; however, it is difficult to understand how central metabolism contributes to pathogen fitness by these approaches because most central pathways are regulated by allosteric mechanisms, e.g., post-translationally. One successful approach has been to construct and utilize central-pathway gene deletions to directly assess fitness in the murine model of ascending UTI (18, 43). Using this approach, it has been possible to assess fitness for a number of mutants in each central pathway in *E. coli* (Fig. 3). The mutants that have been tested in this way are in glycolysis (*pgi*, *pfkA*, *tpiA*, *pykA*), pentose-phosphate pathway (*gnd*, *talA*, *talB*), Entner-Duodoroff (*edd*), TCA cycle (*sdhB*, *fumC*, *frdA*), and gluconeogenesis (*pckA*).

Strains lacking *tpiA* (triose phosphate isomerase) and *pgi* (phosphoglucose isomerase), as well as mutants in irreversible glycolytic steps involving both 6-carbon (*pfkA*; 6-phosphofructokinase transferase) and 3-carbon (*pykA*; pyruvate kinase) demonstrated that neither the preparative or substrate-level phosphorylation stages of glycolysis are required during experimental infection (18, 43). Similarly, it was found that a mutant defective in the oxidative branch of the pentose-phosphate pathway; phosphogluconolactonate (*gnd*), and a mutant defective in gluconate catabolism; 6-phosphoglyconate dehydratase (*edd*), did not demonstrate a fitness defect during experimental UTI (18, 43). *E. coli* encode genes for two transaldolase enzymes, *talA* and *talB*, which function in the non-oxidative pentose-phosphate pathway. TalB is the major transaldolase in *E. coli* that transfers a three-carbon moiety from a C_7 molecule to glyceraldehyde-3-P (C_3) to form erythrose-4-P (C_4) and fructose-P (C_6). This stage of the pentose-phosphate pathway is reversible, and thus, can be uncoupled from the oxidative decarboxylation reactions that produce nicotinamide adenine dinucleotide phosphate-oxidase (NADPH). Interestingly, loss of TalA created a slight fitness advantage and loss of the major transaldolase, TalB, did not affect

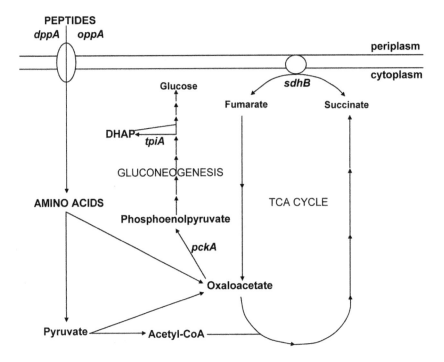

FIGURE 2 UPEC acquires amino acids and requires gluconeogenesis and the TCA cycle for fitness *in vivo*. Peptide substrate-binding protein genes *dppA* and *oppA* are required to import di- and oligopeptides into the cytoplasm from the periplasm. Short peptides are degraded into amino acids in the cytoplasm and converted into pyruvate and oxaloacetate. Pyruvate is converted into acetyl-CoA and enters the TCA cycle to replenish intermediates and generate oxaloacetate. Oxaloacetate is converted to phospho-enolpyruvate by the *pckA* gene product during gluconeogenesis. Mutations in the indicated genes *dppA*, *oppA*, *pckA*, *sdhB*, and *tpiA* demonstrated fitness defects *in vivo*. (18) doi:10.1128/microbiolspec.MBP-0016-2015.f2

E. coli fitness during UTI; however, loss of both TalA and TalB in a double-mutant strain created a modest fitness defect (18, 43), suggesting that isomerization of sugars in the pentose phosphate are important during UTI, presumably for nucleoside biosynthesis.

During bacterial growth on gluconeogenic substrates, peptides and certain amino acids that are present in the urinary tract are broken-down into pyruvate, which can be oxidized in the TCA cycle or reduced to fermentative end-products. The resulting oxaloacetate can fuel gluconeogenesis as the substrate for pyruvate carboxykinase (*pckA*) that generates phophoenolpyruvate and bypasses the irreversible glycolytic reaction catalyzed by pyruvate kinase (*pykA*). Mutation of *pckA*, which disrupts gluconeogenesis,

results in a significant fitness defect for *E. coli* in both bladder and kidneys (18). The aerobic TCA cycle has been proposed to be required for *E. coli* fitness during growth on gluconeogenic substrates present in the urinary tract (18). Specifically, *E. coli* *sdhB*-mutant bacteria have been shown to have fitness defects during UTI (18, 65), suggesting that the reductive TCA cycle may not be operating during host colonization. Mutants lacking fumarate dehydratase (fumarase); *fumC*, and fumarate reductase; *frdA*, have also been tested. It is generally believed that the production of reduced flavin adenine dinucleotide ($FADH_2$) during the conversion of succinate to fumarate by succinate dehydrogenase is avoided during fermentation by modification of the TCA cycle to an incom-

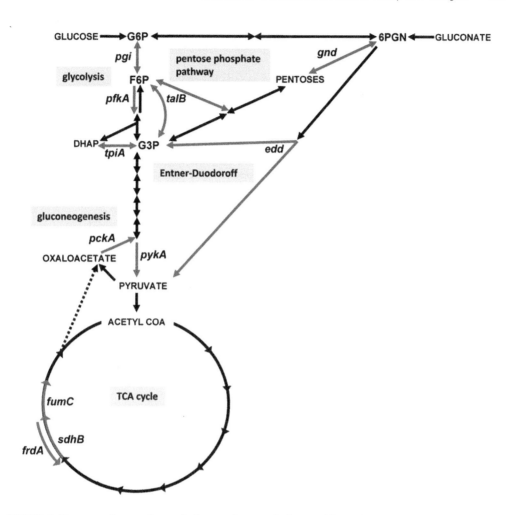

FIGURE 3 **Diagram of central metabolism and map of the specific pathways disrupted by targeted mutations in uropathogenic *E. coli*. Carbon sources or biochemical intermediates shared between pathways are indicated in capital letters or abbreviated: G6P, glucose-6-phosphate; F6P, fructose-6-phosphate; G3P, glyceraldehyde-3-phosphate; 6PGN, 6-phosphogluconate. Reactions are denoted with arrows. Specific reactions (red arrows) were targeted by deletion or insertion in *E. coli* CFT073. In glycolysis: *pgi*, glucose-6-phosphate isomerase; *pfkA*, 6-phosphofructokinase transferase; *tpiA*, triosephosphate isomerase; *pykA*, pyruvate kinase; in pentose phosphate pathway: gnd, 6-phosphogluconate dehydrogenase; *talB*, transaldolase; in Entner-Duodoroff pathway: *edd*, 6-phosphogluconate dehydratase; in gluconeogenesis: *pckA*, phosphoenolpyruvate carboxykinase; and in the TCA cycle: *sdhB*, succinate dehydrogenase; *fumC*, fumarate hydratase; *frdA*, fumarate reductase. (43) doi:10.1128/microbiolspec.MBP-0016-2015.f3**

plete reductive pathway where fumarate conversion to succinate by fumarate reductase replaces succinate-dehydrogenase activity. The loss of FrdA had no effect on *E. coli* colonization of the urinary tract, while loss of FumC created a fitness defect during bladder colonization (43). Finding that FumC contributes to UPEC fitness during UTI suggests that the TCA cycle may be operating as a tool to cope with oxidative stress. It is known that FumC is a fumarase that lacks an iron-sulfur cluster and is part of the superoxide response (SoxRS) regulon (66). Thus, it is possible that the flexibility of the TCA cycle and related energy pathways, as discussed above, may reflect a countermeasure against host defense rather than a *bona fide* indication of aerobic or anaerobic metabolism (13).

Flexible-Energy Pathways

The TCA cycle, in *E. coli* and in nearly all living cells, is an amphibolic pathway because TCA intermediates are important for anabolic processes in addition to catabolism of acetogenic and gluconeogenic carbon sources like amino acids. Both the complete aerobic TCA cycle and the incomplete reductive pathway are intimately linked to energy metabolism and respiration; pyruvate from glycolysis is oxidized to carbon dioxide (CO_2) concomitant with production of nicotinamide adenine dinucleotide (NAD^+) + hydrogen (NADH) and $FADH_2$, both of which must be re-oxidized because all pathways would cease in that absence of NAD^+. During respiration, the re-oxidation of NADH is coupled to the generation of a proton gradient (ΔpH) that is used, among many processes, to drive ATP production via proton influx through the F_1F_0 ATPase stator. Much like the flexibility of *E. coli* carbon metabolism in the intestine, the modular respiratory chains of *E. coli* can be assembled in a variety of configurations depending on the available terminal-electron acceptor and the energy needs of the bacterium (67). In the intestine, the ability to respire aerobically and anaerobically provides a substantial fitness advantage for *E. coli* (68). In the extraintestinal environment, loss of the TCA-cycle enzyme succinate dehydrogenase, SdhB, results in a UPEC strain that has reduced fitness *in vivo* (18, 43), suggesting that a complete TCA cycle and aerobic respiration are important in the urinary tract (13). At first approximation, aerobic respiration and acetogenic growth (fermentation) during UTI appear contradictory; however, during rapid growth in the host urinary tract it is likely that acetate metabolism reflects a certain degree of reductive overflow, or an electron sink, to relieve

carbon flux through the TCA cycle when the respiratory capacity has been reached (46).

The flexibility of energy metabolism and the modular respiratory chains for *E. coli* are also critical for adapting to respiratory stress (69). Utilization of alternative respiratory chains not only confers flexibility dependent upon the available energy source or terminal-electron acceptor (67), but also allows for modulation of the proton-motive force (μH^+), the gradient of charge and protons across the cytoplasmic membrane (70). By reducing membrane potential ($\Delta \psi$), *E. coli* can limit or prevent uptake of antibiotics and exist as a bacterial "persister" that can only be eradicated by antibiotics when specific metabolites are present to stimulate respiratory generation of μH^+ (71). This strategy is also utilized by *E. coli* to prevent killing by the bactericidal activities of mammalian phagocytes. In the urinary tract, extraintestinal *E. coli* infection elicits a massive inflammatory response characterized by neutrophil influx (72, 73) and host production of the antimicrobial peptide LL-37 cathelicidin (74). In response to antimicrobial peptides and acidic pH generated by neutrophils, the PhoPQ-regulatory system becomes activated and up-regulates genes that encode electroneutral respiratory-chain components, cation-import systems, and anion-efflux channels to limit respiratory generation of μH^+ and reverse membrane polarity from inside-negative to inside-positive, respectively (75). This control of membrane potential promotes *E. coli* resistance to cationic antimicrobial peptides and acidic pH, which potentiates bacterial survival against bactericidal activities of neutrophils during UTI. Indeed, UPEC lacking PhoP are exquisitely sensitive to polymyxin and acidic pH, and are severely attenuated during acute infection of the bladder (75).

Sensing and Responding to Environmental Cues

Nutritional sensing is important for competing with other bacteria for limiting nutrients in the intestine (23) and could represent a mechanism signaling the arrival into a different host environment such as the urinary tract. Signal transduction by two-component regulatory systems, which contain a trans-membrane-sensor kinase and a cognate-response-regulator transcription factor, are the best-described mechanism used by bacteria to coordinate gene expression in response to specific external stimuli. In UPEC, disruption of the QseBC two-component regulatory system attenuates virulence in *E. coli* within the urinary tract by altering carbon flux through central pathways, which negatively affects the expression of multiple adhesins that are involved in bladder colonization (65). The horizontally acquired *frz* operon has also been shown to link the metabolic capacity of ExPEC with expression of genes required for adherence to the bladder epithelium; the presence of the *frz* operon favors the ON orientation of the invertible type 1-fimbriae promoter (32). UPEC catabolism of D-serine in the urinary tract is also an important signaling mechanism to trigger virulence-gene expression (44, 76, 77). These studies suggest that movement of ExPEC from a commensal-like lifestyle in the nutrient- and carbon-rich intestine to the nutritionally poor, nitrogen-rich lower urinary tract would be quickly sensed by changes in metabolism, which would trigger expression of genes required for pathogen colonization of an extraintestinal microenvironment.

The nutrient-sensing regulatory system KguSR is a two-component system that has been shown to be important for UPEC fitness during UTI (78). This system appears to be involved in the control of utilization of α-ketoglutarate by regulating target genes that encode an α-ketoglutarate dehydrogenase and a succinyl-CoA synthetase (78). The KguSR system was also suggested to be partially controlled by anaerobiosis (78). Interestingly, the canonical anaerobiosis regulator, Fnr, has been shown to play a key role in UPEC fitness during UTI (79). Fnr impacts UPEC fitness by modulating the expression of type 1 and P fimbriae, as well as affecting motility. It is also possible that Fnr is responsible for the apparent effect of anaerobiosis on the KguSR system (79). Another nutritional response that has been observed in UPEC is the regulation of type 1 fimbriae by sialic acid. It has been shown that NanR and NagC inhibit FimB switching from phase off to phase on, resulting in a decrease in the production of type 1 fimbriae (80). Taken together, these findings all provide examples of nutritional mechanisms whereby UPEC sense the urinary tract environment to modulate expression of virulence factors.

SUMMARY

Future studies aimed at how the presence or acquisition of genomic islands and PAIs affect the metabolome and metabolic capacity of *E. coli* that infect the urinary tract will be important to better define rational drug targets. Investigating changes in carbon flux through central metabolism caused by transcriptional regulators, nutrient-uptake systems, and carbohydrate-utilization systems encoded on PAIs will help to identify traits that confer a fitness advantage for UPEC when faced with bacterial competition for nutrients in the human intestine. For *E. coli* transitioning from the carbon-rich intestine to the urinary tract, it is likely that the ratio of carbon:nitrogen (C:N) or fluctuations in the intracellular C:N levels plays an important role controlling pathogenicity (Fig. 4) (13, 43). Nitrogen metabolism and ammonia generated by catabolism of amino acids can act as a form of long-range interbacterial communication to induce oxidative-stress responses and increase resistance to antibiotics (81).

Basic principles of physiology, shared by nearly all living cells, are beginning to be appreciated as playing key roles in processes that are essential for pathogenesis. Bacterial respiration coordinates type-three secretion-system (TTSS) assembly with movement of *Shigella* from the intestinal lumen to the

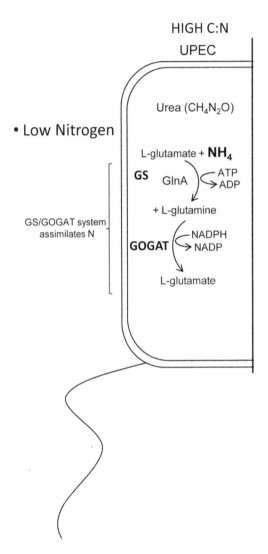

HIGH C:N

UPEC

• Low Nitrogen

Urea (CH$_4$N$_2$O)

L-glutamate + **NH$_4$**

GS   GlnA   ⟨ ATP
                   ⟩ ADP

+ L-glutamine

GS/GOGAT system
assimilates N

GOGAT ⟨ NADPH
          ⟩ NADP

L-glutamate

FIGURE 4 Model describing the C/N ratio within the urinary tract for *E. coli*. The urinary tract environment has a low C/N ratio due to the dilute mixture of amino acids and peptides as the primary carbon source and the abundance of urea in urine providing a substantial nitrogen contribution. *E. coli* is unable to utilize or sense the nitrogen sequestered in urea because it lacks urease, which liberates ammonia from urea. This results in *E. coli* activation of the glutamine synthetase and glutamate oxo-glutarate aminotransferase system (GS/GOGAT) to assimilate nitrogen. (43) doi:10.1128/microbiolspec.MBP-0016-2015.f4

mucosa (82). *E. coli* modulate their respiratory activity and limit or reverse membrane potential as a protective counter-measure against mammalian host defenses during UTI (75). These studies demonstrate the important link between fundamental bioenergetics and pathogenesis and suggest that energy metabolism might be an important signal used by bacterial pathogens to identify specific host microenvironments. By studying carbon metabolism, key differences between *E. coli* growing in the intestine and colonizing the urinary tract have been identified. Commensal *E. coli* require the Entner-Doudoroff pathway and glycolysis for intestinal colonization, while the pentose-phosphate pathway and gluconeogenesis are dispensable (21). In contrast, for *E. coli* infecting the urinary tract, the pathways required for commensal colonization are dispensable while the TCA cycle and gluconeogenesis are required (18). Other aspects of UPEC physiology, such as determining if the numerous well-considered or presently uncharacterized fimbriae that predominate in ExPEC strains (83) function as adhesive organelles to promote bacterial adherence in the intestine or in the urinary tract, will be beneficial to understand the selective pressures that actively shape and maintain the UPEC pan-genome. Once we better understand how ExPEC are able to transition and adapt to both the intestinal and extraintestinal host microenvironments within the same individual, it will be feasible to develop antimicrobials that target pathogenic strains and avoid eradicating beneficial commensal bacteria.

ACKNOWLEDGMENTS

Conflicts of interest: We declare no conflicts.

CITATION

Alteri CJ, Mobley HLT. 2015. Metabolism and fitness of urinary tract pathogens. Microbiol Spectrum 3(3):MBP-0016-2015.

REFERENCES

1. **Foxman B, Barlow R, D'Arcy H, Gillespie B, Sobel JD.** 2000. Urinary tract infection: self-reported incidence and associated costs. *Ann Epidemiol* **10:**509–515.

2. **Litwin MS, Saigal CS, Yano EM, Avila C, Geschwind SA, Hanley JM, Joyce GF, Madison R, Pace J, Polich SM, Wang M.** 2005. Urologic diseases in America Project: analytical methods and principal findings. *J Urol* **173:**933–937.

3. **Kaper JB, Nataro JP, Mobley HL.** 2004. Pathogenic *Escherichia coli. Nat Rev Microbiol* **2:**123–140.

4. **Touchon M, Hoede C, Tenaillon O, Barbe V, Baeriswyl S, Bidet P, Bingen E, Bonacorsi S, Bouchier C, Bouvet O, Calteau A, Chiapello H, Clermont O, Cruveiller S, Danchin A, Diard M, Dossat C, Karoui ME, Frapy E, Garry L, Ghigo JM, Gilles AM, Johnson J, Le Bouguenec C, Lescat M, Mangenot S, Martinez-Jehanne V, Matic I, Nassif X, Oztas S, Petit MA, Pichon C, Rouy Z, Ruf CS, Schneider D, Tourret J, Vacherie B, Vallenet D, Medigue C, Rocha EP, Denamur E.** 2009. Organised genome dynamics in the Escherichia coli species results in highly diverse adaptive paths. *PLoS Genet* **5:**e1000344. doi:10.1371/journal.ppat.1000344

5. **Land M, Hauser L, Jun SR, Nookaew I, Leuze MR, Ahn TH, Karpinets T, Lund O, Kora G, Wassenaar T, Poudel S, Ussery DW.** 2015. Insights from 20 years of bacterial genome sequencing. *Funct Integr Genomics* **15:**141–161.

6. **Croxen MA, Finlay BB.** 2010. Molecular mechanisms of *Escherichia coli* pathogenicity. *Nat Rev Microbiol* **8:**26–38.

7. **Johnson JR, Kuskowski MA, Gajewski A, Soto S, Horcajada JP, Jimenez de Anta MT, Vila J.** 2005. Extended virulence genotypes and phylogenetic background of *Escherichia coli* isolates from patients with cystitis, pyelonephritis, or prostatitis. *J Infect Dis* **191:**46–50.

8. **Johnson JR, Russo TA.** 2005. Molecular epidemiology of extraintestinal pathogenic (uropathogenic) *Escherichia coli. Int J Med Microbiol* **295:**383–404.

9. **Russo TA, Johnson JR.** 2003. Medical and economic impact of extraintestinal infections due to *Escherichia coli*: focus on an increasingly important endemic problem. *Microbes Infect* **5:**449–456.

10. **Chen SL, Hung CS, Pinkner JS, Walker JN, Cusumano CK, Li Z, Bouckaert J, Gordon JI, Hultgren SJ.** 2009. Positive selection identifies an *in vivo* role for FimH during urinary tract infection in addition to mannose binding. *Proc Natl Acad Sci U S A* **106:**22439–22444.

11. **Hommais F, Gouriou S, Amorin C, Bui H, Rahimy MC, Picard B, Denamur E.** 2003. The FimH A27V mutation is pathoadaptive for urovirulence in *Escherichia coli* B2 phylogenetic group isolates. *Infect Immun* **71:**3619–3622.

12. **Johnson JR, Russo TA.** 2002. Extraintestinal pathogenic *Escherichia coli*: "the other bad *E. coli*". *J Lab Clin Med* **139:**155–162.

13. **Alteri CJ, Mobley HL.** 2012. Escherichia coli physiology and metabolism dictates adaptation to diverse host microenvironments. *Curr Opin Microbiol* **15:**3–9.

14. **Hacker J, Blum-Oehler G, Mühldorfer I, Tschäpe H.** 1997. Pathogenicity islands of virulent bacteria: structure, function and impact on microbial evolution. *Mol Microbiol* **23:**1089–1097.

15. **Lloyd AL, Rasko DA, Mobley HL.** 2007. Defining genomic islands and uropathogen-specific genes in uropathogenic *Escherichia coli. J Bacteriol* **189:**3532–3546.

16. **Le Gall T, Clermont O, Gouriou S, Picard B, Nassif X, Denamur E, Tenaillon O.** 2007. Extraintestinal virulence is a coincidental by-product of commensalism in B2 phylogenetic group *Escherichia coli* strains. *Mol Biol Evol* **24:**2373–2384.

17. **Diard M, Garry L, Selva M, Mosser T, Denamur E, Matic I.** 2010. Pathogenicity-associated islands in extraintestinal pathogenic *Escherichia coli* are fitness elements involved in intestinal colonization. *J Bacteriol* **192:**4885–4893.

18. **Alteri CJ, Smith SN, Mobley HL.** 2009. Fitness of *Escherichia coli* during urinary tract infection requires gluconeogenesis and the TCA cycle. *PLoS Pathog* **5:**e1000448. doi:10.1371/journal.ppat.1000448

19. **Moller AK, Leatham MP, Conway T, Nuijten PJ, de Haan LA, Krogfelt KA, Cohen PS.** 2003. An *Escherichia coli* MG1655 lipopolysaccharide deep-rough core mutant grows and survives in mouse cecal mucus but fails to colonize the mouse large intestine. *Infect Immun* **71:**2142–2152.

20. **Meador JP, Caldwell ME, Cohen PS, Conway T.** 2014. *Escherichia coli* pathotypes occupy distinct niches in the mouse intestine. *Infect Immun* **82:**1931–1938.

21. **Chang DE, Smalley DJ, Tucker DL, Leatham MP, Norris WE, Stevenson SJ, Anderson AB, Grissom JE, Laux DC, Cohen PS, Conway T.** 2004. Carbon nutrition of *Escherichia coli* in the mouse intestine. *Proc Natl Acad Sci U S A* **101:**7427–7432.

22. Fabich AJ, Jones SA, Chowdhury FZ, Cernosek A, Anderson A, Smalley D, McHargue JW, Hightower GA, Smith JT, Autieri SM, Leatham MP, Lins JJ, Allen RL, Laux DC, Cohen PS, Conway T. 2008. Comparison of carbon nutrition for pathogenic and commensal *Escherichia coli* strains in the mouse intestine. *Infect Immun* **76**:1143–1152.

23. Martens EC, Roth R, Heuser JE, Gordon JI. 2009. Coordinate regulation of glycan degradation and polysaccharide capsule biosynthesis by a prominent human gut symbiont. *J Biol Chem* **284**:18445–18457.

24. Peekhaus N, Conway T. 1998. What's for dinner?: Entner-Doudoroff metabolism in *Escherichia coli*. *J Bacteriol* **180**:3495–3502.

25. Bonafonte MA, Solano C, Sesma B, Alvarez M, Montuenga L, Garcia-Ros D, Gamazo C. 2000. The relationship between glycogen synthesis, biofilm formation and virulence in *Salmonella enteritidis*. *FEMS Microbiol Lett* **191**:31–36.

26. Jones SA, Jorgensen M, Chowdhury FZ, Rodgers R, Hartline J, Leatham MP, Struve C, Krogfelt KA, Cohen PS, Conway T. 2008. Glycogen and maltose utilization by *Escherichia coli* O157:H7 in the mouse intestine. *Infect Immun* **76**:2531–2540.

27. Miranda RL, Conway T, Leatham MP, Chang DE, Norris WE, Allen JH, Stevenson SJ, Laux DC, Cohen PS. 2004. Glycolytic and gluconeogenic growth of *Escherichia coli* O157:H7 (EDL933) and *E. coli* K-12 (MG1655) in the mouse intestine. *Infect Immun* **72**:1666–1676.

28. Nowrouzian FL, Wold AE, Adlerberth I. 2005. *Escherichia coli* strains belonging to phylogenetic group B2 have superior capacity to persist in the intestinal microflora of infants. *J Infect Dis* **191**:1078–1083.

29. Nowrouzian FL, Adlerberth I, Wold AE. 2006. Enhanced persistence in the colonic microbiota of *Escherichia coli* strains belonging to phylogenetic group B2: role of virulence factors and adherence to colonic cells. *Microbes Infect* **8**:834–840.

30. Schouler C, Taki A, Chouikha I, Moulin-Schouleur M, Gilot P. 2009. A genomic island of an extraintestinal pathogenic *Escherichia coli* strain enables the metabolism of fructo-oligosaccharides, which improves intestinal colonization. *J Bacteriol* **191**:388–393.

31. Porcheron G, Kut E, Canepa S, Maurel MC, Schouler C. 2011. Regulation of fructooligosaccharide metabolism in an extra-intestinal pathogenic *Escherichia coli* strain. *Mol Microbiol* **81**:717–733.

32. Rouquet G, Porcheron G, Barra C, Répérant M, Chanteloup NK, Schouler C, Gilot P.

2009. A metabolic operon in extraintestinal pathogenic *Escherichia coli* promotes fitness under stressful conditions and invasion of eukaryotic cells. *J Bacteriol* **191**:4427–4440.

33. Hagberg L, Engberg I, Freter R, Lam J, Olling S, Svanborg Edén C. 1983. Ascending, unobstructed urinary tract infection in mice caused by pyelonephritogenic *Escherichia coli* of human origin. *Infect Immun* **40**:273–283.

34. Alteri CJ, Mobley HL. 2007. Quantitative profile of the uropathogenic *Escherichia coli* outer membrane proteome during growth in human urine. *Infect Immun* **75**:2679–2688.

35. Asscher AW, Sussman M, Waters WE, Davis RH, Chick S. 1966. Urine as a medium for bacterial growth. *Lancet* **2**:1037–1041.

36. Russo TA, Jodush ST, Brown JJ, Johnson JR. 1996. Identification of two previously unrecognized genes (*guaA* and *argC*) important for uropathogenesis. *Mol Microbiol* **22**:217–229.

37. Brooks T, Keevil CW. 1997. A simple artificial urine for the growth of urinary pathogens. *Lett Appl Microbiol* **24**:203–206.

38. Aubron C, Glodt J, Matar C, Huet O, Borderie D, Dobrindt U, Duranteau J, Denamur E, Conti M, Bouvet O. 2012. Variation in endogenous oxidative stress in *Escherichia coli* natural isolates during growth in urine. *BMC Microbiol* **12**:120.

39. Snyder JA, Haugen BJ, Buckles EL, Lockatell CV, Johnson DE, Donnenberg MS, Welch RA, Mobley HL. 2004. Transcriptome of uropathogenic *Escherichia coli* during urinary tract infection. *Infect Immun* **72**:6373–6381.

40. Hancock V, Vejborg RM, Klemm P. 2010. Functional genomics of probiotic *Escherichia coli* Nissle 1917 and 83972, and UPEC strain CFT073: comparison of transcriptomes, growth and biofilm formation. *Mol Genet Genomics* **284**:437–454.

41. Vejborg RM, de Evgrafov MR, Phan MD, Totsika M, Schembri MA, Hancock V. 2012. Identification of genes important for growth of asymptomatic bacteriuria *Escherichia coli* in urine. *Infect Immun* **80**:3179–3188.

42. Alteri CJ, Hagan EC, Sivick KE, Smith SN, Mobley HL. 2009. Mucosal immunization with iron receptor antigens protects against urinary tract infection. *PLoS Pathog* **5**:e1000586. doi:10.1371/journal.ppat.1000586

43. Alteri CJ, Himpsl SD, Mobley HL. 2015. Preferential use of central metabolism *in vivo* reveals a nutritional basis for polymicrobial infection. *PLoS Pathog* **11**:e1004601. doi:10.1371/journal.ppat.1004601

44. Roesch PL, Redford P, Batchelet S, Moritz RL, Pellett S, Haugen BJ, Blattner FR, Welch RA. 2003. Uropathogenic *Escherichia coli* use

d-serine deaminase to modulate infection of the murine urinary tract. *Mol Microbiol* **49:**55–67.

45. **Bahrani-Mougeot FK, Buckles EL, Lockatell CV, Hebel JR, Johnson DE, Tang CM, Donnenberg MS.** 2002. Type 1 fimbriae and extracellular polysaccharides are preeminent uropathogenic *Escherichia coli* virulence determinants in the murine urinary tract. *Mol Microbiol* **45:**1079–1093.

46. **Hagan EC, Lloyd AL, Rasko DA, Faerber GJ, Mobley HL.** 2010. *Escherichia coli* global gene expression in urine from women with urinary tract infection. *PLoS Pathog* **6:**e1001187. doi:10.1371/journal.ppat.1001187

47. **Bielecki P, Muthukumarasamy U, Eckweiler D, Bielecka A, Pohl S, Schanz A, Niemeyer U, Oumeraci T, von Neuhoff N, Ghigo JM, Häussler S.** 2014. *In vivo* mRNA profiling of uropathogenic *Escherichia coli* from diverse phylogroups reveals common and group-specific gene expression profiles. *MBio* **5:**e01075-01014. doi:10.1128/mBio.01075-14

48. **Subashchandrabose S, Hazen TH, Brumbaugh AR, Himpsl SD, Smith SN, Ernst RD, Rasko DA, Mobley HL.** 2014. Host-specific induction of *Escherichia coli* fitness genes during human urinary tract infection. *Proc Natl Acad Sci U S A* **111:**18327–18332.

49. **Wandersman C, Delepelaire P.** 2004. Bacterial iron sources: from siderophores to hemophores. *Annu Rev Microbiol* **58:**611–647.

50. **Kammler M, Schön C, Hantke K.** 1993. Characterization of the ferrous iron uptake system of *Escherichia coli. J Bacteriol* **175:**6212–6219.

51. **Torres AG, Redford P, Welch RA, Payne SM.** 2001. TonB-dependent systems of uropathogenic *Escherichia coli*: aerobactin and heme transport and TonB are required for virulence in the mouse. *Infect Immun* **69:**6179–6185.

52. **Welch RA, Burland V, Plunkett G III, Redford P, Roesch P, Rasko D, Buckles EL, Liou SR, Boutin A, Hackett J, Stroud D, Mayhew GF, Rose DJ, Zhou S, Schwartz DC, Perna NT, Mobley HL, Donnenberg MS, Blattner FR.** 2002. Extensive mosaic structure revealed by the complete genome sequence of uropathogenic *Escherichia coli. Proc Natl Acad Sci U S A* **99:**17020–17024.

53. **Hagan EC, Mobley HL.** 2009. Haem acquisition is facilitated by a novel receptor Hma and required by uropathogenic *Escherichia coli* for kidney infection. *Mol Microbiol* **71:**79–91.

54. **Johnson JR, Jelacic S, Schoening LM, Clabots C, Shaikh N, Mobley HL, Tarr PI.** 2005. The IrgA homologue adhesin Iha is an *Escherichia coli* virulence factor in murine urinary tract infection. *Infect Immun* **73:**965–971.

55. **Russo TA, Carlino UB, Johnson JR.** 2001. Identification of a new iron-regulated virulence gene, ireA, in an extraintestinal pathogenic isolate of *Escherichia coli. Infect Immun* **69:**6209–6216.

56. **Russo TA, McFadden CD, Carlino-MacDonald UB, Beanan JM, Barnard TJ, Johnson JR.** 2002. IroN functions as a siderophore receptor and is a urovirulence factor in an extraintestinal pathogenic isolate of *Escherichia coli. Infect Immun* **70:**7156–7160.

57. **Watts RE, Totsika M, Challinor VL, Mabbett AN, Ulett GC, De Voss JJ, Schembri MA.** 2012. Contribution of siderophore systems to growth and urinary tract colonization of asymptomatic bacteriuria *Escherichia coli. Infect Immun* **80:**333–344.

58. **Carbonetti NH, Boonchai S, Parry SH, Väisänen-Rhen V, Korhonen TK, Williams PH.** 1986. Aerobactin-mediated iron uptake by *Escherichia coli* isolates from human extraintestinal infections. *Infect Immun* **51:**966–968.

59. **Johnson JR, Moseley SL, Roberts PL, Stamm WE.** 1988. Aerobactin and other virulence factor genes among strains of *Escherichia coli* causing urosepsis: association with patient characteristics. *Infect Immun* **56:**405–412.

60. **Vigil PD, Stapleton AE, Johnson JR, Hooton TM, Hodges AP, He Y, Mobley HL.** 2011. Presence of putative repeat-in-toxin gene tosA in *Escherichia coli* predicts successful colonization of the urinary tract. *MBio* **2:**e00066-00011. doi:10.1128/mBio.00066-11

61. **Henderson JP, Crowley JR, Pinkner JS, Walker JN, Tsukayama P, Stamm WE, Hooton TM, Hultgren SJ.** 2009. Quantitative metabolomics reveals an epigenetic blueprint for iron acquisition in uropathogenic *Escherichia coli. PLoS Pathog* **5:**e1000305. doi:10.1371/journal.ppat.1000305

62. **Melican K, Sandoval RM, Kader A, Josefsson L, Tanner GA, Molitoris BA, Richter-Dahlfors A.** 2011. Uropathogenic *Escherichia coli* P and Type 1 fimbriae act in synergy in a living host to facilitate renal colonization leading to nephron obstruction. *PLoS Pathog* **7:**e1001298. doi:10.1371/journal.ppat.1001298

63. **Anfora AT, Halladin DK, Haugen BJ, Welch RA.** 2008. Uropathogenic *Escherichia coli* CFT073 is adapted to acetatogenic growth but does not require acetate during murine urinary tract infection. *Infect Immun* **76:**5760–5767.

64. **Zdziarski J, Brzuszkiewicz E, Wullt B, Liesegang H, Biran D, Voigt B, Grönberg-Hernandez J, Ragnarsdottir B, Hecker M,**

Ron EZ, Daniel R, Gottschalk G, Hacker J, Svanborg C, Dobrindt U. 2010. Host imprints on bacterial genomes--rapid, divergent evolution in individual patients. *PLoS Pathog* 6: e1001078. doi:10.1371/journal.ppat.1001078

65. **Hadjifrangiskou M, Kostakioti M, Chen SL, Henderson JP, Greene SE, Hultgren SJ.** 2011. A central metabolic circuit controlled by QseC in pathogenic *Escherichia coli. Mol Microbiol* 80:1516–1529.

66. **Park SJ, Gunsalus RP.** 1995. Oxygen, iron, carbon, and superoxide control of the fumarase fumA and fumC genes of *Escherichia coli*: role of the arcA, fnr, and soxR gene products. *J Bacteriol* 177:6255–6262.

67. **Unden G, Bongaerts J.** 1997. Alternative respiratory pathways of *Escherichia coli*: energetics and transcriptional regulation in response to electron acceptors. *Biochim Biophys Acta* 1320:217–234.

68. **Jones SA, Chowdhury FZ, Fabich AJ, Anderson A, Schreiner DM, House AL, Autieri SM, Leatham MP, Lins JJ, Jorgensen M, Cohen PS, Conway T.** 2007. Respiration of *Escherichia coli* in the mouse intestine. *Infect Immun* 75:4891–4899.

69. **Shepherd M, Sanguinetti G, Cook GM, Poole RK.** 2010. Compensations for diminished terminal oxidase activity in *Escherichia coli*: cytochrome bd-II-mediated respiration and glutamate metabolism. *J Biol Chem* 285:18464–18472.

70. **Kralj JM, Hochbaum DR, Douglass AD, Cohen AE.** 2011. Electrical spiking in *Escherichia coli* probed with a fluorescent voltage-indicating protein. *Science* 333:345–348.

71. **Allison KR, Brynildsen MP, Collins JJ.** 2011. Metabolite-enabled eradication of bacterial persisters by aminoglycosides. *Nature* 473: 216–220.

72. **Bergsten G, Samuelsson M, Wullt B, Leijonhufvud I, Fischer H, Svanborg C.** 2004. PapG-dependent adherence breaks mucosal inertia and triggers the innate host response. *J Infect Dis* 189:1734–1742.

73. **Godaly G, Bergsten G, Hang L, Fischer H, Frendéus B, Lundstedt AC, Samuelsson M, Samuelsson P, Svanborg C.** 2001. Neutrophil recruitment, chemokine receptors, and resistance to mucosal infection. *J Leukoc Biol* 69:899–906.

74. **Chromek M, Slamová Z, Bergman P, Kovács L, Podracká L, Ehrén I, Hökfelt T, Gudmundsson GH, Gallo RL, Agerberth B, Brauner A.** 2006. The antimicrobial peptide cathelicidin protects the urinary tract against invasive bacterial infection. *Nat Med* 12:636–641.

75. **Alteri CJ, Lindner JR, Reiss DJ, Smith SN, Mobley HL.** 2011. The broadly conserved regulator PhoP links pathogen virulence and membrane potential in *Escherichia coli. Mol Microbiol* 82:145–163.

76. **Anfora AT, Haugen BJ, Roesch P, Redford P, Welch RA.** 2007. Roles of serine accumulation and catabolism in the colonization of the murine urinary tract by *Escherichia coli* CFT073. *Infect Immun* 75:5298–5304.

77. **Haugen BJ, Pellett S, Redford P, Hamilton HL, Roesch PL, Welch RA.** 2007. *In vivo* gene expression analysis identifies genes required for enhanced colonization of the mouse urinary tract by uropathogenic *Escherichia coli* strain CFT073 *dsdA. Infect Immun* 75:278–289.

78. **Cai W, Wannemuehler Y, Dell'anna G, Nicholson B, Barbieri NL, Kariyawasam S, Feng Y, Logue CM, Nolan LK, Li G.** 2013. A novel two-component signaling system facilitates uropathogenic *Escherichia coli*'s ability to exploit abundant host metabolites. *PLoS Pathog* 9:e1003428. doi:10.1371/journal.ppat.1003428

79. **Barbieri NL, Nicholson B, Hussein A, Cai W, Wannemuehler YM, Dell'Anna G, Logue CM, Horn F, Nolan LK, Li G.** 2014. FNR regulates expression of important virulence factors contributing to pathogenicity of uropathogenic *Escherichia coli. Infect Immun* 82: 5086–5098.

80. **Sohanpal BK, El-Labany S, Lahooti M, Plumbridge JA, Blomfield IC.** 2004. Integrated regulatory responses of fimB to N-acetylneuraminic (sialic) acid and GlcNAc in *Escherichia coli* K-12. *Proc Natl Acad Sci U S A* 101:16322–16327.

81. **Bernier SP, Létoffé S, Delepierre M, Ghigo JM.** 2011. Biogenic ammonia modifies antibiotic resistance at a distance in physically separated bacteria. *Mol Microbiol* 81:705–716.

82. **Marteyn B, West NP, Browning DF, Cole JA, Shaw JG, Palm F, Mounier J, Prévost MC, Sansonetti P, Tang CM.** 2010. Modulation of *Shigella* virulence in response to available oxygen *in vivo. Nature* 465:355–358.

83. **Spurbeck RR, Stapleton AE, Johnson JR, Walk ST, Hooton TM, Mobley HL.** 2011. Fimbrial profiles predict virulence of uropathogenic *Escherichia coli* strains: contribution of ygi and yad fimbriae. *Infect Immun* 79:4753–4763.

Bacterial Metabolism in the Host Environment: Pathogen Growth and Nutrient Assimilation in the Mammalian Upper Respiratory Tract

11

SANDRA K. ARMSTRONG[1]

INTRODUCTION

Pathogens evolve in specific host niches and microenvironments that provide the physical and nutritional requirements conducive to their growth. In addition to using the host as a source of food, bacterial pathogens must avoid the immune response to their presence. The mammalian upper respiratory tract is a site that is exposed to the external environment, and is readily colonized by bacteria that live as resident flora or as pathogens. These bacteria can remain localized, descend to the lower respiratory tract, or traverse the epithelium to disseminate throughout the body. By virtue of their successful colonization of the respiratory epithelium, these bacteria obtain the nutrients needed for growth, either directly from host resources or from other microbes. This chapter describes the upper respiratory tract environment, including its tissue and mucosal structure, prokaryotic biota, and biochemical composition that would support microbial life. *Neisseria meningitidis* and the *Bordetella* species are discussed as examples of bacteria that have no known external reservoirs but have evolved to obligately colonize the mammalian upper respiratory tract.

[1]Department of Microbiology, University of Minnesota Medical School, Minneapolis, MN 55455-0312.
Metabolism and Bacterial Pathogenesis
Edited by Tyrrell Conway and Paul Cohen
© 2015 American Society for Microbiology, Washington, DC
doi:10.1128/microbiolspec.MBP-0007-2014

THE UPPER RESPIRATORY TRACT ENVIRONMENT

Structure of the Upper Airway: Tissue and Cells

Hairlike nasal vibrissae filter large particles from inhaled air that is then warmed, humidified, and conducted into the respiratory tract (Fig 1). Pathogenic microbes that enter the nasal cavity are able to bypass this filtration, and can make contact with the airway epithelium. The anterior nasal cavity has a keratinized stratified squamous epithelium that becomes less keratinized posteriorly (1, 2). Air flows through both sides of the nasal septum through scroll-shaped turbinates that have increased surface area, allowing inspired air to be humidified (to nearly 100% relative humidity) and warmed (to ~34°C). From the turbinate region, posteriorly throughout the rest of the nasal cavity, the surface consists of a pseudostratified columnar epithelium comprised of secretory goblet cells, columnar cells, and ciliated columnar cells (1, 2). The olfactory region is located in the posterior nasal cavity and is interspersed with chemoreceptive neurons. Submucosal glands produce the majority of respiratory secretions that are moved to the epithelial surface via ducts. In humans, these glands are found in the nose, trachea, and

bronchi; in mice they are restricted to the nose and laryngeal region of the trachea. The nasopharynx, oropharynx, and larynx have pseudostratified columnar epithelia similar to that found in the posterior nasal cavity; the vocal cords have a squamous epithelium that lacks ciliated cells. The ciliated pseudostratified columnar epithelium extends throughout the respiratory tract, but not to the smallest bronchioles and alveoli. On regional surfaces that routinely encounter friction (e.g., food movement, swallowing), there is stratified squamous epithelium. These regions of the upper respiratory tract provide a moist, warm environment for microbial growth, but to effectively interact with those host cells, organisms attempting to colonize the airway must encounter the appropriate epithelial cells for which its adhesins and colonization factors have evolved.

The mammalian respiratory tract contains local respiratory lymphoid tissues such as the nasopharynx-associated and bronchus-associated lymphoid tissues (3). Other airway mucosal cells include macrophages and dendritic cells, some of which are resident intra-epithelial dendritic cells that are positioned for luminal antigen sampling. T cells can also localize to intraepithelial spaces and to the lamina propria, along with IgA-producing plasma cells, mast cells, and B cells (3). Resident memory T cells may also be present in the mucosa after their migration from local lymphoid tissues (4), and the nasal mucosa may harbor M cells that are involved in antigen sampling at localized lymphoid tissues (4, 5). Inflammation of the airway increases vascular permeability, promoting immune-cell migration and transudation of plasma components onto the respiratory epithelial surface (6, 7).

The Mucosa

The respiratory epithelial surface is bathed in a mucus blanket that effectively traps particulates that are then moved toward the esophagus by mucociliary clearance for sub-

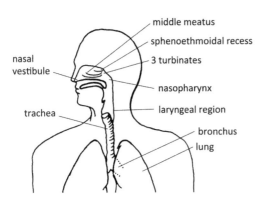

FIGURE 1 Diagram showing regions of the human upper respiratory tract. doi:10.1128/microbiolspec. MBP-0007-2014.f1

sequent elimination by swallowing or expectoration (8, 9). Components of this mucus blanket contribute to its high osmolarity, but also constitute potential food sources for pathogens. Mucins abundant on mucosal surfaces are high molecular mass glycoproteins comprised of a protein core, with carbohydrate (80–90% of the mass) and sulfate (1–2%) moieties (1, 2, 10). Mucins have characteristic Pro-Thr-Ser repeats as well as Cys-rich domains that provide key disulfide-bonding capability, important for mucin multimerization and mucus function. There are two main types of mucins: membrane-bound and secreted (gel-forming). Respiratory epithelial cells produce the membrane-bound mucins MUC1, MUC4, and MUC16, which can be released from the cell surface (10) and are proposed to have important cell regulatory roles as well as key roles in mucociliary function. In the respiratory tract, the most abundant gel-forming mucins are MUC5AC and MUC5B, with MUC2 and MUC19 found in lower abundance (10, 11). In humans, epithelial goblet cells produce primarily MUC5AC, whereas MUC5B is produced by the submucosal glands and other surface secretory-cell types (1). In the mouse respiratory tract, very little Muc5ac is made, whereas secretory cells produce Muc5b (1). Within the secretory cell, newly synthesized mucin is dehydrated and then enclosed within membrane-bound secretory granules. ATP and UTP are potent extracellular signals that trigger the secretion process, during which the granules fuse with the cytoplasmic membrane to allow mucin release onto the epithelium and subsequent hydration in the airway surface liquid (1, 10, 12).

The mucus layer closest to the lumen, often called the gel phase, is 0.5–5 μm thick and consists of a network (pore size, approx. 500 nm) (1, 3, 14) of gel-forming mucins. Beneath this highly viscous layer and in direct contact with the epithelium, is the periciliary layer, with a depth of approximately 7 μm. This periciliary layer, often referred to as the sol phase, has traditionally

been thought to be the less viscous watery layer of the mucus blanket, allowing free movement of the cilia to achieve mucus clearance. Recent *ex vivo* studies using human airway tissue have provided detailed compositional analyses of the mucus layers, revealing that the periciliary layer is not "watery", but filled with a mesh network of membrane-bound mucins and mucopolysaccharides (14). These network molecules fill the interciliary spaces and are also connected to the cilia and microvilli of airway epithelial cells. This complex periciliary layer is proposed to function as a brush and stabilize the two mucus layers, in part by prevention of soluble mucins in the luminal gel layer from extending into the periciliary layer and disrupting ciliary movement.

Human Upper Respiratory Tract Microbiota

Microbial inhabitants of the nasopharynx and other regions of the respiratory tract are likely to not only modify that environment, but may also produce metabolites and other factors that influence the growth of pathogenic microbes. There have been some recent efforts to map the biogeography of the respiratory bacterial population using 16S rRNA metagenomic approaches (15–22). Sampling of specific regions of the human respiratory tract can be challenging, since obtaining specimens can involve invasive procedures and certain sampling methods are prone to cross contamination of samples from differing sites. The anterior nares (nasal vestibule), where there is keratinized squamous epithelium, generally yield Staphylococci and other skin inhabitants such as *Propionibacterium* species (15–17). Recently, Yan and colleagues characterized the bacterial biota from three distinct regions of the human nasal cavity, comparing persistent carriers of *Staphylococcus aureus* with non-persistent carriers (22). The study sampled not only the anterior nares, but also the more posterior middle meatus and sphenoethmoidal recess (Fig 1), both of which are covered in the ciliated, pseudo-stratified, columnar epithelium characteristic of the respiratory mucosa. In general, the three nasal sites of both *S. aureus* carriers and non-carriers were characterized by significant populations of members of the Actinobacteria and α-Proteobacteria (Table 1) (22). Subjects persistently colonized with *S. aureus* also appeared to be co-colonized with larger numbers of other different taxa than non-persistent carriers. Genera, including those from the *Firmicutes*, Actinobacteria, γ- and β- Proteobacteria, and other *Staphylococcus* species, were variably prevalent in the sample set from non-carriers. Of note, the microbiota from the middle meatus and sphenoethmoidal recess nasal sites of individuals from both subject groups exhibited more taxonomic diversity compared with the nasal vestibule site. Since the sphenoethmoidal recess is the closest geographically of the three sampled regions to the nasopharynx, these results may inform our understanding of the nasopharyngeal and tracheal microbiota. In a different study, Charlson and colleagues performed a comprehensive biogeographical analysis of the healthy human respiratory tract, using methods designed to minimize cross-contamination of epithelial site samples (19). For the nasopharynx, they reported that *Staphylococcaceae* and *Propionibacteriaceae* were predominant, and noted significant relative abundance of members of families listed in Table 1. Their results comparing multiple respiratory tract regions indicated more overall diversity in the bacterial populations from nasopharyngeal samples than from oral, oropharynx, and bronchoalveolar lavage fluid (BALF) samples.

Hilty et al examined the bacterial populations of the human nose, oropharynx, the left upper lung lobe, and BALF from healthy controls and patients with asthma and chronic obstructive pulmonary disease (18). Populations found in the left upper lobe of all of the adult donors were similar and included taxa listed in Table 1. Compared with controls,

patients with chronic obstructive pulmonary disease had significantly fewer *Prevotella* species and more *Haemophilus* and other Proteobacteria species. BALF from children with asthma also were colonized with fewer *Prevotella* species and more *Staphylococcus*, *Haemophilus*, and other Proteobacteria. Huang and colleagues examined bronchial epithelial brushings from healthy controls and patients with asthma (20). Compared with controls, asthmatics had a significantly higher bacterial burden. Importantly, a diverse microbiota was identified in both subject groups, but several taxa, in particular, members of the *Comamonadaceae*, *Oxalobacteraceae*, and *Sphingomonadaceae*, were associated with asthma patients experiencing bronchial hyperresponsiveness.

Although these analyses have yielded useful information about the human respiratory microbiota, the majority of the identified bacterial taxa have not been studied to any extent that would allow predictions of their metabolites or compounds that could be produced or excreted *in vivo* to promote the growth of infecting pathogens. The aforementioned study of the human microbiota from three nasal sites revealed that in per sistent carriers of *S. aureus*, there was a significant association with co-colonization by *Corynebacterium accolens* (22). In non-carriers, there was instead, colonization by *Corynebacterium pseudodiphtheriticum*. In vitro co-culture experiments showed that growth of *C. accolens* was enhanced by the presence of *S. aureus*, and conversely, *S. aureus* growth was stimulated by *C. accolens*. There was no growth enhancement when *C. pseudodiphtheriticum* was substituted for *C. accolens*. Bacteria that have evolved to be obligate pathogens may be auxotrophic for vitamins and cofactors that other bacterial species can produce endogenously. For example, it is well known that some respiratory *Haemophilus* species are incapable of *de novo* production of heme and nicotinamide adenine dinucleotide (NAD), and on solid culture media these organisms will form satellite

TABLE 1 **Human respiratory tract microbiota**

Location	Donor	Major taxonomic groups identified (listed in decreasing order of abundance)	Reference
Nasal cavity: 3 sites	*S. aureus* non-carriers	*Corynebacterium* *Propionibacterium* *Escherichia-Shigella* *Dolosigranulum* *Tomitella* *Moraxella* uncultured β-proteobacteria other *Staphylococcus* species	22
Nasal cavity: 3 sites	*S. aureus* carriers	*Corynebacterium* *Propionibacterium* *Escherichia-Shigella* non-*aureus Staphylococcus* *Streptococcus* *Prevotella* *Anaerococcus* *Peptoniphilus*	22
Nasopharynx	healthy controls	*Propionibacteriaceae* *Staphyloccoaceae* *Corynebacteriaceae* *Micrococcaceae* *Streptococcaceae* *Prevotellaceae* *Veillonellaceae* *Moraxellaceae* *Lachnospiraceae* *Acintomycetaceae* *Flavobacteriaceae* *Enterobacteriaceae* *Flexibacteriaceae* *Microbacteriaceae*	19
Upper left lung lobe	healthy controls, patients with asthma or chronic obstructive pulmonary disease	*Corynebacterium* and other Actinobacteria *Prevotella* and other Bacterioidetes *Staphylococcus* *Streptococcus* *Veillonella* and other Firmicutes *Neisseria flava/sicca* *Haemophilus* sp and other Proteobacteria *Fusobacterium* *Megasphaera*	18
Bronchoalveolar lavage fluid	healthy controls, patients with asthma	*Prevotella* other Bacterioidetes *Staphylococcus* *Streptococcus* *Veillonella* other Firmicutes *Haemophilus* *Neisseria* other Proteobacteria Fusobacteria	18

colonies surrounding bacterial species that produce these classical X (heme) and V (NAD) factors (23, 24). *S. aureus* was once used in clinical microbiology laboratories as the source of V factor for the satellite test for identification of *Haemophilus* species (25). In the host-airway environment, the growth factors needed by *Haemophilus* species must be provided by exogenous sources, so co-colonizing organisms, such as staphylococci, might supply needed heme, NAD, and other nutrients. Secreted enzymes such as proteases, mucinases, lipases, and nucleases from resident microflora would be predicted to degrade host resources to yield soluble nutrients. *Staphylococcus* species are known to excrete a number of such enzymes that also include hyaluronidase, DNase, coagulase, lipase, and staphylokinase (26–28). *Streptococcus* species can produce hemolysins, cytolytic toxins, and pyrogenic toxins (29). Since these enzymes and toxins may lyse host cells, making their cytosolic contents available for microbial degradation and consumption, commensal *Staphylococcus* and *Streptococcus* species may play important roles in modulating airway nutrient sources available to invading pathogens.

Nutrients in Airway Mucus

Respiratory mucus contains a number of host molecules that contribute to resistance to infection, but may also be degraded and utilized by microbes as nutrient sources. Lysozyme, immunoglobulins, complement components, cytokines, secretory leukoprotease inhibitor, defensins, cathelicidins, and lactoferrin are examples of potential sources of sugars or amino acids (30–32). Within the mucus, debris from sloughed and dead respiratory cells (1, 2) may provide additional sources of nutrition such as lipids, nucleic acids, and proteins. The respiratory microbiota, including pathogens, are also likely to use small, readily assimilated metabolites that are in the airway surface liquid. Using nuclear magnetic resonance (NMR) and mass spectrometry, metabolomic analyses of BALF from mice

and humans have identified airway metabolites that may serve as nutrients for respiratory commensals and pathogens.

The BALF of healthy mice were found to contain a variety of amino acids, carbohydrates, acids, alcohols, and other metabolites (Table 2) (33). Mouse respiratory *Aspergillus fumigatus* infections yielded BALF that contained these metabolites but also contained increased ethanol concentrations over those in BALF from uninfected mice (34). BALF from mice exposed to airborne silica dust demonstrated higher levels of lactate, glutamate, creatine, lysine, glycogen/glucose, glycine, proline, and 4-hydroxyproline compared with BALF from unexposed mice (35).

Using BALF from healthy humans and patients with acute respiratory distress syndrome or acute lung injury, Evans and colleagues used liquid chromatography–mass spectrometry (LC-MS) to identify the metabolites listed in Table 2, all of which were found in increased concentrations in diseased patients (36). These authors also identified the metabolites palmitic acid, phosphatidylcholines, and stearic acid in both healthy and diseased subjects. A study by Rai and colleagues used NMR to identify airway metabolites in patients with acute respiratory distress syndrome and/or acute lung injury and control patients that did not have those respiratory diseases (37). In both patient populations, they identified the metabolites listed in Table 2. In patients diagnosed with acute respiratory disease, elevated levels of branched-chain amino acids arginine, glutamate, glycine, aspartate, acetate, taurine, threonine, lactate, and succinate were observed and the proline concentration was reduced, compared with control samples.

In sum, these mouse and human studies indicate that in the respiratory tract of healthy animals there exists a baseline level of metabolites available for microbial consumption. As an infection initiates and inflammation ensues, levels of certain metabolites would increase due to transudation of plasma components onto the epithelial surface. This would serve to further increase the levels of potential nutrients for pathogens. Based on these analyses and the known composition of airway fluids, a microbe's carbon and nitrogen sources might be readily satisfied. Potential sulfur sources would include taurine, cysteine, and sulfate derived from mucins, and inorganic sulfur compounds. Phosphorus requirements could be met by host phospholipids, phosphocholine, glycerophosphocholine, nucleic acids, and other phosphate sources. Airway surface liquid is known to contain ascorbic acid (38) as well as high concentrations of glutathione that are critical for defense against oxidative damage (39); therefore, microbes may use these as carbon, nitrogen, and sulfur sources. Since bacterial auxotrophs such as *Haemophilus* species inhabit the respiratory tract, and excreted metabolites and the contents of dead epithelial and microbial cells exist in airway fluids, one may predict that requisite vitamins such as B12, NAD precursors, and other cofactors and ions exist in airway surface liquid and their concentrations may be enhanced during inflammation or by production by commensal biota.

Iron and Its Acquisition by Microbes

Most living cells, including prokaryotes, require nutritional iron. In the mammalian host, there is exceedingly little iron available for microbial growth due to host iron-homeostatic mechanisms that function to limit the amount of free iron (40–42). The peptide hormone hepcidin is a key regulator of erythropoiesis and it also represses the uptake of dietary iron in the gut and the liberation of iron from macrophages and hepatocytes (43–45). Hepcidin inhibits iron transport to plasma and tissues by binding the cellular ferroportin iron exporter and inducing its breakdown within the lysosome. The bone morphogenetic protein pathway transcriptionally regulates hepcidin production, and iron abundance and inflammation also upregulate expression of the hepcidin gene (45). Another arm of the host

iron-withholding defense is that exerted by the iron-binding glycoproteins transferrin and lactoferrin, each of which can coordinate two atoms of ferric iron (43, 44). Transferrin is found primarily in plasma and is the primary means by which the cells of the body receive nutritional iron. Lactoferrin exists on mucosal surfaces and is also stored within neutrophil granules that can be deployed upon encounter with microbes (43, 46). The N-terminal region of lactoferrin (termed lactoferricin) has been shown *in vitro* to have antimicrobial activities that are distinct from its iron-scavenging functions. In human respiratory epithelial fluids, lactoferrin concentrations are approximately 100–1000 µg/ml (46); transferrin is found in very low abundance here, but inflammation promotes transudation of serum molecules onto the airway epithelium, thus potentially increasing its concentration at that site.

Pathogenic bacteria must overcome host iron restriction for successful *in vivo* growth, and when starved for iron, they employ several general mechanisms for retrieval of iron from host sources. Under conditions of iron-replete growth (a rare circumstance *in vivo*), expression of iron-acquisition genes is usually repressed by Fur (or an analogous regulator) that requires ferrous iron as a corepressor (47). Iron starvation leads to decreased intracellular iron corepressor concentrations and subsequent derepression of genes involved in iron uptake. For numerous pathogens, the predominant classes of *in vivo*-expressed genes are involved in nutrient acquisition, and of those, iron-acquisition genes dominate the transcription profile (48). Therefore, microbial iron-acquisition gene expression is regarded as a transcriptional signature of the host microenvironment. In addition to regulating the systems required for iron assimilation, iron starvation is a major signal controlling the expression of bacterial virulence factor genes (49, 50).

Bacteria may produce and excrete siderophores that chelate ferric iron and bind cognate surface receptors for iron uptake (51).

TABLE 2 **Metabolites in airway fluids**

Airway sample	Donor	Major metabolites identified	Reference
Bronchoalveolar lavage fluid	mice: infected or treated mice, healthy controls	glutamate glycine proline lysine taurine 4-hydroxyproline glucose glycogen glycerol acetate formate lactate succinate acetone isopropanol ethane ethanol glycerophosphocholine phosphocholine choline creatine creatine phosphate creatinine	33, 34, 35
Bronchoalveolar lavage fluid	humans: healthy controls, patients with acute respiratory distress syndrome or acute lung injury	L-glutamate L-leucine methionine L-phenylalanine L-proline L-tyrosine L-tryptophan L-threonate D-glucose alpha ketoglutarate cis-aconitate L-lactate citrate creatinine creatine O-acetylcarnitine guanosine hypoxanthine bis(2-ethylhexyl)phthalate uridine hippurate inosine palmitoleic acid arachidonic acid linoleic acid	36

(Continued on next page)

TABLE 2 Metabolites in airway fluids (Continued)

Airway sample	Donor	Major metabolites identified	Reference
Bronchoalveolar lavage fluid	humans: patients with acute respiratory distress syndrome and/or acute lung injury undiseased control patients	alanine arginine aspartate glutamate glycine leucine isoleucine lysine phenylalanine proline threonine valine taurine acetate lactate succinate choline creatine betaine adenine ethanol	37

Bacteria may also have evolved transport systems allowing utilization of xenosiderophores (siderophores produced by other microbes in the vicinity). Siderophores produced by commensal organisms inhabiting the respiratory tract may provide iron to invading pathogens that can use those xenosiderophores. *Corynebacterium, Burkholderia,* and *Actinomyces* species have been identified as respiratory inhabitants (Table 1) and members of these genera have been reported to produce siderophores (52–54); however, it is not known whether the described siderophore-producing species can be found in the respiratory tract. *Staphylococcus* species (55), *Pseudomonas aeruginosa* (56, 57), and *Escherichia coli* (41, 51) each can produce multiple siderophores that could be used by other bacterial species, and have been identified as respiratory microbiota. The nasopharyngeal pathogens *Bordetella pertussis* (58), *Neisseria meningitidis* (59, 60), and *Haemophilus* species (61) are known to use *E. coli* siderophores; thus commensal *E. coli* may promote the *in vivo* growth of these pathogens.

Some organisms have surface receptors allowing hemin uptake or the direct, siderophore-independent uptake of host iron from transferrin and lactoferrin (42, 62). Bacteria may also have systems to transport inorganic ferrous or ferric iron sources (63, 64). Bacterial growth stimulation by host catecholamine neuroendocrine hormones such as epinephrine and norepinephrine has been described (65, 66). Studies indicate that norepinephrine can remove the iron from transferrin and lactoferrin, making it available for growth of *E. coli* and *Salmonella enteriditis* (67–69). Their use of norepinephrine required production of their endogenous siderophores enterobactin and salmochelin, respectively, or their breakdown products. Differently, the pathogens *Bordetella bronchiseptica* (70), *S. aureus* (55), and *Campylobacter jejuni* (71) are able to use host catecholamines to retrieve iron from transferrin and lactoferrin in the *absence* of a siderophore. Thus, these catecholamines function as pseudosiderophores. Although the concentrations of neuroendocrine hormones in the airway have not been well documented, sensory neurons of the airway control reflexes such as coughing, bronchodilation, and mucociliary action, as well as plasma leakage, through the function of neuroendocrine hormones (72). Cells of the human bronchus-associated lymphoid tissue have catecholamine neurotransmitter receptors (73) and rat nasal mucus contains norepinephrine and dopamine (74). Therefore, catecholamines may be available to bacterial cells in the host respiratory tract.

Members of the lipocalin family are small proteins characterized by a binding cleft for hydrophobic ligands (75–77). There are a number of lipocalin types in mammals, including lipocalin 1 found in tears and lipocalin 2 that is a biomarker of kidney damage. These two lipocalins have also been shown to be present on respiratory mucosal surfaces (77) and have been demonstrated to bind microbial siderophores. Lipocalin 1 was shown to bind to the enterobactin siderophore produced by members of the *Enterobacteriaceae*, as well as other structurally different siderophores (78). Lipocalin 2 (also known as siderocalin) binds the ferric and apo- forms of enterobactin in addition to other catechol siderophores (79) and norepinephrine (80). This lipocalin also interacts with an endogenous mammalian catechol that may function as a siderophore in host iron trafficking and homeostasis (81–83). Respiratory lipocalin 2 production in mice was induced after infection by *Streptococcus pneumoniae*, *Haemophilus influenzae* (84), and *Klebsiella pneumoniae* (85) and lipocalin 2 was identified in the nasal secretions from healthy humans (84). Pathogens such as *E. coli* that rely on catechol siderophores for iron acquisition *in vivo* may encounter significant iron restriction due to loss of that siderophore to lipocalin 2 sequestration, and to general competition for ferric iron by lactoferrin. Bacteria that produce and/or use siderophores that are not lipocalin ligands may have an *in vivo* growth advantage. Further, if there is a true mammalian iron-binding catechol siderophore that naturally interacts with lipocalin 2, it is plausible

that microbes will have evolved mechanisms to obtain that complexed iron.

BACTERIAL PATHOGENS OF THE NASOPHARYNX

Neisseria meningitidis

Members of the genus *Neisseria* are Gram-negative β-proteobacteria that include the closely genetically related obligate human pathogens *Neisseria gonorrhoeae*, the bacterial agent of gonorrhea, and *Neisseria meningitidis*, a cause of fulminant meningitis, as well as a number of nonpathogenic commensal species (86–90). Although *N. gonorrhoeae* can colonize the pharynx and other mucosal sites in addition to the reproductive tract, the site of initial *N. meningitidis* colonization is the human upper respiratory tract epithelium. Most commonly, *N. meningitidis* lives as a commensal of the nasopharynx, oropharynx, buccal epithelium, and tonsils. Approximately 10% of healthy humans are colonized by *N. meningitidis* and experience no disease symptoms. Rarely, the organism can enter host respiratory epithelial cells, transit and exit those cells, and reach the submucosal tissues to disseminate via the bloodstream, causing septicemia, and pass the blood-brain barrier to cause meningitis. Invasive disease is associated with high mortality, even when treated with appropriate supportive and antibiotic therapy, as well as neurological sequelae and lifelong disability.

Like other Neisseriae, *N. meningitidis* is a diplococcus that lacks flagellar swimming motility (86). An antiphagocytic polysaccharide capsule is produced and it provides the basis for serotyping *N. meningitidis* strains. There are 12 capsular serogroups, but the majority of disease is caused by serogroups A, B, and C (86). Meningococcal vaccines based on capsular polysaccharides or protein conjugates are in wide use to prevent disease by serogroups A, C, Y, and W135 (86–90). *Neisseria* species are well known for their ability to take up exogenous DNA and undergo natural transformation (86). These organisms are capable of conjugation and resident plasmids exist in some strains. Pathogenic *Neisseria* species are also known for their phase variation of virulence factors such as the opacity proteins (91, 92), pili (93, 94), lipooligosaccharide (95), and heme-uptake transporters HpuAB and HmbR (86, 90, 96).

Colonization

N. meningitidis is transmitted between human hosts by airborne droplets and by respiratory secretions, and it has been hypothesized that the polysaccharide capsule may prevent desiccation during transit (90). Epithelial cell culture and *ex vivo* respiratory tissue models of infection have yielded useful information about meningococcal colonization that may translate to infection of the human nasopharynx (97, 98). The major adhesins are the type IV pili, also involved in twitching motility, and the opacity proteins Opa and Opc (99). Type IV pili are equipped with the PilC tip adhesin and are required for initial interaction with the host epithelial surface after which the organisms then become more intimately associated with the host cells (100). CrgA is a regulator present in the pathogenic Neisseriae that is activated upon contact with epithelial cells (101, 102); in *N. meningitidis*, it is reported to play an important role in the controlled expression of pilus and capsule synthesis genes. The PhoP-PhoQ two component regulatory system is also induced on host cell contact and is involved in adherence of *N. meningitidis* to the epithelium. The Pho system is also involved in the ability of the organism to transit epithelial cells to reach the submucosa (103, 104). The outer membrane porins PorA and PorB (105), lipooligosaccharide, factor H-binding protein (106), the NadA (group B strains) (107) and TspA (108) adhesins, the antiphagocytic NhhA adhesin/complement resistance protein (109), and the HrpA-HrpB two-partner secretion system (110) also play key roles in adherence and colonization. Opa, Opc, and PorB have been reported as important for uptake of *N. meningitidis* into host cells, a prerequisite to dissemination (99). Few secreted products are produced, but the organism does excrete an immunoglobulin A1 (IgA) protease that cleaves the antibody molecule at the hinge region (111), and can also degrade host lysosomal LAMP1, enabling the organism to prevent lysosome maturation (112).

Colonization of the nasopharyngeal epithelium by *N. meningitidis* is hypothesized to involve biofilm formation and studies of biofilm growth have used abiotic surfaces or cultured cells as experimental systems (113–115). Since there is a paucity of nonhuman animal models of meningococcal nasopharyngeal infection, there is little evidence that these organisms live in biofilm communities in the host. However, one *ex vivo* study of human tonsillar tissue revealed what appeared to be microcolonies of *N. meningitidis* present either within or beneath the epithelial surface, suggestive of subsurface biofilm formation (116). *N. meningitidis* strains isolated from the bloodstream during invasive disease are uniformly encapsulated but capsule production by *N. meningitidis* living in the nasopharynx potentially as a biofilm, is downregulated (89). Biofilm formation depends on twitching motility conferred by the type IV pili and the pilus protein PilX, and maintenance of the structure also involves HrpA-HrpB (116). Extracellular DNA is an important component of the extracellular biofilm matrix (116). When *N. meningitidis* contacts the host cell surface, the CrgA regulator represses capsular gene expression and the *ampD* and *mltB* genes exhibit increased expression (101). The products of *ampD* and *mltB* are involved in peptidoglycan recycling and in lytic functions, respectively, and are needed for initial biofilm formation; thus, they may provide a means for bacterial autolysis and release of DNA into the biofilm matrix (116). Experimental results indicate that a major contributor to the elaboration of extracellular DNA and biofilm structural stability is the outer membrane autolysin, phospholipase A.

Metabolism

N. meningitidis has complete Entner-Doudoroff and pentose phosphate pathways and a tricarboxylic acid (TCA) cycle (86). It can use glucose and maltose as carbon sources, but apparently does not use other commonly metabolized sugars. The meningococcus can also use pyruvate and lactate as carbon sources (117, 118). Pyruvate is oxidatively decarboxylated by pyruvate dehydrogenase, to produce acetyl CoA that is oxidized to CO_2 in the TCA cycle. There are no *N. meningitidis* genes specifying the isocitrate-lyase or malate-synthase activities characteristic of a glyoxalate shunt. Growth studies revealed that *N. meningitidis* can use lactate as a sole carbon source more avidly than it uses glucose, and low lactate concentrations stimulate growth on glucose (118). The *N. meningitidis* lactate permease is required for utilization of lactate, but importantly, it was shown to be required for *ex vivo* colonization of human nasopharyngeal tissues (118). *N. meningitidis* has three lactate dehydrogenases that yield pyruvate that can be fed into the TCA cycle (119). Since both D and L isomers of lactate can be used, it was proposed that since mammalian cells do not produce D-lactate, that which is produced by bacterial flora, such as lactic acid bacteria, may provide this carbon source to the meningococcus (119). *N. meningitidis* has a variety of aminopeptidases for the cleavage and liberation of N-terminal amino acids from proteins and peptides. Amino acids are assimilated to form other amino acids and can also be oxidatively deaminated and fed into the TCA cycle (120). Glutamate and proline, which are present in airway surface fluids of humans (Table 2), are two preferred amino acids catabolized by *N. meningitidis*. Glutamate is processed through the activities of one or both *N. meningitidis* glutamate dehydrogenases (121). Glutamate uptake has

also been implicated in *N. meningitidis* internalization and survival in epithelial and other host cells (122).

In vivo sources of nitrogen for the meningococcus most likely include amino acids like glutamate, or other nitrogenous compounds. *N. meningitidis* is an aerobe, preferring an *in vitro* atmosphere with increased CO_2. *N. gonorrhoeae* growth is similarly optimal with increased CO_2 but it can also grow anaerobically using nitrite as an electron acceptor (123). The *Neisseria* AniA nitrite reductase yields nitric oxide (NO), which is then converted to nitrous oxide by the Nor enzyme that uses electrons from quinones (124). *N. gonorrhoeae* and commensal *Neisseria* species have the *aniA* and *norB* genes. Commensal Neisseriae also have a functional nitrate reductase and can complete the denitrification process to take nitrous oxide to N_2 via a nitrous-oxide reductase (124). In many *N. meningitidis* isolates, the *aniA* gene is no longer intact, indicating loss of the nitrite-reductase function. The meningococcal nitric oxide reductase, however, is functional and predicted to provide resistance to host nitric oxide-mediated killing. *N. meningitidis*, but not *N. gonorrhoeae*, has a functional superoxide dismutase that confers additional protection from oxidative killing. *N. meningitidis* also produce SodB (125), SodC (126), catalase, peroxidase (127), and glutathione peroxidase (128). *N. meningitidis* has genes encoding a MntABC manganese transport system that, in *N. gonorrhoeae*, also supplies the manganese involved in protection from oxidative stress (129, 130).

In vivo, *N. meningitidis* is likely to obtain sulfur from cysteine, sulfates, or thiosulfate, although some strains have an absolute requirement for cysteine (86, 131). No reports could be found that described *Neisseria* mucinases or desulfating enzymes that could be used to liberate cysteine and sulfate from respiratory mucins. *N. meningitidis* Z2491 (serogroup A), MC58 (serogroup B), FAM18 (serogroup C), and other serogroup strains each have apparently intact genes encoding a sulfate-transport apparatus and the Cys enzymes required for cysteine biosynthesis from sulfate (120). Absent from the genomes was a *cysC* gene encoding an adenosine phosphosulfate (APS) phosphokinase activity that would produce phosphoadenosine phosphosulfate (PAPS) from APS. This activity could be provided by the product of *cysN* which, in *Burkholderia cenocepacia*, was named *cysNC*, the product of which was noted to be a fusion of the ATP-sulfurylase and APS-kinase domains and proposed to have dual activities in cysteine synthesis from sulfate (132).

Likely sources of phosphorus for *N. meningitidis* in the nasopharynx include compounds such as phospholipids, phosphocholine, and nucleic acids (Table 2). Little has been published on phosphorus metabolism in *Neisseria*, but *N. meningitidis* and *N. gonorrhoeae* have polyphosphate-kinase genes, the products of which promote formation of polyphosphate, and they also have the Ppx enzyme for cleavage of this stored phosphate (133, 134). For zinc acquisition, *N. meningitidis* uses the ZnuD outer membrane receptor-mediated transport system for both zinc and heme (135). *N. meningitidis* has the genetic capability for *de novo* NAD synthesis as well as having genes for the Preiss-Handler salvage pathway for NAD synthesis from nicotinamide or nicotinic acid. These organisms also have the genes encoding enzymes for the synthesis of flavins and thiamine (120).

Iron

All strains of *N. meningitidis* studied to date can obtain iron from transferrin, lactoferrin, and hemoglobin (90, 96, 136–138). Both *N. meningitidis* and *N. gonorrhoeae* produce the TbpA-TbpB cell surface-receptor complex that is specific for human transferrin. TbpA has a structure characteristic of members of the TonB-dependent outer membrane-transporter family (139, 140). In most Gram-negative bacteria, the TonB system is needed to supply the energy from the proton-motive force for transport of iron and other nutrient sources across the outer membrane (141). TbpA and the TbpB lipoprotein form an outer membrane-receptor complex that recognizes and binds the C-lobe of ferric transferrin, removes its iron at the cell surface, and transports that iron to the periplasm for subsequent binding by the FbpA periplasmic protein and uptake across the cytoplasmic membrane by the FbpBC transporter (142). After iron extraction, the apotransferrin is released from the receptor. Crystallographic and structural analyses allowed the meningococcal TbpA and TbpB interactive surfaces to be mapped onto human transferrin and provided data to support a model of iron extraction and release of spent transferrin from the receptor (139). The TbpAB system is found in all *N. meningitidis* and *N. gonorrhoeae* strains that have been examined (90). *N. meningitidis* and *N. gonorrhoeae* produce analogous dipartite-surface receptors for lactoferrin (LbpA-LbpB) and hemoglobin (HpuA-HpuB), although they are less well characterized than TbpA-TbpB (90). Recent studies implicated the LbpB protein in resistance of *N. meningitidis* to the N-terminal lactoferricin portion of lactoferrin that has antimicrobial activity (143). Although there are *N. gonorrhoeae* clinical isolates that lack *lbpBA* genes, all *N. meningitidis* strains that have been examined have them. A *N. gonorrhoeae tbpBA* mutant was avirulent in human volunteers, whereas an *lbpBA* mutant retained virulence (144, 145). Since *N. gonorrhoeae* initially infects human mucosae, one may surmise that lactoferrin would be a primary iron source rather than transferrin. It is unknown whether Tbp (or Lbp) is important for *N. meningitidis* nasopharyngeal colonization or dissemination in humans.

N. meningitidis (and *N. gonorrhoeae*) have a second transport system for heme. HmbR is a TonB-dependent outer membrane receptor for hemin and hemoglobin (90, 96). In the bloodstream, the amount of free hemin or hemoglobin is limited due to binding by hemopexin and haptoglobin, respectively

(62). The amount of available heme in the nasopharynx is unknown. However, since other nasopharyngeal bacterial pathogens such as *H. influenzae* (62) and *Bordetella pertussis* (146) possess heme-utilization systems, and damage to the respiratory epithelium would make heme more available, it is reasonable to hypothesize that heme compounds can be obtained by bacteria living in the nasopharynx.

Neisseria species do not produce siderophores; however, they are capable of using some xenosiderophores, presumably excreted by microbiota of the host. The hydroxamate siderophore aerobactin, produced by members of the *Enterobacteriaceae*, can be used by *N. meningitidis* and *N. gonorrhoeae* (59, 60). Both *Neisseria* species also use ferric enterobactin, which relies on the FetABCD transport system for uptake (147, 148). FetA is the TonB-dependent outer membrane receptor and the FetBCD proteins comprise a periplasmic binding protein-dependent transporter for iron uptake across the cytoplasmic membrane. FetA-specific antibodies have been detected in patients convalescing from disseminated meningococcal infection and it has been studied as a potential vaccine candidate (90). Sequencing the *fetA* genes from a number of *N. meningitidis* strains, both invasive and other, revealed a great deal of allelic variation and antigenic diversity in the protein, perhaps limiting its potential utility as a vaccine antigen. The *fetABCD* genes are Fur repressed and are also subject to genetic variation that affects transcription and therefore, Fet protein production, in both pathogenic *Neisseria* species (147).

In *N. meningitidis*, expression of *fur, tbpBA, lbpBA, fbpABC, hmbR,* and *fetA* is under control of the Fur repressor that uses ferrous iron as co-repressor to limit transcription under conditions of iron plenty (90, 149). Other *N. meningitidis* genes predicted to encode TonB-dependent receptors that may play a role in the acquisition of iron or other nutrients have been identified. These receptors include TdfF, TdfH, and TdfJ, although they remain uncharacterized and their substrates unknown. TdfF and TdfJ are most similar to ferric siderophore receptors and TdfH is similar to other bacterial heme receptors. A global microarray study of *N. meningitidis* genes that were expressed under iron-replete versus iron-limiting conditions identified a number of genes upregulated during iron starvation (149). Known iron-acquisition genes were directly repressed by Fur, as were genes encoding the FrpA virulence factor, DnaK, the ClpB chaperone, the DsbA protein disulfide oxidoreductase, fumarase C, L-lactate dehydrogenase, certain NAD biosynthesis enzymes, and the *nrrF* small RNA. Similar to the positive Fur regulation of *E. coli sodA*, bacterioferritin, and succinate dehydrogenase genes via the *ryhB* regulatory sRNA (150), in *N. meningitidis*, expression of succinate dehydrogenase subunit genes *sdhA* and *sdhC* is repressed by the regulatory *nrrF* sRNA, that itself is iron and Fur repressed (151). Under iron-replete conditions, the TCA cycle genes *sdhA* and *sdhC* are expressed and iron limitation decreases their transcription. In *N. meningitidis*, Fur can directly activate expression of genes involved in resistance to oxidative stress including *katA, sodB, norB,* and *aniA* in addition to genes encoding a fumarate hydratase, and Nuo subunits of the NADH dehydrogenase (149).

Gene-expression profiling examined iron-starved *N. meningitidis* cultured in the presence of hemoglobin, transferrin, or lactoferrin (152). Growth on hemoglobin induced expression of genes including those encoding FrpC and surface proteins FhaB and PilE, in addition to Fur and the NorB nitric oxide reductase. Transferrin or lactoferrin utilization led to increased transcription of genes such as those encoding the PorB and Hsf surface proteins, TCA cycle enzymes, cytochrome C, and alcohol dehydrogenase. Lactoferrin also upregulated genes specifying PilP and LgtG, which is a lipopolysaccharide-modifying enzyme that impacts serum resistance. Since hemoglobin, transferrin, and lactoferrin are primarily localized to different anatomical regions, it was proposed that because these iron compounds elicited such differences in gene expression, they essentially function as host niche indicators for *N. meningitidis*.

The *Bordetella* Respiratory Pathogens

The Bordetellae are Gram-negative β-proteo-bacteria of the family *Alcaligenaceae*. There are presently nine described *Bordetella* species, three of which are highly genetically related respiratory pathogens of mammals, considered to be subspecies, and often called the classical *Bordetella* species (153, 154). *Bordetella pertussis* is a strictly human-adapted species and is the agent of whooping cough or pertussis. *Bordetella parapertussis* causes whooping cough-type infections in humans and can also infect sheep. *Bordetella bronchiseptica* infects a broad range of mammalian hosts and causes respiratory diseases such as canine kennel cough and swine atrophic rhinitis. Genomic sequence analyses suggest that a *B. bronchiseptica*-like organism is the ancestor of *B. pertussis* and *B. parapertussis*, which have sustained significant gene loss and rearrangement compared to the genome of *B. bronchiseptica* (155). For example, *B. bronchiseptica* expresses flagellar genes and is motile, but *B. pertussis* and *B. parapertussis* are not motile and many of their flagellar genes are mutated. This type of genetic loss is thought to be the result of their evolution toward obligate and host species-specific colonization. These three classical pathogenic *Bordetella* species colonize the cilia of the host respiratory epithelium where they multiply and produce a variety of virulence factors, resulting in disease symptoms such as the uncontrollable coughing characteristic of pertussis. Although intracellular bacteria can be observed in infected immune-competent animals, including humans, the classical Bordetellae are thought to primarily inhabit the epithelial surface and do not disseminate throughout the body (153, 154). Despite active immunization, pertussis remains an endemic disease that has been increasing in frequency.

The respiratory Bordetellae produce a variety of virulence factors. *B. pertussis* is unique in its production of pertussis toxin; *B. bronchiseptica* and *B. parapertussis* have the *ptx* genes, but mutation has silenced them (155). Pertussis toxin is a classical A-B type toxin that enters the host cell and covalently modifies G protein-inhibitory subunits by ADP ribosylation, rendering the host protein locked in a dysregulated state, and altering normal cellular function (156). Pertussis toxin inhibits airway macrophage function, interferes with neutrophil chemotaxis, and suppresses the adaptive immune response to infection (157). All three classic *Bordetella* species secrete an adenylate cyclase that also has pore-forming lytic activity (158, 159). This adenylate cyclase is a member of the RTX family of toxins and targets a variety of cell types, using host calmodulin in a cytoplasmic reaction that generates supraphysiologic concentrations of cAMP. Bordetellae produce a dermonecrotic toxin that is a trans-glutaminase that acts on host Rho GTPases (160); its role during infection is not known. The classic *Bordetella* species have a type III secretion system that causes necrotic death of cultured cells that is independent of caspase activity (161). *B. bronchiseptica* type III secretion-system mutants exhibit attenuation in the lower respiratory tract of infected rodents (162) and in swine (163). The type III system BteA-secreted effector has cytotoxic activity on cultured cells, although the mechanism has yet to be determined (153, 154).

Bordetellae produce a peptidoglycan-derived disaccharide tetrapeptide that is usually recycled in bacteria during cell wall synthesis and turnover (164). In hamster tracheal explant cultures this peptidoglycan fragment, tracheal cytotoxin, synergizes with lipopolysaccharide to induce an inflammatory response and increase host nitric oxide synthase activity. As a result, host cells hyperproduce NO, ultimately resulting in the death and extrusion of ciliated cells from the epithelium (164). *N. gonorrhoeae* produces similar peptidoglycan fragments that cause inflammation and tissue destruction in fallopian tube cultures (165). *N. meningitidis* appears to recycle its peptidoglycan fragments more efficiently and those that are not recycled have only modest proinflammatory activity. Interestingly, the marine bacterium *Vibrio fischeri* produces a peptidoglycan-derived tracheal cytotoxin that causes remodeling of its squid host's ciliated tissues for the effective establishment of a host-symbiont relationship (166). *Bordetella, N. gonorrhoeae*, and *V. fischeri* interact with host ciliated epithelia and may have evolved a mechanism to modify those tissues to promote their colonization or to access nutrients released from destroyed host cells.

Most known *Bordetella* virulence genes are transcriptionally activated by a two-component system consisting of the BvgS membrane-sensor kinase and the DNA-binding response regulator BvgA (167–169). Bvg$^+$ phase bacteria express BvgA-activated genes, such as those encoding pertussis toxin, adenylate cyclase, the type III secretion system, adhesins, and certain metabolic functions. If the NaCl in *Bordetella* growth medium is replaced with sulfate salts, there is a reversible loss of virulence-associated traits; upon return to standard medium lacking sulfate, production of the virulence traits resumes (170). This phenomenon is known as antigenic (or phenotypic) modulation. Millimolar concentrations of nicotinic acid in the growth medium cause similar reversible modulation (171), as does growth at low temperature (e.g., 25°C) (170). The transcriptional response to modulating stimuli requires the BvgAS sensory system (172). *bvg* mutants and modulated wild-type cells are considered Bvg$^-$ phase organisms and do not express BvgAS-activated genes, but instead express genes (such as motility genes) that are often denoted as Bvg-repressed. A *B. bronchiseptica bvgS* mutant can survive remarkably better in nutrient-free phosphate-buffered saline compared with the Bvg$^+$ parent strain and provision of a modulating

concentration of sulfate rescues the survival of the wild-type strain (173). The metabolism underlying these experimental observations is not known. Bacteria grown at intermediate modulator concentrations are characterized by expression of the BipA surface protein and adhesin genes but not genes encoding secreted virulence factors (154, 169). It is unknown whether the sulfate, 25°C growth, or nicotinic acid modulators are natural *in vivo* signals detected by BvgS. The Bvg$^+$ phase is required for infection in experimental hosts, but no *in vivo* role for the Bvg$^-$ phase has been demonstrated (169, 174, 175). However, it was shown that *B. bronchiseptica* can encounter modulating conditions while living in the mouse respiratory tract (176).

Colonization

Pertussis is an acute disease during which the organisms are effectively cleared from the host, although coughing can persist for months (153), whereas infection of nonhuman mammals by *B. bronchiseptica* can be chronic (177). A study of the contribution of the mouse nasal microbiota to *Bordetella* colonization demonstrated that at a very low inoculum, *B. bronchiseptica* could readily displace the resident flora and establish nasal colonization. For *B. pertussis*, even very high inocula could not compete with the nasal biota, but pretreatment of mice with antibiotics significantly reduced their infectious dose to levels similar to those of *B. bronchiseptica* (177). Pertussis patients occasionally present with primary or secondary pneumonia, but *B. pertussis* is generally localized to the upper respiratory tract (153). Histologic studies examining respiratory tissues from patients who died from pertussis, specimens from necropsied *Bordetella*-infected experimental animals, and from infected tissue explants, show that *Bordetella* cells adhere preferentially to cilia (178–180). These observations suggest that all ciliated respiratory epithelial surfaces are targets of *Bordetella* colonization. The ciliary receptors to which *Bordetella* cells adhere are not known.

B. bronchiseptica infection of dog tracheal explants demonstrated remarkably rapid induction of ciliostasis upon bacterial contact (181). Microarray studies of cultured BEAS-2B human bronchial epithelial cells infected with *B. pertussis* showed upregulated proinflammatory cytokine responses and increased transcription of the *MUC2* and *MUC5AC* genes (182). In addition, *B. pertussis* was shown to bind these respiratory mucins, which is a trait that may be highly relevant to airway colonization. Ciliostasis, along with mucus accumulation and damage to the epithelium, may interfere with effective clearing of the organisms and may provide nutrients that promote continued bacterial growth.

B. pertussis experimental studies have used a number of animal models, with the mouse being the most often used, although it does not display symptoms typical of human whooping cough (153, 154). Recently, a baboon model of pertussis was developed that recapitulates many of the symptoms of this coughing disease (183). Experimental models of natural *B. bronchiseptica* respiratory infection include the mouse, rat, rabbit, guinea pig, and swine. *Bordetella* surface molecules implicated in colonization include lipopolysaccharide, type I fimbriae, the autotransporter protein pertactin, and filamentous hemagglutinin (FHA) (153, 154). Currently used acellular pertussis vaccines consist of detoxified pertussis toxin, FHA, pertactin and fimbrial proteins, depending on the manufacturer (153). On rabbit tracheal explants, *B. bronchiseptica* mutants lacking pertactin, fimbriae, or FHA exhibited decreased binding to cilia, suggesting that all three collectively function as adhesins (179). The fimbrial subunits can undergo genetic-phase variation, and fimbriae were shown to be important in colonizing the rodent lower respiratory tract (184, 185). A two-partner secretion system processes and transports FHA to the bacterial surface where it is tethered, although a significant amount is also released (186). FHA has heparin-binding and carbohydrate-recognition domains needed for *Bordetella* adherence to cells in culture. *B. bronchiseptica* FHA-deficient mutants colonized only the nose but not the trachea or lower respiratory tract of rats (187) and swine (188). In mouse lower respiratory tract infections, *B. bronchiseptica* mutants lacking FHA elicited a stronger inflammatory response than did an isogenic FHA$^+$ strain, suggesting that this adhesin may have a role in influencing the inflammatory response that may promote persistence of the infection (187, 189). *In vitro* biofilm formation by *B. bronchiseptica* was reported to involve FHA and fimbriae, and both Bvg$^+$ and Bvg intermediate-phase cells optimally produced biofilms (190). The Bps exopolysaccharide (191) and extracellular DNA (192) have been identified as components of *Bordetella* biofilms. Importantly, the Bps polysaccharide was required for *B. bronchiseptica* biofilm formation in the nasal cavities of mice (191). *B. pertussis* also formed nasal biofilms in infected mice and *bps* mutants poorly colonized the nose and trachea compared with biofilm-proficient strains (193). The Bps polysaccharide has a role in complement resistance (194), as do the BrkA and Vag8 complement-resistance proteins (195, 196). Flagellar motility is required for the early stages of *in vitro B. bronchiseptica* biofilm formation, which is also characterized by expression of Bvg$^-$ phase genes (197). As the biofilm structure matures, the Bvg-regulated gene-expression program shifts to a Bvg$^+$ response, with decreased flagellar gene transcription. The stringent response signal ppGpp was shown to be involved in biofilm formation *in vitro* (198) and cyclic-di-GMP and a putative diguanylate cyclase were reported to affect *B. bronchiseptica* motility and biofilm formation (199). It has long been thought that *Bordetella* species produced a capsule, which has been infrequently observed and not well characterized (200). Capsule production has not been reported to be involved in biofilm development or maintenance.

Metabolism

The classic *Bordetella* subspecies are obligate aerobes, and based on growth experiments and analysis of the *Bordetella* genomic databases, they do not have a glycolytic pathway for utilization of carbohydrates (155, 201). Their main mode of metabolism is the oxidative deamination of amino acids. These Bordetellae have a functional TCA cycle as well as an apparently intact glyoxalate bypass, consistent with the reported metabolism of acetate (202). *B. pertussis* cultured in standard minimal media having imbalanced nitrogen:carbon levels was shown to excrete ammonia as well as acetate, acetoacetate, and β-hydroxybutyrate, and intracellular poly-β-hydroxybutyrate storage granules were observed (203). Glutamate was identified in early research as a critical nutrient for *B. pertussis* (204) and can serve as the main amino acid for growth. *B. parapertussis* and *B. bronchiseptica* have genes encoding a urease, but in *B. pertussis* several of these genes are mutated (155). The urease has no apparent role in virulence (205) and the genes are repressed by BvgAS (206). Of the three classic subspecies, only *B. bronchiseptica* has an intact set of genes encoding components of a periplasmic nitrate reductase. No other putative denitrification gene candidates are present in the genome databases of the three classic subspecies.

Cysteine or cystine are preferred sulfur sources for the Bordetellae, and glutathione can also be used. For *B. pertussis*, neither methionine nor inorganic sulfate can serve as a substitute for cysteine (207, 208). Concordantly, *B. pertussis* has mutations in the *cys* genes involved in sulfate transport and cysteine biosynthesis from sulfate (120, 201). *B. parapertussis* and *B. bronchiseptica* have intact *cys* genes, but not a *cysC* homolog encoding an APS kinase. Similar to *N. meningitidis* that also lacks a *cysC*, *Bordetella* spp. have a CysN that is a predicted bifunctional sulfate adenylyltransferase–adenylylsulfate kinase protein (CysNC), as described for *B. cenocepacia* (132), that demonstrates high similarity with *N. meningitidis* CysN (53% identity; $9 e^{-148}$). *In vitro*, sulfate is a known inducer of the *Bordetella*-modulation response, abrogating expression of Bvg-activated genes (170). It was demonstrated that in *B. pertussis* stationary-phase cultures, metabolism of cysteine in the medium resulted in production and excretion of sulfate and pyruvate (209). The accumulated sulfate resulted in modulation of the bacteria, with decreased production of pertussis toxin. In the nasopharynx, one main host source of sulfur for *B. pertussis* is likely cysteine. The Bordetellae produce no known mucinases. That *B. pertussis* has lost the capacity to synthesize cysteine indicates it has evolved to utilize the cysteine reliably available in the host, obviating the need to take up sulfate for cysteine synthesis. In the host, *B. parapertussis* and *B. bronchiseptica* likely use cysteine as the preferred sulfur source, but may also utilize sulfate and perhaps related sulfates. For sulfur acquisition, all three subspecies have genes predicted to allow the use of taurine (120, 155), which is found in airway surface fluids (Table 2).

Genomic-sequence analyses predict the Bordetellae can synthesize their required vitamins and cofactors including heme, thiamine, flavins, biotin, and folate. The *Bordetella* also have genes predicted to be involved in magnesium and zinc uptake (155, 201). The classical Bordetellae require NAD precursors such as nicotinic acid or nicotinamide and lack obvious *nadA* and *nadB* genes encoding NAD-biosynthesis enzymes, indicating the absence of a *de novo* NAD-biosynthesis pathway (155, 201, 210). *In vitro*, millimolar concentrations of nicotinic acid cause the Bvg-modulation response, but similar concentrations of nicotinamide have no modulating effect. That Bordetellae and pathogens such as *H. influenzae* lack *de novo* NAD-biosynthesis pathways, yet still colonize the respiratory tract, implies that NAD precursors are in sufficient abundance at this site to support microbial growth. Host sources of phosphorus in the upper respiratory tract likely include nucleic acids, phospholipids, phosphocholine, and inorganic phosphate. The classic *Bordetella* strains have a gene locus encoding putative phosphate transport permease proteins (*pstSCAB*) as well as a polyphosphate kinase and exopolyphosphatase. These strains also have a gene predicted to encode a low-affinity inorganic phosphate transporter.

Iron

To obtain iron, *B. pertussis, B. parapertussis,* and *B. bronchiseptica* produce and utilize the alcaligin siderophore (211, 212, 213), and *B. pertussis* and *B. bronchiseptica* are also capable of using the xenosiderophores enterobactin (58), ferrichrome, and desferrioxamine B (214). *B. bronchiseptica* was also reported to use aerobactin, ferrichrysin, ferricrocin, ferrirubin, protochelin, schizokinen, vicibactin, and pyoverdins (215). *B. pertussis, B. parapertussis*, and *B. bronchiseptica* each have at least 13 genes predicted to encode TonB-dependent outer-membrane receptors for iron uptake. The systems for four iron sources have been characterized to date: alcaligin, enterobactin, heme, and catecholamines.

Alcaligin is a dihydroxamate siderophore produced by the classical Bordetellae for iron uptake *via* the TonB-dependent outer-membrane receptor, FauA (216). Upon iron depletion and Fur derepression, the alcaligin-system genes are positively regulated by the AraC family transcriptional regulator, AlcR, in a process that requires activation by alcaligin (217). Interestingly, the alcaligin gene cluster of the opportunistic pathogen *Bordetella holmesii* was identified on a pathogenicity island apparently acquired by horizontal transfer, a process hypothesized to have contributed to its emergence and adaptation to the human host (218). Bordetellae do not produce the catechol siderophores enterobactin, salmochelin, and corynebactin, but can use them to obtain iron *via* the BfeA outer-membrane receptor (58, 219). After Fur derepression, transcription of *bfeA* is induced by the BfeR AraC-type regulator that requires enterobactin or other catechols

for its function (211, 219). The classical *Bordetella* subspecies use heme, presumably both as an iron and a heme source. Hemin is transported to the periplasm by the BhuR TonB-dependent receptor and subsequent uptake occurs via the ABC family BhuTUV transporter; BhuS bears similarity to known heme degrading or utilization proteins (146). The *Bordetella bhuRSTUV* heme genes are positively regulated by a surface-signaling mechanism that involves the adjacent Fur- and iron-repressible *hurI* and *hurR* genes. Under iron-depleted conditions, *hurIR-bhuRSTUV* are derepressed, resulting in basal levels of protein production (220, 221). When extracellular hemin binds the BhuR receptor, the receptor-occupancy signal initiates a regulatory cascade. The periplasmic N-terminal extension of BhuR transmits the signal to the cytoplasmic membrane-bound HurR anti-sigma factor with its bound HurI sigma factor (221). This activates the HurP protease, resulting in HurR cleavage, release of HurI, and dramatically increased *bhuRSTUV* transcription and Bhu protein production (222).

Infection studies have demonstrated the importance of *Bordetella* iron acquisition systems to growth in a host. A *B. pertussis tonB* mutant was attenuated in mouse infections (223) and a *B. bronchiseptica* alcaligin-biosynthesis mutant exhibited significantly reduced respiratory tract colonization in neonatal swine (224). Studies using a *B. pertussis* alcaligin-receptor mutant showed that alcaligin utilization was critical for colonization in a mouse model of respiratory infection (225) and the BfeA enterobactin receptor was important for *B. pertussis* growth in the early stages of mouse infection (226). In contrast, a *B. pertussis bhuR* heme-receptor mutant showed an *in vivo* growth defect only in the later stages of infection, suggesting that persistence in the host depends on heme utilization (227). During late infection, with increased inflammation and tissue damage, heme may become available to *Bordetella* cells on the epithelial surface. The ability of the heme-receptor mutant to produce and use its alcaligin siderophore could not compensate for its growth defect during late infection, suggesting that alcaligin may not effectively scavenge iron once extensive tissue damage by *B. pertussis* has occurred. *In vivo* gene-expression studies demonstrated that *B. pertussis* increases expression of the alcaligin, enterobactin, and heme utilization genes during mouse respiratory infection (226). Since these iron-uptake systems are positively regulated in a manner requiring the presence of the cognate iron source, these results indicate these iron sources are present in the host during infection. Furthermore, the temporal patterns of gene expression were correlated with the infection stage during which the iron systems have their greatest impact on *in vivo* fitness: alcaligin gene expression was high throughout infection, the enterobactin-receptor gene was most highly expressed during early infection, and heme-receptor gene transcription was low during initial infection and significantly increased after the peak of infection (226). These mutant and gene-expression studies confirmed that *Bordetella* can discriminate between iron sources in the changing host environment and this ability contributes to its success as a pathogen.

The iron-acquisition systems used by *B. pertussis* during human infection are unknown. However, a study of human sera from uninfected controls and *B. pertussis* culture-positive patients showed antiserum responses to the receptors for alcaligin, enterobactin, and heme (226). This indicates that *B. pertussis* produces these proteins during natural human infection of the upper respiratory tract. These results also underscore the concept that *B. pertussis* is iron-starved *in vivo*, and implies that the cognate iron sources that induce expression of the alcaligin, enterobactin, and heme-system genes are available for consumption and perceived by *B. pertussis* in the human host environment.

Transcriptional-profiling studies of *B. pertussis* identified genes that were up-regulated in response to iron starvation (228). In addition to identifying alcaligin, heme, and enterobactin-system genes, increased transcription of the type III secretion-system genes was observed as well as increased type III secretion-protein production. Genes encoding the previously unknown *fbpABC* periplasmic-binding protein-dependent transporter were also identified. Importantly, the Fbp ferric iron-uptake system was not specific for a single iron source but was required for utilization of inorganic ferric iron and ferric complexes of alcaligin, enterobactin, ferrichrome, and desferrioxamine B in both *B. pertussis* and *B. bronchiseptica*. Since the prototypic iron-uptake system often has a dedicated periplasmic and cytoplasmic membrane-transport apparatus, the finding that the Fbp system was needed for utilization of four structurally distinct siderophores is novel. The *B. pertussis* cDNA microarray study (228) also identified *ftrA*, encoding a cytoplasmic membrane permease similar to the *Saccharomyces* Ftr1 (230) and *E. coli* EfeU (231) proteins required for ferrous iron utilization. *Brucella abortus* has a FtrABCD ferrous iron-utilization system that is very similar to the *Bordetella* system (232). In *B. pertussis* and *B. bronchiseptica*, *ftrABCD* are needed for ferrous iron utilization during growth at pH conditions of <7.0. Siderophores bind ferric iron, but pH values less than 7.0 favor ferrous iron stability, and siderophores like alcaligin lose their effectiveness. In the mammalian respiratory tract, the pH values range from ~5.5 to 7.9 (233, 234), suggesting that iron acquisition by pathogens in the airway requires the flexibility to utilize not only a variety of iron sources, but both oxidized and reduced forms of iron.

Host catecholamine neuroendocrine hormones such as epinephrine, norepinephrine, and dopamine have overall structural similarity to enterobactin and can remove iron from transferrin and lactoferrin, making it available to bacterial pathogens (65–69). *B. bronchiseptica* can obtain iron bound to transferrin or lactoferrin by a mechanism that uses these catecholamines (and L-DOPA) and any of three TonB-dependent outer membrane receptors, BfrA, BfrD, and BfrE (70). This iron-retrieval mechanism functions in the absence of any siderophore, indicating the receptors bind ferric complexes of the catecholamines. With ferric transferrin as the only source of iron, norepinephrine can synergize with alcaligin or enterobactin and significantly stimulate *B. bronchiseptica* growth. On infecting a host, *B. bronchiseptica* may not yet produce useful amounts of alcaligin and may rely on xeno-siderophores or catecholamine hormones for iron uptake from lactoferrin or transferrin. That *Bordetella* cells can use host catecholamines to obtain iron from lactoferrin and transferrin in the absence of a siderophore suggests that these organisms have evolved to exploit host stress hormones for nutritional gain.

CONCLUDING REMARKS

N. meningitidis is thought to have evolved from nonpathogenic *Neisseria* mucosal-commensal species and itself grows as a commensal most of the time. Despite its paucity of toxins and other obvious virulence factors, *N. meningitidis* cells that colonize the nasopharynx can enter the epithelial cells and gain access to the submucosa where they can disseminate, leading to catastrophic septicemia and meningitis. *B. pertussis, B. parapertussis*, and *B. bronchiseptica* likely evolved from *Achromobacter* species that live in the external environment. Bordetellae such as *Bordetella petrii* (235) and *Bordetella* strain 10d (236) are also inhabitants of the outside world, living in soil and river sediments. In contrast to *N. meningitidis*, the classical *Bordetella* species evolved to produce a variety of secreted toxins that modify or damage host cells. Unlike *N. meningitidis*, these pathogenic *Bordetella* species remain localized to the upper respiratory tract of immune-competent hosts. Despite these differences, both *N. meningitidis* and *B. pertussis* have evolved in their hosts to become obligate human pathogens. They are sensitive to desiccation, are similarly transmitted between hosts, and share metabolic traits such as an obligately aerobic respiration, and the ability to deaminate amino acids for use as carbon sources. Microorganisms that live in the nasopharynx have a close relationship with mucus, and infecting pathogens must penetrate the mucus layers to colonize respiratory epithelial cells. Flagellar motility is not necessary for transit through the mucus blanket since *N. meningitidis* and *B. pertussis* do not produce flagella. These bacteria may be able to interact uniquely with airway mucin networks, and effectively surf the mucus layers to reach the epithelium. For *N. meningitidis*, its twitching motility may have a role in this process.

Components of mucus are sources of nutrition for *N. meningitidis* and the classical *Bordetella* species. Whiteley and colleagues' review of the host as a growth medium (237) noted Louis Pasteur's early model of how the host, in effect, serves as a culture vessel for microbial pathogens (238). Edward D. Garber was a plant geneticist who also studied fungal and bacterial plant pathogens. His studies led him to further crystallize the fundamental concept that microbes that are successful are those that can obtain required nutrients from the host (239, 240). A host niche that has those nutrients and also lacks inhibitory factors will be effectively colonized. Recent advances in genetic and analytical technologies have provided critical knowledge of host-niche environments. Determination of the contributions of resident airway microbiota to pathogen colonization and nutrient acquisition would be especially informative. It is also important to know what specific mucosal metabolites are taken up by pathogens and metabolized in different airway microenvironments, during different stages of infection, and in healthy, chronically diseased, and stressed hosts. Identification of the molecules that bacterial pathogens assimilate while living in the respiratory tract will allow development of more biologically relevant growth media for their *in vitro* cultivation and study. These studies will provide a more comprehensive understanding of the connections between bacterial metabolism and virulence.

ACKNOWLEDGMENTS

I thank Timothy Brickman and Vaiva Vezys for helpful discussions and critical review of sections of this chapter. Research in my laboratory has been supported by Public Health Service Grant AI-31088 from the National Institute of Allergy and Infectious Diseases.

CITATION

Armstrong SK. 2015. Bacterial metabolism in the host environment: pathogen growth and nutrient assimilation in the mammalian upper respiratory tract. Microbiol Spectrum 3(3):MBP-0007-2014

REFERENCES

1. **Fahy JV, Dickey BF.** 2010. Airway mucus function and dysfunction. *N Engl J Med* **363:**2233–2247.
2. **Sahin-Yilmaz A, Naclerio RM.** 2011. Anatomy and physiology of the upper airway. *Proc Am Thorac Soc* **8:**31–39.
3. **Holt PG, Strickland DH, Wikstrom ME, Jahnsen FL.** 2008. Regulation of immunological homeostasis in the respiratory tract. *Nat Rev Immunol* **8:**142–152.
4. **Cauley LS, Lefrançois L.** 2013. Guarding the perimeter: protection of the mucosa by tissue-resident memory T cells. *Mucosal Immunol* **6:**14–23.
5. **Kim DY, Sato A, Fukuyama S, Sagara H, Nagatake T, Kong IG, Goda K, Nochi T, Kunisawa J, Sato S, Yokota Y, Lee CH, Kiyono H.** 2011. The airway antigen sampling system: respiratory M cells as an alternative gateway for inhaled antigens. *J Immunol* **186:**4253–4262.

6. **Serikov VB, Choi H, Chmiel KJ, Wu R, Widdicombe JH.** 2004. Activation of extracellular regulated kinases is required for the increase in airway epithelial permeability during leukocyte transmigration. *Am J Respir Cell Mol Biol* **30:**261–270.
7. **Persson CG, Erjefalt JS, Greiff L, Andersson M, Erjefält I, Godfrey RW, Korsgren M, Linden M, Sundler F, Svensson C.** 1998. Plasma-derived proteins in airway defence, disease and repair of epithelial injury. *Eur Respir J* **11:**958–970.
8. **Knowles MR, Boucher RC.** 2002. Mucus clearance as a primary innate defense mechanism for mammalian airways. *J Clin Invest* **109:**571–577.
9. **Rubin BK.** 2002. Physiology of airway mucus clearance. *Respir Care* **47:**761–768.
10. **Evans CM, Koo JS.** 2009. Airway mucus: the good, the bad, the sticky. *Pharmacol Ther* **121:**332–348.
11. **Ali MS, Pearson JP.** 2007. Upper airway mucin gene expression: a review. *Laryngoscope* **117:**932–938.
12. **Davis CW, Dickey BF.** 2008. Regulated airway goblet cell mucin secretion. *Annu Rev Physiol* **70:**487–512.
13. **Cone RA.** 2009. Barrier properties of mucus. *Adv Drug Deliv Rev* **61:**75–85.
14. **Button B, Cai LH, Ehre C, Kesimer M, Hill DB, Sheehan JK, Boucher RC, Rubinstein M.** 2012. A periciliary brush promotes the lung health by separating the mucus layer from airway epithelia. *Science* **337:**937–941.
15. **Frank DN, Feazel LM, Bessesen MT, Price CS, Janoff EN, Pace NR.** 2010. The human nasal microbiota and *Staphylococcus aureus* carriage. *PLoS ONE* **5:**e10598. doi:10.1371/journal.pone.0010598.
16. **Lemon KP, Klepac-Ceraj V, Schiffer HK, Brodie EL, Lynch SV, Kolter R.** 2010. Comparative analyses of the bacterial microbiota of the human nostril and oropharynx. *mBio* **1**(3):e00129-10. doi:10.1128/mBio.00129-10.
17. **Pettigrew MM, Laufer AS, Gent JF, Kong Y, Fennie KP, Metlay JP.** 2012. Upper respiratory tract microbial communities, acute otitis media pathogens, and antibiotic use in healthy and sick children. *Appl Environ Microbiol* **78:**6262–6270.
18. **Hilty M, Burke C, Pedro H, Cardenas P, Bush A, Bossley C, Davies J, Ervine A, Poulter L, Pachter L, Moffatt MF, Cookson WOC.** 2010. Disordered microbial communities in asthmatic airways. *PLoS ONE* **5:**e8578. doi:10.1371/journal.pone.0008578.

19. **Charlson ES, Bittinger K, Haas AR, Fitzgerald AS, Frank I, Yadav A, Bushman FD, Collman RG.** 2011. Topographical continuity of bacterial populations in the healthy human respiratory tract. *Am J Respir Crit Care Med* **184:**957–963.

20. **Huang YJ, Nelson CE, Brodie EL, DeSantis TZ, Baek MS, Liu J, Woyke T, Allgaier M, Bristow J, Wiener-Kronish JP, Sutherland ER, King TS, Icitovic N, Martin RJ, Calhoun WJ, Castro M, Denlinger LC, Dimango E, Kraft M, Peters SP, Wasserman SI, Wechsler ME, Boushey HA, Lynch SV; National Heart, Lung, and Blood Institute's Asthma Clinical Research Network.** 2011. Airway microbiota and bronchial hyperresponsiveness in patients with suboptimally controlled asthma. *J Allergy Clin Immunol* **127:**372–381.

21. **Sze MA, Dimitriu PA, Hayashi S, Elliott WM, McDonough JE, Gosselink JV, Cooper J, Sin DD, Mohn WW, Hogg JC.** 2012. The lung tissue microbiome in chronic obstructive pulmonary disease. *Am J Respir Crit Care Med* **185:**1073–1080.

22. **Yan M, Pamp SJ, Fukuyama J, Hwang PH, Cho DY, Holmes S, Relman DA.** 2013. Nasal microenvironments and interspecific interactions influence nasal microbiota complexity and *S. aureus* carriage. *Cell Host Microbe* **14:**631–640.

23. **Thjötta T, Avery OT.** 1921. Studies on bacterial nutrition: II. Growth accessory substances in the cultivation of hemophilic bacilli. *J Exp Med* **34:**97–114.

24. **Thjötta T, Avery OT.** 1921. Studies on bacterial nutrition: III. Plant tissue, as a source of growth accessory substances, in the cultivation of *Bacillus influenzae. J Exp Med* **34:**455–466.

25. **Evans NM, Smith DD.** 1972. The effect of the medium and source of growth factors on the satellitism test for *Haemophilus* species. *J Med Microbiol* **5:**509–514.

26. **Dinges MM, Orwin PM, Schlievert PM.** 2000. Exotoxins of *Staphylococcus aureus. Clin Microbiol Rev* **13:**16–34.

27. **Vandenesch F, Lina G, Henry T.** 2012. *Staphylococcus aureus* hemolysins, bi-component leukocidins, and cytolytic peptides: a redundant arsenal of membrane-damaging virulence factors? *Front Cell Infect Microbiol* **2:**12. doi:10.3389/fcimb.2012.00012.

28. **Cole JN, Barnett TC, Nizet V, Walker MJ.** 2011. Molecular insight into invasive group A streptococcal disease. *Nat Rev Microbiol* **9:** 724–736.

29. **Spaulding AR, Salgado-Pabón W, Kohler PL, Horswill AR, Leung DY, Schlievert PM.** 2013. Staphylococcal and streptococcal superantigen exotoxins. *Clin Microbiol Rev* **26:**422–447.

30. **Cole AM, Dewan P, Ganz T.** 1999. Innate antimicrobial activity of nasal secretions. *Infect Immun* **67:**3267–3275.

31. **Ganz T.** 2003. Defensins: antimicrobial peptides of innate immunity. *Nat Rev Immunol* **3:** 710–720.

32. **Ganz T.** 2004. Antimicrobial polypeptides. *J Leukoc Biol* **75:**34–38.

33. **Hong JH, Lee WC, Hsu YM, Liang HJ, Wan CH, Chien CL, Lin CY.** 2014. Characterization of the biochemical effects of naphthalene on the mouse respiratory system using NMR-based metabolomics. *J Appl Toxicol* **34:**1379–1388.

34. **Grahl N, Puttikamonkul S, Macdonald JM, Gamcsik MP, Ngo LY, Hohl TM, Cramer RA.** 2011. In vivo hypoxia and a fungal alcohol dehydrogenase influence the pathogenesis of invasive pulmonary aspergillosis. *PLoS Pathog* **7:**e1002145. doi:10.1371/journal.ppat.1002145.

35. **Hu JZ, Rommereim DN, Minard KR, Woodstock A, Harrer BJ, Wind RA, Phipps RP, Sime PJ.** 2008. Metabolomics in lung inflammation: a high-resolution (1) h NMR study of mice exposed to silica dust. *Toxicol Mech Methods* **18:**385–398.

36. **Evans CR, Karnovsky A, Kovach MA, Standiford TJ, Burant CF, Stringer KA.** 2014. Untargeted LC-MS metabolomics of bronchoalveolar lavage fluid differentiates acute respiratory distress syndrome from health. *J Proteome Res* **13:**640–649.

37. **Rai RK, Azim A, Sinha N, Sahoo JN, Singh C, Ahmed A, Saigal S, Baronia AK, Gupta D, Gurjar M, Poddar B, Singh RK.** 2013. Metabolic profiling in human lung injuries by high-resolution nuclear magnetic resonance spectroscopy of bronchoalveolar lavage fluid (BALF). *Metabolomics* **9:**667–676.

38. **Schock BC, Koostra J, Kwack S, Hackman RM, van der Vliet A, Cross CE.** 2004. Ascorbic acid in nasal and tracheobronchial airway lining fluids. *Free Radic Biol Med* **37:**1393–1401.

39. **Gould NS, Min E, Gauthier S, Martin RJ, Day BJ.** 2011. Lung glutathione adaptive responses to cigarette smoke exposure. *Respir Res* **12:**133.

40. **Lankford CE.** 1973. Bacterial assimilation of iron. *Crit Rev Microbiol* **2:**273–331.

41. **Neilands JB.** 1981. Iron absorption and transport in microorganisms. *Annu Rev Nutr* **1:**27–46.

42. **Ratledge C, Dover LG.** 2000. Iron metabolism in pathogenic bacteria. *Annu Rev Microbiol* **54:**881–941.

43. **Hentze MW, Muckenthaler MU, Andrews NC.** 2004. Balancing acts: molecular control of mammalian iron metabolism. *Cell* **117:**285–297.

44. **Andrews NC.** 2008. Forging a field: the golden age of iron biology. *Blood* **112:**219–230.

45. **Ganz T, Nemeth E.** 2012. Hepcidin and iron homeostasis. *Biochim Biophys Acta* **1823:**1434–1443.

46. **Alexander DB, Iigo M, Yamauchi K, Suzui M, Tsuda H.** 2012. Lactoferrin: an alternative view of its role in human biological fluids. *Biochem Cell Biol* **90:**279–306.

47. **Hantke K.** 1981. Regulation of ferric iron transport in *Escherichia coli* K12: isolation of a constitutive mutant. *Mol Gen Genet* **182:**288–292.

48. **Mahan MJ, Heithoff DM, Sinsheimer RL, Low DA.** 2000. Assessment of bacterial pathogenesis by analysis of gene expression in the host. *Annu Rev Genet* **34:**139–164.

49. **Murphy JR, Michel JL, Teng M.** 1978. Evidence that the regulation of diphtheria toxin production is directed at the level of transcription. *J Bacteriol* **135:**511–516.

50. **Grant C, Vasil ML.** 1986. Analysis of transcription of the exotoxin A gene of *Pseudomonas aeruginosa. J Bacteriol* **168:**1112–1119.

51. **Neilands JB.** 1995. Siderophores: structure and function of microbial iron transport compounds. *J Biol Chem* **270:**26723–26726.

52. **Budzikiewicz H, Bössenkamp A, Taraz K, Pandey A, Meyer JM.** 1997. Corynebactin, a cyclic catecholate siderophore from *Corynebacterium glutamicum* ATCC 14067 (*Brevibacterium* sp DSM 20411). *Z Naturforsch C* **52:**551–554.

53. **Zajdowicz S, Haller JC, Krafft AE, Hunsucker SW, Mant CT, Duncan MW, Hodges RS, Jones DN, Holmes RK.** 2012. Purification and structural characterization of siderophore (corynebactin) from *Corynebacterium diphtheriae*. *PLoS ONE* **7:**e34591. doi:10.1371/journal.pone.0034591.

54. **Moelling C, Oberschlacke R, Ward P, Karijolich J, Borisova K, Bjelos N, Bergeron L.** 2007. Metal-dependent repression of siderophore and biofilm formation in *Actinomyces naeslundii. FEMS Microbiol Lett* **275:**214–220.

55. **Beasley FC, Marolda CL, Cheung J, Buac S, Heinrichs DE.** 2011. *Staphylococcus aureus* transporters Hts, Sir, and Sst capture iron liberated from human transferrin by Staphyloferrin A, Staphyloferrin B, and catecholamine stress hormones, respectively, and contribute to virulence. *Infect Immun* **79:**2345–2355.

56. **Cox CD, Rinehart KL, Moore ML, Cook JC Jr.** 1981. Pyochelin: novel structure of an iron-chelating growth promoter for *Pseudomonas aeruginosa. Proc Natl Acad Sci U S A* **78:**4256–4260.

57. **Cox CD, Adams P.** 1985. Siderophore activity of pyoverdin for *Pseudomonas aeruginosa. Infect Immun* **48:**130–138.

58. **Beall B, Sanden GN.** 1995. A *Bordetella pertussis fepA* homologue required for utilization of exogenous ferric enterobactin. *Microbiology* **141**(Pt 12):3193–3205.

59. **Yancey RJ, Finkelstein RA.** 1981. Assimilation of iron by pathogenic *Neisseria* spp. *Infect Immun* **32:**600–608.

60. **Dyer DW, West EP, McKenna W, Thompson SA, Sparling PF.** 1988. A pleiotropic iron-uptake mutant of *Neisseria meningitidis* lacks a 70-kilodalton iron-regulated protein. *Infect Immun* **56:**977–983.

61. **Williams P, Morton DJ, Towner KJ, Stevenson P, Griffiths E.** 1990. Utilization of enterobactin and other exogenous iron sources by *Haemophilus influenzae, H. parainfluenzae* and *H. paraphrophilus. J Gen Microbiol* **136:**2343–2350.

62. **Genco CA, Dixon DW.** 2001. Emerging strategies in microbial haem capture. *Mol Microbiol* **39:**1–11.

63. **Hantke K.** 1987. Ferrous iron transport mutants in *Escherichia coli* K12. *FEMS Microbiol Lett* **44:**53–57.

64. **Andrews SC, Robinson AK, Rodríguez-Quiñones F.** 2003. Bacterial iron homeostasis. *FEMS Microbiol Rev* **27:**215–237.

65. **Freestone PP, Sandrini SM, Haigh RD, Lyte M.** 2008. Microbial endocrinology: how stress influences susceptibility to infection. *Trends Microbiol* **16:**55–64.

66. **Lyte M.** 2004. Microbial endocrinology and infectious disease in the 21st century. *Trends Microbiol* **12:**14–20.

67. **Burton CL, Chhabra SR, Swift S, Baldwin TJ, Withers H, Hill SJ, Williams P.** 2002. The growth response of *Escherichia coli* to neurotransmitters and related catecholamine drugs requires a functional enterobactin biosynthesis and uptake system. *Infect Immun* **70:**5913–5923.

68. **Freestone PP, Haigh RD, Williams PH, Lyte M.** 2003. Involvement of enterobactin in norepinephrine-mediated iron supply from transferrin to enterohaemorrhagic *Escherichia coli. FEMS Microbiol Lett* **222:**39–43.

69. **Sandrini SM, Shergill R, Woodward J, Muralikuttan R, Haigh RD, Lyte M, Freestone PP.** 2010. Elucidation of the mechanism by which catecholamine stress hormones liberate iron from the innate immune defense proteins transferrin and lactoferrin. *J Bacteriol* **192:**587–594.

70. **Armstrong SK, Brickman TJ, Suhadolc RJ.** 2012. Involvement of multiple distinct *Bordetella* receptor proteins in the utilization of iron liberated from transferrin by host catecholamine stress hormones. *Mol Microbiol* **84:**446–462.

71. **Zeng X, Xu F, Lin J.** 2009. Molecular, antigenic, and functional characteristics of ferric enterobactin receptor CfrA in *Campylobacter jejuni*. *Infect Immun* **77:**5437–5448.

72. **Barnes PJ.** 2001. Neurogenic inflammation in the airways. *Respir Physiol* **125:**145–154.

73. **Cavallotti C, Bruzzone P, Tonnarini G, Cavallotti D.** 2004. Distribution of catecholaminergic neurotransmitters and related receptors in human bronchus-associated lymphoid tissue. *Respiration* **71:**635–640.

74. **Lucero MT, Squires A.** 1998. Catecholamine concentrations in rat nasal mucus are modulated by trigeminal stimulation of the nasal cavity. *Brain Res* **807:**234–236.

75. **Flower DR, North AC, Sansom CE.** 2000. The lipocalin protein family: structural and sequence overview. *Biochim Biophys Acta* **1482:**9–24.

76. **Kjeldsen L, Cowland JB, Borregaard N.** 2000. Human neutrophil gelatinase-associated lipocalin and homologous proteins in rat and mouse. *Biochim Biophys Acta* **1482:**272–283.

77. **Dittrich AM, Meyer HA, Hamelmann E.** 2013. The role of lipocalins in airway disease. *Clin Exp Allergy* **43:**503–511.

78. **Fluckinger M, Haas H, Merschak P, Glasgow BJ, Redl B.** 2004. Human tear lipocalin exhibits antimicrobial activity by scavenging microbial siderophores. *Antimicrob Agents Chemother* **48:**3367–3372.

79. **Goetz DH, Holmes MA, Borregaard N, Bluhm ME, Raymond KN, Strong RK.** 2002. The neutrophil lipocalin NGAL is a bacteriostatic agent that interferes with siderophore-mediated iron acquisition. *Mol Cell* **10:**1033–1043.

80. **Miethke M, Skerra A.** 2010. Neutrophil gelatinase-associated lipocalin expresses antimicrobial activity by interfering with L-norepinephrine-mediated bacterial iron acquisition. *Antimicrob Agents Chemother* **54:**1580–1589.

81. **Bao G, Clifton M, Hoette TM, Mori K, Deng SX, Qiu A, Viltard M, Williams D, Paragas N, Leete T, Kulkarni R, Li X, Lee B, Kalandadze A, Ratner AJ, Pizarro JC, Schmidt-Ott KM, Landry DW, Raymond KN, Strong RK, Barasch J.** 2010. Iron traffics in circulation bound to a siderocalin (Ngal)-catechol complex. *Nat Chem Biol* **6:**602–609.

82. **Devireddy LR, Hart DO, Goetz DH, Green MR.** 2010. A mammalian siderophore synthesized by an enzyme with a bacterial homolog involved in enterobactin production. *Cell* **141:**1006–1017.

83. **Liu Z, Ciocea A, Devireddy L.** 2014. Endogenous siderophore 2,5-dihydroxybenzoic acid deficiency promotes anemia and splenic iron overload in mice. *Mol Cell Biol* **34:**2533–2546.

84. **Nelson AL, Barasch JM, Bunte RM, Weiser JN.** 2005. Bacterial colonization of nasal mucosa induces expression of siderocalin, an iron-sequestering component of innate immunity. *Cell Microbiol* **7:**1404–1417.

85. **Chan YR, Liu JS, Pociask DA, Zheng M, Mietzner TA, Berger T, Mak TW, Clifton MC, Strong RK, Ray P, Kolls JK.** 2009. Lipocalin 2 Is required for pulmonary host defense against *Klebsiella* infection. *J Immunol* **182:**4947–4956.

86. **Stein DC.** 2006. The Genus *Neisseria*, p 602–647. *In* Dworkin M, Falkow S, Rosenberg E, Schleifer KH, Stackebrandt E (eds), The Prokaryotes: **vol. 5:** *Proteobacteria: alpha and beta subclasses.* Springer-Verlag, Berlin/Heidelberg.

87. **Stephens DS.** 2007. Conquering the meningococcus. *FEMS Microbiol Rev* **31:**3–14.

88. **Hill DJ, Griffiths NJ, Borodina E, Virji M.** 2010. Cellular and molecular biology of *Neisseria meningitidis* colonization and invasive disease. *Clin Sci* **118:**547–564.

89. **Trivedi K, Tang CM, Exley RM.** 2011. Mechanisms of meningococcal colonisation. *Trends Microbiol* **19:**456–463.

90. **Cornelissen CN, Sparling PF.** 2004. *Neisseria*, p 256–272. *In* Crosa JH, Mey AR, Payne SM (eds), *Iron transport in bacteria.* ASM Press, Washington, DC.

91. **Hobbs MM, Seiler A, Achtman M, Cannon JG.** 1994. Microevolution within a clonal population of pathogenic bacteria: recombination, gene duplication and horizontal genetic exchange in the opa gene family of *Neisseria meningitidis*. *Mol Microbiol* **12:**171–180.

92. **Jerse AE, Cohen MS, Drown PM, Whicker LG, Isbey SF, Seifert HS, Cannon JG.** 1994. Multiple gonococcal opacity proteins are expressed during experimental urethral infection in the male. *J Exp Med* **179:**911–920.

93. **Rayner CF, Dewar A, Moxon ER, Virji M, Wilson R.** 1995. The effect of variations in the expression of pili on the interaction of *Neisseria meningitidis* with human nasopharyngeal epithelium. *J Infect Dis* **171:**113–121.

94. **Chamot-Rooke J, Mikaty G, Malosse C, Soyer M, Dumont A, Gault J, Imhaus AF, Martin P, Trellet M, Clary G, Chafey P, Camoin L, Nilges M, Nassif X, Duménil G.** 2011. Posttranslational modification of pili upon cell contact triggers *N. meningitidis* dissemination. *Science* **331:**778–782.

95. **Jennings MP, Hood DW, Peak IR, Virgi M, Moxon ER.** 1995. Molecular analysis of a locus for the biosynthesis and phase-variable expression of the lacto-N-neotetraose terminal lipopolysaccharide structure in *Neisseria meningitidis*. *Mol Microbiol* **18:**729–740.

96. Lewis LA, Gipson M, Hartman K, Ownbey T, Vaughn J, Dyer DW. 1999. Phase variation of HpuAB and HmbR, two distinct haemoglobin receptors of *Neisseria meningitidis* DNM2. *Mol Microbiol* **32**:977–989.

97. Pujol C, Eugène E, de Saint Martin L, Nassif X. 1997. Interaction of *Neisseria meningitidis* with a polarized monolayer of epithelial cells. *Infect Immun* **65**:4836–4842.

98. Exley RM, Goodwin L, Mowe E, Shaw J, Smith H, Read RC, Tang CM. 2005. *Neisseria meningitidis* lactate permease is required for nasopharyngeal colonization. *Infect Immun* **73**:5762–5766.

99. Virji M, Makepeace K, Ferguson DJ, Achtman M, Moxon ER. 1993. Meningococcal Opa and Opc proteins: their role in colonization and invasion of human epithelial and endothelial cells. *Mol Microbiol* **10**:499–510.

100. Rudel T, Scheurerpflug I, Meyer TF. 1995. *Neisseria* PilC protein identified as type-4 pilus tip-located adhesin. *Nature* **373**:357–359.

101. Deghmane AE, Giorgini D, Larribe M, Alonso JM, Taha MK. 2002. Down-regulation of pili and capsule of *Neisseria meningitidis* upon contact with epithelial cells is mediated by CrgA regulatory protein. *Mol Microbiol* **43**:1555–1564.

102. Ieva R, Alaimo C, Delany I, Spohn G, Rappuoli R, Scarlato V. 2005. CrgA is an inducible LysR-type regulator of *Neisseria meningitidis*, acting both as a repressor and as an activator of gene transcription. *J Bacteriol* **187**:3421–3430.

103. Jamet A, Rousseau C, Monfort JB, Frapy E, Nassif X, Martin P. 2009. A two-component system is required for colonization of host cells by meningococcus. *Microbiology* **155**:2288–2295.

104. Johnson CR, Newcombe J, Thorne S, Borde HA, Eales-Reynolds LJ, Gorringe AR, Funnell SG, McFadden JJ. 2001. Generation and characterization of a PhoP homologue mutant of *Neisseria meningitidis*. *Mol Microbiol* **39**:1345–1355.

105. Hey A, Li MS, Hudson MJ, Langford PR, Kroll JS. 2013. Transcriptional profiling of *Neisseria meningitidis* interacting with human epithelial cells in a long-term *in vitro* colonization model. *Infect Immun* **81**:4149–4159.

106. Madico G, Welsch JA, Lewis LA, McNaughton A, Perlman DH, Costello CE, Ngampasutadol J, Vogel U, Granoff DM, Ram S. 2006. The meningococcal vaccine candidate GNA1870 binds the complement regulatory protein factor H and enhances serum resistance. *J Immunol* **177**:501–510.

107. Comanducci M, Bambini S, Brunelli B, Adu-Bobie J, Aricò B, Capecchi B, Giuliani MM, Masignani V, Santini L, Savino S, Granoff DM, Caugant DA, Pizza M, Rappuoli R, Mora M. 2002. NadA, a novel vaccine candidate of *Neisseria meningitidis*. *J Exp Med* **195**:1445–1454.

108. Oldfield NJ, Bland SJ, Taraktsoglou M, Dos Ramos FJ, Robinson K, Wooldridge KG, Ala'Aldeen DA. 2007. T-cell stimulating protein A (TspA) of *Neisseria meningitidis* is required for optimal adhesion to human cells. *Cell Microbiol* **9**:463–478.

109. Sjölinder H, Eriksson J, Maudsdotter L, Aro H, Jonsson AB. 2008. Meningococcal outer membrane protein NhhA is essential for colonization and disease by preventing phagocytosis and complement attack. *Infect Immun* **76**:5412–5420.

110. Talà A, Progida C, De Stefano M, Cogli L, Spinosa MR, Bucci C, Alifano P. 2008. The HrpB-HrpA two-partner secretion system is essential for intracellular survival of *Neisseria meningitidis*. *Cell Microbiol* **10**:2461–2482.

111. Plaut AG, Gilbert JV, Artenstein MS, Capra JD. 1975. *Neisseria gonorrhoeae* and *Neisseria meningitidis*: extracellular enzyme cleaves human immunoglobulin A. *Science* **190**:1103–1105.

112. Ayala P, Lin L, Hopper S, Fukuda M, So M. 1998. Infection of epithelial cells by pathogenic Neisseriae reduces the levels of multiple lysosomal constituents. *Infect Immun* **66**:5001–5007.

113. Neil RB, Apicella MA. 2009. Clinical and laboratory evidence for *Neisseria meningitidis* biofilms. *Future Microbiol* **4**:555–563.

114. Sim RJ, Harrison MM, Moxon ER, Tang CM. 2000. Underestimation of meningococci in tonsillar tissue by nasopharyngeal swabbing. *Lancet* **356**:1653–1654.

115. Yi K, Rasmussen AW, Gudlavalleti SK, Stephens DS, Stojiljkovic I. 2004. Biofilm formation by *Neisseria meningitidis*. *Infect Immun* **72**:6132–6138.

116. Lappann M, Vogel U. 2010. Biofilm formation by the human pathogen *Neisseria meningitidis*. *Med Microbiol Immunol* **199**:173–183.

117. Leighton MP, Kelly DJ, Williamson MP, Shaw JG. 2001. An NMR and enzyme study of the carbon metabolism of *Neisseria meningitidis*. *Microbiology* **147**:1473–1482.

118. Exley RM, Shaw J, Mowe E, Sun YH, West NP, Williamson M, Botto M, Smith H, Tang CM. 2005. Available carbon source influences the resistance of *Neisseria meningitidis* against complement. *J Exp Med* **201**:1637–1645.

119. Erwin AL, Gotschlich EC. 1993. Oxidation of D-lactate and L-lactate by *Neisseria meningitidis*: purification and cloning of meningococcal D-lactate dehydrogenase. *J Bacteriol* **175**:6382–6391.

120. **KEGG: Kyoto Encyclopedia of Genes and Genomes.** (http://www.genome.jp/kegg/).

121. **Pagliarulo C, Salvatore P, De Vitis LR, Colicchio R, Monaco C, Tredici M, Talà A, Bardaro M, Lavitola A, Bruni CB, Alifano P.** 2004. Regulation and differential expression of *gdhA* encoding NADP-specific glutamate dehydrogenase in *Neisseria meningitidis* clinical isolates. *Mol Microbiol* **51:**1757–1772.

122. **Talà A, Monaco C, Nagorska K, Exley RM, Corbett A, Zychlinsky A, Alifano P, Tang CM.** 2011. Glutamate utilization promotes meningococcal survival *in vivo* through avoidance of the neutrophil oxidative burst. *Mol Microbiol* **81:**1330–1342.

123. **Knapp JS, Clark VL.** 1984. Anaerobic growth of *Neisseria gonorrhoeae* coupled to nitrite reduction. *Infect Immun* **46:**176–181.

124. **Moir JW.** 2011. A snapshot of a pathogenic bacterium mid-evolution: *Neisseria meningitidis* is becoming a nitric oxide-tolerant aerobe. *Biochem Soc Trans* **39:**1890–1894.

125. **Norrod P, Morse SA.** 1979. Absence of superoxide dismutase in some strains of *Neisseria gonorrhoeae*. *Biochem Biophys Res Com* **90:**1287–1294.

126. **Wilks KE, Dunn KL, Farrant JL, Reddin KM, Gorringe AR, Langford PR, Kroll JS.** 1998. Periplasmic superoxide dismutase in meningococcal pathogenicity. *Infect Immun* **66:**213–217.

127. **Archibald FS, Duong MN.** 1986. Superoxide dismutase and oxygen toxicity defenses in the genus *Neisseria*. *Infect Immun* **51:**631–641.

128. **Moore TD, Sparling PF.** 1995. Isolation and identification of a glutathione peroxidase homolog gene, *gpxA*, present in *Neisseria meningitidis* but absent in *Neisseria gonorrhoeae*. *Infect Immun* **63:**1603–1607.

129. **Tettelin H, Saunders NJ, Heidelberg J, Jeffries AC, Nelson KE, Eisen JA, Ketchum KA, Hood DW, Peden JF, Dodson RJ, Nelson WC, Gwinn ML, DeBoy R, Peterson JD, Hickey EK, Haft DH, Salzberg SL, White O, Fleischmann RD, Dougherty BA, Mason T, Ciecko A, Parksey DS, Blair E, Cittone H, Clark EB, Cotton MD, Utterback TR, Khouri H, Qin H, Vamathevan J, GILL J, Scarlato V, Masignani V, Pizza M, Grandi G, Sun L, Smith HO, Fraser CM, Moxon ER, Rappuoli R, Venter JC.** 2000. Complete genome sequence of *Neisseria meningitidis* serogroup B strain MC58. *Science* **287:**1809–1815.

130. **Seib KL, Tseng H-J, McEwan AG, Apicella MA, Jennings MP.** 2004. Defenses against oxidative stress in *Neisseria gonorrhoeae* and *Neisseria meningitidis*: distinctive systems for different lifestyles. *J Infect Dis* **190:**136–147.

131. **Le Faou A.** 1984. Sulphur nutrition and metabolism in various species of *Neisseria*. *Ann Microbiol (Paris)* **135B:**3–11.

132. **Iwanicka-Nowicka R, Zielak A, Cook AM, Thomas MS, Hryniewicz MM.** 2007. Regulation of sulfur assimilation pathways in *Burkholderia cenocepacia*: identification of transcription factors CysB and SsuR and their role in control of target genes. *J Bacteriol* **189:**1675–1688.

133. **Tinsley CR, Manjula BN, Gotschlich EC.** 1993. Purification and characterization of polyphosphate kinase from *Neisseria meningitidis*. *Infect Immun* **61:**3703–3710.

134. **Zhang Q, Li Y, Tang CM.** 2010. The role of the exopolyphosphatase PPX in avoidance by *Neisseria meningitidis* of complement-mediated killing. *J Biol Chem* **285:**34259–34268.

135. **Stork M, Bos MP, Jongerius I, de Kok N, Schilders I, Weynants VE, Poolman JT, Tommassen J.** 2010. An outer membrane receptor of *Neisseria meningitidis* involved in zinc acquisition with vaccine potential. *PLoS Pathog* **6:**e1000969. doi:10.1371/journal.ppat.1000969.

136. **Perkins-Balding D, Ratliff-Griffin M, Stojiljkovic I.** 2004. Iron transport systems in *Neisseria meningitidis*. *Microbiol Mol Biol Rev* **68:**154–171.

137. **Schryvers AB, Morris LJ.** 1988. Identification and characterization of the human lactoferrin-binding protein from *Neisseria meningitidis*. *Infect Immun* **56:**1144–1149.

138. **Tsai J, Dyer DW, Sparling PF.** 1988. Loss of transferrin receptor activity in *Neisseria meningitidis* correlates with inability to use transferrin as an iron source. *Infect Immun* **56:**3132–3138.

139. **Noinaj N, Easley NC, Oke M, Mizuno N, Gumbart J, Boura E, Steere AN, Zak O, Aisen P, Tajkhorshid E, Evans RW, Gorringe AR, Mason AB, Steven AC, Buchanan SK.** 2012. Structural basis for iron piracy by pathogenic *Neisseria*. *Nature* **483:**53–58.

140. **Noinaj N, Cornelissen CN, Buchanan SK.** 2013. Structural insight into the lactoferrin receptors from pathogenic *Neisseria*. *J Struct Biol* **184:**83–92.

141. **Postle K, Larsen RA.** 2007. TonB-dependent energy transduction between outer and cytoplasmic membranes. *Biometals* **20:**453–465.

142. **Adhikari P, Berish SA, Nowalk AJ, Veraldi KL, Morse SA, Mietzner TA.** 1996. The *fbpABC* locus of *Neisseria gonorrhoeae* functions in the periplasm-to-cytosol transport of iron. *J Bacteriol* **178:**2145–2149.

143. **Morgenthau A, Beddek A, Schryvers AB.** 2014. The negatively charged regions of lactoferrin binding protein B, an adaptation against anti-microbial peptides. *PLoS ONE* **9:**e86243. doi:10.1371/journal.pone.0086243.

144. **Cornelissen CN, Kelley M, Hobbs MM, Anderson JE, Cannon JG, Cohen MS, Sparling PF.** 1998. The transferrin receptor expressed by gonococcal strain FA1090 is required for the experimental infection of human male volunteers. *Mol Microbiol* **27:**611–616.

145. **Anderson JE, Hobbs MM, Biswas GD, Sparling PF.** 2003. Opposing selective forces for expression of the gonococcal lactoferrin receptor. *Mol Microbiol* **48:**1325–1337.

146. **Vanderpool CK, Armstrong SK.** 2001. The *Bordetella bhu* locus is required for heme iron utilization. *J Bacteriol* **183:**4278–4287.

147. **Thompson EA, Feavers IM, Maiden MC.** 2003. Antigenic diversity of meningococcal enterobactin receptor FetA, a vaccine component. *Microbiology* **149:**1849–1858.

148. **Marsh JW, O'Leary MM, Shutt KA, Harrison LH.** 2007. Deletion of *fetA* gene sequences in serogroup B and C *Neisseria meningitidis* isolates. *J Clin Microbiol* **45:**1333–1335.

149. **Yu C, Genco CA.** 2012. Fur-mediated global regulatory circuits in pathogenic *Neisseria* species. *J Bacteriol* **194:**6372–6381.

150. **Massé E, Gottesman S.** 2002. A small RNA regulates the expression of genes involved in iron metabolism in *Escherichia coli. Proc Natl Acad Sci U S A* **99:**4620–4625.

151. **Mellin JR, Goswami S, Grogan S, Tjaden B, Genco CA.** 2007. A novel fur-and iron-regulated small RNA, NrrF, is required for indirect fur-mediated regulation of the *sdhA* and *sdhC* genes in *Neisseria meningitidis. J Bacteriol* **189:**3686–3694.

152. **Jordan PW, Saunders NJ.** 2009. Host iron binding proteins acting as niche indicators for *Neisseria meningitidis. PLoS ONE* **4:**e5198. doi:10.1371/journal.pone.0005198.

153. **Mattoo S, Cherry JD.** 2005. Molecular pathogenesis, epidemiology, and clinical manifestations of respiratory infections due to *Bordetella pertussis* and other *Bordetella* subspecies. *Clin Microbiol Rev* **18:**326–382.

154. **Melvin JA, Scheller EV, Miller JF, Cotter PA.** 2014. *Bordetella pertussis* pathogenesis: current and future challenges. *Nat Rev Microbiol* **12:**274–288.

155. **Parkhill J, Sebaihia M, Preston A, Murphy LD, Thomson N, Harris DE, Holden MT, Churcher CM, Bentley SD, Mungall KL, Cerdeño-Tárraga AM, Temple L, James K, Harris B, Quail MA, Achtman M, Atkin R, Baker S, Basham D, Bason N, Cherevach I, Chillingworth T, Collins M, Cronin A, Davis P, Doggett J, Feltwell T, Goble A, Hamlin N, Hauser H, Holroyd S, Jagels K, Leather S,** Moule S, Norberczak H, O'Neil S, Ormond D, Price C, Rabbinowitsch E, Rutter S, Sanders M, Saunders D, Seeger K, Sharp S, Simmonds M, Skelton J, Squares R, Squares S, Stevens K, Unwin L, Whitehead S, Barrell BG, Maskell DJ. 2003. Comparative analysis of the genome sequences of *Bordetella pertussis, Bordetella parapertussis* and *Bordetella bronchiseptica. Nat Genet* **35:**32–40.

156. **Katada T, Tamura M, Ui M.** 1983. The A protomer of islet-activating protein, pertussis toxin, as an active peptide catalyzing ADP-ribosylation of a membrane protein. *Arch Biochem Biophys* **224:**290–298.

157. **Carbonetti NH.** 2010. Pertussis toxin and adenylate cyclase toxin: key virulence factors of *Bordetella pertussis* and cell biology tools. *Future Microbiol* **5:**455–469.

158. **Hewlett E, Wolff J.** 1976. Soluble adenylate cyclase from the culture medium of *Bordetella pertussis:* purification and characterization. *J Bacteriol* **127:**890–898.

159. **Eby JC, Gray MC, Warfel JM, Paddock CD, Jones TF, Day SR, Bowden J, Poulter MD, Donato GM, Merkel TJ, Hewlett EL.** 2013. Quantification of the adenylate cyclase toxin of *Bordetella pertussis in vitro* and during respiratory infection. *Infect Immun* **81:**1390–1398.

160. **Horiguchi Y, Senda T, Sugimoto N, Katahira J, Matsuda M.** 1995. Bordetella bronchiseptica dermonecrotizing toxin stimulates assembly of actin stress fibers and focal adhesions by modifying the small GTP-binding protein rho. *J Cell Sci* **108:**3243–3251.

161. **Stockbauer KE, Foreman-Wykert AK, Miller JF.** 2003. *Bordetella* type III secretion induces caspase 1-independent necrosis. *Cell Microbiol* **5:**123–132.

162. **Yuk MH, Harvill ET, Miller JF.** 1998. The BvgAS virulence control system regulates type III secretion in *Bordetella bronchiseptica. Mol Microbiol* **28:**945–959.

163. **Nicholson TL, Brockmeier SL, Loving CL, Register KB, Kehrli ME Jr, Shore SM.** 2014. The *Bordetella bronchiseptica* type III secretion system Is required for persistence and disease severity but not transmission in swine. *Infect Immun* **82:**1092–1103.

164. **Flak TA, Goldman WE.** 1999. Signalling and cellular specificity of airway nitric oxide production in pertussis. *Cell Microbiol* **1:**51–60.

165. **Woodhams KL, Chan JM, Lenz JD, Hackett KT, Dillard JP.** 2013. Peptidoglycan fragment release from *Neisseria meningitidis. Infect Immun* **81:**3490–3498.

166. **Koropatnick TA, Engle JT, Apicella MA, Stabb EV, Goldman WE, McFall-Ngai MJ.**

2004. Microbial factor-mediated development in a host-bacterial mutualism. *Science* **306:**1186–1188.

167. **Weiss AA, Hewlett EL, Myers GA, Falkow S.** 1983. Tn5-induced mutations affecting virulence factors of *Bordetella pertussis*. *Infect Immun* **42:**33–41.

168. **Aricó B, Miller JF, Roy C, Stibitz S, Monack D, Falkow S, Gross R, Rappuoli R.** 1989. Sequences required for expression of *Bordetella pertussis* virulence factors share homology with prokaryotic signal transduction proteins. *Proc Nat Acad Sci U S A* **86:**6671–6675.

169. **Cotter PA, Jones AM.** 2003. Phosphorelay control of virulence gene expression in *Bordetella. Trends Microbiol* **11:**367–373.

170. **Lacey BW.** 1960. Antigenic modulation of *Bordetella pertussis. J Hyg* **58:**57–93.

171. **Pusztai Z, Joó I.** 1967. Influence of nicotinic acid on the antigenic structure of *Bordetella pertussis. Ann Immunol Hung* **10:**63–67.

172. **Miller JF, Johnson SA, Black WJ, Beattie DT, Mekalanos JJ, Falkow S.** 1992. Constitutive sensory transduction mutations in the *Bordetella pertussis bvgS* gene. *J Bacteriol* **174:**970–979.

173. **Cotter PA, Miller JF.** 1994. BvgAS-mediated signal transduction: analysis of phase-locked regulatory mutants of *Bordetella bronchiseptica* in a rabbit model. *Infect Immun* **62:**3381–3390.

174. **Martinez de Tejada G, Cotter PA, Heininger U, Camilli A, Akerley BJ, Mekalanos JJ, Miller JF.** 1998. Neither the Bvg- phase nor the *vrg6* locus of *Bordetella pertussis* is required for respiratory infection in mice. *Infect Immun* **66:**2762–2768.

175. **Nicholson TL, Brockmeier SL, Loving CL, Register KB, Kehrli ME Jr, Stibitz SE, Shore SM.** 2012. Phenotypic modulation of the virulent Bvg phase is not required for pathogenesis and transmission of *Bordetella bronchiseptica* in swine. *Infect Immun* **80:**1025–1036.

176. **Mason E, Henderson MW, Scheller EV, Byrd MS, Cotter PA.** 2013. Evidence for phenotypic bistability resulting from transcriptional interference of *bvgAS* in *Bordetella bronchiseptica. Mol Microbiol* **90:**716–733.

177. **Weyrich LS, Feaga HA, Park J, Muse SJ, Safi CY, Rolin OY, Young SE, Harvill ET.** 2014. Resident microbiota affect *Bordetella pertussis* infectious dose and host specificity. *J Infect Dis* **209:**913–921.

178. **Mallory FB, Hornor AA.** 1912. Pertussis: the histological lesion in the respiratory tract. *J Med Res* **27:**115–1243.

179. **Edwards JA, Groathouse NA, Boitano S.** 2005. *Bordetella bronchiseptica* adherence to cilia is mediated by multiple adhesin factors and blocked by surfactant protein A. *Infect Immun* **73:**3618–3626.

180. **Paddock CD, Sanden GN, Cherry JD, Gal AA, Langston C, Tatti KM, Wu KH, Goldsmith CS, Greer PW, Montague JL, Eliason MT, Holman RC, Guarner J, Shieh WJ, Zaki SR.** 2008. Pathology and pathogenesis of fatal *Bordetella pertussis* infection in infants. *Clin Infect Dis* **47:**328–338.

181. **Anderton TL, Maskell DJ, Preston A.** 2004. Ciliostasis is a key early event during colonization of canine tracheal tissue by *Bordetella bronchiseptica. Microbiology* **150:**2843–2855.

182. **Belcher CE, Drenkow J, Kehoe B, Gingeras TR, McNamara N, Lemjabbar H, Basbaum C, Relman DA.** 2000. The transcriptional responses of respiratory epithelial cells to *Bordetella pertussis* reveal host defensive and pathogen counter-defensive strategies. *Proc Natl Acad Sci U S A* **97:**13847–13852.

183. **Warfel JM, Beren J, Kelly VK, Lee G, Merkel TJ.** 2012. Nonhuman primate model of pertussis. *Infect Immun* **80:**1530–1536.

184. **Geuijen CA, Willems RJ, Bongaerts M, Top J, Gielen H, Mooi FR.** 1997. Role of the *Bordetella pertussis* minor fimbrial subunit, FimD, in colonization of the mouse respiratory tract. *Infect Immun* **65:**4222–4228.

185. **Mattoo S, Miller JF, Cotter PA.** 2000. Role of *Bordetella bronchiseptica* fimbriae in tracheal colonization and development of a humoral immune response. *Infect Immun* **68:**2024–2033.

186. **Mazar J, Cotter PA.** 2006. Topology and maturation of filamentous haemagglutinin suggest a new model for two-partner secretion. *Mol Microbiol* **62:**641–654.

187. **Julio SM, Inatsuka CS, Mazar J, Dieterich C, Relman DA, Cotter PA.** 2009. Natural-host animal models indicate functional interchangeability between the filamentous haemagglutinins of *Bordetella pertussis* and *Bordetella bronchiseptica* and reveal a role for the mature C-terminal domain, but not the RGD motif, during infection. *Mol Microbiol* **71:**1574–1590.

188. **Nicholson TL, Brockmeier SL, Loving CL.** 2009. Contribution of *Bordetella bronchiseptica* filamentous hemagglutinin and pertactin to respiratory disease in swine. *Infect Immun* **77:**2136–2146.

189. **Henderson MW, Inatsuka CS, Sheets AJ, Williams CL, Benaron DJ, Donato GM, Gray MC, Hewlett EL, Cotter PA.** 2012. Contribution of *Bordetella* filamentous hemagglutinin and adenylate cyclase toxin to suppression and evasion of interleukin-17-mediated inflammation. *Infect Immun* **80:**2061–2075.

190. **Irie Y, Mattoo S, Yuk MH.** 2004. The Bvg virulence control system regulates biofilm formation in *Bordetella bronchiseptica*. *J Bacteriol* **186:**5692–5698.

191. **Sloan GP, Love CF, Sukumar N, Mishra M, Deora R.** 2007. The *Bordetella* Bps polysaccharide is critical for biofilm development in the mouse respiratory tract. *J Bacteriol* **189:**8270–8276.

192. **Conover MS, Mishra M, Deora R.** 2011. Extracellular DNA is essential for maintaining *Bordetella* biofilm integrity on abiotic surfaces and in the upper respiratory tract of mice. *PLoS ONE* **6:**e16861. doi:10.1371/journal.pone.0016861.

193. **Conover MS, Sloan GP, Love CF, Sukumar N, Deora R.** 2010. The Bps polysaccharide of *Bordetella pertussis* promotes colonization and biofilm formation in the nose by functioning as an adhesin. *Mol Microbiol* **77:**1439–1455.

194. **Ganguly T, Johnson JB, Kock ND, Parks GD, Deora R.** 2014. The *Bordetella pertussis* Bps polysaccharide enhances lung colonization by conferring protection from complement-mediated killing. *Cell Microbiol* **16:**1105–1118.

195. **Fernandez RC, Weiss AA.** 1994. Cloning and sequencing of a *Bordetella pertussis* serum resistance locus. *Infect Immun* **62:**4727–4738.

196. **Marr N, Shah NR, Lee R, Kim EJ, Fernandez RC.** 2011. *Bordetella pertussis* autotransporter Vag8 binds human C1 esterase inhibitor and confers serum resistance. *PLoS ONE* **6:**e20585. doi:10.1371/journal.pone.0020585.

197. **Nicholson TL, Conover MS, Deora R.** 2012. Transcriptome profiling reveals stage-specific production and requirement of flagella during biofilm development in *Bordetella bronchiseptica*. *PLoS ONE* **7:**e49166. doi:10.1371/journal.pone.0049166.

198. **Sugisaki K, Hanawa T, Yonezawa H, Osaki T, Fukutomi T, Kawakami H, Yamamoto T, Kamiya S.** 2013. Role of (p)ppGpp in biofilm formation and expression of filamentous structures in *Bordetella pertussis*. *Microbiology* **159:**1379–1389.

199. **Sisti F, Ha DG, O'Toole GA, Hozbor D, Fernández J.** 2013. Cyclic-di-GMP signaling regulates motility and biofilm formation in *Bordetella bronchiseptica*. *Microbiology* **159:**869–879.

200. **Neo Y, Li R, Howe J, Hoo R, Pant A, Ho S, Alonso S.** 2010. Evidence for an intact polysaccharide capsule in *Bordetella pertussis*. *Microbes Infect* **12:**238–245.

201. **Armstrong SK, Gross R.** 2007. Primary metabolism and physiology of *Bordetella* species, p 167–190. *In* Locht C (ed), *Bordetella: Molecular Biology*. Horizon Scientific Press, Norwich, UK.

202. **Bundeally AE, Rao SS.** 1966. Stimulation of growth of *Bordetella pertussis* by acetate. *Indian J Exp Biol* **4:**23–24.

203. **Thalen M, van den IJssel J, Jiskoot W, Zomer B, Roholl P, de Gooijer C, Beuvery C, Tramper J.** 1999. Rational medium design for *Bordetella pertussis*: basic metabolism. *J Biotechnol* **75:**147–159.

204. **Jebb WH, Tomlinson AH.** 1951. The catabolic activity of washed suspensions of *Haemophilus pertussis*. *J Gen Microbiol* **5:**951–965.

205. **Monack DM, Falkow S.** 1993. Cloning of *Bordetella bronchiseptica* urease genes and analysis of colonization by a urease-negative mutant strain in a guinea-pig model. *Mol Microbiol* **10:**545–553.

206. **McMillan DJ, Shojaei M, Chhatwal GS, Guzman CA, Walker MJ.** 1996. Molecular analysis of the *bvg*-repressed urease of *Bordetella bronchiseptica*. *Microb Pathog* **21:**379–394.

207. **Jebb WH, Tomlinson AH.** 1957. The minimal amino acid requirements of *Haemophilus pertussis*. *J Gen Microbiol* **17:**59–63.

208. **Fukumi H, Sayama E, Tomizawa JI, Uchida T.** 1953. Nutritional requirements and respiratory pattern of *pertussis-parapertusis-bronchisepticus* group of microorganisms. *Jpn J Med Sci Biol* **6:**587–601.

209. **Bogdan JA, Nazario-Larrieu J, Sarwar J, Alexander P, Blake MS.** 2001. *Bordetella pertussis* autoregulates pertussis toxin production through the metabolism of cysteine. *Infect Immun* **69:**6823–6830.

210. **Hornibrook JW.** 1940. Nicotinic acid as a growth factor for *Haemophilus pertussis*. *Proc Soc Exp Biol Med U S A* **45:**598–599.

211. **Brickman TJ, Anderson MT, Armstrong SK.** 2007. *Bordetella* iron transport and virulence. *Biometals* **20:**303–322.

212. **Moore CH, Foster LA, Gerbig DG Jr, Dyer DW, Gibson BW.** 1995. Identification of alcaligin as the siderophore produced by *Bordetella pertussis* and *B. bronchiseptica*. *J Bacteriol* **177:**1116–1118.

213. **Brickman TJ, Hansel JG, Miller MJ, Armstrong SK.** 1996. Purification, spectroscopic analysis and biological activity of the macrocyclic dihydroxamate siderophore alcaligin produced by *Bordetella pertussis* and *Bordetella bronchiseptica*. *Biometals* **9:**191–203.

214. **Beall B, Hoenes T.** 1997. An iron-regulated outer-membrane protein specific to *Bordetella bronchiseptica* and homologous to ferric siderophore receptors. *Microbiology* **143**(Pt 1):135–145.

215. **Pradel E, Locht C.** 2001. Expression of the putative siderophore receptor gene *bfrZ* is

controlled by the extracytoplasmic-function sigma factor BupI in *Bordetella bronchiseptica*. *J Bacteriol* **183**:2910–2917.

216. **Brickman TJ, Armstrong SK.** 1999. Essential role of the iron-regulated outer membrane receptor FauA in alcaligin siderophore-mediated iron uptake in *Bordetella* species. *J Bacteriol* **181**:5958–5966.

217. **Brickman TJ, Kang HY, Armstrong SK.** 2001. Transcriptional activation of *Bordetella* alcaligin siderophore genes requires the AlcR regulator with alcaligin as inducer. *J Bacteriol* **183**:483–489.

218. **Diavatopoulos DA, Cummings CA, van der Heide HG, van Gent M, Liew S, Relman DA, Mooi FR.** 2006. Characterization of a highly conserved island in the otherwise divergent *Bordetella holmesii* and *Bordetella pertussis* genomes. *J Bacteriol* **188**:8385–8394.

219. **Anderson MT, Armstrong SK.** 2004. The BfeR regulator mediates enterobactin-inducible expression of *Bordetella* enterobactin utilization genes. *J Bacteriol* **186**:7302–7311.

220. **Vanderpool CK, Armstrong SK.** 2003. Heme-responsive transcriptional activation of *Bordetella bhu* genes. *J Bacteriol* **185**:909–917.

221. **Vanderpool CK, Armstrong SK.** 2004. Integration of environmental signals controls expression of *Bordetella* heme utilization genes. *J Bacteriol* **186**:938–948.

222. **King-Lyons ND, Smith KF, Connell TD.** 2007. Expression of *hurP*, a gene encoding a prospective site 2 protease, is essential for heme-dependent induction of *bhuR* in *Bordetella bronchiseptica*. *J Bacteriol* **189**:6266–6275.

223. **Pradel E, Guiso N, Menozzi FD, Locht C.** 2000. *Bordetella pertussis* TonB, a Bvg-independent virulence determinant. *Infect Immun* **68**:1919–1927.

224. **Register KB, Ducey TF, Brockmeier SL, Dyer DW.** 2001. Reduced virulence of a *Bordetella bronchiseptica* siderophore mutant in neonatal swine. *Infect Immun* **69**:2137–2143.

225. **Brickman TJ, Armstrong SK.** 2007. Impact of alcaligin siderophore utilization on *in vivo* growth of *Bordetella pertussis*. *Infect Immun* **75**:5305–5312.

226. **Brickman TJ, Hanawa T, Anderson MT, Suhadolc RJ, Armstrong SK.** 2008. Differential expression of *Bordetella pertussis* iron transport system genes during infection. *Mol Microbiol* **70**:3–14.

227. **Brickman TJ, Vanderpool CK, Armstrong SK.** 2006. Heme transport contributes to in vivo fitness of *Bordetella pertussis* during primary infection in mice. *Infect Immun* **74**:1741–1744.

228. **Brickman TJ, Cummings CA, Liew SY, Relman DA, Armstrong SK.** 2011. Transcriptional profiling of the iron starvation response in *Bordetella pertussis* provides new insights into siderophore utilization and virulence gene expression. *J Bacteriol* **193**:4798–4812.

229. **Brickman TJ, Armstrong SK.** 2012. Iron and pH-responsive FtrABCD ferrous iron utilization system of *Bordetella* species. *Mol Microbiol* **86**:580–593.

230. **Singh A, Severance S, Kaur N, Wiltsie W, Kosman DJ.** 2006. Assembly, activation, and trafficking of the Fet3p.Ftr1p high affinity iron permease complex in *Saccharomyces cerevisiae*. *J Biol Chem* **281**:13355–13364.

231. **Cao J, Woodhall MR, Alvarez J, Cartron ML, Andrews SC.** 2007. EfeUOB (YcdNOB) is a tripartite, acid-induced and CpxAR-regulated, low-pH Fe^{2+} transporter that is cryptic in *Escherichia coli* K-12 but functional in *E. coli* O157:H7. *Mol Microbiol* **65**:857–875.

232. **Elhassanny AE, Anderson ES, Menscher EA, Roop RM II.** 2013. The ferrous iron transporter FtrABCD is required for the virulence of *Brucella abortus* 2308 in mice. *Mol Microbiol* **88**:1070–1082.

233. **Fischer H, Widdicombe JH.** 2006. Mechanisms of acid and base secretion by the airway epithelium. *J Membr Biol* **211**:139–150.

234. **Ng AW, Bidani A, Herring TA.** 2004. Innate host defense of the lung: effects of lung-lining fluid pH. *Lung* **182**:297–317.

235. **von Wintzingerode F, Schattke A, Siddiqui RA, Rösick U, Göbel UB, Gross R.** 2001. *Bordetella petrii* sp *nov.*, isolated from an anaerobic bioreactor, and amended description of the genus *Bordetella*. *Int J Syst Evol Microbiol* **51**:1257–1265.

236. **Takenaka S, Asami T, Orii C, Murakami S, Aoki K.** 2002. A novel meta-cleavage dioxygenase that cleaves a carboxyl-group substituted 2-aminophenol. Purification and characterization of 4-amino-3-hydroxybenzoate 2,3-dioxygenase from *Bordetella* sp strain 10d. *Eur J Biochem* **269**:5871–5877.

237. **Brown SA, Palmer KL, Whiteley M.** 2008. Revisiting the host as a growth medium. *Nat Rev Microbiol* **6**:657–666.

238. **Pasteur L.** 1878. La theorie des germes. *Comptes Rendus. l'Academie des Sciences* **86**:1037–1043.

239. **Garber ED.** 1954. The role of nutrition in the host-parasite relationship. *Proc Natl Acad Sci U S A* **40**:1112–1118.

240. **Garber ED.** 1960. The host as a growth medium. *Ann N Y Acad Sci* **88**:1187–1194.

Saliva as the Sole Nutritional Source in the Development of Multispecies Communities in Dental Plaque

12

NICHOLAS S. JAKUBOVICS[1]

INTRODUCTION

Despite major improvements in oral hygiene over the last few decades, dental caries and periodontitis remain two of the most common diseases in industrialized countries. For example, in the most recent National Health and Nutrition Examination Survey conducted between 2009–2010 in the U.S., almost one in four children between 3–9 years living in poverty had untreated dental caries (1). In the U.K., almost 50% of children have obvious decay experience in permanent teeth by the age of 15 years (2). Periodontitis, an inflammatory condition that involves the loss of the supporting structures of teeth, is more common in older age groups. In the U.S., over one-third of adults ≥30 years have moderate to severe periodontitis (3). Both dental caries and periodontitis are caused by the growth of microbial biofilms on tooth surfaces, known as dental plaque.

The human mouth is an open environment that is exposed to microorganisms in the air and in foods and drinks that are consumed. Some of the bacteria that enter the mouth will remain only transiently as they are not adapted to thrive in this environment. It is currently estimated that around 1,000 species of bacteria are able to exist stably in the human mouth, and

[1]School of Dental Sciences, Newcastle University, UK.
Metabolism and Bacterial Pathogenesis
Edited by Tyrrell Conway and Paul Cohen
© 2015 American Society for Microbiology, Washington, DC
doi:10.1128/microbiolspec.MBP-0013-2014

that individuals typically maintain between 50–200 species from this wider pool (4, 5). Although there are relatively large inter-individual variations in the oral microbiota, approximately 15 genera are almost universally found in supragingival dental plaque (above the gumline), including *Actinomyces, Campylobacter, Capnocytophaga, Corynebacterium, Fusobacterium, Granulicatella, Neisseria, Prevotella, Streptococcus,* and *Veillonella* (5–7). Bacteria growing below the gumline in subgingival dental plaque are cut off from access to saliva and dietary nutrients and instead obtain nutrients from gingival crevicular fluid, a serum exudate. As a consequence, the subgingival microbiota tends to become enriched in anaerobic, proteolytic bacteria such as *Filifactor, Fusobacterium, Parvimonas, Porphyromonas, Prevotella, Tannerella,* and *Treponema* (8, 9).

Many of the proteolytic bacteria in subgingival dental plaque produce virulence factors that promote disease in animal models, and it is generally accepted that these organisms are involved in periodontitis in humans (10). Both dental caries and periodontitis are probably best explained by the ecological plaque hypothesis that was developed in the 1990s by Marsh and coworkers (11). This concept states that supragingival dental plaque develops naturally on tooth surfaces and reaches a state of microbial equilibrium, where many different microbial species coexist without causing harm to the host. Disease occurs when the equilibrium is perturbed by changes in the host, the microbiota, or the environment. In the case of dental caries, frequent intake of sugars imposes a selective pressure that promotes the growth of highly acidogenic species such as *Streptococcus mutans* or *Lactobacillus* species. Risk factors for periodontitis include smoking or the accumulation of plaque or calculus at the gum margins, which irritate the gingivae and lead to the formation of periodontal pockets.

According to the ecological plaque hypothesis, low levels of supragingival dental plaque are not harmful. Indeed, oral health-care relies largely on preventing dental plaque overgrowth through regular brushing and flossing. These procedures result in repetitive cycles of colonization and growth of microorganisms on the tooth, followed by removal of the maturing dental plaque. Evidence from animal studies indicates that dietary influences, such as fasting or ingesting high levels of simple sugars, have little effect on the initial accumulation of dental plaque (12). Instead, nascent dental plaque relies almost exclusively on salivary constituents for nutrition. This review will provide an overview of how bacterial consortia develop to obtain the maximum nutritional benefit from the catabolism of salivary substrates, and will explore the nutritional interactions between oral bacteria that contribute to dental plaque growth.

THE COMPOSITION OF SALIVA

Saliva is a complex biological fluid that originates from the major and minor salivary glands. Whole saliva also contains components from non-salivary origins such as gingival crevicular fluid, exfoliated epithelial cells, oral bacteria, and oral wounds. The relative contribution of each source varies between individuals and in an individual at different times. Stimulation of saliva production, for example by smell, taste, or mastication, increases flow from the parotid gland. Saliva flow is also affected by many other factors including age, hydration, physical exercise, drugs, circadian rhythms, nutrition, and systemic diseases (13). The contribution from gingival crevicular fluid is influenced by the presence and extent of periodontal disease (14). Therefore, the composition of saliva and, hence, the nutrients available to oral bacteria will vary greatly in different people and at different times. Nevertheless, there are many consistent characteristics of saliva, and a number of bacterial enzymes and scavenging systems have been identified that are critical for catabolism and uptake of salivary substrates.

On average, saliva is approximately 99% water, with the balance made up by organic and inorganic ions, peptides, proteins, and glycoproteins (Table 1). The major cations in saliva are sodium and potassium, whilst ammonium, calcium, and magnesium ions are present at lower concentrations (15). Chloride and phosphate are the most abundant anions, and nitrate, nitrite, sulfate, and thiosulfate tend to be present at sub-millimolar levels. In addition, organic anions including lactate, acetate, propionate, and formate have been detected in whole saliva, possibly as a consequence of the metabolism of dental-plaque bacteria (16). Trace metals are essential for microbial growth and are often a limiting factor within body fluids. The total concentration of trace metals can be measured directly by inductively coupled-plasma mass spectroscopy ICP-MS and it has been shown that the secretion rates of different trace metals vary dramatically day-to-day (17). Importantly, most trace metals are tightly bound to host proteins, and are not readily available to microorganisms.

The protein and glycoprotein content of human saliva has been studied in great detail. The total protein content of whole saliva is approximately 3,000 mg L^{-1}, whereas the concentration of stimulated saliva varies between approximately 500–1,000 mg L^{-1}. Free amino acids are present at low concentrations, and examples are given in Table 1. At least 400 different proteins have been detected in whole human saliva (18). Over 95% of salivary polypeptides belong to the major salivary protein families including acidic and basic proline-rich proteins, amylases, mucins MUC5B (MG1) and MUC7 (MG2), salivary agglutinin gp340, cystatin, histatins, and statherin (19). By weight, the majority of polypeptides in saliva are glycosylated, by N-linked or O-linked glycosylation. To release carbohydrate moieties from these glycoproteins, bacteria must produce a wide array of glycosidases that together can cleave many different glycosidic linkages. Free sugar levels in saliva are low, typically <100 µM glucose in non-diabetics for example (20), and therefore glycoproteins are an important source of energy for bacteria in the oral cavity.

The Acquired Enamel Pellicle

Salivary proteins and glycoproteins adsorb to the tooth surface and form a conditioning layer, termed the acquired enamel pellicle (AEP), which acts as a substrate for bacterial attachment. Some constituents of saliva adsorb to enamel more readily than others and therefore the AEP is enriched in certain components compared with whole saliva. The AEP is formed when teeth erupt during childhood and is never fully removed even by professional dental hygiene. However, the AEP is continually remodelled and replenished by salivary polypeptides (21). During the formation of AEP, proteins with affinity for calcium and phosphate, such as histatins and statherin, tend to accumulate early (22). Subsequently, histatins, but not statherin, decrease in abundance, probably due to proteolysis. Proteins that interact with other proteins accumulate in the AEP over time, including mucin MUC5B, amylase, lysozyme, and lactoperoxidase (22). Some of these enzymes remain stable within the AEP and potentially may influence bacterial colonization (21). For example, amylase in the AEP may help to degrade starch into shorter maltosaccharides that can be utilized by dental-plaque bacteria.

NUTRIENT REQUIREMENTS OF ORAL BACTERIA

Broadly, oral bacteria can be divided into saccharolytic strains, which include many of the early colonizers and cariogenic bacteria, such as *Streptococcus* spp., *Neisseria* spp., *Lactobacillus* spp., *Scardovia* spp., *Actinomyces* spp., and asaccharolytic, strictly anaerobic

TABLE 1 Concentrations of major components of saliva

Component	Approximate concentration (mg L^{-1})	Comments	Reference(s)
Total protein	500–3,000	Higher concentrations in whole saliva than parotid; concentration decreases with higher stimulated flow rates	(88–91)
Urea	120–200		(88)
Glucose	5–14	Approximately 2-fold lower in stimulated vs resting saliva	(20, 88)
Cations			
Sodium	0–800	Approximately 4-fold higher in stimulated saliva	(15, 88)
Potassium	430–1,000		(15, 88)
Ammonium	18–72		(15)
Calcium	20–110		(15, 88)
Magnesium	0.17–7.2		(15, 17, 92)
Anions			
Chloride	280–690	Unstimulated saliva	(16)
Phosphate	140–330	Unstimulated saliva	(16)
Bicarbonate	0–1,220	Concentration increases with saliva flow rate	(88, 90)
Nitrate	4.3–100	Unstimulated saliva	(16)
Nitrite	0–18	Unstimulated saliva	(16)
Sulfate	6.7–21	Unstimulated saliva	(16)
Thiocyanate	7.6–34	Unstimulated saliva	(16)
Fluoride	0.01–0.048	Increases transiently following fluoride mouthrinse/dentifrice	(88, 91)
Lactate	0–5.4	Unstimulated saliva	(16)
Acetate	1.8–48	Unstimulated saliva	(16)
Propionate	0–10	Unstimulated saliva	(16)
Formate	0–1.4	Unstimulated saliva	(16)
*Free amino acids**			
Alanine	0.57–1.9		(93, 94)
Arginine	0.59–5.1		(93, 94)
Glutamine	0–2.8		(93, 94)
Glutamate	1.6–3.8		(93, 94)
Glycine	1.1–5.8		(93, 94)
Serine	0.92–3.0		(93, 94)
Tyrosine	1.1–10		(93, 94)
Valine	0–1.4		(93, 94)
Trace metals			
Aluminium	0.0014		(17)
Copper	0.0015–0.034		(17, 92)
Iron	0.12		(92)
Manganese	0.0029–0.0045		(17, 92)
Rubidium	0.064		
Strontium	0.0022		
Zinc	0.014–0.08		(17, 92)

*Only amino acids that were detected in both of the referenced studies are shown.

proteolytic bacteria commonly associated with subgingival dental plaque, such as *Porphyromonas* spp., *Prevotella* spp., and *Parvimonas* species. It is important to note, however, that our understanding of oral bacterial metabolism comes exclusively from analysis of the bacteria that have been cultured in the laboratory, and approximately 50% of oral bacteria have not been cultured to date (4). Most oral bacteria have relatively complex nutritional requirements, and the development of chemically defined media capable of supporting growth has been instrumental in defining their nutrient requirements. There are now chemically defined media available that support the growth of at least some strains of several different oral bacterial genera including *Streptococcus*, *Actinomyces*, *Fusobacterium*, *Treponema*, and *Porphyromonas* (23–26). Some of the key nutrients required for the growth of bacteria in isolation are discussed below. It is likely that many of the key insights into oral bacterial metabolism in the future will begin from *in silico* predictions of metabolic pathways and potentials.

Carbohydrates

The production of organic acids from simple carbohydrates by microorganisms is central to the dental caries process. *Streptococcus mutans* and *Lactobacillus* spp. are able to metabolize a wide range of carbohydrates and to survive and grow at the low pH levels generated by carbohydrate metabolism. These bacteria are frequently elevated in carious lesions (27). Sucrose is thought to be particularly important as it is converted to high molecular weight exopolysaccharides by *Streptococcus* spp. and other oral microorganisms, and these polysaccharides provide protection and attachment sites for dental-plaque bacteria (28). However, the concentrations of simple sugars such as glucose are very low in saliva, and it is only once acidogenic bacteria are established within dental plaque that they are able to metabolize sufficient quantities of dietary carbohydrates to drive the progression of dental caries. The integration of highly acidogenic bacteria into dental plaque may be facilitated by the production of low levels of organic acids by moderately acidogenic early colonizers such as *Streptococcus* spp. (for example, *S. mitis*, *S. oralis*, *S. sanguinis*, and *S. gordonii*) and *Actinomyces* spp. (29). *In vitro*, these early colonizers can metabolize a relatively broad range of simple sugars (30, 31). However, simple sugars are present only at low concentrations in saliva and during the development of dental plaque it is likely that salivary glycoproteins are the most important source of carbohydrates.

The degradation of salivary glycoproteins requires a battery of enzyme activities to release the carbohydrate moieties and cleave the peptide backbone. The complete degradation of glycoproteins requires a consortium of microorganisms that each contribute different enzyme activities (32). Glycosidases produced by oral *Streptococcus* spp. and *Actinomyces* spp. include N-acetyl-β-D-glucosaminidase, β-D-galactosidase, α-L-fucosidase, α- and β-mannosidase, and sialidase (neuraminidase) (32–35). Gram-negative periodontal pathogens, such as *Porphyromonas gingivalis* and *Tannerella forsythia*, also produce a range of glycosidases including α- and β-mannosidases, sialidases, and β-hexosaminidases (36, 37). Together these enzymes can release oligosaccharides from complex host substrates such as MUC5B or IgA1 (32, 38).

The catabolism of sugars results in organic acids that, in turn, may be utilized as substrates by certain oral bacteria. For example, many strains of *Actinomyces* spp. can grow on lactate as the sole nutrient source (39). The periodontal species *Aggregatibacter actinomycetemcomitans* grows preferentially on lactate even when glucose and fructose are available (40), and lactate is a key nutrient for oral *Veillonella* spp., *Neisseria* spp., and *Haemophilus* spp. (41, 42). Some of the potential metabolic exchanges between oral bacteria are discussed in more detail below.

Proteins and Amino Acids

Subgingival dental plaque is isolated from the salivary environment of the oral cavity, and instead is fed by the serum exudate gingival crevicular fluid (GCF). In comparison to saliva, GCF is a relatively rich source of proteins; the total protein concentration in GCF is approximately 20,000–130,000 mg L^{-1} (43). Proteins provide a critical source of nutrients for subgingival dental plaque bacteria. The bacterial species most strongly associated with periodontitis, *P. gingivalis*,

Treponema denticola, and *T. forsythia*, each produce extracellular proteases that degrade the synthetic trypsin analogue N-benzoyl-DL-arginine-2-naphthylamide (BANA), and BANA has been employed (with limited success) as a test for the presence of these organisms (44). Free amino acids may also be an important source of nutrients for oral bacteria. For example, *F. nucleatum* has been shown to ferment glutamate, lysine, and histidine (45), glycine is catabolized by *T. denticola* (46), and proline is metabolized by *Neisseria* spp. (42). In supragingival dental plaque, the catabolism of arginine by oral streptococci releases ammonia, which elevates the pH in the local environment and may help to protect against dental caries (47).

In addition to energy production, amino acids are essential for bacterial growth processes. Most oral bacteria are auxotrophic for multiple amino acids and chemically defined growth media typically contain all 20 natural amino acids. The development of a chemically defined medium for oral streptococci in the 1970s provided a resource for determining amino acid requirements through 'leave-one-out' experiments (48, 49). These studies identified a number of auxotrophies that were shared between species, as well as some that were specific to individual strains of a species. For example, cysteine was universally required by streptococci, and arginine was needed by all *S. sanguinis* strains, whereas glutamate, lysine, and histidine were required by only some Mutans Streptococcus strains (48, 49). Interestingly, certain amino acids were only required by Mutans Streptococci in aerobic conditions, and particularly in the absence of sodium carbonate. More recently, it has been shown that *S. gordonii* requires arginine for aerobic growth but not in the absence of oxygen (50). In *S. sanguinis* SK36, a comprehensive knockout mutagenesis has confirmed the activity of enzymes encoded by genes in each of the amino acid biosynthetic pathways that are present in this strain (51).

Nucleic Acids

Extracellular nucleic acids, and extracellular DNA (eDNA) in particular, are a major source of phosphorous for microorganisms in some environments, such as deep-sea sediments (52). Evidence is accumulating that oral bacteria release DNA into the surrounding biofilm environment (53). In periodontal pockets, eDNA accumulates from the release of neutrophil extracellular traps (54). At the same time, many oral bacteria produce extracellular deoxyribonuclease (DNase) enzymes. It is possible that eDNA provides a nutrient store for oral bacteria and that the production of DNase enzymes allows the mobilization of this store. However, at present there is little evidence regarding the nutritional importance of eDNA for oral bacteria.

Vitamins and Cofactors

Requirements for vitamins among different streptococci are relatively uniform, and pantothenate and nicotinic acid were essential or growth stimulatory for all 18 strains tested in one study (55). Culture media for anaerobic oral bacteria typically include menadione (vitamin K) and hemin to facilitate the growth of strict anaerobes such as *P. gingivalis*. Hemin is critical for the growth of black-pigmented strains (*Porphyromonas* spp. and *Prevotella* spp.); however, requirements for menadione are strain-dependent (26). Oral treponemes are dependent on volatile fatty acids and require either isobutyric acid or 2-methylbutyric acid for growth (56). The growth of periodontal bacteria may be enhanced by hormones from the host. For example, cortisol, which is elevated in saliva under stressful conditions, has been shown to promote the growth of *P. gingivalis* (57). Steroid hormones have also been shown to promote growth of *A. actinomycetemcomitans* (58). Stress, obesity, and systemic diseases such as diabetes affect the host's hormonal status and each of these is considered a risk factor for periodontitis (59). It is possible that the concentrations of specific hormones in the oral environment may impact on the growth of subgingival bacteria, and the pathogenicity of dental plaque.

Trace Metals

Sequestration of trace metals is a common strategy for restricting the growth of invading pathogens. Iron, in particular, is tightly bound by host proteins such as transferrin and lactoferrin. The acquisition of heme iron is essential for the growth of black-pigmented anaerobes, and the mechanisms of iron acquisition from the host by *P. gingivalis* have been studied in detail (60). A combination of hemagglutinin and proteases (gingipains) are deployed for binding to host erythrocytes and releasing hemoglobin. The hemin is then imported into *P. gingivalis* cells by the iron-heme transport system encoded by the *ihtABCDE* genes, a TonB-linked receptor with an ATP-binding cassette (ABC) uptake system encoded by hemin-transport genes *htrABCD*, or a heme-uptake system encoded by the *hmu* locus (60).

In *S. mutans* the uptake of iron is mediated by an ABC transporter designated SloABC (61). In addition to iron (Fe^{3+}), SloABC also transports Mn^{2+}. Homologous ABC transporters are present in other oral streptococci including *S. gordonii* (ScaCBA), *S. parasanguinis* (FimCBA), and *S. sanguinis* (SsaACB), although the ScaCBA system in *S. gordonii* appears to be more selective for Mn^{2+} ions than Fe^{3+} (62–64). In each case, the expression of the ABC transporter is repressed by a DtxR family regulator primarily in the presence of Mn^{2+} (63, 65–67). In the case of *S. gordonii*, the ABC transporter is strongly up-regulated in serum or saliva, indicating that bioavailable Mn^{2+} is scarce in these body fluids (67). The precise intracellular functions of Mn^{2+} within streptococcal cells are not yet fully clear, but one important role for this metal is in protection against oxidative stress (63, 68). It is noteworthy that the

P. gingalivis Mn^{2+} transporter FeoB2 is up-regulated following a switch from anaerobic to microaerophilic conditions, which is also consistent with a role for Mn^{2+} in protection against oxidative stress in this organism (69).

Many different metal ions are essential co-factors for bacterial enzymes, including cobalt (Co^{2+}), copper (Cu^{2+}), and zinc (Zn^{2+}). Nickel (Ni^{2+}) is rarely required by bacteria, but in *Streptococcus salivarius* and *Actinomyces johnsonii* (previously *Actinomyces* WVA 693) Ni^{2+} ions are required for the activity of urease (70). A high-affinity Ni^{2+} transporter has been identified in *S. salivarius*, encoded by the genes *ureMQO* (71).

METABOLIC CO-OPERATION BETWEEN ORAL BACTERIA

Overall, the nutrient content of saliva is sufficient to provide all essential nutrients and lacks only a strong source of energy for strains of *Streptococcus* spp. and *Actinomyces* spp., and these bacteria grow well when an exogenous energy source such as glucose is added (72). It is estimated that humans produce 500–600 mL of saliva per day (73), and the continual replenishment of nutrients may enable growth even when the overall concentration of glucose is low. Nevertheless, there is strong evidence that interspecies interactions play a key role in growth on saliva within the oral cavity and that the derivation of energy from host salivary substrates is maximised by the co-operative degradation of macromolecules and the recycling of waste products between oral bacteria.

Food Webs

Many of the waste products of oral bacterial species are important nutrients for other oral bacteria and several synergistic interactions have been demonstrated *in vitro* (Fig. 1A). Early studies on *P. gingivalis* and *T. denticola* identified a synergy based on the exchange of isobutyric acid, produced by *P. gingivalis*,

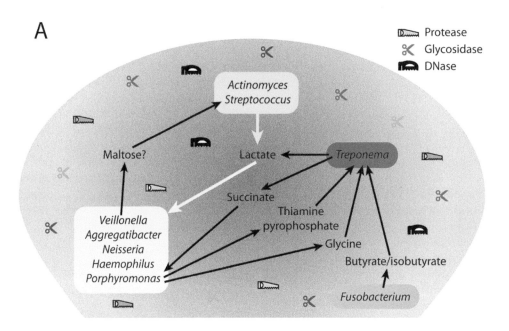

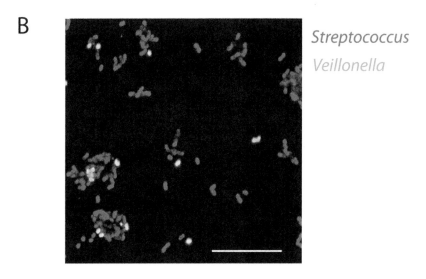

FIGURE 1 Metabolic interactions between oral bacteria A. Food web based on observed nutrient interactions between oral bacteria (see text for details). Black arrows indicate nutrients produced or consumed by a single organism; colored lines show nutrient feeding between groups of organisms. The production of maltose from *Veillonella* has not yet been confirmed. The extracellular environment contains a pool of digestive enzymes produced by different microorganisms. Extracellular enzymes capable of degrading proteins, carbohydrates, or DNA are indicated, and different types of protease or glycosidase are indicated by different colors. B. Proximity of *Veillonella* sp. cells (green, labelled with antibodies against *Veillonella* sp. PK1910) and *Streptococcus* cells (red, labelled with antibodies against receptor polysaccharide) in 8 h dental plaque formed on the surface of an enamel chip in the mouth of a volunteer. Scale bar = 20 μm. Part B is from Palmer et al. (79) with kind permission of the lead author. doi:10.1128/microbiolspec.MBP-0013-2014.f1

for *T. denticola*-secreted succinate. More recently, interactions between this pair have been further investigated using DNA microarrays to assess global gene expression in mixed-species chemostat cultures (46). Here, it was found that *P. gingivalis* also produces thiamine pyrophosphate and free glycine, which are utilized by *T. denticola*. Thiamine pyrophosphate is upregulated by *P. gingivalis* in response to interactions with *T. denticola*. By contrast, glycine is liberated from glycine-containing peptides by *P. gingivalis* proteases, and is rapidly catabolized by *T. denticola* into lactate and acetate. In turn, *P. gingivalis* is able to utilize lactate for growth, particularly under strictly anaerobic conditions (69). Short-chain fatty acids, including butyric acid, are also produced by *F. nucleatum* and it is possible that *T. denticola* benefits from the presence of *F. nucleatum* (74). Organic acids, and lactate in particular, are central to many of the metabolic interactions between oral bacteria (Fig. 1A).

Following the widespread introduction of genomic and metagenomic sequencing, attempts have been made to develop *in silico* models to predict the key metabolic interactions between bacteria in communities such as dental plaque and to link them to co-occurrence patterns. Network-interaction models have been developed to assess co-occurrence relationships within different body environments, and it is interesting to note that the genera *Streptococcus.*, *Porphyromonas*, and *Veillonella*, which have all been shown to undergo metabolic interactions with partner oral bacteria (Fig. 1A), form network 'hubs' with large numbers of co-occurrence relationships (75). On the other hand, *Streptococcus* (facultatively anaerobic, saccharolytic) has a very different metabolism from *Tannerella* (obligate anaerobe, proteolytic) and these genera display a negative co-occurrence relationship (75). Co-occurrence patterns were also identified that involve oral microorganisms that are not well-characterized metabolically. For example, TM7 phylum co-occurred with *Moreyella*

on the tongue dorsum and a positive relationship was observed between Synergistetes phylum and *Treponema* in subgingival dental plaque. One member of the Synergistetes phylum, *Fretibacterium fastidiosum*, has been cultured in the laboratory and shown to be asaccharolytic and to produce acetate, propionate, and hydrogen sulfide (76). However, the full nutrient requirements of this species are not yet understood and growth requires extracts from other oral bacteria. Recently, a representative of the TM7 phylum was grown in axenic laboratory culture on Fastidious Anaerobe Agar supplemented with 5% defibrinated horse blood (77). Whilst the metabolic potential of this strain has not yet been characterized in detail, synergistic biofilm growth was observed in co-cultures with *S. gordonii*, *Parvimonas micra*, or *F. nucleatum* compared with monocultures.

Metabolic Exchange as a Driver for Spatial Organization

Many oral bacteria bind to phylogenetically distinct cells through specific adhesin-receptor interactions in a process known as coaggregation (78). One function of coaggregation may be to bring cells into close proximity where metabolic co-operation is maximized. *S. gordonii* and *Veillonella* spp. coaggregate *in vitro*, and are often found in close proximity in early dental plaque [Fig. 1B; (79)]. *Veillonella* spp. benefit from lactate produced by streptococcal metabolism of carbohydrates. In turn, *Veillonella* produces a signal or nutrient, possibly maltose, that induces the up-regulation of *S. gordonii amyB* gene expression in neighboring cells (80). A more complex spatial relationship has been identified between *S. gordonii* and *A. actinomycetemcomitans* (81). In this case, *A. actinomycetemcomitans* obtains lactate from *S. gordonii*. However, H_2O_2 from *S. gordonii* is inhibitory towards *A. actinomycetemcomitans*. It appears that the production of the biofilm matrix-degrading glycosidase Dispersin B by *A. actinomycetemcomitans* is fine-tuned to facilitate the positioning of *A. actinomycetemcomitans* at a safe distance (≥4 μm) from *S. gordonii*, but still close enough to scavenge lactate.

Dental plaque typically contains 50–200 phylogenetically distinct microorganisms, and the potential for pairwise interactions is vast. However, dental plaque develops in a relatively reproducible spatiotemporal pattern and certain microorganisms, primarily *Streptococcus* spp., are almost always found in high numbers at the first stages of plaque development. Using a simplified model of dental biofilm development, it was found that metabolically similar bacteria with overlapping enzyme activities tend to be found in association with one another at different stages in biofilm formation and that gradients of metabolic functions extend across the different layers of the biofilm (82). At the same time, there appears to be a trade-off between the presence of similar enzymes in close proximity and the potential for synergistic interactions, which require complementary enzyme functions. Overall, the analysis is consistent with a major role for metabolism in driving the spatiotemporal organization of dental plaque.

METABOLISM OF ORAL BACTERIA *IN VIVO*

The '-omics' technologies, and particularly those applied to whole communities (metagenomics, metaproteomics, metatranscriptomics, metabolomics, and others) are already beginning to revolutionize our understanding of oral microbial metabolism in the host. Proteomics has been applied in a variety of different approaches, such as to analyse changes in the metabolic potential of *P. gingivalis* following uptake into host cells (83). In the absence of significant growth, pathways for transcription and protein production were generally elevated in internalized *P. gingivalis*, indicating that the host cytoplasm provides a relatively nutrient-rich environment. In mixed-species cultures the

S. gordonii proteome responds in a partner-specific manner (84). Thus, with *F. nucleatum*, the pathway for *S. gordonii* acetate production was decreased, whereas in interactions with *P. gingivalis*, the pathway for conversion of pyruvate to acetate was increased. With both species, lactate-production pathways were increased compared with monocultures. *P. gingivalis* has been shown to utilize lactate (69), and therefore interactions with *S. gordonii* may provide a source of nutrients. Similar analyses have recently been reported that center on the *F. nucleatum* proteome in two- or three-species communities (85). Overall, interspecies interactions led to reductions in amino acid fermentation and increases in butanoate production. It is possible that *P. gingivalis* competes with *F. nucleatum* for amino acids, although it is less clear why amino acid fermentation should be decreased in co-culture with *S. gordonii*.

Metabolomics technologies have been applied to investigate microbial metabolism in supragingival dental plaque, and largely appear to reflect the presence of *Streptococcus* spp. and *Actinomyces* spp. in this environment (86). Metatranscriptomics provides a very sensitive, although somewhat indirect, measure of metabolic potential in the mouth. The metatranscriptomic analysis of subgingival dental plaque showed differences in metabolism between healthy and periodontally affected sites (87). Specifically, diseased sites were enriched in pathways for lysine fermentation to butyrate, catabolism of histidine, nucleotide biosynthesis, and pyruvate fermentation. These changes were concordant with shifts in the microbial population. In the case of butyrate production, for example, *F. nucleatum* was the only organism found to have a pathway for the conversion of lysine to butyrate (87).

SUMMARY

Oral biofilms represent highly complex microbial communities that can exist in harmony with the host, but are also prone to cause disease following shifts in the microbial balance within the community. Metabolism appears to be a central driver for intermicrobial interactions that lead to the establishment of dental plaque and, in some instances, lead to changes in the plaque composition that result in dental caries or periodontitis. Computational-modelling methods and analysis of genes, proteins, transcripts, or metabolites at the microbial community level are starting to yield exciting new insights about the key metabolic pathways within oral microbial communities. At the same time, further studies of more simplified two- or three-species interactions are required to elucidate the many different synergistic and competitive interactions between bacteria and to place them in the context of oral health or disease. Ultimately, it is anticipated that modulating microbial metabolism in the oral cavity may prove an effective means for controlling oral disease.

CITATION

Jakubovics NS. 2015. Saliva as the sole nutritional source in the development of multispecies communities in dental plaque, Microbiol Spectrum 3(3):MBP-0013-2014.

REFERENCES

1. **Dye BA, Li X, Beltran-Aguilar ED.** 2012. Selected oral health indicators in the United States, 2005-2008. *NCHS Data Brief* **96**:1–8.
2. **Pitts NB, Chestnutt IG, Evans D, White D, Chadwick B, Steele JG.** 2006. The dentinal caries experience of children in the United Kingdom, 2003. *Br Dent J* **200**:313–320.
3. **Thornton-Evans G, Eke P, Wei L, Palmer A, Moeti R, Hutchins S, Borrell LN, Centers for Disease Control and Prevention.** 2013. Periodontitis among adults aged ≥30 years - United States, 2009–2010. *MMWR Surveill Summ* **62**(Suppl 3):129–135.
4. **Dewhirst FE, Chen T, Izard J, Paster BJ, Tanner AC, Yu WH, Lakshmanan A, Wade WG.** 2010. The human oral microbiome. *J Bacteriol* **192**:5002–5017.
5. **Bik EM, Long CD, Armitage GC, Loomer P, Emerson J, Mongodin EF, Nelson KE, Gill SR, Fraser-Liggett CM, Relman DA.** 2010. Bacterial diversity in the oral cavity of 10 healthy individuals. *ISME J* **4**:962–974.
6. **Li K, Bihan M, Methè BA.** 2013. Analyses of the stability and core taxonomic memberships of the human microbiome. *PLoS One* **8**:e63139. doi:10.1371/journal.pone.0063139
7. **Xu X, He J, Xue J, Wang Y, Li K, Zhang K, Guo Q, Liu X, Zhou Y, Cheng L, Li M, Li Y, Li Y, Shi W, Zhou X.** 2014. Oral cavity contains distinct niches with dynamic microbial communities. *Environ Microbiol*: In press. doi:10.1111/1462-2920.12502.
8. **Ge X, Rodriguez R, Trinh M, Gunsolley J, Xu P.** 2013. Oral microbiome of deep and shallow dental pockets in chronic periodontitis. *PLoS One* **8**:e65520. doi:10.1371/journal.pone.0065520
9. **Li Y, He J, He Z, Zhou Y, Yuan M, Xu X, Sun F, Liu C, Li J, Xie W, Deng Y, Qin Y, VanNostrand JD, Xiao L, Wu L, Zhou J, Shi W, Zhou X.** 2014. Phylogenetic and functional gene structure shifts of the oral microbiomes in periodontitis patients. *ISME J* **8**:1879–1891.
10. **Teles RP, Haffajee AD, Socransky SS.** 2006. Microbiological goals of periodontal therapy. *Periodontol 2000* **42**:180–218.
11. **Marsh PD.** 2003. Are dental diseases examples of ecological catastrophes? *Microbiology* **149**:279–294.
12. **Bowden GH, Li YH.** 1997. Nutritional influences on biofilm development. *Adv Dent Res* **11**:81–99.
13. **de Almeida Pdel V, Grégio AM, Machado MA, de Lima AA, Azevedo LR.** 2008. Saliva composition and functions: a comprehensive review. *J Contemp Dent Pract* **9**:72–80.
14. **Goodson JM.** 2003. Gingival crevice fluid flow. *Periodontol 2000* **31**:43–54.
15. **Mori M, Kaseda M, Yamamoto T, Yamada S, Itabashi H.** 2012. Capillary ion electrophoresis-capacitively coupled contactless conductivity detection of inorganic cations in human saliva on a polyvinyl alcohol-coated capillary. *Anal Bioanal Chem* **402**:2425–2430.
16. **Chen Z, Feng S, Pow EH, Lam OL, Mai S, Wang H.** 2014. Organic anion composition of human whole saliva as determined by ion chromatography. *Clin Chim Acta* **438**:231–235.
17. **Kim YJ, Kim YK, Kho HS.** 2010. Effects of smoking on trace metal levels in saliva. *Oral Dis* **16**:823–830.
18. **Jehmlich N, Golatowski C, Murr A, Salazar G, Dhople VM, Hammer E, Volker U.** 2014. Comparative evaluation of peptide desalting

methods for salivary proteome analysis. *Clin Chim Acta* **434**:16–20.

19. **Helmerhorst EJ, Oppenheim FG.** 2007. Saliva: a dynamic proteome. *J Dent Res* **86:** 680–693.

20. **Jurysta C, Bulur N, Oguzhan B, Satman I, Yilmaz TM, Malaisse WJ, Sener A.** 2009. Salivary glucose concentration and excretion in normal and diabetic subjects. *J Biomed Biotechnol* **2009**:430426.

21. **Hannig C, Hannig M, Attin T.** 2005. Enzymes in the acquired enamel pellicle. *Eur J Oral Sci* **113**:2–13.

22. **Lee YH, Zimmerman JN, Custodio W, Xiao Y, Basiri T, Hatibovic-Kofman S, Siqueira WL.** 2013. Proteomic evaluation of acquired enamel pellicle during *in vivo* formation. *PLoS One* **8**:e67919. doi:10.1371/journal.pone.0067919

23. **Lawson JW.** 1971. Growth of cariogenic streptococci in chemically defined medium. *Arch Oral Biol* **16**:339–342.

24. **Socransky SS, Dzink JL, Smith CM.** 1985. Chemically defined medium for oral microorganisms. *J Clin Microbiol* **22**:303–305.

25. **Terleckyj B, Willett NP, Shockman GD.** 1975. Growth of several cariogenic strains of oral streptococci in a chemically defined medium. *Infect Immun* **11**:649–655.

26. **Wyss C.** 1992. Growth of *Porphyromonas gingivalis, Treponema denticola, T. pectinovorum, T. socranskii,* and *T. vincentii* in a chemically defined medium. *J Clin Microbiol* **30**:2225–2229.

27. **Schulze-Schweifing K, Banerjee A, Wade WG.** 2014. Comparison of bacterial culture and 16S rRNA community profiling by clonal analysis and pyrosequencing for the characterization of the dentine caries-associated microbiome. *Front Cell Infect Microbiol* **4**:164.

28. **Koo H, Falsetta ML, Klein MI.** 2013. The exopolysaccharide matrix: a virulence determinant of cariogenic biofilm. *J Dent Res* **92**:1065–1073.

29. **Takahashi N, Nyvad B.** 2011. The role of bacteria in the caries process: ecological perspectives. *J Dent Res* **90**:294–303.

30. **de Soet JJ, Nyvad B, Kilian M.** 2000. Strain-related acid production by oral streptococci. *Caries Res* **34**:486–490.

31. **Johnson JL, Moore LV, Kaneko B, Moore WE.** 1990. *Actinomyces georgiae* sp. nov., *Actinomyces gerencseriae* sp. nov., designation of two genospecies of *Actinomyces naeslundii,* and inclusion of *A. naeslundii* serotypes II and III and *Actinomyces viscosus* serotype II in *A. naeslundii* genospecies 2. *Int J Syst Bacteriol* **40**:273–286.

32. **Bradshaw DJ, Homer KA, Marsh PD, Beighton D.** 1994. Metabolic cooperation in oral microbial communities during growth on mucin. *Microbiology* **140**:3407–3412.

33. **Homer KA, Roberts G, Byers HL, Tarelli E, Whiley RA, Philpott-Howard J, Beighton D.** 2001. Mannosidase production by viridans group streptococci. *J Clin Microbiol* **39**:995–1001.

34. **Yeung MK.** 1993. Complete nucleotide sequence of the *Actinomyces viscosus* T14V sialidase gene: presence of a conserved repeating sequence among strains of *Actinomyces* spp. *Infect Immun* **61**:109–116.

35. **Kilian M, Mikkelsen L, Henrichsen J.** 1989. Taxonomic study of viridans streptococci - description of *Streptococcus gordonii* sp. nov and emended descriptions of *Streptococcus sanguis* (White and Niven 1946), *Streptococcus oralis* (Bridge and Sneath 1982), and *Streptococcus mitis* (Andrewes and Horder 1906). *Int J Syst Evol Microbiol* **39**:471–484.

36. **Rangarajan M, Aduse-Opoku J, Hashim A, Paramonov N, Curtis MA.** 2013. Characterization of the α- and β-mannosidases of *Porphyromonas gingivalis. J Bacteriol* **195:** 5297–5307.

37. **Roy S, Phansopa C, Stafford P, Honma K, Douglas CW, Sharma A, Stafford GP.** 2012. Beta-hexosaminidase activity of the oral pathogen *Tannerella forsythia* influences biofilm formation on glycoprotein substrates. *FEMS Immunol Med Microbiol* **65**:116–120.

38. **Reinholdt J, Tomana M, Mortensen SB, Kilian M.** 1990. Molecular aspects of immunoglobulin A1 degradation by oral streptococci. *Infect Immun* **58**:1186–1194.

39. **van der Hoeven JS, van den Kieboom CW.** 1990. Oxygen-dependent lactate utilization by *Actinomyces viscosus* and *Actinomyces naeslundii. Oral Microbiol Immunol* **5**:223–225.

40. **Brown SA, Whiteley M.** 2007. A novel exclusion mechanism for carbon resource partitioning in *Aggregatibacter actinomycetemcomitans. J Bacteriol* **189**:6407–6414.

41. **Aujoulat F, Bouvet P, Jumas-Bilak E, Jean-Pierre H, Marchandin H.** 2014. *Veillonella seminalis* sp. nov., a novel anaerobic Gram-stain-negative coccus from human clinical samples, and emended description of the genus *Veillonella. Int J Syst Evol Microbiol* **64**:3526–3531.

42. **Traudt M, Kleinberg I.** 1996. Stoichiometry of oxygen consumption and sugar, organic acid and amino acid utilization in salivary sediment and pure cultures of oral bacteria. *Arch Oral Biol* **41**:965–978.

43. Curtis MA, Griffiths GS, Price SJ, Coulthurst SK, Johnson NW. 1988. The total protein concentration of gingival crevicular fluid. Variation with sampling time and gingival inflammation. *J Clin Periodontol* **15**:628–632.

44. Andrade JA, Feres M, Figueiredo LC, Salvador SL, Cortelli SC. 2010. The ability of the BANA Test to detect different levels of *P. gingivalis, T. denticola* and *T. forsythia*. *Braz Oral Res* **24**:224–230.

45. Rogers AH, Zilm PS, Gully NJ, Pfennig AL, Marsh PD. 1991. Aspects of the growth and metabolism of *Fusobacterium nucleatum* ATCC 10953 in continuous culture. *Oral Microbiol Immunol* **6**:250–255.

46. Tan KH, Seers CA, Dashper SG, Mitchell HL, Pyke JS, Meuric V, Slakeski N, Cleal SM, Chambers JL, McConville MJ, Reynolds EC. 2014. *Porphyromonas gingivalis* and *Treponema denticola* exhibit metabolic symbioses. *PLoS Pathog* **10**:e1003955. doi:10.1371/journal.ppat.1003955

47. Liu YL, Nascimento M, Burne RA. 2012. Progress toward understanding the contribution of alkali generation in dental biofilms to inhibition of dental caries. *Int J Oral Sci* **4**:135–140.

48. Cowman RA, Perrella MM, Adams BO, Fitzgerald RJ. 1975. Amino acid requirements and proteolytic activity of *Streptococcus sanguis*. *Appl Microbiol* **30**:374–380.

49. Terleckyj B, Shockman GD. 1975. Amino acid requirements of *Streptococcus mutans* and other oral streptococci. *Infect Immun* **11**:656–664.

50. Jakubovics NS, Gill SR, Iobst SE, Vickerman MM, Kolenbrander PE. 2008. Regulation of gene expression in a mixed-genus community: stabilized arginine biosynthesis in *Streptococcus gordonii* by coaggregation with *Actinomyces naeslundii*. *J Bacteriol* **190**:3646–3657.

51. Xu P, Ge X, Chen L, Wang X, Dou Y, Xu JZ, Patel JR, Stone V, Trinh M, Evans K, Kitten T, Bonchev D, Buck GA. 2011. Genome-wide essential gene identification in *Streptococcus sanguinis*. *Sci Rep* **1**:125.

52. Dell'Anno A, Danovaro R. 2005. Extracellular DNA plays a key role in deep-sea ecosystem functioning. *Science* **309**:2179.

53. Liao S, Klein MI, Heim KP, Fan Y, Bitoun JP, Ahn SJ, Burne RA, Koo H, Brady LJ, Wen ZT. 2014. *Streptococcus mutans* extracellular DNA is upregulated during growth in biofilms, actively released via membrane vesicles, and influenced by components of the protein secretion machinery. *J Bacteriol* **196**:2355–2366.

54. Vitkov L, Klappacher M, Hannig M, Krautgartner WD. 2009. Extracellular neutrophil traps in periodontitis. *J Periodontal Res* **44**:664–672.

55. Rogers AH. 1973. The vitamin requirements of some oral streptococci. *Arch Oral Biol* **18**: 227–232.

56. Wyss C. 2007. Fatty acids synthesized by oral treponemes in chemically defined media. *FEMS Microbiol Lett* **269**:70–76.

57. Akcali A, Huck O, Buduneli N, Davideau JL, Köse T, Tenenbaum H. 2014. Exposure of *Porphyromonas gingivalis* to cortisol increases bacterial growth. *Arch Oral Biol* **59**:30–34.

58. Sreenivasan PK, Meyer DH, Fives-Taylor PM. 1993. Factors influencing the growth and viability of *Actinobacillus actinomycetemcomitans*. *Oral Microbiol Immunol* **8**:361–369.

59. AlJehani YA. 2014. Risk factors of periodontal disease: review of the literature. *Int J Dent* **2014**:182513.

60. Lewis JP. 2010. Metal uptake in host-pathogen interactions: role of iron in *Porphyromonas gingivalis* interactions with host organisms. *Periodontol 2000* **52**:94–116.

61. Paik S, Brown A, Munro CL, Cornelissen CN, Kitten T. 2003. The *sloABCR* operon of *Streptococcus mutans* encodes an Mn and Fe transport system required for endocarditis virulence and its Mn-dependent repressor. *J Bacteriol* **185**:5967–5975.

62. Kolenbrander PE, Andersen RN, Baker RA, Jenkinson HF. 1998. The adhesion-associated *sca* operon in *Streptococcus gordonii* encodes an inducible high-affinity ABC transporter for Mn^{2+} uptake. *J Bacteriol* **180**:290–295.

63. Crump KE, Bainbridge B, Brusko S, Turner LS, Ge X, Stone V, Xu P, Kitten T. 2014. The relationship of the lipoprotein SsaB, manganese and superoxide dismutase in *Streptococcus sanguinis* virulence for endocarditis. *Mol Microbiol* **92**:1243–1259.

64. Oetjen J, Fives-Taylor P, Froeliger EH. 2002. The divergently transcribed *Streptococcus parasanguis* virulence-associated *fimA* operon encoding an Mn(2+)-responsive metal transporter and *pepO* encoding a zinc metallopeptidase are not coordinately regulated. *Infect Immun* **70**:5706–5714.

65. Haswell JR, Pruitt BW, Cornacchione LP, Coe CL, Smith EG, Spatafora GA. 2013. Characterization of the functional domains of the SloR metalloregulatory protein in *Streptococcus mutans*. *J Bacteriol* **195**:126–134.

66. Chen YYM, Shieh HR, Chang YC. 2013. The expression of the *fim* operon is crucial for the

survival of *Streptococcus parasanguinis* FW213 within macrophages but not acid tolerance. *PLoS One* 8:e66163. doi:10.1371/journal.pone.0066163

67. **Jakubovics NS, Smith AW, Jenkinson HF.** 2000. Expression of the virulence-related Sca (Mn^{2+}) permease in *Streptococcus gordonii* is regulated by a diphtheria toxin metallorepressor-like protein ScaR. *Mol Microbiol* 38:140–153.

68. **Jakubovics NS, Smith AW, Jenkinson HF.** 2002. Oxidative stress tolerance is manganese (Mn^{2+}) regulated in *Streptococcus gordonii*. *Microbiology* 148:3255–3263.

69. **Lewis JP, Iyer D, Anaya-Bergman C.** 2009. Adaptation of *Porphyromonas gingivalis* to microaerophilic conditions involves increased consumption of formate and reduced utilization of lactate. *Microbiology* 155:3758–3774.

70. **Barboza-Silva E, Castro AC, Marquis RE.** 2005. Mechanisms of inhibition by fluoride of urease activities of cell suspensions and biofilms of *Staphylococcus epidermidis*, *Streptococcus salivarius*, *Actinomyces naeslundii* and of dental plaque. *Oral Microbiol Immunol* 20: 323–332.

71. **Chen YY, Burne RA.** 2003. Identification and characterization of the nickel uptake system for urease biogenesis in *Streptococcus salivarius* 57. I. *J Bacteriol* 185:6773–6779.

72. **de Jong MH, van der Hoeven JS, van OS JH, Olijve JH.** 1984. Growth of oral *Streptococcus* species and *Actinomyces viscosus* in human saliva. *Appl Environ Microbiol* 47:901–904.

73. **Watanabe S, Dawes C.** 1988. The effects of different foods and concentrations of citric acid on the flow rate of whole saliva in man. *Arch Oral Biol* 33:1–5.

74. **Kurita-Ochiai T, Fukushima K, Ochiai K.** 1995. Volatile fatty acids, metabolic by-products of periodontopathic bacteria, inhibit lymphocyte proliferation and cytokine production. *J Dent Res* 74:1367–1373.

75. **Faust K, Sathirapongsasuti JF, Izard J, Segata N, Gevers D, Raes J, Huttenhower C.** 2012. Microbial co-occurrence relationships in the human microbiome. *PLoS Comput Biol* 8: e1002606. doi:10.1371/journal.pcbi.1002606

76. **Vartoukian SR, Downes J, Palmer RM, Wade WG.** 2013. *Fretibacterium fastidiosum* gen. nov., sp. nov., isolated from the human oral cavity. *Int J Syst Evol Microbiol* 63:458–463.

77. **Soro V, Dutton LC, Sprague SV, Nobbs AH, Ireland AJ, Sandy JR, Jepson MA, Micaroni M, Splatt PR, Dymock D, Jenkinson HF.** 2014. Axenic culture of a candidate division TM7 bacterium from the human oral cavity and biofilm interactions with other oral bacteria. *Appl Environ Microbiol* 80:6480–6489.

78. **Katharios-Lanwermeyer S, Xi C, Jakubovics NS, Rickard AH.** 2014. Mini-review: Microbial coaggregation: ubiquity and implications for biofilm development. *Biofouling* 30:1235–1251.

79. **Palmer RJ Jr, Diaz PI, Kolenbrander PE.** 2006. Rapid succession within the *Veillonella* population of a developing human oral biofilm in situ. *J Bacteriol* 188:4117–4124.

80. **Johnson BP, Jensen BJ, Ransom EM, Heinemann KA, Vannatta KM, Egland KA, Egland PG.** 2009. Interspecies signaling between *Veillonella atypica* and *Streptococcus gordonii* requires the transcription factor CcpA. *J Bacteriol* 191:5563–5565.

81. **Stacy A, Everett J, Jorth P, Trivedi U, Rumbaugh KP, Whiteley M.** 2014. Bacterial fight-and-flight responses enhance virulence in a polymicrobial infection. *Proc Natl Acad Sci U S A* 111:7819–7824.

82. **Mazumdar V, Amar S, Segrè D.** 2013. Metabolic proximity in the order of colonization of a microbial community. *PLoS One* 8:e77617. doi:10.1371/journal.pone.0077617

83. **Hendrickson EL, Xia Q, Wang T, Lamont RJ, Hackett M.** 2009. Pathway analysis for intracellular *Porphyromonas gingivalis* using a strain ATCC 33277 specific database. *BMC Microbiol* 9:185.

84. **Hendrickson EL, Wang TS, Dickinson BC, Whitmore SE, Wright CJ, Lamont RJ, Hackett M.** 2012. Proteomics of *Streptococcus gordonii* within a model developing oral microbial community. *BMC Microbiol* 12:211.

85. **Hendrickson EL, Wang T, Beck DA, Dickinson BC, Wright CJ, Lamont RJ, Hackett M.** 2014. Proteomics of *Fusobacterium nucleatum* within a model developing oral microbial community. *Microbiologyopen* 3:729–751.

86. **Takahashi N, Washio J, Mayanagi G.** 2010. Metabolomics of supragingival plaque and oral bacteria. *J Dent Res* 89:1383–1388.

87. **Jorth P, Turner KH, Gumus P, Nizam N, Buduneli N, Whiteley M.** 2014. Metatranscriptomics of the human oral microbiome during health and disease. *MBio* 5:e01012-14. doi:10.1128/mBio.01012-14

88. **Edgar WM.** 1992. Saliva: its secretion, composition and functions. *Br Dent J* 172:305–312.

89. **Kejriwal S, Bhandary R, Thomas B, Kumari S.** 2014. Estimation of levels of salivary mucin, amylase and total protein in gingivitis and chronic periodontitis patients. *J Clin Diagn Res* 8:ZC56–ZC60.

90. **Neyraud E, Bult JH, Dransfield E.** 2009. Continuous analysis of parotid saliva during resting and short-duration simulated chewing. *Arch Oral Biol* 54:449–456.

91. **Naumova EA, Sandulescu T, Bochnig C, Al Khatib P, Lee WK, Zimmer S, Arnold WH.** 2014. Dynamic changes in saliva after acute mental stress. *Sci Rep* **4:**4884.

92. **Watanabe M, Asatsuma M, Ikui A, Ikeda M, Yamada Y, Nomura S, Igarashi A.** 2005. Measurements of several metallic elements and matrix metalloproteinases (MMPs) in saliva from patients with taste disorder. *Chem Senses* **30:**121–125.

93. **Brand HS, Jörning GG, Chamuleau RA, Abraham-Inpijn L.** 1997. Effect of a protein-rich meal on urinary and salivary free amino acid concentrations in human subjects. *Clin Chim Acta* **264:**37–47.

94. **Syrjanen SM, Alakuijala L, Alakuijala P, Markkanen SO, Markkanen H.** 1990. Free amino acid levels in oral fluids of normal subjects and patients with periodontal disease. *Arch Oral Biol* **35:**189–193.

Enteric Pathogens Exploit the Microbiota-generated Nutritional Environment of the Gut

13

ALLINE R. PACHECO[1] and VANESSA SPERANDIO[1]

DYNAMICS OF INTESTINAL COLONIZATION BY PATHOGENIC BACTERIA

The mammalian gastrointestinal (GI) tract harbors a diverse collection of indigenous bacteria known as the microbiota. The number of bacterial cells within our bodies exceeds the number of our cells by one order of magnitude (3). Homeostasis of the microbiota is maintained by differential nutrient utilization and physical separation from the gut mucosa (4). However, environmental perturbations such as antibiotic treatment, changes in diet, and infection lead to substantial alterations in composition and structure of the microbiota, referred to as *dysbiosis* (5–8).

Efficient use of nutrient sources in the gut has a major impact on colonization by pathogenic bacterial species given that nutrient sources are scarce, and they compete with the exquisitely adapted commensal bacteria for these nutrients. According to Freter's hypothesis, the ability of a pathogen to thrive during intestinal colonization depends on its ability to efficiently utilize nutrient sources and find a suitable niche for colonization (9). Competition for nutrient acquisition between enteric pathogens and the microbiota constitutes a protective mechanism against

[1]University of Texas Southwestern Medical Center, Dept. of Microbiology.
Metabolism and Bacterial Pathogenesis
Edited by Tyrrell Conway and Paul Cohen
© 2015 American Society for Microbiology, Washington, DC
doi:10.1128/microbiolspec.MBP-0001-2014

infection and is an important aspect of colonization resistance. Therefore, evolution of new nutrient acquisition mechanisms and metabolic diversification contributes to a pathogen's survival and persistence and is an important determinant of the course of bacterial infections. Two important strategies employed by enteric pathogens are the alternative use of carbon sources (10) and utilization of byproducts of the microbiota's metabolism.

Linking metabolism to the precise coordination of virulence gene expression is a key step in the adaptation of pathogens towards their colonization niches. In this chapter, we will discuss nutrient detection, acquisition, and utilization by enteric pathogens and the barriers against intestinal infection, highlighting the vital role played by the gut microbiota in these processes.

INTESTINAL BARRIERS AGAINST INFECTION

Enteric pathogenic bacteria face a series of barriers to colonizing the GI tract. The human gut is a very complex ecosystem that harbors a high number of commensal bacteria that compete with pathogens for nutrients and space. In addition, the intestinal epithelium is covered by a protective viscous mucus layer that impairs easy bacterial access to the epithelium (10, 11). Suffice it to say, tropism for the mammalian intestine co-evolved with several virulence traits that helped bacterial pathogens cross the aforementioned barriers. Some crucial virulence traits comprise the ability to attach to mucus and cell surface receptors; production of proteases and toxins; expression of flagella to swim across the mucus layer; invasion of epithelial cells; and quorum sensing (12–17). In addition to strict *pro quo* virulence factors, pathogens also require suitable nutrients. Therefore, nutrient-sensing systems play a major role in both early and late phases of infection.

The Intestinal Microbiota

The mammalian GI tract microbiota plays a fundamental role in human health. Ten trillion to 100 trillion microbes inhabit the distal segment of the human gut, with most belonging to the Bacteroidetes (Gram-negative) and Firmicutes (Gram-positive) phyla (3, 18, 19). Metagenomic studies have shown that the species composition of the microbiota is very diverse; nonetheless, there is conservation in the microbial phyla shared by all individuals (20, 21). In addition, the composition of the intestinal microbial community can vary according to the host genetic background (22, 23). The gut microbiota plays many roles in the host homeostasis and has been referred to as the "forgotten organ" (24). The genetic repertoire of this community is referred to as the *microbiome* and gives the human host metabolic capabilities not encoded in our genome (25, 26).

The Mucus Layer: A Source of Protection and Nutrients

The single layer of epithelial cells that separates the luminal contents from the GI mucosa is the target of many pathogenic bacteria (27, 28). Nonetheless, not all bacteria can directly interact with enterocytes. A gel-like mucus layer overlays the intestinal epithelial cells, shielding the colonic epithelium from bacteria (29). The mucus layer is in a dynamic state, being constantly synthesized and secreted by specialized goblet cells and degraded to a large extent by indigenous intestinal microbes (30, 31). In fact, mucus utilization has recently been proposed as a co-evolved adaptation of gut resident bacteria and their host (32).

The mucus is composed of mucin, antimicrobial peptides, glycoproteins, glycolipids and epithelial cell debris but 50% of it is made of polysaccharides (33). The major structural component of the mucus is mucin, a glycoprotein that has a protein backbone connected to hydrophilic and hygroscopic oligosaccharide side-chains, which form a gel-like tridimensional structure (34). Also, as part of the mucus composition, there are other goblet cell products such as trefoil peptides (TFF), resistin-like molecule β (RELMβ), and Fc-γ binding protein (Fcgbp), antimicrobial peptides (beta-defensin) and lysozymes secreted by the Paneth cells, and IgA secreted by enterocytes (34, 35). Moreover, the microbiota modulates mucin synthesis and secretion (36). O-linked glycans comprise 80% of the total weight of mucins and are a major nutrient source for bacteria (29). In addition, they provide attachment sites for commensal and pathogenic bacteria (29, 37). A diverse collection of 13 monosaccharides is part of the mucus composition: arabinose, fucose, galactose, gluconate, glucuronate, galacturonate, mannose, glucosamine, N-acetyl-glucosamine, galactosamine, N-acetyl-galactosamine, N-acetylneuramic acid, and ribose. All of these sugars are made available to pathogenic bacteria due to host epithelial cell turnover and the polysaccharide-degrading activity of commensal anaerobes. Hence, the mucus layer is an important habitat and source of nutrients for bacterial communities that colonize mucosal surfaces.

The highly glycosylated MUC2 mucin is synthesized by goblet cells in the small and large intestines and is a major component of the mucus layer (33). The intestinal epithelium also expresses membrane-bound mucins: MUC1, MUC3, MUC4, MUC12, MUC13, and MUC17. However, MUC2 is the predominant mucin in human and murine colons (33), and these mucins are constantly being degraded by the action of glycosidases produced by the anaerobic bacteria that dominate the colonic microbiota (38).

The important role of the mucus as a defensive barrier against infection and injury can be illustrated by the differential length and thickness of its structure along the GI tract. The mucus layer is progressively thicker towards the large intestine, the site where the highest numbers of commensal bacteria reside. The mucus is the front line of the innate host defense against pathogenic microbes (34). Alterations of the mucus layer, such as the lack of MUC2, renders mice susceptible to bacterial adhesion to the intestinal epithelium, resulting in disturbances of the intestinal physiology (33). It is noteworthy that deficiency in mucus production causes changes in the normal localization of commensal bacteria in the colon (39). In addition, recent evidence highlights the importance of mucin in the defense against bacterial infection, with MUC2-deficient mice developing severe life-threatening colitis when infected with the murine enteric pathogen *Citrobacter rodentium* (33). MUC2 has highly fucosylated glycans, and MUC17, a membrane-bound mucin expressed in the large intestine, protects against invasion by enteroinvasive *E. coli* (EIEC) (41).

The stratified mucus layer that protects the intestinal epithelium displays a highly complex structure. The mucus layer is spatially divided in two layers: an outer loose mucus layer facing the intestinal lumen and a thick inner mucus layer firmly attached to the epithelial cells (29). Using fluorescent *in situ* hybridization (FISH) staining for bacteria, Johansson et al elegantly demonstrated that commensal bacteria communities reside in the outer mucus layer, while the inner layer is virtually devoid of bacteria (39).

Gut commensals are not found in close contact with the epithelium lining, in contrast to pathogenic bacteria, which employ virulence factors to reach close proximity to enterocytes. EHEC produces the plasmid-encoded metalloprotease StcE, which targets intestinal mucins, therefore contributing to bacterial penetration towards the colonic epithelium (42). In fact, one of the theories of how commensal bacteria do not cause disease, even when sharing several common features with pathogenic bacteria, is their localization in the GI tract. Commensal bacteria are associated with the outer mucus layer but not at the interface of the intestinal epithelium, and this physical separation is thought to prevent dysbiosis (43). Both pathogenic and commensal bacteria can consume carbohydrates from the mucus as a carbon and energy source (44–47). In addition to being a nutrient source for indigenous and foreign bacteria, the mucus components can also be exploited as cues to trigger production of virulence factors by pathogens. MUC2 triggers virulence expression in *Campylobacter jejuni*, with MUC2 exposure leading *C. jejuni* to express cytolethal distending toxin JlpA, and flagellin (28).

The stratification of the mucus layer is important to sustain the symbiotic relationship between the microbiota and the host. It has been demonstrated that the lectin RegIII gamma, produced by Paneth cells of the intestinal epithelium, is important to the physical separation of the microbiota and the intestinal epithelium (4, 48). The physical separation of the intestinal microbiota and the colonic epithelium represents evidence that the mucus layers act as an effective barrier (11).

MICROBIOTA AND NUTRIENT GENERATION IN THE GUT

The consortium of the resident microbes that inhabit the human gut possess an incredible arsenal of glycolytic enzymes that allow the utilization of complex polysaccharides originated from the host itself or its diet (48–50). These polysaccharides are hydrolyzed into monosaccharides, which are subsequently utilized as carbon and energy sources after being released as free sugars in the intestinal lumen (51). Therefore, the metabolic activity of the microbiota is intrinsically related to the generation of the nutritional environment of the gut, which also impacts the survival and persistence of pathogenic bacteria. In this section, we will discuss the role of nutrient acquisition in the gut by pathogenic bacteria and how

the microbiota modulates the availability of these nutrient sources.

Commensal Bacteria and Polysaccharide Degradation in the Gut

One of the major roles played by the microbiota is the manipulation of carbohydrate sources in the gut. The members of the microbiota community are fermenters with a broad range of metabolic capacity, being able to digest complex glycan structures originated from the host diet, host structures such as mucus, and cell-associated glycans otherwise not digested by invading microbes (50, 52–54). This plural metabolic capacity allows the microbiota to explore unique niches and survive in the human gut in homeostasis. Hence, the gut microbiota functions as a metabolic organ, helping the human host obtain energy from otherwise indigestible dietary sources (55).

The ability to degrade complex polysaccharides highlights the crucial role displayed by the gut commensals in metabolic pathways in the gut. Most of the knowledge of polysaccharide utilization by the microbiota derives from investigations of the glycophagic symbiont *Bacteroides thetataiotaomicron*. *B. theta*, a strict anaerobe, is the most abundant resident of the distal small intestine and colon of mice and humans (56, 57). The *B. theta* genome encodes an impressive arsenal of 246 glycolitic enzymes (58). This prominent commensal bacterium can degrade plant- and animal-derived complex glycans into monosaccharides, providing nutrient sources for other commensals, consequently playing a fundamental role in the supply of energy sources in the gut. Pathogenic bacteria exploit this *B. theta*-dependent nutrient availability. Therefore, the polysaccharide-degrading activity of commensal bacteria might affect colonization by pathogens by representing a source of competition but also a nutrient-generating machine capable of supporting pathogens to successfully grow and find a niche in the host. The ability of

Bacteroides sp to use a diverse range of glycans is dependent on genes in its polysaccharide-utilization loci (PULs) (59).

Mucus utilization by primary fermenters such as *B. theta* involves the production of glycosyl hydrolases secreted into the environment. Consequently, monosaccharides are released from the complex glycan structures into the lumen, where they are accessible to other microbial species. *B. theta* dedicates about 18% of its genome to polysaccharide utilization, the PULs (58). Conversely, gut pathogens are not equally equipped to consume complex host-derived glycans. Although some pathogens, such as EHEC, are able to produce a mucus-degrading protease, StcE, it does not encode glycosidases. This means that gut pathogens rely on the glycosidase activity of commensal bacteria to access and import free monosaccharides for catabolism.

Certain members of the microbiota, such as *Bifidobacterium sp*, may differ in their effects on nutrient generation to pathogenic bacteria. In contrast with *B. theta*, *B. longum subspecies infantis* (*B.infantis*) typically produces intracellular glycosidases to import complex glycans into the cell for digestion into monosaccharides (60, 61). *B. bifidum*, however, secretes glycosidases similarly to *B. theta* (62). Consequently, the access to free monosaccharides by microbial pathogens during polysaccharide degradation by certain species of *Bifidobacteria* is different, which may have interesting biological implications because *Bifidobacteria* are used as probiotics and may be able to control or prevent enteric infections (63).

A special relationship takes place between *B. theta* and its host. *B. theta* consumption of fucose may have a major impact on enteropathogens able to utilize this sugar as a carbon and energy source (64). *B. theta* induces the expression of fucosylated glycoconjugates by the host intestinal epithelium (65). Then *B. theta* produces fucosidases that harvest fucose from mucosal glycans (66). Fucose is abundant is intestinal glyco-

conjugates, and it is usually a terminal α-linked sugar (67, 68). The triggering of intestinal fucosylation by *B. theta* depends on the bacterial density and the production of a *B. theta*-derived signal that remains elusive (69). In addition to using host-derived fucose as a carbon and energy source, *B. theta* also incorporates fucose into a capsular polysaccharide via an O-glycosylation system, which is believed to be important for competitive colonization of the gut (70). Nutrient utilization by *B. theta* can be modulated by diet: a diet rich in plant glycans triggers expression of genes involved in metabolism of dietary substrates by *B. theta*; conversely, on exposure to a diet devoid of complex glycans, *B. theta* switches its metabolism towards host glycans (50). The metabolic switch that *B. theta* undergoes during a change in diet might affect pathogen access to fucose, which could directly impact the outcome of bacterial infections.

Primary fermenters differ in their capacity to utilize carbohydrates, which is relevant *in vivo* because dietary changes cause alterations in the community structure of the microbiota, and likely, in the outcome of end-fermented products (71). Polysaccharide degradation has been recognized as a core function encoded within the microbiome (72). The prominent adult gut symbiont, *B. theta* encodes an arsenal of 261 glycosyl hydrolases (73). Other distal gut residents such as *Akkermansia muciniphila* have a mucin-degrading ability that may also lead to release of free monosaccharides that can be utilized by pathogenic enterobacteria (74).

A generation of free monosaccharides in the gut, an end product of extensive polysaccharide degradation by members of the microbiota, is a major modulator of the nutrient environment accessible to invading pathogens. Most pathogenic bacteria do not encode glycosyl hydrolases in their genomes and rely on simple monosaccharides or disaccharides as substrates for growth *in vivo* (32). Therefore, the enzymatic activity of the gut microbiota, to a certain degree, may render the host particularly susceptible to different infections. This concept could be further explored to design customized diets or probiotic interventions aimed at improving pathogen exclusion based on nutrient competition. This concept is exemplified by Deriu et al, who demonstrated that administration of the probiotic *E.coli* Nissle 1917 strain reduced murine colonization by *Salmonella enterica Typhimiurium* (75).

INTERPLAY BETWEEN COMMENSAL AND PATHOGENIC BACTERIA

Given the high content of commensal bacteria residing in the gut, the colonization site of several bacterial pathogens, it is not surprising that a complex relationship might arise from these interactions. Nonetheless, little is known of the mechanisms that govern the crosstalk between pathogens and the microbiota or the impact of the commensal bacteria on pathogenesis and infection outcomes. Elucidation of the processes involved in interactions among host, microbiota, and pathogens is of major importance in the design of novel therapeutic interventions (76–78).

Gut pathogens harbor several traits to maximize proliferation in the lumen, including motility, chemotaxis, and iron-scavenging and nutrient-sensing systems. In addition, pathogenic species can hijack carbohydrate utilization pathways of resident microbes for their own advantage by exploring the end-product of glycosidases produced by anaerobes from the microbiota to obtain monossacharides as carbon sources (38).

Competition for Nutrients and Colonization Resistance

While primary fermenters have a major impact on nutrient generation for bacterial pathogens, commensal bacteria displaying similar nutrient requirements pose a threat against pathogen survival during intestinal infection. By consuming similar carbohydrates, commensal *E. coli* competes with EHEC O157:H7 for nutrients, leading this pathogen to explore different niches to proliferate in the gut (79–81).

Commensal and pathogenic *E. coli* differs in the types of carbohydrates it preferably utilizes *in vivo* as carbon sources. EHEC can grow on mucus (82). Studies indicate that EHEC can grow *in vitro* on cecal mucus prepared from mice but cannot grow in the luminal content, suggesting that these bacteria colonize the mouse intestine by growing in the mucus layer that overlays the cecal epithelium (43). In addition, other studies provide evidence that carbohydrates derived from the mucus can support the growth of *E. coli* during murine colonization (44–47).

Commensal and pathogenic *E. coli* share their preferences for particular carbon sources they utilize during intestinal colonization, but they also present differences that reflect their spatial segregation inside the human gut. Commensal *E. coli* strains are found in the lumen and attached to the mucus layer, while pathogenic *E. coli* strains are able to cross the mucus layer and reach proximity to the IECs, which also exposes them to nutrients exclusively available at the epithelium interface (44, 83). Among the nutrient sources consumed by both commensal and pathogenic *E. coli* are monossaccharides and disaccharides. The fact that they consume similar carbon sources *in vivo* is the basis of colonization resistance that commensal *E. coli* imposes on pathogenic *E. coli* species.

Colonization of the mammalian intestine by EHEC requires precise coordination of metabolic and virulence factors. The infectious dose of EHEC is remarkably low compared with other enteric pathogens, highlighting important adaptations of EHEC to the human intestine. EHEC must expand its population to high numbers and find a niche in the colon, which is a major challenge considering the immense number of residing commensal bacteria adapted to live in the colon during millions of years of co-evolution with the human host (84).

An investigation of the carbohydrate utilization profile of EHEC in the bovine gut has revealed that this strain can catabolize mucus-derived carbohydrates inside the cattle gut and do so more rapidly than resident microbes, including commensal *E. coli* (85). Cattle are the major reservoir of EHEC O157:H7, which has tropism to the recto-anal junction (RAJ) (86). *In vitro* growth competition assays using WT and the EHEC sugar utilization mutant strains Δ*manA*, Δ*nagE*, Δ*nanAT,* and Δ*galK*, which are deleted for genes involved in catabolism of six major mucus-derived monosaccharides (galactose, N-acetyl-glucosamine [GlcNAc], N-acetylgalactosamine [GalNAc], fucose, mannose and N-acetyl neuraminic acid [Neu5Ac]), showed that the ability to consume mannose, GlcNAc, Neu5Ac, and galactose is important for EHEC growth, suggesting that metabolism of the aforementioned carbohydrates confers a growth advantage to EHEC in the bovine intestine (85).

In vivo carbon consumption was investigated using the streptomycin-treated mouse model and elucidated many aspects of the competition and nutrient utilization that allows EHEC to successfully colonize the mammalian intestine. Commensal and pathogenic *E. coli* share the ability to consume arabinose, fucose, and N-acetylglucosamine in the mouse intestine. EHEC is able to catabolize galactose, hexuronates, mannose, and ribose, while commensal *E. coli* exclusively catabolize gluconate and N-acetyl-neuramic acid (79). These data indicate that differential carbon nutrition in the gut contributes to niche adaptation of EHEC and helps avoid competition with commensal *E. coli*. According to the nutrient-niche hypothesis elaborated by Freter et al, better consumption of a limiting nutrient source than an organism's competitors is imperative for successful colonization of the intestine (87, 89).

Recent evidence indicates that the ability to consume similar carbohydrate sources is an important factor that may influence bacterial infection. Kamada et al reported that competition with members of the gut microbiota for the same nutrients is necessary for pathogen clearance (88). This study represents a great advance on the investigation of the relationship among commensal and pathogenic bacteria, which has also shown that classic virulence traits such as the type 3 secretion system (T3SS), which is known to be required for host cell contact by EHEC, is also important for competition with gut microbiota (88).

EHEC does not significantly compete with *B. theta* for nutrient utilization during growth in mucus but competes with commensal *E. coli* for the same carbon sources during growth in the mammalian intestine (44, 47, 79, 89). One such carbon source is fucose, which is released into the lumen by glycophagic bacteria such as *B. theta* and can be utilized by *E. coli*, which itself cannot hydrolyze complex mucus carbohydrates (44, 47, 79). Because both EHEC and commensal *E. coli* compete for fucose utilization in the lumen, it would be counterproductive for EHEC to invest a lot of resources in the utilization of this carbon source in this compartment, where commensal *E. coli* are present (79, 89). However, EHEC can efficiently use other carbon sources, such as galactose, hexorunates, mannose, and ribose, which are not used by commensal *E. coli* in the intestine (79).

NUTRIENT SENSING IN THE GUT

Differential utilization of limiting nutrients is the basis for the coexistence of members of the gut microbiota. It has also a major impact on bacterial infection, as pathogens explore alternative nutrient sources to avoid competition with commensals. In cases of non-cronical infections such as EHEC or EPEC, nutrient competition among commensal and pathogenic bacteria impacts the outcome of infection, leading to resolution of this infection and elimination of the intruder.

Fucose Sensing Regulates Intestinal Colonization by EHEC

EHEC is the causative agent of outbreaks of bloody diarrhea worldwide, with about 5% to 7% of the cases in any given outbreak developing a life-threatening complication known as *hemolytic uremic syndrome* (HUS) (90, 91). EHEC colonizes the human large intestine through the formation of attaching and effacing (AE) lesions on intestinal epithelial cells (92). Most genes necessary for AE lesion formation are clustered in a pathogenicity island (PAI) named the locus of enterocyte effacement (LEE) (93). The LEE region contains five major operons: *LEE1-5* (94–96), which encodes a type III secretion system (TTSS) (12), an adhesin (intimin) (97) and its receptor (Tir) (98), and effector proteins (99–103). The LEE genes and the non–LEE-encoded effector, EspFu, are both required for the formation of AE lesions (104).

Cell-to-cell communication among bacteria in the intestine is a major mechanism that shapes bacteria-host relationships. Pathogenic bacteria such as EHEC can also cross-communicate with the host by detecting mammalian hormones (105). By virtue of its remarkably low infectious dose (50 CFU) (106), successful colonization of the human colon by EHEC relies largely on sensing multiple signals to coordinate the expression of virulence genes. EHEC exploits the autoinducer-3 (AI-3)/epinephrine (Epi)/norepinephrine (NE) interkingdom signaling cascade to trigger expression of motility and AE lesion genes, two pathogenic traits that are crucial for colonization but are required at different time points during infection (105). The host hormones Epi/NE are specifically sensed by two histidine sensor kinases: QseC and QseE (107, 108). QseE is downstream of QseC in this signaling cascade, given that transcription of QseE is activated through QseC (109). In addition to sensing these host hormones, QseC also senses the bacterial signal AI-3 (110). QseE, however, does not sense AI-3, thereby discriminating between host- and bacterial-derived signals (108). QseC and QseE activate virulence gene expression and pathogenesis *in vitro* and *in vivo* in EHEC (108, 110, 111). On sensing these signals, QseC and QseE autophosphorylation increases, initiating a signaling cascade that promotes virulence gene expression. QseE exclusively phosphorylates the QseF response regulator (112). QseC, however, phosphorylates three response regulators: QseB, QseF, and KdpE.

Signal sensing by EHEC is crucial for colonization of the mammalian colon due to the orchestration of multiple virulence pathways that aim to promote intimate attachment of the bacteria to the apical portion of enterocytes (113). Also of major importance is the activation of pathways that allow suitable nutrition during infection. Recently, Pacheco et al (64) demonstrated that EHEC encodes for a two-component system (TCS) named FusKR, in which FusK is the sensor kinase and FusR is the response regulator (64). The FusKR TCS is repressed by the adrenergic-sensing QseBC and QseEF TCSs.

Investigation of the signal triggering *fusKR* transcription indicated that it was a component of the mucus. Using a combination of biochemical and genetic approaches, L-fucose was identified as the signal that activates the FusKR signaling cascade in EHEC. In fact, FusK specifically increases its autophosphorylation in response to fucose (64). FusKR signaling leads to repression of LEE and fucose utilization gene expression, allowing the pathogen to save energy by preventing unnecessary virulence gene expression while crossing the mucus layer and avoiding competition with the commensal *E. coli* for carbon sources, given that commensal *E. coli* preferentially catabolize fucose in the mammalian intestine (64).

In vitro competition assays have demonstrated that the modulation of carbon availability by the prominent gut symbiont *B. theta* alters the effect of fucose utilization by EHEC on the expression of *ler*, the master regulator of the LEE genes. In the presence of free fucose in the media, *B. theta* has no effect on *ler* transcription, but this scenario changed on co-culture of *B. theta* and EHEC on mucin. During growth on mucin, EHEC relies on *B. theta* to access free fucose, and as the result of this relationship, expression of *ler* is reduced (64). Therefore, the interaction between EHEC and the commensal bacterium *B. theta* is able to change the pathogen's virulence due to the nutrient modulatory activity of *B. theta*. *In vivo* competition assays using the infant rabbit model (114), which is able to reproduce several aspects of EHEC-mediated disease, show that the EHEC *fusK* mutant is attenuated for virulence, and regulation of *ler* by FusK plays a determinant role on EHEC fitness during intestinal colonization (Fig. 1) (64).

The genes encoding *fusKR* are clustered in a pathogenicity island (OI-20) only found in EPEC O55:H7 (the *E. coli* lineage that gave rise to EHEC O157:H7) (115, 116), EHEC O157:H7, and *C. rodentium*, AE GI pathogens that colonize the colon. EHEC's ancestor, EPEC O55:H7 (116), is the only other serotype of *E. coli* to harbor *fusKR*, suggesting that acquisition of these genes is recent. The recent acquisition of OI-20 on EHEC evolution provided this pathogen with a novel signal transduction system, suggesting that expression of this TCS in mucus facilitates EHEC adaptation to the mammalian intestine.

The modulation of the nutrient supply in the gut by commensal microbes and its effects on bacterial virulence supports the use of probiotic interventions to control bacterial infections. It was demonstrated that the probiotic strain *Lactobacillus casei* consumes fucosyl-α-1,3- N-acetylglucosamine (Fuc-α-1,3-GlcNAc) as a carbon source and releases free L-fucose into the media (117). Given that fucose can repress EHEC virulence, it would be interesting to investigate the effects of a symbiotic approach (i.e., combination of a probiotic strain and a prebiotic diet) on preventing or treating EHEC infections.

Sensing of Glycolytic versus Gluconeogenic Environments

Glycolytic environments inhibit the expression of the LEE genes. Conversely, growth in a gluconeogenic environment activates expression of these genes. Part of this sugar-dependent regulation is achieved through two transcription factors: KdpE and Cra. KdpE and Cra interact to optimally and directly activate expression of the LEE genes in a metabolite-dependent fashion (118). EHEC competes with commensal *E. coli* (the predominant species in the γ-Proteobacteria) for the same carbon sources during growth in the mammalian intestine (1, 44, 47, 79, 89). EHEC uses glycolytic substrates for initial growth but is unable to effectively compete for these carbon sources beyond the first few days and begins to utilize gluconeogenic substrates to stay within the intestine (44). Hence, it is advantageous to coordinate expression of the LEE with these environmental conditions. Commensal *E. coli* can be found in the lumen, which is glycolytic due to the abundant sugar sources supplied by the glycophagic microbiota, while the interface with the epithelium is a more gluconeogenic environment. Hence, the KdpE/Cra-dependent activation of the LEE under gluconeogenic conditions ensures that these genes are optimally expressed only at the epithelium interface and not in the lumen (Fig. 1).

Ethanolamine Utilization

Ethanolamine is a breakdown product of phosphatidylethanolamine, which is an abundant phospholipid of mammalian and bacterial cell membranes (119–121). Epithelial cell turnover and the gut microbiota are important sources of ethanolamine in the gut, which can be taken up and utilized as a carbon and/or nitrogen source by a number of bacterial species, including pathogenic bacteria such as EHEC and *Salmonella*. Exfoliation of intestinal cells also releases

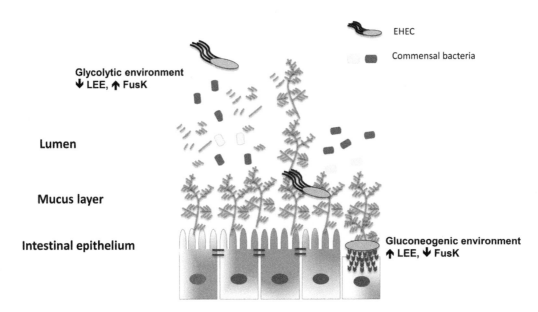

FIGURE 1 Nutritional cues regulate the locus of enterocyte effacement (LEE) gene expression in enterohemorrhagic *E. coli* (EHEC). Glycophagic members of the microbiota such as *B. theta* make fucose from mucin accessible to EHEC, and EHEC interprets this information to recognize that it is in the lumen, where expression of its LEE-encoded type III secretion system (TTSS) is onerous and not advantageous. Using yet another nutrient-based environmental cue, EHEC also times LEE expression through recognition of glycolytic and gluconeogenic environments. The lumen is more glycolytic due to predominant glycophagic members of the microbiota degrading complex polysaccharides into monosaccharides that can be readily utilized by nonglycophagic bacterial species such as *E. coli* and *C. rodentium*. In contrast, the tight mucus layer between the lumen and the epithelial interface in the gastrointestinal (GI) tract is devoid of microbiota; it is known as a "zone of clearance." At the epithelial interface, the environment is regarded as gluconeogenic. Hence, the coupling of LEE regulation to optimal expression under gluconeogenic and low-fucose conditions mirrors the interface with the epithelial layer environment in the GI tract, ensuring that EHEC will express only LEE at optimal levels to promote attaching and effacing lesion formation at the epithelial interface. doi:10.1128/microbiolspec.MBP-0001-2014.f1

ethanolamine into the intestine (122–125). Ethanolamine is broken down into acetaldehyde and ammonia by the enzyme complex ethanolamine ammonia lyase, encoded by genes *eutB* and *eutC*. Ammonia is used as a nitrogen source, while acetaldehyde is converted into acetyl-CoA (126).

Ethanolamine consumption supports the growth of EHEC *in vivo* and also confers EHEC a competitive advantage over the indigenous microbiota in the bovine intestine (127). EHEC can utilize ethanolamine as a nitrogen source in the bovine small intestine, in contrast with commensal *E. coli* strains. Ethanolamine metabolism allows EHEC to flourish in the bovine gut (its main reservoir),

which contributes to the spreading of EHEC infections (127).

In addition to its role as a nitrogen source, ethanolamine functions as a signaling molecule that triggers EHEC virulence expression. Kendall et al demonstrated that EHEC senses ethanolamine partially via EutR, a previously known receptor for ethanolamine. On EHEC growth on M9 minimal medium with ethanolamine as the sole nitrogen source, expression of the LEE and Stx increases markedly, indicating that ethanolamine is a signal that triggers EHEC virulence gene expression (123). It was also shown that ethanolamine triggers transcription of the EHEC adrenergic sensors *qseC* and *qseE*, which are involved in cell-to-cell signaling and bacteria-host communication (123). These studies suggest that ethanolamine sensing may contribute to EHEC persistence in the mammalian gut, not only by supporting EHEC growth but also by controlling transcription of major virulence factors. The research conducted by Kendall et al also indicates that EHEC encodes an additional, yet unidentified, ethanolamine sensor (123).

While currently available data leave no doubt of the pivotal role played by ethanolamine during gut colonization by enteric pathogens, little is known of the sensory systems employed in ethanolamine detection. Future research is necessary to unravel the receptors involved in early ethanolamine sensing, which are critical steps in infection.

The Effects of Inflammation

Pathogen-promoted inflammation during enteric infection is now appreciated as a strategy to promote rather than a consequence of bacterial infection. Enteric pathogens such as *C. rodentium* and *Salmonella* can benefit from the inflammatory environment or the overall changes in the bacterial community that result from inflammation. By provoking intestinal inflammation, the murine pathogen *C. rodentium* reduces the overall number of commensal bacteria in the microbiota, which gives the pathogen a colonization advantage (76, 128). Although later stages of inflammation result in pathogen clearance from the gut, inflammation in the early stages of infection helps *C. rodentium* replicate and increase its population when competing with commensal microbes.

Destruction of intestinal integrity by inflammation promotes *Salmonella* Typhimurium persistence in the gut. Inflammation triggered by *Salmonella* releases a new electron acceptor, tetrathionate, which allows *Salmonella* to outcompete the gut microbiota and proliferate in the gut lumen (129). In addition, tetrathionate allows *Salmonella*

Typhimurium to use ethanolamine as a carbon source in the inflamed intestine (124). The *eutC* mutant, which cannot grow anaerobically on ethanolamine as a carbon source, was outcompeted by the wild-type (WT) strain only in the presence of tetrathionate, indicating that ethanolamine utilization and tetrathionate respiration likely occur concomitantly. Ethanolamine levels present in colons of mice infected and uninfected were similar, indicating that ethanolamine consumption in the inflamed intestine was not due to release of ethanolamine due to epithelial cell destruction caused by inflammation. Interestingly, the growth advantage conferred by the ability to consume ethanolamine *in vivo* relies on the ability to respire tetrathionate (124). In the absence of the electron acceptor tetrathionate, respiration of ethanolamine does not support *Salmonella* growth in the mouse intestine. Therefore, the inflammatory response orchestrated by *Salmonella* during infection creates a nutritional environment that supports its replication in the gut lumen.

Nutrient Competition

Kamada et al (88) have demonstrated that the combined effects of virulence gene expression and competition with the microbiota are both crucial for pathogen clearance using the murine pathogen *C. rodentium*. *C. rodentium* is a natural mouse pathogen that causes colonic hyperplasia, and similarly to EHEC and EPEC, forms AE lesions on IECs. *C. rodentium* has been extensively used as a model for EHEC and EPEC infections (88). During infection of conventional mice, *C. rodentium* requires expression of the LEE to compete with indigenous microbes, while LEE expression is not necessary for *C. rodentium* colonization of germ-free mice. It was also shown that virulence gene expression (LEE) was triggered early but was reduced during late stages of infection, causing relocation of *C. rodentium* from the epithelium to the gut lumen, where the pathogen was exposed to commensal bacteria and had to compete for similar carbon sources for luminal growth. This shows that virulence and metabolism act in concert during bacterial infection, and both nutrient utilization and production of virulence traits are required to establish a successful colonization by pathogenic bacteria. A closer look at the nutrient competition between *C. rodentium* and commensal *E. coli* and *B. tetha* indicated that *E. coli* can outcompete *C. rodentium* due its ability to grow on monosaccharides, while *B. theta* does not outcompete *C. rodentium* because it can grow on polysaccharides. This work demonstrated that competition for similar nutrient sources is an important determinant of the outcome of bacterial infections of the mammalian intestine, reinforces Freter's concept, and raises the possibility that shifting the commensal microbiota towards nutrient competition with pathogens may be an alternative to fight bacterial infections.

A recent study shows that EHEC colonization could be prevented by the probiotic strains *E. coli* Nissle 1917 and *E. coli* HS, based on the ability of these combined commensal strains to compete for the carbohydrate niches occupied by EHEC to colonize the mammalian gut. EHEC utilizes arabinose, galactose, and gluconate, carbohydrates also consumed by *E. coli* Nissle 1917 and *E. coli* HS. EHEC also competes with *E. coli* HS for ribose and N-acetylglucosamine, while it competes with *E. coli* Nissle 1917 for mannose (81).

CONCLUSIONS

Nutrient scavenging by pathogenic bacteria from microbiota-derived products is an emerging theme in bacterial pathogenesis. Dietary changes causing shifts in gut microbial populations are well established, although the consequences regarding infection by pathogenic agents are mostly unknown. The use of probiotic strains to reinforce colonization resistance may be a better alternative to treatment of infections for which antibiotic treatment is not advisable, such as EHEC and *Salmonella*. Future investigations on the relationships between indigenous microbiota members and pathogenic microorganisms are crucial for the development of new effective preventive and curative strategies for enteric infections.

CITATION

Pacheco AR, Sperandio V. 2014. Enteric pathogens exploit the microbiota-generated nutritional environment of the gut. Microbiol Spectrum 3(3):MBP-0001-2014.

REFERENCES

1. **Kamada N, Kim YG, Sham HP, Vallance BA, Puente JL, Martens EC, Nunez G.** 2012. Regulated virulence controls the ability of a pathogen to compete with the gut microbiota. *Science* **336:**1325–1329.
2. **Thiennimitr P, Winter SE, Baumler AJ.** 2011. Salmonella, the host and its microbiota. *Curr Opin Microbiol* **15:**108–114.
3. **Gill SR, Pop M, Deboy RT, Eckburg PB, Turnbaugh PJ, Samuel BS, Gordon GL, Relman DA, Fraser-Liggett CM, Nelson KE.** 2006. Metagenomic analysis of the human distal gut microbiome. *Science* **312:**1355–1359.
4. **Vaishnava S, Yamamoto M, Severson KM, Ruhn KA, Yu X, Koren O, Ley R, Wakeland EK, Hooper LV.** 2011. The antibacterial lectin RegIIIgamma promotes the spatial segregation of microbiota and host in the intestine. *Science* **334:**255–258.
5. **Stecher B, Maier L, Hardt WD.** 2013. 'Blooming' in the gut: how dysbiosis might contribute to pathogen evolution. *Nat Rev Microbiol* **11:**277–284.
6. **Lupp C, Robertson ML, Wickham ME, Sekirov I, Champion OL, Gaynor EC, Finlay BB.** 2007. Host-mediated inflammation disrupts the intestinal microbiota and promotes the overgrowth of Enterobacteriaceae. *Cell Host Microbe* **2:**204.
7. **Stecher B, Robbiani R, Walker AW, Westendorf AM, Barthel M, Kremer M, Chaffron S, Macpherson AJ, Buer J, Parkhill J, Dougan G, von Mering C, Hardt WD.** 2007. Salmonella enterica serovar typhimurium exploits

inflammation to compete with the intestinal microbiota. *PLoS Biol* **5:**2177–2189.

8. **Sekirov I, Tam NM, Jogova M, Robertson ML, Li Y, Lupp C, Finlay BB.** 2008. Antibiotic-induced perturbations of the intestinal microbiota alter host susceptibility to enteric infection. *Infect Immun* **76:**4726–4736.

9. **Freter R, Brickner H, Botney M, Cleven D, Aranki A.** 1983. Mechanisms that control bacterial populations in continuous-flow culture models of mouse large intestinal flora. *Infect Immun* **39:**676–685.

10. **Lamont JT.** 1992. Mucus: the front line of intestinal mucosal defense. *Ann N Y Acad Sci* **664:**190–201.

11. **McGuckin MA, Linden SK, Sutton P, Florin TH.** 2011. Mucin dynamics and enteric pathogens. *Nat Rev Microbiol* **9:**265–278.

12. **Jarvis KG, Giron JA, Jerse AE, McDaniel EK, Donnenberg MS, Kaper JB.** 1995. Enteropathogenic *Escherichia coli* contains a putative type III secretion system necessary for the export of proteins involved in attaching and effacing lesion formation. *Proc Natl Acad Sci U S A* **92:**7996–8000.

13. **Sperandio V, Mellies JL, Nguyen W, Shin S, Kaper JB.** 1999. Quorum sensing controls expression of the type III secretion gene transcription and protein secretion in enterohemorrhagic and enteropathogenic Escherichia coli. *Proc Natl Acad Sci U S A* **96:**15196–15201.

14. **Harrington SM, Sheikh J, Henderson LR, Ruiz-Perez F, Cohen PS, Nataro JP.** 2009. The Pic protease of enteroaggregative *Escherichia coli* promotes intestinal colonization and growth in the presence of mucin. *Infect Immun* **77:**2465–2473.

15. **O'Brien AD, LaVeck GD, Griffin DE, Thompson MR.** 1980. Characterization of Shigella dysenteriae 1 (Shiga) toxin purified by anti-Shiga toxin affinity chromatography. *Infect Immun* **30:**170–179.

16. **Weinstein DL, Carsiotis M, Lissner CR, O'Brien AD.** 1984. Flagella help *Salmonella typhimurium* survive within murine macrophages. *Infect Immun* **46:**819–825.

17. **Finlay BB, Falkow S.** 1990. Salmonella interactions with polarized human intestinal Caco-2 epithelial cells. *J Infect Dis* **162:**1096–1106.

18. **Ley RE, Hamady M, Lozupone C, Turnbaugh PJ, Ramey RR, Bircher JS, Schlegel ML, Tucker TA, Schrenzel MD, Knight R, Gordon JL.** 2008. Evolution of mammals and their gut microbes. *Science* **320:**1647–1651.

19. **Savage DC.** 1977. Microbial ecology of the gastrointestinal tract. *Annu Rev Microbiol* **31:**107–133.

20. **Eckburg PB, Bik EM, Bernstein CN, Purdom E, Dethlefsen L, Sargent M, Gill SR, Nelson KE, Relman DA.** 2005. Diversity of the human intestinal microbial flora. *Science* **308:**1635–1638.

21. **Qin J, Li R, Raes J, Arumugam M, Burgdorf KS, Manichanh C, Nielsen T, Pons N, Levenez F, Yamada T, Mende DR, Li J, Xu J, Li S, Li D, Cao J, Wang B, Liang H, Zheng H, Xie Y, Tap J, Lepage P, Bertalan M, Batto JM, Hansen T, Le Paslier D, Linneberg A, Nielsen HB, Pelletier E, Renault P, Sicheritz-Ponten T, Turner K, Zhu H, Yu C, Jian M, Zhou Y, Li Y, Zhang X, Qin N, Yang H, Wang J, Brunak S, Dore J, Guarner F, Kristiansen K, Pedersen O, Parkhill J, Weissenbach J, Bork P, Ehrlich SD.** 2010. A human gut microbial gene catalogue established by metagenomic sequencing. *Nature* **464:**59–65.

22. **Stewart JA, Chadwick VS, Murray A.** 2005. Investigations into the influence of host genetics on the predominant eubacteria in the faecal microflora of children. *J Med Microbiol* **54:**1239–1242.

23. **Benson AK, Kelly SA, Legge R, Ma F, Low SJ, Kim J, Zhang M, Oh PL, Nehrenberg D, Hua K, Kachman SD, Moriyama EN, Walter J, Peterson DA, Pomp D.** 2010. Individuality in gut microbiota composition is a complex polygenic trait shaped by multiple environmental and host genetic factors. *Proc Natl Acad Sci U S A* **107:**18933–18938.

24. **O'Hara AM, Shanahan F.** 2006. The gut flora as a forgotten organ. *EMBO Rep* **7:**688–693.

25. **Peterson J, Garges S, Giovanni M, McInnes P, Wang L, Schloss JA, Bonazzi V, McEwen JE, Wetterstrand KA, Deal C, Baker CC, Di Francesco V, Howcroft TK, Karp RW, Lunsford RD, Wellington CR, Belachew T, Wright M, Giblin C, David H, Mills M, Salomon R, Mullins C, Akolkar B, Begg L, Davis C, Grandison L, Humble M, Khalsa J, Little AR, Peavy H, Pontzer C, Portnoy M, Sayre MH, Starke-Reed P, Zakhari S, Read J, Watson B, Guyer M.** 2009. The NIH Human Microbiome Project. *Genome Res* **19:**2317–2323.

26. **Turnbaugh PJ, Ley RE, Hamady M, Fraser-Liggett CM, Knight R, Gordon JI.** 2007. The human microbiome project. *Nature* **449:**804–810.

27. **Png CW, Linden SK, Gilshenan KS, Zoetendal EG, McSweeney CS, Sly LI, McGuckin MA, Florin TH.** 2010. Mucolytic bacteria with increased prevalence in IBD mucosa augment in vitro utilization of mucin

by other bacteria. *Am J Gastroenterol* **105:** 2420–2428.

28. **Tu QV, McGuckin MA, Mendz GL.** 2008. Campylobacter jejuni response to human mucin MUC2: modulation of colonization and pathogenicity determinants. *J Med Microbiol* **57:**795–802.

29. **Johansson ME, Larsson JM, Hansson GC.** 2011. The two mucus layers of colon are organized by the MUC2 mucin, whereas the outer layer is a legislator of host-microbial interactions. *Proc Natl Acad Sci U S A* **108** (Suppl 1):4659–4665.

30. **Larson G, Falk P, Hoskins LC.** 1988. Degradation of human intestinal glycosphingolipids by extracellular glycosidases from mucin-degrading bacteria of the human fecal flora. *J Biol Chem* **263:**10790–10798.

31. **Hoskins LC, Agustines M, McKee WB, Boulding ET, Kriaris M, Niedermeyer G.** 1985. Mucin degradation in human colon ecosystems. Isolation and properties of fecal strains that degrade ABH blood group antigens and oligosaccharides from mucin glycoproteins. *J Clin Invest* **75:**944–953.

32. **Marcobal A, Southwick AM, Earle KA, Sonnenburg JL.** 2013. A refined palate: Bacterial consumption of host glycans in the gut. *Glycobiology* **23:**1038–1046.

33. **Bergstrom KS, Kissoon-Singh V, Gibson DL, Ma C, Montero M, Sham HP, Ryz N, Huang T, Velcich A, Finlay BB, Chadee K, Vallance BA.** 2010. Muc2 protects against lethal infectious colitis by disassociating pathogenic and commensal bacteria from the colonic mucosa. *PLoS Pathog* **6:**e1000902.

34. **Kim YS, Ho SB.** 2010. Intestinal goblet cells and mucins in health and disease: recent insights and progress. *Curr Gastroenterol Rep* **12:**319–330.

35. **Salzman N, Underwood MA, Bevins CL.** 2007. Paneth cells, defensins, and the commensal microbiota: a hypothesis on intimate interplay at the intestinal mucosa. *Semin Immunol* **19:**70–83.

36. **Mack DR, Michail S, Wei S, McDougall L, Hollingsworth MA.** 1999. Probiotics inhibit enteropathogenic E. coli adherence in vitro by inducing intestinal mucin gene expression. *Am J Physiol* **276:**G941–G950.

37. **Robbe C, Capon C, Flahaut C, Michalski JC.** 2003. Microscale analysis of mucin-type O-glycans by a coordinated fluorophore-assisted carbohydrate electrophoresis and mass spectrometry approach. *Electrophoresis* **24:**611–621.

38. **Corfield AP, Wagner SA, Clamp JR, Kriaris MS, Hoskins LC.** 1992. Mucin degradation in the human colon: production of sialidase, sialate O-acetylesterase, N-acetylneuraminate lyase, arylesterase, and glycosulfatase activities by strains of fecal bacteria. *Infect Immun* **60:**3971–3978.

39. **Johansson ME, Phillipson M, Petersson J, Velcich A, Holm L, Hansson GC.** 2008. The inner of the two Muc2 mucin-dependent mucus layers in colon is devoid of bacteria. *Proc Natl Acad Sci USA* **105:**15064–15069.

40. **Robbe C, Capon C, Coddeville B, Michalski JC.** 2004. Structural diversity and specific distribution of O-glycans in normal human mucins along the intestinal tract. *Biochem J* **384:**307–316.

41. **Resta-Lenert S, Das S, Batra SK, Ho SB.** 2011. Muc17 protects intestinal epithelial cells from enteroinvasive E. coli infection by promoting epithelial barrier integrity. *Am J Physiol Gastrointest Liver Physiol* **300:**G1144–G1155.

42. **Grys TE, Walters LL, Welch RA.** 2006. Characterization of the StcE protease activity of Escherichia coli O157:H7. *J Bacteriol* **188:** 4646–4653.

43. **Wadolkowski EA, Laux DC, Cohen PS.** 1988. Colonization of the streptomycin-treated mouse large intestine by a human fecal Escherichia coli strain: role of growth in mucus. *Infect Immun* **56:**1030–1035.

44. **Miranda RL, Conway T, Leatham MP, Chang DE, Norris WE, Allen JH, Stevenson SJ, Laux DC, Cohen PS.** 2004. Glycolytic and gluconeogenic growth of Escherichia coli O157:H7 (EDL933) and E. coli K-12 (MG1655) in the mouse intestine. *Infect Immun* **72:**1666–1676.

45. **Peekhaus N, Conway T.** 1998. What's for dinner?: Entner-Doudoroff metabolism in Escherichia coli. *J Bacteriol* **180:**3495–3502.

46. **Montagne L, Toullec R, Lalles JP.** 2000. Calf intestinal mucin: isolation, partial characterization, and measurement in ileal digesta with an enzyme-linked immunosorbent assay. *J Dairy Sci* **83:**507–517.

47. **Chang DE, Smalley DJ, Tucker DL, Leatham MP, Norris WE, Stevenson SJ, Anderson AB, Grissom JE, Laux DC, Cohen PS, Conway T.** 2004. Carbon nutrition of Escherichia coli in the mouse intestine. *Proc Natl Acad Sci U S A* **101:**7427–7432.

48. **Martens EC, Roth R, Heuser JE, Gordon JI.** 2009. Coordinate regulation of glycan degradation and polysaccharide capsule biosynthesis by a prominent human gut symbiont. *J Biol Chem* **284:**18445–18457.

49. **McNulty NP, Wu M, Erickson AR, Pan C, Erickson BK, Martens EC, Pudlo NA,**

Muegge BD, Henrissat B, Hettich RL, Gordon JI. 2013. Effects of Diet on Resource Utilization by a Model Human Gut Microbiota Containing Bacteroides cellulosilyticus WH2, a Symbiont with an Extensive Glycobiome. *PLoS Biol* **11**:e1001637.

50. Sonnenburg JL, Xu J, Leip DD, Chen CH, Westover BP, Weatherford J, Buhler JD, Gordon JI. 2005. Glycan foraging in vivo by an intestine-adapted bacterial symbiont. *Science* **307**:1955–1959.

51. Salyers AA, West SE, Vercellotti JR, Wilkins TD. 1977. Fermentation of mucins and plant polysaccharides by anaerobic bacteria from the human colon. *Appl Environ Microbiol* **34**:529–533.

52. Martens EC, Chiang HC, Gordon JI. 2008. Mucosal glycan foraging enhances fitness and transmission of a saccharolytic human gut bacterial symbiont. *Cell Host Microbe* **4**:447–457.

53. Martens EC, Lowe EC, Chiang H, Pudlo NA, Wu M, McNulty NP, Abbott DW, Henrissat B, Gilbert HJ, Bolam DN, Gordon JI. 2011. Recognition and degradation of plant cell wall polysaccharides by two human gut symbionts. *PLoS Biol* **9**:e1001221.

54. Marcobal A, Barboza M, Sonnenburg ED, Pudlo N, Martens EC, Desai P, Lebrilla CB, Weimer BC, Mills DA, German JB, Sonnenburg JL. 2011. Bacteroides in the infant gut consume milk oligosaccharides via mucus-utilization pathways. *Cell Host Microbe* **10**:507–514.

55. Flint HJ, Bayer EA, Rincon MT, Lamed R, White BA. 2008. Polysaccharide utilization by gut bacteria: potential for new insights from genomic analysis. *Nat Rev Microbiol* **6**:121–131.

56. Moore WE, Holdeman LV. 1974. Human fecal flora: the normal flora of 20 Japanese-Hawaiians. *Appl Microbiol* **27**:961–979.

57. Ushijima T, Takahashi M, Tatewaki K, Ozaki Y. 1983. A selective medium for isolation and presumptive identification of the Bacteriodes fragilis group. *Microbiol Immunol* **27**:985–993.

58. Xu J, Mahowald MA, Ley RE, Lozupone CA, Hamady M, Martens EC, Henrissat B, Coutinho PM, Minx P, Latreille P, Cordum H, Van Brunt A, Kim K, Fulton RS, Fulton LA, Clifton SW, Wilson RK, Knight RD, Gordon JI. 2007. Evolution of symbiotic bacteria in the distal human intestine. *PLoS Biol* **5**:e156.

59. Bjursell MK, Martens EC, Gordon JI. 2006. Functional genomic and metabolic studies of the adaptations of a prominent adult human gut symbiont, Bacteroides thetaiotaomicron, to the suckling period. *J Biol Chem* **281**:36269–36279.

60. Garrido D, Ruiz-Moyano S, Mills DA. 2012. Release and utilization of N-acetyl-D-glucosamine from human milk oligosaccharides by Bifidobacterium longum subsp. infantis. *Anaerobe* **18**:430–435.

61. Garrido D, Dallas DC, Mills DA. 2013. Consumption of human milk glycoconjugates by infant-associated bifidobacteria: mechanisms and implications. *Microbiology* **159**:649–664.

62. Garrido D, Barile D, Mills DA. 2012. A molecular basis for bifidobacterial enrichment in the infant gastrointestinal tract. *Adv Nutr* **3**:415S–421S.

63. Collins MD, Gibson GR. 1999. Probiotics, prebiotics, and synbiotics: approaches for modulating the microbial ecology of the gut. *Am J Clin Nutr* **69**:1052S–1057S.

64. Pacheco AR, Curtis MM, Ritchie JM, Munera D, Waldor MK, Moreira CG, Sperandio V. 2012. Fucose sensing regulates bacterial intestinal colonization. *Nature* **492**:113–117.

65. Bry L, Falk PG, Midtvedt T, Gordon JI. 1996. A model of host-microbial interactions in an open mammalian ecosystem. *Science* **273**:1380–1383.

66. Coyne MJ, Reinap B, Lee MM, Comstock LE. 2005. Human symbionts use a host-like pathway for surface fucosylation. *Science* **307**:1778–1781.

67. Bjork S, Breimer ME, Hansson GC, Karlsson KA, Leffler H. 1987. Structures of blood group glycosphingolipids of human small intestine. A relation between the expression of fucolipids of epithelial cells and the ABO, Le and Se phenotype of the donor. *J Biol Chem* **262**:6758–6765.

68. Finne J, Breimer ME, Hansson GC, Karlsson KA, Leffler H, Vliegenthart JF, van Halbeek H. 1989. Novel polyfucosylated N-linked glycopeptides with blood group A, H, X, and Y determinants from human small intestinal epithelial cells. *J Biol Chem* **264**:5720–5735.

69. Hooper LV, Xu J, Falk PG, Midtvedt T, Gordon JI. 1999. A molecular sensor that allows a gut commensal to control its nutrient foundation in a competitive ecosystem. *Proc Natl Acad Sci U S A* **96**:9833–9838.

70. Fletcher CM, Coyne MJ, Villa OF, Chatzidaki-Livanis M, Comstock LE. 2009. A general O-glycosylation system important to the physiology of a major human intestinal symbiont. *Cell* **137**:321–331.

71. Fischbach MA, Sonnenburg JL. 2011. Eating for two: how metabolism establishes interspecies interactions in the gut. *Cell Host Microbe* **10**:336–347.

72. **Lozupone CA, Hamady M, Cantarel BL, Coutinho PM, Henrissat B, Gordon JI, Knight R.** 2008. The convergence of carbohydrate active gene repertoires in human gut microbes. *Proc Natl Acad Sci U S A* **105:**15076–15081.

73. **Xu J, Bjursell MK, Himrod J, Deng S, Carmichael LK, Chiang HC, Hooper LV, Gordon JI.** 2003. A genomic view of the human-Bacteroides thetaiotaomicron symbiosis. *Science* **299:**2074–2076.

74. **Derrien M, Vaughan EE, Plugge CM, de Vos WM.** 2004. Akkermansia muciniphila gen. nov., sp. nov., a human intestinal mucin-degrading bacterium. *Int J Syst Evol Microbiol* **54:**1469–1476.

75. **Deriu E, Liu JZ, Pezeshki M, Edwards RA, Ochoa RJ, Contreras H, Libby SJ, Fang FC, Raffatellu M.** 2013. Probiotic bacteria reduce salmonella typhimurium intestinal colonization by competing for iron. *Cell Host Microbe* **14:**26–37.

76. **Stecher B, Hardt WD.** 2008. The role of microbiota in infectious disease. *Trends Microbiol* **16:**107–114.

77. **Curtis MM, Sperandio V.** 2011. A complex relationship: the interaction among symbiotic microbes, invading pathogens, and their mammalian host. *Mucosal Immunol* **4:**133–138.

78. **Keeney KM, Finlay BB.** 2011. Enteric pathogen exploitation of the microbiota-generated nutrient environment of the gut. *Curr Opin Microbiol* **14:**92–98.

79. **Fabich AJ, Jones SA, Chowdhury FZ, Cernosek A, Anderson A, Smalley D, McHargue JW, Hightower GA, Smith JT, Autieri SM, Leatham MP, Lins JJ, Allen RL, Laux DC, Cohen PS, Conway T.** 2008. Comparison of carbon nutrition for pathogenic and commensal Escherichia coli strains in the mouse intestine. *Infect Immun* **76:**1143–1152.

80. **Momose Y, Hirayama K, Itoh K.** 2008. Competition for proline between indigenous Escherichia coli and E. coli O157:H7 in gnotobiotic mice associated with infant intestinal microbiota and its contribution to the colonization resistance against E. coli O157:H7. *Antonie Van Leeuwenhoek* **94:**165–171.

81. **Maltby R, Leatham-Jensen MP, Gibson T, Cohen PS, Conway T.** 2013. Nutritional basis for colonization resistance by human commensal Escherichia coli strains HS and Nissle 1917 against E. coli O157:H7 in the mouse intestine. *PLoS One* **8:**e53957.

82. **Snider TA, Fabich AJ, Conway T, Clinkenbeard KD.** 2009. E. coli O157:H7 catabolism of intestinal mucin-derived carbohydrates and colonization. *Vet Microbiol* **136:**150–154.

83. **Moller AK, Leatham MP, Conway T, Nuijten PJ, de Haan LA, Krogfelt KA, Cohen PS.** 2003. An Escherichia coli MG1655 lipopolysaccharide deep-rough core mutant grows and survives in mouse cecal mucus but fails to colonize the mouse large intestine. *Infect Immun* **71:**2142–2152.

84. **Ley RE, Lozupone CA, Hamady M, Knight R, Gordon JI.** 2008. Worlds within worlds: evolution of the vertebrate gut microbiota. *Nat Rev Microbiol* **6:**776–788.

85. **Bertin Y, Chaucheyras-Durand F, Robbe-Masselot C, Durand A, de la Foye A, Harel J, Cohen PS, Conway T, Forano E, Martin C.** 2013. Carbohydrate utilization by enterohaemorrhagic Escherichia coli O157:H7 in bovine intestinal content. *Environ Microbiol* **15:**610–622.

86. **Naylor SW, Roe AJ, Nart P, Spears K, Smith DG, Low JC, Gally DL.** 2005. Escherichia coli O157 : H7 forms attaching and effacing lesions at the terminal rectum of cattle and colonization requires the LEE4 operon. *Microbiology* **151:**2773–2781.

87. **Freter R, Stauffer E, Cleven D, Holdeman LV, Moore WE.** 1983. Continuous-flow cultures as in vitro models of the ecology of large intestinal flora. *Infect Immun* **39:**666–675.

88. **Kamada N, Kim YG, Sham HP, Vallance BA, Puente JL, Martens EC, Nunez G.** 2012. Regulated Virulence Controls the Ability of a Pathogen to Compete with the Gut Microbiota. *Science* **336:**1325–1329.

89. **Autieri SM, Lins JJ, Leatham MP, Laux DC, Conway T, Cohen PS.** 2007. L-fucose stimulates utilization of D-ribose by Escherichia coli MG1655 DeltafucAO and E. coli Nissle 1917 DeltafucAO mutants in the mouse intestine and in M9 minimal medium. *Infect Immun* **75:**5465–5475.

90. **Karmali MA.** 2004. Prospects for preventing serious systemic toxemic complications of Shiga toxin-producing Escherichia coli infections using Shiga toxin receptor analogues. *J Infect Dis* **189:**355–359.

91. **Tarr PI, Gordon CA, Chandler WL.** 2005. Shiga-toxin-producing Escherichia coli and haemolytic uraemic syndrome. *Lancet* **365:**1073–1086.

92. **Kaper JB, Nataro JP, Mobley HL.** 2004. Pathogenic Escherichia coli. *Nat Rev Microbiol* **2:**123–140.

93. **McDaniel TK, Jarvis KG, Donnenberg MS, Kaper JB.** 1995. A genetic locus of enterocyte

effacement conserved among diverse entero-bacterial pathogens. *Proc Natl Acad Sci U S A* **92:**1664–1668.

94. **Elliott SJ, Hutcheson SW, Dubois MS, Mellies JL, Wainwright LA, Batchelor M, Frankel G, Knutton S, Kaper JB.** 1999. Identification of CesT, a chaperone for the type III secretion of Tir in enteropathogenic Escherichia coli. *Mol Microbiol* **33:**1176–1189.

95. **Mellies JL, Elliott SJ, Sperandio V, Donnenberg MS, Kaper JB.** 1999. The Per regulon of entero-pathogenic Escherichia coli : identification of a regulatory cascade and a novel transcriptional activator, the locus of enterocyte effacement (LEE)-encoded regulator (Ler). *Mol Microbiol* **33:**296–306.

96. **Elliott SJ, Wainwright LA, McDaniel TK, Jarvis KG, Deng YK, Lai LC, McNamara BP, Donnenberg MS, Kaper JB.** 1998. The complete sequence of the locus of enterocyte effacement (LEE) from enteropathogenic Escherichia coli E2348/69. *Mol Microbiol* **28:**1–4.

97. **Jerse AE, Yu J, Tall BD, Kaper JB.** 1990. A genetic locus of enteropathogenic Escherichia coli necessary for the production of attaching and effacing lesions on tissue culture cells. *Proc Natl Acad Sci U S A* **87:**7839–7843.

98. **Kenny B, DeVinney R, Stein M, Reinscheid DJ, Frey EA, Finlay BB.** 1997. Enteropatho-genic E. coli (EPEC) transfers its receptor for intimate adherence into mammalian cells. *Cell* **91:**511–520.

99. **McNamara BP, Donnenberg MS.** 1998. A novel proline-rich protein, EspF, is secreted from enteropathogenic Escherichia coli via the type III export pathway. *FEMS Microbiol Lett* **166:**71–78.

100. **Kenny B, Jepson M.** 2000. Targeting of an enteropathogenic Escherichia coli (EPEC) effector protein to host mitochondria. *Cell Microbiol* **2:**579–590.

101. **Elliott SJ, Krejany EO, Mellies JL, Robins-Browne RM, Sasakawa C, Kaper JB.** 2001. EspG, a novel type III system-secreted protein from enteropathogenic Escherichia coli with similarities to VirA of Shigella flexneri. *Infect Immun* **69:**4027–4033.

102. **Tu X, Nisan I, Yona C, Hanski E, Rosenshine I.** 2003. EspH, a new cytoskeleton-modulating effector of enterohaemorrhagic and entero-pathogenic Escherichia coli. *Mol Microbiol* **47:**595–606.

103. **Kanack KJ, Crawford JA, Tatsuno I, Karmali MA, Kaper JB.** 2005. SepZ/EspZ is secreted and translocated into HeLa cells by the enteropathogenic Escherichia coli type III secretion system. *Infect Immun* **73:**4327–4337.

104. **Campellone KG, Robbins D, Leong JM.** 2004. EspFU is a translocated EHEC effector that interacts with Tir and N-WASP and promotes Nck-independent actin assembly. *Developmental cell* **7:**217–228.

105. **Sperandio V, Torres AG, Jarvis B, Nataro JP, Kaper JB.** 2003. Bacteria-host communica-tion: the language of hormones. *Proc Natl Acad Sci U S A* **100:**8951–8956.

106. **Tilden J Jr, Young W, McNamara AM, Custer C, Boesel B, Lambert-Fair MA, Majkowski J, Vugia D, Werner SB, Hollingsworth J, Morris GJ Jr.** 1996. A new route of transmission for Escherichia coli: infection from dry fermented salami. *Am J Public Health* **86:**1142–1145.

107. **Clarke MB, Hughes DT, Zhu C, Boedeker EC, Sperandio V.** 2006. The QseC sensor kinase: A bacterial adrenergic receptor. *Proc Natl Acad Sci USA* **103:**10420–10425.

108. **Reading NC, Rasko DA, Torres AG, Sperandio V.** 2009. The two-component sys-tem QseEF and the membrane protein QseG link adrenergic and stress sensing to bacterial pathogenesis. *Proc Natl Acad Sci U S A* **106:**5889–5894.

109. **Reading NC, Torres AG, Kendall MM, Hughes DT, Yamamoto K, Sperandio V.** 2007. A novel two-component signaling system that activates transcription of an enterohemorrhagic Esche-richia coli effector involved in remodeling of host actin. *J Bacteriol* **189:**2468–2476.

110. **Clarke MB, Hughes DT, Zhu C, Boedeker EC, Sperandio V.** 2006. The QseC sensor kinase: A bacterial adrenergic receptor. *Proc Natl Acad Sci USA* **103:**10420–10425.

111. **Rasko DA, Moreira CG, Li de R, Reading NC, Ritchie JM, Waldor MK, Williams N, Taussig R, Wei S, Roth M, Hughes DT, Huntley JF, Fina MW, Falck JR, Sperandio V.** 2008. Targeting QseC signaling and viru-lence for antibiotic development. *Science* **321:**1078–1080.

112. **Yamamoto K, Hirao K, Oshima T, Aiba H, Utsumi R, Ishihama A.** 2005. Functional characterization in vitro of all two-component signal transduction systems from Escherichia coli. *J Biol Chem* **280:**1448–1456.

113. **Hughes DT, Sperandio V.** 2008. Inter-king-dom signalling: communication between bac-teria and their hosts. *Nat Rev Microbiol* **6:**111–120.

114. **Ritchie JM, Thorpe CM, Rogers AB, Waldor MK.** 2003. Critical roles for stx2, eae, and tir in enterohemorrhagic Escherichia coli-induced diarrhea and intestinal inflamma-tion in infant rabbits. *Infect Immun* **71:**7129–7139.

115. **Reid SD, Herbelin CJ, Bumbaugh AC, Selander RK, Whittam TS.** 2000. Parallel evolution of virulence in pathogenic Escherichia coli. *Nature* **406:**64–67.

116. **Wick LM, Qi W, Lacher DW, Whittam TS.** 2005. Evolution of genomic content in the stepwise emergence of Escherichia coli O157: H7. *J Bacteriol* **187:**1783–1791.

117. **Rodriguez-Diaz J, Rubio-del-Campo A, Yebra MJ.** 2012. Lactobacillus casei ferments the N-Acetylglucosamine moiety of fucosyl-alpha-1,3-N-acetylglucosamine and excretes L-fucose. *Appl Environ Microbiol* **78:**4613–4619.

118. **Njoroge JW, Nguyen Y, Curtis MM, Moreira CG, Sperandio V.** 2012. Virulence Meets Metabolism: Cra and KdpE Gene Regulation in Enterohemorrhagic Escherichia coli. *MBio* 3:e00280-00212.

119. **Bakovic M, Fullerton MD, Michel V.** 2007. Metabolic and molecular aspects of ethanolamine phospholipid biosynthesis: the role of CTP:phosphoethanolamine cytidylyltransferase (Pcyt2). *Biochem Cell Biol* **85:**283–300.

120. **Dowhan W.** 1997. Molecular basis for membrane phospholipid diversity: why are there so many lipids? *Annu Rev Biochem* **66:**199–232.

121. **Dowhan W.** 1997. Phosphatidylserine decarboxylases: pyruvoyl-dependent enzymes from bacteria to mammals. *Methods Enzymol* **280:**81–88.

122. **Cotton PB.** 1972. Non-dietary lipid in the intestinal lumen. *Gut* **13:**675–681.

123. **Kendall MM, Gruber CC, Parker CT, Sperandio V.** 2012. Ethanolamine controls expression of genes encoding components involved in inter-kingdom signaling and virulence in enterohemorrhagic Escherichia coli O157:H7. *MBio* **3:**e00050-12.

124. **Thiennimitr P, Winter SE, Winter MG, Xavier MN, Tolstikov V, Huseby DL, Sterzenbach T, Tsolis RM, Roth JR, Baumler AJ.** 2011. Intestinal inflammation allows Salmonella to use ethanolamine to compete with the microbiota. *Proc Natl Acad Sci U S A* **108:**17480–17485.

125. **Stojiljkovic I, Baumler AJ, Heffron F.** 1995. Ethanolamine utilization in Salmonella typhimurium: nucleotide sequence, protein expression, and mutational analysis of the cchA cchB eutE eutJ eutG eutH gene cluster. *J Bacteriol* **177:**1357–1366.

126. **Garsin DA.** 2010. Ethanolamine utilization in bacterial pathogens: roles and regulation. *Nat Rev Microbiol* **8:**290–295.

127. **Bertin Y, Girardeau JP, Chaucheyras-Durand F, Lyan B, Pujos-Guillot E, Harel J, Martin C.** 2011. Enterohaemorrhagic Escherichia coli gains a competitive advantage by using ethanolamine as a nitrogen source in the bovine intestinal content. *Environ Microbiol* **13:**365–377.

128. **Lupp C, Robertson ML, Wickham ME, Sekirov I, Champion OL, E Gaynor EC, Finlay BB.** 2007. Host-mediated inflammation disrupts the intestinal microbiota and promotes the overgrowth of Enterobacteriaceae. *Cell Host Microbe* **2:**119–129.

129. **Winter SE, Thiennimitr P, Winter MG, Butler BP, Huseby DL, Crawford RW, Russell JM, Bevins CL, Adams LG, Tsolis RM, Roth JR, Baumler AJ.** 2010. Gut inflammation provides a respiratory electron acceptor for Salmonella. *Nature* **467:**426–429.

The Roles of Inflammation, Nutrient Availability and the Commensal Microbiota in Enteric Pathogen Infection

14

BÄRBEL STECHER[1][2]

CHARACTERISTICS AND MECHANISMS OF INFLAMMATION-MEDIATED DYSBIOSIS

Characteristics of the Inflammation-Associated Gut Microbiota

Genetic susceptibility, the mucosal immune system, and environmental factors such as the microbiota, stress, and diet, contribute to the pathogenesis of inflammatory bowel disease (IBD) (15). Involvement of the microbiota has been proposed early on, as microbiota manipulation by antibiotics or probiotics can treat or alleviate IBD symptoms in humans. In experimental animal models, the gut luminal microbiota is required for the induction of chronic inflammation (16). Different theories about how the microbiota is involved in the pathogenesis of IBD have been proposed. (1) Mutations that lead to a defective mucosal barrier function (e.g., mucus layer, innate killing, antimicrobial peptides) involve excessive translocation of commensal bacteria and triggering of proinflammatory signalling cascades. (2) Abnormal host immune regulation induces an overshooting immune response against intrinsic commensal bacteria. (3) The presence of an unidentified pathogen leads to induction of the disease or (4), a dysbiotic microbiota, characterized

[1]Max von Pettenkofer Institute, LMU Munich, Pettenkoferstr. 9a, 80336 Munich, Germany; [2]German Center for Infection Research (DZIF), Partner Site LMU Munich, Pettenkoferstr. 9a, 80336 Munich, Germany.
Metabolism and Bacterial Pathogenesis
Edited by Tyrrell Conway and Paul Cohen
© 2015 American Society for Microbiology, Washington, DC
doi:10.1128/microbiolspec.MBP-0008-2014

by an imbalance between "beneficial" and "potentially harmful" commensal bacteria, acts as a trigger or driver of the disease. The latter theory has been challenged by studies conducted in experimental rodent models: inflammatory disease conditions in the course of chronic colitis or enteropathogen infection can disrupt normal microbiota composition, induce dysbiosis, and favor overgrowth of pathogens and commensals with an increased virulence potential (10, 14, 17). Therefore, dysbiosis may not only be considered as a cause but also a consequence of gut inflammation.

What are the characteristics of IBD associated dysbiosis? Before methods for culture-independent microbiota analysis became available, several studies revealed alterations in mucosal and fecal microbiota composition in IBD patients. Some surveys already revealed an increase in the aerobic members of the microbiota (18–20). These early investigations using bacterial culturing methods were confirmed and extended by numerous recent studies in patient and control cohorts as well as in animal models using molecular, culture-independent techniques such as 16S rRNA gene-sequencing (21) and metagenomics (22). A general picture on the nature of inflammation-induced dysbiosis is gradually emerging: Most studies report a significantly decreased microbial diversity in active IBD, decreased abundance of particular obligate anaerobic Gram-positive bacteria (e.g., *Ruminococcaceae*, *Lachnospiraceae*), and a concomitant increase in facultative anaerobic bacteria such as *Enterococci* and *Streptococci* (23, 24) as well as Gram-negative Proteobacteria (in particular members of the *Enterobacteriaceae*) (21, 25, 26). A decrease of *Faecalibacterium prausnitzii* (27, 28), an abundant butyrate-producer in the human gut (29), has been proposed as a microbial marker for active disease. Reports on changes in other bacterial groups are somewhat contradictory. Some studies observe depletion of *Bacteroidetes* (21, 25), while others report enrichment (30, 31). No consistent changes in the abundance of the "probiotic"

Lactobacilli and Actinobacteria (i.e., Bifidobacteria) in IBD were observed. In any case, most studies agree in terms of a significant enrichment of *Enterobacteriaceae*, particularly *E. coli*, in mucosa-associated and fecal microbiota of IBD patients. In Crohn's disease patients, a disease-specific novel *E. coli* pathotype, termed adherent *invasive E. coli* (AIEC) was described (32, 33). Numbers of AIEC are not only elevated in the inflamed gut lumen, but they were also found to exacerbate the disease (second section). Similarly, enrichment of *E. coli* is also observed in experimental rodent colitis models based on genetic defects (34, 35), chemical induction (36, 37), or pathogen infection (38–40). This condition is termed *Enterobacterial* "blooms" (11).

A number of enteric pathogens are also able to exploit inflammatory responses for their own benefit. In the healthy intestine, the complex anaerobic microbiota efficiently blocks colonization and infection of the major human enteric pathogens. This "colonization resistance" is alleviated in the presence of gut inflammation and enables pathogen overgrowth. Prominent examples include *Citrobacter rodentium*, the agent of murine transmissible colonic hyperplasia; *Campylobacter jejuni;* and *Salmonella enterica* serovar Typhimurium (*S.* Typhimurium), which causes enterocolitis in humans (36, 38). A number of studies suggest that other pathogens such as *Klebsiella pneumoniae*, *Proteus mirabilis* (35), *Vibrio cholerae* (41), *Clostridium difficile* (42), and *Enterococcus* spp. (38, 43), may also benefit from an inflammatory milieu in the gut. Thus, an inflammatory milieu in the gut can alter survival, attachment, or growth of enteric pathogens and related commensal species.

Mechanisms triggering inflammation-induced dysbiosis and Enterobacterial "blooms"

Why do inflammatory conditions in the gut provide an advantage to facultative anaerobes and suppress many of the normally abundant anaerobic members of the microbiota? Different explanations have already been proposed: The "food hypothesis" suggests that inflamed intestine offers an altered nutrient spectrum or adhesion receptor sites that can be exploited only by a subset of bacteria. Alternatively (or additionally), antibacterial effector mechanisms released by the inflamed mucosa may selectively inhibit or kill a large part of the intrinsic microbiota, while the more "hardy" enteric pathogens and their close relatives among the commensals would remain unaffected ("differential killing hypothesis"). Over the past years, key studies in the field uncovered a number of underlying mechanisms about how mucosal inflammation supports overgrowth of resident or experimentally introduced facultative anaerobic members, in particular the family members of the *Enterobacteriaceae*. These mechanisms are illustrated later in the chapter.

Physiology of facultative anaerobic members of the microbiota

Enterobacteriaceae, *Enterococcus* spp., *Streptococcus* spp., and *Staphylococcus* spp. are prominent facultative anaerobes of the human fecal microbiota. Right after birth, these species are among the first colonizers of the infant gut and continue to be dominant during early life (44–46). In contrast, they represent only a minor fraction of the microbiota of adults, because facultative anaerobes are suppressed by the obligate anaerobic microbiota (47, 48). Interestingly, enrichment of facultative anaerobes is not only observed upon inflammation-induced dysbiosis but also upon broad-spectrum antimicrobial chemotherapy (antibiotic treatment). Antibiotic-caused disruption of the microbiota and depletion of its obligate anaerobic members lead to overgrowth of γ-Proteobacteria and Gram positive cocci (49, 50). This suggests that facultative anaerobic members of the microbiota share common physiologic properties that allow them to overgrow if the microbiota is disturbed.

What are the characteristics of these facultative anaerobic bacteria? *E. coli* can be considered the "work horse" of fundamental and molecular microbiology. This model organism enabled the discovery of a number of metabolic pathways of which homologues exist in many other species (51). The species *E. coli* includes commensal and different pathogenic strains (pathotypes) exhibiting different lifestyles. Accordingly, more than 20% of an *E. coli* genome represents strain-specific genes (degradation processes, virulence factors) (52, 53). Variable genes show to a great part an ancestry of horizontal gene transfer (i.e., plasmids, genomic islands, prophages and transposons). The conserved genes (core genome) comprise the pathways for central biosynthetic metabolism. *E. coli* can use a wide variety of carbon sources that are abundantly available in the mammalian intestine (54). In terms of energy metabolism, *E. coli* is a facultative anaerobic, chemoorganotrophic organism with the ability to switch between fermentative and respiratory metabolism. Under aerobic conditions, *E. coli* uses oxygen as a terminal electron acceptor. NADH, formate, and hydrogen can be used as electron donors. In the absence of oxygen, *E. coli* can switch to anaerobic respiration and use (in the order of preference) nitrate, nitrite, trimethylamine-N-oxide (TMAO), dimethyl sulfoxide (DMSO) and fumarate as electron acceptors (55). The expression of terminal oxidases is, at the transcriptional level, regulated by the O_2 and NO_3^- sensing regulators FNR, ArcA/B, NarX/L, and NarP/Q. In the absence of exogenous electron acceptors *E. coli* shifts to mixed acid fermentation, producing acetate, formic acid, succinic acid, lactic acid, and ethanol at different stoichiometric ratios (56).

In terms of the characteristics discussed above, *Enterococcus* spp. can be considered a Gram-positive "equivalent" of *E. coli*. Enterococci are opportunistic pathogens and cause a great portion of fatal nosocomial infections in humans in Europe and the United States. The dominant isolated species are *E. faecalis* and *E. faecium*. Often, strains exhibit multiple resistances to antibiotics. As *E. coli*, *Enterococcus* spp. exhibits an open pangenome structure featuring a large fraction of horizontally acquired DNA (57). In terms of energy metabolism, *Enterococci* belong to the lactic acid bacteria (Lactobacillales) and use the homolactic fermentative pathway for energy production (58). Some species were reported to harbor cytochrome-oxidases and have the capacity to perform oxidative phosphorylation and respire O_2. This property seems to depend on the external supplementation of heme (59). Besides, *E. faecalis* may be able to perform anaerobic respiration as suggested by the presence of a periplasmic nitrate reductase (*napA*) in the genome of strain OG1RF (ATCC 47077) (60).

In summary, the facultative anaerobic *Enterobacteriaceae* and *Enterococci* exhibit similar physiologic and metabolic properties that may enable them to bloom in the inflamed gut.

The "Oxygen-Hypothesis"

Degradation and fermentation of complex carbohydrates is the main metabolic strategy employed by anaerobic microbial communities (i.e., Bacteroidetes, Firmicutes) in the mammalian large intestine. Levels of facultative anaerobic *Enterobacteriaceae* and *Enterococcaceae* are only minor constituents of a healthy microbiota (47), which might be due to a relative deficiency in anaerobic e^--acceptors in the lower gut. Inspired by the observation that, under inflammatory conditions, the microbial community is characterized by a shift from obligate to facultative anaerobes, it was suggested that an increased oxygen tension may be a feature of the inflamed gut mucosa. O_2 would serve facultative anaerobes as electron acceptor and at the same time inhibit growth of highly oxygen-sensitive commensals ("The Oxygen-Hypothesis" (61)). A steep O_2 gradient reaches from the vascularised intestinal epithelium to the anaerobic gut lumen (62). Intestinal

inflammation was shown to lead to mucosal tissue hypoxia. It is technically challenging to accurately determine pO_2 in the gut lumen. Spectral electron paramagnetic resonance (EPR) imaging has been established to measure oxygenation in living mice (63). As an alternative method, Pd-porphyrin can be used to quantify local O_2-tension using an intravital phosphorescence assay (64). Because oxidative (O_2-dependent) respiration is thermodynamically more favorable than anaerobic respiration or fermentative metabolism, it is conceivable that increased oxygen levels, derived from blood and hemoglobin, reach the gut lumen via the damaged mucosa and foster growth of the facultative anaerobes. As a matter of fact, oxygen intake as a consequence of small bowel transplantation leads to a local increase in abundance of facultative anaerobic *Enterobacteriales* and *Lactobacillales*, which rapidly decrease after surgical closure (65). Clearly, an increase of oxygen concentration may play a role in intestinal dysbiosis. However, no studies have measured O_2 levels in the gut lumen in the course of acute or chronic colitis to confirm the "Oxygen Hypothesis."

Electron donors for anaerobic respiration

Apparently, introduction of a respiratory electron donor supports growth of facultative anaerobic bacteria in the gut lumen. While an elevated abundance of oxygen in the gut lumen upon acute or chronic colitis has not yet been experimentally verified, other electron donors supporting anaerobic respiration were shown to play a role. *E. coli* and *S.* Typhimurium can use a variety of oxidized substrates as terminal electron acceptors for anaerobic respiration. The substrate spectrum includes (in the order of preference): nitrate (NO_3^-), nitrite (NO_2^-), DMSO, TMAO, and fumarate (55). In addition, *S.* Typhimurium can respire on tetrathionate, a product of the oxidation of thiosulfate (66).

Reactive oxygen species (ROS) and reactive nitrogen species (RNS) play a major role in the antibacterial immune defense of mice and men (67). Proinflammatory cytokine signalling induces the activation of intestinal epithelial ROS-generating enzymes such as Nox-1 and Duox2 (68–69). In addition, neutrophils transmigrating into the gut lumen upon severe inflammation can produce additional superoxide radicals (O_2^-) via the inducible phagocyte NADPH oxidase Phox (70). Neutrophils convert superoxide radicals to hydrogen peroxide and hypochlorite (OCl^-) by the enzymes myeloperoxidase (MPO) and superoxide-dismutase (SOD). RNS are also produced by the inducible nitric oxide synthase (iNOS, *Nos2*) by epithelial cells and neutrophils in the inflamed gut mucosa. Further, NO can react with a superoxide radical to form peroxynitrite ($ONOO^-$). In human IBD patients, elevated production of ROS and RNS was reported (71, 72). Likewise, in a mouse model for acute *S.* Typhimurium-induced colitis, increased expression of NADPH oxidase complex (*Cybb*) and the inducible nitric oxide synthase (*Nos2*) were measured (73).

Seminal work by Winter et al. first revealed that gut inflammation leads to the generation of electron acceptors, which boost anaerobic respiration of *Enterobacteriaceae* (74). They demonstrated that, in the course of *S.* Typhimurium-induced colitis, tetrathionate ($S_4O_6^{2-}$) is generated from thiosulfate ($S_2O_3^{2-}$) via ROS-mediated oxidation. Thiosulfate is produced by epithelial cells upon detoxification of hydrogen sulphide (H_2S), which is a product of anaerobic sulphate-reducing bacteria (SRB), which are members of the normal microbiota (75).

In contrast with many commensal *Enterobacteriaceae*, *S.* Typhimurium harbours the genes for tetrathionate respiration (*ttrBCA* operon and the two-component regulatory system *ttrRS*), which are encoded on the *Salmonella* pathogenicity island 2 (SPI-2) (66). By respiring on tetrathionate, *S.* Typhimurium can gain a competitive advantage over the competing microbiota in the inflamed gut (74). In contrast, in the absence of inflammation, no tetrathionate is produced,

and tetrathionate respiration is dispensable for the fitness of *S.* Typhimurium. In the same way, the *ttr* operon does not confer any benefit for systemic infection (66). Later, it was shown that other electron acceptors such as NO_3^- but also DMSO and TMAO are formed as a side effect of ROS and RNS production in the gut. NO_3^- is generated in an iNOS-dependent fashion by conversion of peroxynitrite (76), while DMSO and TMAO are produced by oxidation of organic sulfides and tertiary amines derived from host cell building blocks. *S.* Typhimurium, as well as commensal *E. coli*, produce NO_3^-, NO_2^-, DMSO, and TMAO reductases (section 2.2.1). All these respiratory reductases contain a molybdopterin cofactor (77). Therefore, growth of an *E. coli moaA* mutant, which is deficient in molybdopterin biosynthesis and anaerobic respiration, is severely attenuated in the gut upon dextrane-sulfate sodium (DSS)-induced colitis (78). In the absence of colitis, gut colonization of *E. coli moaA* is not attenuated. This data revealed that anaerobic respiration of diverse substrates confers a major fitness benefit to *E. coli*—and other related facultative anaerobic bacteria—in the inflamed gut. Presumably, successful competition for external anaerobic electron acceptors against host intrinsic commensal *E. coli* strains defines the fitness of the pathogen *S.* Typhimurium. Both commensal and pathogenic *E. coli* strains as well as *S.* Typhimurium often harbour even several genes encoding NO_3^-, NO_2^-, DMSO, and TMAO reductases. Apparently, the ability to reduce tetrathionate also gives *S.* Typhimurium a competitive edge against *E. coli*. However, localization of the *ttr*-operon on a mobile genetic element makes it likely to be transmitted horizontally to other *Enterobacteriaceae*, which may also benefit from this function in inflammation-induced blooms. Indeed, similar sequences (~70% identity) are present in the genomes of other *Enterobacteriaceae*, namely *Klebsiella* spp., *Morganella* spp., *Enterobacter* spp. and *Serratia* spp.

Alternative nutrient sources in the inflamed gut

Besides anaerobic electron acceptors, the inflamed gut offers several other nutrient niches that could selectively foster bacterial growth. Inflammatory mediators lead to disruption of the mucosal epithelium and an increased shedding of dead epithelial cells. Epithelial cell membranes contain lipids and phospholipids such as phosphatidylcholine and phosphatidylethanolamine. The latter is converted to ethanolamine by microbial activity. Ethanolamine is an abundant substrate in the bovine and murine intestine (79, 80), and S. Typhimurium can use it as sole source of carbon and nitrogen. Ethanolamine is converted to ammonia, acetaldehyde (ethanol), and acetyl-coA in a vitamin B_{12}-dependent manner; in *Salmonella*, the pathway is encoded by a 17-gene cluster, the *eut* operon (81). Yet, ethanolamine can only be used as a sole energy source in the presence of oxygen or upon anaerobic tetrathionate respiration (82, 83). Conversely, S. Typhimurium benefits from ethanolamine-degradation in the inflamed gut upon concomitant use of tetrathionate (80). The ability to use ethanolamine is restricted to certain bacteria among the Firmicutes, Actinobacteria, and Proteobacteria (84, 85). Of note, *Enterococcus faecalis*, *E. coli*, and *Clostridium difficile*, which all "bloom" in an inflamed gut, can also degrade ethanolamine (86–88).

Intestinal mucins form a gel-like structure that contributes to protection against invasion of both commensals and enteric pathogens. On top, mucins shield the epithelium from damage by intestinal contents (89, 90). The goblet cell is a specialized cell type of the intestinal epithelium that mediates production and secretion of intestinal mucins. MUC2, the major colonic mucin in humans and mice, is a large protein rich in proline, serine, and threonine, which is processed by extensive O-glycosylation. N-actetyl-D-galactosamine, N-acetyl-D-glucosamine, N-acetylneuraminic acid (sialic acid), L-fucose, and D-galactose are the five major mucin-derived sugars (91). This makes the mucus layer an attractive bacterial habitat and nutrient source. Increased mucus production and its secretion is another hallmark of enteric pathogen infection (92–94). Mucus secretion by intestinal goblet cells in response to *Salmonella* infection is controlled IFN-γR-signalling (95). Mucus-associated bacteria are increased in IBD, and a number of known mucin-degrading bacteria were shown to be enriched in the inflamed gut. For instance, *Ruminococcus gnavus* and *Ruminococcus torques* were more abundant in UC/CD patients, while *Akkermansia muciniphila* was decreased (96). In mice, bacterial families like the *Enterobacteriaceae, Verrucomicrobiaceae* (mainly *Akkermansia* spp.), *Erysipelotrichaceae, Deferribacteracae* (mainly *Mucispirillum* spp.), and *Bacteroides acidofaciens* were enriched in DSS colitis (37). Moreover, an enrichment of transcripts for genes related to mucin-degradation was detected upon development of DSS colitis (97). *Mucispirillum schaedleri* and *Akkermansia muciniphila* are two species that have been isolated from mouse and human intestinal mucus, respectively (98, 99). A recent study confirmed *in vivo* mucin-degrading activity for both species, as well as for *B. acidifaciens* (100). Thus, enrichment of mucolytic bacteria supports the notion that increased mucin production in the course of intestinal inflammation expands this ecological niche and positively selects for mucin-degrading bacteria. Accordingly, in human CD patients, glycosidase activity in fecal samples was significantly higher than in healthy controls (101). Along the same lines, accumulation of sugar moieties (lactose, galactinol, melibiose, raffinose) was detected in the *Salmonella*-infected, inflamed gut (102). Accumulation of these metabolites may be causally linked to depletion of commensal bacteria. Further, increased levels of sugars might be exploited by the pathogen for growth. A similar scenario has been described after antibiotic-mediated disruption of the microbiota: Disturbance of commensal microbial food webs gives rise to free microbiota-liberated monosaccharides in the gut, which can promote growth of enteropathogenic bacteria such as *S.* Typhimurium and *Clostridium difficile* (103). Motility and chemotaxis were essential for *S.* Typhimurium to move towards the epithelium and benefit from mucin-derived sugars such as D-galactose (92). Moreover, *S.* Typhimurium employs aerotaxis to benefit from the anaerobic electron acceptors nitrate and tetrathionate, which are generated in close proximity to the mucosa (104). Lastly, blood, serum, and erythrocytes are leaking into the gut lumen from damaged mucosa. Besides antimicrobial effector molecules (e.g., complement, antimicrobial peptides), serum contains proteins, ions, and glucose, which may also promote growth of many "serum-resistant" bacteria (e.g., *E. coli* and *S.* Typhimurium) (105).

Antimicrobial effector mechanisms of the inflamed mucosa

The purpose of an inflammatory response is to eliminate infectious agents or attenuate their replication. Therefore, the release of antimicrobial mediators is a hallmark of intestinal inflammation in order to kill pathogens, impede their growth in close vicinity to the epithelium, and prevent invasion. As a matter of fact, the innate immune system can, for the most part, not distinguish between co-colonizing pathogens and commensals in the gut lumen. Therefore, acute inflammation in the gut may damage both beneficial and harmful members of the microbiota (106).

An effective instrument of the host's innate response is restriction of iron availability. Free iron in blood and mucosal secretions are kept at low levels, which is mediated by host proteins such as lactoferrin, transferrin, ferritin, and heme (107). Upon infection, the host can limit free iron concentration in the blood by different mechanisms, including secretion of the hormone hepcidin, which downregulates the iron transporter ferroportin 1 in the intestinal epithelium and prevents discharge of iron into the bloodstream (108, 109). To overcome iron starvation, bacteria can produce iron-scavenging molecules, termed siderophores, which are released from the bacteria and have a high affinity for iron (mostly Fe^{III}). Specific surface receptors serve the reabsorption of iron-siderophore complexes. *Enterobacteriaceae* can produce an arsenal of different catecholate-type and hydroxamate-type siderophores (110). Prominent examples are the siderophore enterochelin and the hydroxamate-type siderophore aerobactin. Enterochelin is synthesized by all *Enterobacteriaceae*, including *E. coli* and *S.* Typhimurium. The host, in turn, counteracts enterochelin by expressing lipocalin-2 (LCN2), an antibacterial protein that tightly binds the ferric enterochelin and thus blocks siderophore-mediated iron uptake. *S.* Typhimurium on the other hand, can evade LCN2-mediated inhibition by synthesizing the enterochelin variant salmochelin. Salmochelin-synthesis is mediated by the *iroBCDE iroN* gene cluster. *IroN* encodes a salmochelin-specific outer membrane receptor. *IroB* codes for a glucosyltransferase that modifies enterobactin to produce salmochelin; *iroC* encodes an ABC transporter required for transport of salmochelins; and *iroD*, and *iroE* code for a cytoplasmic esterase and a periplasmic hydrolase, respectively (111). Salmochelin-mediated resistance to LCN2 conferred a competitive advantage to *S.* Typhimurium when colonizing the inflamed intestine of wild-type but not of LCN2-deficient mice (112). LCN2 is produced by epithelial cells but also neutrophils, which transmigrate into the gut lumen upon *S.* Typhimurium-triggered inflammation and engulf bacteria (113, 114).

Moreover, the antimicrobial repertoire of neutrophils includes proteases and ROS produced by the NADPH oxidase complex. Futher, neutrophils produce myeloperoxidase, calprotectin, and elastase and release lactoferrin at the site of infection (70). Neutrophil elastase mediates changes of microbiota composition observed in response to *S.* Typhimurium infection in the mouse colitis model (115). Specifically, neutrophil elastase activity is linked to loss of bacterial families such as *Lachnospiraceae* and *Ruminococcaceae* and an increase in *Barnsiellae* and *S.* Typhimurium. Antibody-mediated depletion of neutrophils reverts these changes, while application of recombinant elastase to mice partially recapitulates *Salmonella*-induced dysbiosis (115). This suggests that elastase may differentially inhibit/kill *S.* Typhimurium and the commensal microbiota and thereby contribute to inflammation-induced pathogen overgrowth.

Calprotectin is one of the most abundant antimicrobial proteins in neutrophils and is secreted during inflammation at high levels. Calprotectin binds metal ions such as Zn^{2+} and Mn^{2+} and thereby contributes to inhibition of bacterial replication by inducing nutrient starvation (116). Interestingly, it was shown that manganese depletion increases bacterial susceptibility to ROS, as Mn^{2+} is an essential cofactor of superoxide-dismutases (117). *S.* Typhimurium is somewhat resistant to calprotectin by expressing a high-affinity Zn^{2+} transporter, *znuABC* (118). Accordingly, an *S.* Typhimurium *znuA* mutant was outcompeted by the microbiota, demonstrating that competition for Zn^{2+} is important for efficient colonization of the inflamed intestine, which is depleted by calprotectin. Indeed, in calprotectin-deficient mice ($S100a9^{-/-}$), *znuABC* did not confer a competitive advantage to *S.* Typhimurium.

Recently, it was shown that neutrophil transmigration into the gut lumen during *Toxoplasma gondii* infection results in the formation of organized intraluminal structures ("casts" or pseudomembranes), which encapsulate bacteria and thus prevent them from contacting the epithelium and invading the mucosa (40). $Fpr1^{-/-}$ mice, lacking the neutrophil N-formyl peptide receptor Fpr1, showed reduced "cast" formation, which suggests that neutrophil transmigration is chemotactically guided by bacterial patterns within the gut lumen, because the N-formyl peptides are products of bacterial growth.

In addition to antibacterial effectors of neutrophils, epithelial and paneth cells produce a number of antimicrobial mediators. In the small intestine, Paneth cells produce α-defensins (cryptdin- 1 to cryptdin- 6 in mice, HD5, and HD6 in humans) and the C-type lectins of the RegIII family (RegIIIβ, RegIIIγ). Several studies have demonstrated that antimicrobial peptides can indeed significantly influence gut microbiota composition: transgenic mice overexpressing α-defensin HD5 show altered microbiota composition in the ileal lumen (119). Similarly, changes in microbiota composition were reported for mice expressing a human alpha-defensin gene (DEFA5) as well as for $MMP7^{-/-}$ mice (lacking alpha-defensins) (120). However, alpha-defensins are not regulated by proinflammatory stimuli. While human beta defensin 1 (HBD-1) is constitutively expressed in the colonic and ileal epithelium, expression of human and mouse beta defensins (beta defensin HBD2-4, mouse beta defensin-3) is induced in the colon of IBD patients in response to infections and chronic colitis (121, 122). Broad-spectrum antimicrobial activity of beta defensins and significant variability in MICs has been reported. Aerobic bacteria generally exhibited higher susceptibility than anaerobes (123). Defensins were shown to effectively target pathogens, including *S.* Typhimurium, *Shigella flexneri*, and *E. coli* (124–126). Yet, their activity spectrum against the vast and diverse number of commensal bacteria has not been investigated in detail. Finally, it remains to be shown whether alpha- or beta-defensins contribute to inflammation-induced differential killing of microbiota and pathogens.

Expression of the C-type lectins RegIIIγ and RegIIIβ is upregulated during mucosal colonization with symbionts as well as upon pathogen infection. The lectins are released into the gut lumen and possess potent antimicrobial activity *in vitro*. Differential bacterial killing has been demonstrated both for RegIIIγ and RegIIIβ (127, 128). *In vitro*, RegIIIβ kills diverse commensal gut bacteria but not *S.* Typhimurium (128). Feeding of RegIIIβ to mice showed suppression of a RegIIIβ-sensitive commensal *E. coli*, while *S.* Typhimurium was not inhibited. These data suggest that RegIIIβ production by the host could promote S. Typhimurium infection by eliminating inhibitory members of the gut microbiota. Afterwards, a growth-phase–dependent antimicrobial activity against *S.* Typhimurium was revealed (129). Therefore, *in vivo* experiments involving RegIII-deficient mice will be necessary to further elucidate the mechanisms of differential killing of microbiota and pathogens in the inflamed gut.

Besides being a nutrient source for intestinal bacteria, mucus can also act as a structural matrix for the retention of antimicrobial molecules (e.g., defensins, cathelicidins, lysozymes, lectins). It was shown that mucin binding of the antibacterial lectin RegIIIγ strengthens the barrier function of the mucus layer in such a way that it is kept free of microbes (130). Therefore, increased mucus production upon inflammation may also augment antimicrobial activity and longevity of antimicrobial proteins by tethering them to mucins.

The Role of Dysbiosis in Enteric Infections and Human IBD

Inflammation creates conditions that can be better exploited or sustained by those bacteria, adapted to or resistant to immune defences (such as pathogens). Thereby, intestinal inflammation affects the microbiota in a specific, stereotypic manner, and a pre-existing beneficial microbial community can be shifted towards a "facultative pathogenic" microbiota. This inflammatory microbiota may further promote disease and result in a "vicious circle" of microbiota alteration and induction of inflammation. In the inflamed gut, bacterial species, which are normally less abundant, become enriched ("blooming"). While these species are generally harmless, apathogenic commensals in a healthy host at low abundance, they can become pathogenic

when enriched in an immune-compromised host. At high density, pathogenesis might be driven by the proinflammatory action of bacterial products (LPS, toxin production), cell adherence, tissue invasion (131), or increased resistance to phagocytosis. Members of the gut microbial community featuring these properties were termed "pathobionts" (132). For example, expansion of a specific multidrug-resistant *Escherichia coli* strain in the gut of immunodeficient mice occurred in the course of antibiotic therapy. Overgrowth in the gut facilitated tissue invasion and triggering of *E. coli*-induced sepsis (133). In fact, there is compelling evidence that bacteria with the ability to "bloom" in the inflamed gut can also trigger inflammation under these conditions.

The *E. coli* AIEC pathotype is a prominent example of such "pathobionts." AIEC are isolated from healthy individuals, in which they do not trigger disease symptoms. In response to inflammatory conditions, AIEC are positively selected, grow to high numbers, and can further promote disease in human IBD. In CD patients, elevated AIEC levels have been reported consistently (25, 134). AIEC strains can invade epithelial cells, and they are detected in mucosal ulcers and granuloma (16). *In vitro*, AIEC show high-level adherence to intestinal epithelial cell lines and can invade and survive within cells (135). IBD patients sometimes harbor systemic antibodies against *E. coli* (136). This suggests that AIEC are "pathobionts" that can benefit from the compromised mucosal immune system of a CD susceptible individual (e.g., polymorphisms in NOD2, ATG16L1, reduced defensin production (137)) or pre-existing inflammation. Overgrowth of AIEC in CD patients might well be related to concomitant depletion of other commensals and their metabolites (e.g., *Faecalibacterium prausnitzii*, butyrate) (28) to pre-existing inflammation or other unknown factors.

In mouse models of chronic colitis, AIEC only triggers/promotes disease in combination with certain host genotypes. Upregulation of carcinoembryonic antigen-related cell adhesion molecules (CEACAMs) on intestinal epithelium was reported for CD patients (138). The AIEC strain *E. coli* LF82 triggered colitis in CEABAC10 transgenic mice, which express human CEACAMs (139), a putative receptor for bacterial type 1 pili. Accordingly, induction of disease was found to depend on expression of type 1 pili by LF82. This supported the idea that CD patients express abnormally high CEACAM-levels, which act as receptor for AIEC and promote bacterial invasion and induction of inflammation. This may lead to a vicious circle of AIEC-induced inflammation and inflammation-induced up-regulation of the AIEC-receptor. Furthermore, AIEC was also shown to trigger disease in a genetically predisposed host. TLR5-deficient mice, lacking the pattern-recognition receptor for bacterial flagellin, are highly susceptible to developing colitis (140). The incidence of colitis in TLR5-deficient mice was associated with transient microbiota instability and increased abundance of Proteobacteria in the post-weaning period (141). When the microbiota was experimentally disrupted (germfree or antibiotic-treated mice), AIEC colonization triggered colitis in TLR5$^{-/-}$ but not WT mice. In conclusion, this data support the idea that AIEC are pathobionts, which are usually kept in check by the microbiota. If, for any reason, they grow to high levels, they can trigger or drive colitis in a genetically predisposed host.

IL-10-deficient mice have become a widely used tool to test the "inflammatogenic" potential of different commensals, such as species with the ability to "bloom" in the inflamed gut. IL-10–deficient mice spontaneously develop intestinal inflammation when colonized with a complex microbiota (142). Interestingly, monoassociation of germfree IL-10–deficient mice with "nonpathogenic" strains of *E. coli* (mouse isolate) and *Enterococcus faecalis* (strain OG1RF) triggered colitis (143). Upon co-colonization of both strains, symptoms of intestinal disease are even more severe (144). What is known about the mechanisms as to how these bacteria induce disease? In the case of *Enterococcus faecalis* OG1RF, it was shown that it expresses a metalloprotease (gelatinase) that compromises epithelial integrity, which in turn triggers colitis in this model (145). *Helicobacter hepaticus* is one of the best characterized pathobionts in mice. It asymptomatically colonizes immune-proficient mice while, in combination with a normal SPF microbiota, it causes disease in colitis-susceptible mouse models (146). In contrast with *E. coli* and *E. faecalis*, however, mono-association of IL-10–deficient mice with *H. hepaticus* does not lead to colitis (147). Under specific pathogen-free (SPF) conditions, mice from different SPF facilities display strong differences in their susceptibility to *H. hepaticus*-induced pathology (148). Therefore, *H. hepaticus* seems to require the presence of a "colitogenic" microbiota to trigger or exacerbate colitis in IL-10–deficient mice. So far, it is unclear which members of a normal microbiota are necessary and how they influence disease progression. The remaining microbiota may affect the mucosal immune system of the mice or modulate environmental conditions in the intestine and, thereby, *H. hepaticus* colonization or virulence factor expression. So far, it is also unclear if an inflamed gut environment would support *H. hepaticus* growth.

In conclusion, bacterial "blooms" occurring as a consequence of intestinal inflammation may contain a high proportion of pathobionts, which engage in disease perpetuation. This might feed into a futile cycle of inflammation-induced pathobiont "blooms" and "bloom"-induced inflammation. Clearly, future work is needed to further characterize the mechanisms of inflammation-induced enrichment of pathobionts and the complex interplay of pathobionts and the host's immune system.

Role of Dysbiosis for Bacterial Competition and Microbiota-Pathogen Evolution

Dysbiosis has a considerable impact on mucosal immune homeostasis and human health.

Also, inflammatory changes affect the microbiota in many ways: on the one hand, the inflammatory milieu in the gut opens up new niches that can be exploited by pathogens, commensals, and pathobionts. The bacteria respond with changes in gene expression and adapt to life in the inflamed gut. Moreover, the small subset of strains of the normal microbiota that are adapted to these conditions intensely compete for resources and produce a set of elaborate competition factors (e.g., siderophores, colicins) to ensure a competitive advantage. By creating such a specific environment, inflammation can give rise to transient high concentrations of closely related bacteria that are normally low-level inhabitants of the gut ecosystem. "Blooming" drastically increases the chance of these related bacteria to horizontally exchange genetic material. Thus, inflammation-induced environmental alterations can act as driving force for bacterial evolution.

Colicin-dependent competition of *Enterobacteriaceae* in the inflamed gut

Enterobacteriaceae comprise many inflammation-adapted species that are well characterized at a molecular level. In the case of *E. coli* and *Salmonella enterica*, a number of studies have addressed the mechanisms and how these species have adapted to growth in the inflamed intestine and to competition against the commensal microbiota. Iron is a limiting nutrient in the inflamed gut and, hence, the ability to compete for iron is a prerequisite for successful growth in this environment (149). Both species produce an array of high-affinity siderophores and cognate iron-siderophore receptors (see section 2.2.5). Competition for iron in the inflamed gut has been shown to be an effective mechanism of action of probiotics to inhibit pathogen growth. The probiotic *E. coli* Nissle strain produces a number of different siderophores (150). Competition for siderophores with *S.* Typhimurium in the inflamed gut leads to a decrease in pathogen titers and explains, at least in part, the mechanism of

protection against *S.* Typhimurium infection conferred by *E. coli* Nissle (149).

Enterobacteriaceae might also "parasitize" on siderophores produced by close relatives cocolonizing in the same gut, respectively (151). Parasitism has to be prevented in order to ensure successful colonization of this niche by the siderophore producer. To this end, *E. coli* and relatives produce an arsenal of different bacteriocins, the colicins, to fight off phylogenetically close competitors (152). Colicin production is a common trait in *E. coli* populations (153). On average, 30% of natural *E. coli* populations produce one or more colicins (154). Colicins can kill bacteria by one of four general mechanisms: inner-membrane pore formation, DNAse or RNAse activity, and interference with cell wall synthesis. For cell entry, colicins first bind to a specific outer-membrane receptor. Thereafter, "group A" colicins (colicin A, E1-E9, K, N, U) pass through the periplasm via the Tol-dependent import pathway, while "group B" colicins (colicin B, D, Ia, Ib, M) exploit the TonB-pathway for entry. The colicin producer protects itself against self-killing by expressing a cognate immunity protein. In addition to the immunity protein, strains can become resistant to the colicin by mutations in the colicin-import pathway or the cellular target (referred to as "colicin-resistant," as opposed to "immune"). Since this import pathways usually serve the uptake of key nutrients (e.g., iron), mutations conferring resistance to colicins might concomitantly decrease bacterial fitness.

In addition to porins (OmpA, OmpF) and the vitamin B12 receptor (BtuB), many colicins bind to iron-siderophore receptors on the outer bacterial membrane. Therefore, these colicins can also act as a "weapon" specifically directed against competitors for siderophores: On the one hand, colicins compete with siderophores for receptor binding. Receptor-binding of the colicin would block siderophore uptake regardless of the competitor exhibiting an additional intrinsic resistance to the colicin (e.g., by carrying a mutation in the import pathway). On the other hand, colicins targeting siderophore-receptors can directly eliminate colicin-sensitive bacteria, which express the respective receptor. Indeed, the degree of sensitivity to colicin Ib increases with the quantity of CirA-receptor molecules on the bacterial surface, which is upregulated in response to iron limitation in a Fur-dependent manner (155). A vast number of colicin/siderophore-receptor couples follow the same principle: Colicin B and D target FepA (enterochelin-receptor); colicin M targets FhuA (ferrichrome-receptor); colicin Ia/Ib and microcin L both target CirA (binds breakdown products of enterochelin); pesticin targets FyuA (Yersiniabactin) (156); and cloacin DF13 targets IutA (aerobactin receptor). Because all the named receptors are induced in response to iron-limiting conditions (157), it is highly likely that increased sensitivity to colicin-mediated killing under iron depletion may also apply to other colicins binding to TonB-dependent outer-membrane transporters. Further studies are needed to experimentally verify that colicins can prevent siderophore "parasitism" *in vivo* in the inflamed gut.

Colicin producers have to pay a certain cost for colicin expression. On one hand, colicin synthesis results in a higher metabolic load, which may affect physiology and growth rates of the bacteria. Moreover, colicin release is mediated by bacterial lysis and is therefore lethal for the bacteria (152). Colicin producers solve this problem by "division of labor": Colicins are expressed by only a small subpopulation that sacrifices itself and releases colicins,which benefits the whole population and contributes to survival of the genotype. Colicin expression is repressed by LexA and is under the control of the bacterial SOS-response. DNA damage (e.g., induced by ultraviolet light, radicals, or antibiotics) leads to activation of the protease RecA, which in turn mediates LexA degradation and triggers the SOS-response (158). The SOS response is induced in a stochastic fashion (159). *In vitro*, in the absence of DNA-

damaging agents, colicins are expressed only by a small fraction (<1%) of the total population (160). The rate of colicin-expression *in vivo* (e.g., in the mammalian gut) has never been addressed. Nevertheless, it is clear that bacteria have to tightly control expression of colicins to prevent excessive cell lysis. This implies that expression should be increased under conditions in which the chances are high to encounter closely related competitors.

Inflammation-inflicted *Enterobacterial* blooms can contain high loads of different closely related *Enterobacteriaceae*, which likely compete for the same resources (e.g., iron, anaerobic electron acceptors). The individual strains may hugely benefit from colicin production under this highly competitive situation. Recently, we have shown in a murine colitis model that colicin Ib (ColIb) provides a competitive advantage to *S.* Typhimurium against commensal *E. coli* strains (155). In fact, the benefit was apparent only in the inflamed gut. Expression of colicin Ib (*cib*), which is under negative control by Fur and LexA, was induced in the inflamed gut. This suggests that, under this condition, *S.* Typhimurium senses iron limitation as well as signals triggering the SOS response, which leads to increased *cib* expression. On the other hand, expression of the ColIb receptor *cirA* in *E. coli* was upregulated as well. Thus, physiological changes of the murine intestine in *Salmonella*-induced colitis are likely to provide the environmental cues required for upregulation of both colicins and their corresponding receptors. These findings shed new light on the role of colicins as important fitness factors providing a competitive advantage for bacterial growth in the inflamed gut.

Horizontal gene transfer and bacterial evolution

Owing to the enormous bacterial density, horizontal gene transfer (HGT) among bacteria in the mammalian gut is thought to occur at relatively high frequencies compared with other microbial ecosystems (161, 162).

HGT enables bacterial evolution in quantum leaps rather than via stepwise adaptation by mutations to changing environments. Mechanisms of HGT occur via the uptake of naked DNA by natural transformation, phage-mediated transduction, and the exchange of plasmids by conjugation. Conjugation is an HGT mechanism which that direct physical contact between the donor and the recipient strain and thereby is a highly efficient way for bacteria to exchange DNA. Conjugative plasmids encode the machinery (conjugative pilus), which serves plasmid transfer from the donor to the recipient strain. When the bacteria involved are present at high densities (e.g., in the gut), conjugation is most effective. Indeed, intestinal metagenomes harbor a large variety of conjugative plasmids and transposons (163).

It was recently shown that inflammation can foster HGT by mediating a transient increase of total *Enterobacteriaceae* in the gut (11, 164). During the course of *Salmonella* infection in the mouse colitis model, pColIb, a conjugative colicin plasmid of the *S.* Typhimurium strain SL1344 (165), was transmitted at high frequency to commensal *E. coli* strains, which are part of the normal microbiota of the mice. Conjugation frequency was increased when the overall levels of donor and recipient strains were elevated, such as within inflammation-inflicted *Enterobacteriaceae* blooms or in gnotobiotic mice, which do not exhibit colonization resistance against *Enterobacteriaceae* (166). In contrast, a complex microbiota supressed growth of *S.* Typhimurium and *E. coli*, alleviating conjugation. Interestingly, colicin loci are often associated with transposons, insertion sequences, and mobilizable or conjugative plasmids (167). This suggests that inflammation-inflicted blooms not only provide the environmental cues for colicin-dependent bacterial competition (section 2.4.1), they also enhance the spread of colicin plasmids among blooming *Enterobacteriaceae*. Therefore, inflammation-inflicted blooms play an important role in colicin ecology.

Enterobacteriaceae in general, and *E. coli* in particular, are highly "promiscuous" bacterial species. Their genomes contain a large proportion of horizontally acquired DNA (168–171). *S.* Typhimurium virulence and fitness factors are frequently associated with pathogenicity islands, prophages, transposons, and plasmids, pointing at a history of HGT. Because *E. coli* and *Salmonella* genomes exhibit high homology (>80%) (172), they presumably share a common ancestor (173) from which *S.* Typhimurium and *E. coli* strains evolved independently. Evolution has been driven in particular by the horizontal acquisition of virulence and fitness factors. Thus, we suggested that virulence of contemporary *Salmonella* spp. may have evolved in two stages (11). Because the property to bloom in the inflamed gut is a characteristic of both commensal *E. coli* and *S.* Typhimurium, the common ancestor of contemporary commensal *E. coli* and *Salmonella* spp., a commensal inhabitant of the mammalian intestine, may successively have acquired factors to profit from a pre-existing intestinal inflammation (e.g., genes for iron acquisition, resistance to antimicrobial mediators). Thereafter, commensal *E. coli* remained at this stage, while *Salmonella* spp. acquired an array of virulence factors (e.g., *Salmonella* pathogenicity islands, Type 3 secretion systems, and effector proteins, fimbriae/pili), which enabled it to trigger the inflammatory response by itself. From then on, overgrowth of *Salmonella* in episodes of inflammation-induced blooms with other *Enterobacteriaceae* led to a continuous exchange of "inflammation fitness factors" by HGT among the different species.

From the beginning of the antibiotic era, excessive and uncontrolled antibiotic use has led to the emergence of antibiotic (AB)-resistant strains. Because a large part of resistance can be transferred horizontally (e.g., conjugative transposons/plasmids), inflammation-induced blooms may also form the playground for exchange of antibiotic-resistance cassettes. Mainly in recent years, a number of multiple AB-resistant *S.* Typhimurium strains have emerged, causing an increasingly serious threat to global public health and leading to elevated morbidity, mortality, and treatment costs. The multidrug-resistant *S.* Typhimurium strain DT104, which caused a global epidemic in the 1990s, carries resistance to ampicillin, chloramphenicol, spectinomycin/streptomycin, sulfonamides, and tetracyclines (174, 175). Most of the AB resistance genes are clustered on a 13kb multidrug-resistant region within *Salmonella* genomic island 1, but some isolates carry additional AB-resistance plasmids (176). A recent study comparing the genome sequences of more than 300 DT104 isolates from humans and animals revealed that the AB resistance profile of the isolates is more diverse than core genome variability, pointing at a rather frequent horizontal exchange of AB-resistances (177). As a consequence of the HIV pandemic, a highly invasive disease caused by nontyphoidal *Salmonella* strains has become a serious public health problem in sub-Saharan Africa. *S.* Typhimurium causes a fatal systemic infection in immune-compromised HIV-infected patients. These African epidemic strains are often multidrug-resistant to ampicillin, cotrimoxazole, and chloramphenicol, complicating the clinical treatment of the disease (178). Here, the incidence of plasmid-encoded extended spectrum beta lactamase (ESBL) or AmpC beta-lactamase–producing strains is on the rise (179). This alarming increase in clinical MDR *Salmonella* isolates suggests that horizontal spread of AB resistances commonly occurs among virulent epidemic *S.* Typhimurium strains. It remains to be shown whether *Enterobacterial* blooms in the gut of infected patients play a role in driving transmission and remixing AB resistances.

Similar to *E. coli*, *Enterococcus* spp. also has an open pangenome, which is shaped by HGT (57, 180). Genomic sequencing data have allowed the identification of the enterococcal "mobilome," which includes plasmids, insertion sequences, transposons, and integrons that can be mobilized between cells (181). As for *E. coli*, Enterococcal plasmids encode AB resistances, virulence factors, and bacteriocins. Interestingly, a recent study revealed that the genomes of clinical *E. faecium* isolates were enriched in mobile elements, virulence factors, and AB resistance genes compared with environmental isolates (180). Although this interesting correlation was not seen for the other human pathogenic species, *E. faecalis*, it suggests that Enterococci adapt to the clinical setting by HGT. One could imagine that the ability of Enterococci to dwell in inflammation-triggered bacterial blooms may contribute to frequent HGT among the clinical strains.

CONCLUSION/OUTLOOK

For many years, the intestinal microbiota has been considered an insignificant ingredient of our body. However, research during the last decade has taught us numerous new functions of our commensals and how bacteria shape and modulate their host's metabolism and immunity. An increasing number of human disease conditions were identified as having a correlative or causative association with specific microbiota alterations (3). Later, it was more and more appreciated that an inflammatory immune response in the gut (both IBD and pathogen-induced) can also directly shape the composition of the microbiota and trigger dysbiosis. This discovery has far-reaching consequences. It fundamentally changes our understanding of the pathogenesis of human IBD. Moreover, it sheds new light on the evolution of bacterial virulence. Currently, we are still getting a clearer picture of the "inflammabiome" which is characterized by the environmental conditions bacteria encounter in an inflamed intestine and their response to it. Future research directed towards an in-depth characterization of this "inflammabiome" will yield valuable insights into the metabolic pathways and virulence factors that are employed by pathogens and members of the microbiota in inflammation-triggered blooms. Interference with "blooming" by specific drugs or probiotics may offer an elegant strategy to revert a dysbiotic microbiota to a normal state. This may lead to the development of new therapies for the treatment of human IBD and infectious diseases. Further, prevention of *"Enterobacterial* blooms" in the hospital setting may decrease the rate of horizontal transfer of AB resistances and virulence factors.

CITATION

Stecher B. 2015. The Roles of Inflammation, Nutrient Availability and the Commensal Microbiota in Enteric Pathogen Infection, Microbiol Spectrum 3(2):MBP-0008-2014.

REFERENCES

1. **Blumberg R, Powrie F.** 2012. Microbiota, disease, and back to health: a metastable journey. *Sci Transl Med* **4:**137rv137.

2. **Nicholson JK, Holmes E, Kinross J, Burcelin R, Gibson G, Jia W, Pettersson S.** 2012. Host-gut microbiota metabolic interactions. *Science* **336:**1262–1267.

3. **Cho I, Blaser MJ.** 2012. The human microbiome: at the interface of health and disease. *Nat Rev Genet* **13:**260–270.

4. **Lozupone CA, Stombaugh JI, Gordon JI, Jansson JK, Knight R.** 2012. Diversity, stability and resilience of the human gut microbiota. *Nature* **489:**220–230.

5. **Dethlefsen L, Relman DA.** 2011. Incomplete recovery and individualized responses of the human distal gut microbiota to repeated antibiotic perturbation. *Proc Natl Acad Sci U S A* **108**(Suppl 1):4554–4561.

6. **Hill DA, Hoffmann C, Abt MC, Du Y, Kobuley D, Kirn TJ, Bushman FD, Artis D.** 2010. Metagenomic analyses reveal antibiotic-induced temporal and spatial changes in intestinal microbiota with associated alterations in immune cell homeostasis. *Mucosal Immunol* **3:**148–158.

7. **Wu GD, Chen J, Hoffmann C, Bittinger K, Chen YY, Keilbaugh SA, Bewtra M, Knights D, Walters WA, Knight R, Sinha R, Gilroy E, Gupta K, Baldassano R, Nessel L, Li H,** Bushman FD, Lewis JD. 2011. Linking long-term dietary patterns with gut microbial enterotypes. *Science* **334:**105–108.

8. **Walker AW, Ince J, Duncan SH, Webster LM, Holtrop G, Ze X, Brown D, Stares MD, Scott P, Bergerat A, Louis P, McIntosh F, Johnstone AM, Lobley GE, Parkhill J, Flint HJ.** 2011. Dominant and diet-responsive groups of bacteria within the human colonic microbiota. *ISME J* **5:**220–230.

9. **Zhang X, Zhao Y, Zhang M, Pang X, Xu J, Kang C, Li M, Zhang C, Zhang Z, Zhang Y, Li X, Ning G, Zhao L.** 2012. Structural changes of gut microbiota during berberine-mediated prevention of obesity and insulin resistance in high-fat diet-fed rats. *PLoS ONE* **7:**e42529.

10. **Stecher B, Berry D, Loy A.** 2013. Colonization resistance and microbial ecophysiology: using gnotobiotic mouse models and single-cell technology to explore the intestinal jungle. *FEMS Microbiol Rev* **37:**793–829.

11. **Stecher B, Maier L, Hardt WD.** 2013. 'Blooming' in the gut: how dysbiosis might contribute to pathogen evolution. *Nat Rev Microbiol* **11:**277–284.

12. **Hooper LV, Littman DR, Macpherson AJ.** 2012. Interactions between the microbiota and the immune system. *Science* **336:**1268–1273.

13. **Manichanh C, Borruel N, Casellas F, Guarner F.** 2012. The gut microbiota in IBD. *Nat Rev Gastroenterol Hepatol* **9:**599–608.

14. **Winter SE, Lopez CA, Baumler AJ.** 2013. The dynamics of gut-associated microbial communities during inflammation. *EMBO Rep* **14:**319–327.

15. **Strober W, Fuss I, Mannon P.** 2007. The fundamental basis of inflammatory bowel disease. *Journal Clin Invest* **117:**514–521.

16. **Sartor RB.** 2008. Microbial influences in inflammatory bowel diseases. *Gastroenterology* **134:**577–594.

17. **Sansonetti PJ.** 2008. Host-bacteria homeostasis in the healthy and inflamed gut. *Current opinion in gastroenterology* **24:**435–439.

18. **Peach S, Lock MR, Katz D, Todd IP, Tabaqchali S.** 1978. Mucosal-associated bacterial flora of the intestine in patients with Crohn's disease and in a control group. *Gut* **19:**1034–1042.

19. **Giaffer MH, Holdsworth CD, Duerden BI.** 1991. The assessment of faecal flora in patients with inflammatory bowel disease by a simplified bacteriological technique. *J Med Microbiol* **35:**238–243.

20. **Wensinck F, Custers-van L, Poppelaars-Kustermans PA, Schroder AM.** 1981. The faecal flora of patients with Crohn's disease. *J Hyg (Lond)* **87:**1–12.

21. Frank DN, St Amand AL, Feldman RA, Boedeker EC, Harpaz N, Pace NR. 2007. Molecular-phylogenetic characterization of microbial community imbalances in human inflammatory bowel diseases. *Proc Natl Acad Sci U S A* **104:**13780–13785.

22. Qin J, Li R, Raes J, Arumugam M, Burgdorf KS, Manichanh C, Nielsen T, Pons N, Levenez F, Yamada T, Mende DR, Li J, Xu J, Li S, Li D, Cao J, Wang B, Liang H, Zheng H, Xie Y, Tap J, Lepage P, Bertalan M, Batto JM, Hansen T, Le Paslier D, Linneberg A, Nielsen HB, Pelletier E, Renault P, Sicheritz-Ponten T, Turner K, Zhu H, Yu C, Jian M, Zhou Y, Li Y, Zhang X, Qin N, Yang H, Wang J, Brunak S, Dore J, Guarner F, Kristiansen K, Pedersen O, Parkhill J, Weissenbach J, Bork P, Ehrlich SD. 2010. A human gut microbial gene catalogue established by metagenomic sequencing. *Nature* **464:**59–65.

23. Fyderek K, Strus M, Kowalska-Duplaga K, Gosiewski T, Wedrychowicz A, Jedynak-Wasowicz U, Sladek M, Pieczarkowski S, Adamski P, Kochan P, Heczko PB. 2009. Mucosal bacterial microflora and mucus layer thickness in adolescents with inflammatory bowel disease. *World J Gastroenterol* **15:**5287–5294.

24. Gosiewski T, Strus M, Fyderek K, Kowalska-Duplaga K, Wedrychowicz A, Jedynak-Wasowicz U, Sladek M, Pieczarkowski S, Adamski P, Heczko PB. 2012. Horizontal distribution of the fecal microbiota in adolescents with inflammatory bowel disease. *J Pediatr Gastroenterol Nutr* **54:**20–27.

25. Baumgart M, Dogan B, Rishniw M, Weitzman G, Bosworth B, Yantiss R, Orsi RH, Wiedmann M, McDonough P, Kim SG, Berg D, Schukken Y, Scherl E, Simpson KW. 2007. Culture independent analysis of ileal mucosa reveals a selective increase in invasive *Escherichia coli* of novel phylogeny relative to depletion of Clostridiales in Crohn's disease involving the ileum. *ISME J* **1:**403–418.

26. Sokol H, Seksik P, Rigottier-Gois L, Lay C, Lepage P, Podglajen I, Marteau P, Dore J. 2006. Specificities of the fecal microbiota in inflammatory bowel disease. *Inflamm Bowel Dis* **12:**106–111.

27. Sokol H, Seksik P, Furet JP, Firmesse O, Nion-Larmurier I, Beaugerie L, Cosnes J, Corthier G, Marteau P, Dore J. 2009. Low counts of *Faecalibacterium prausnitzii* in colitis microbiota. *Inflamm Bowel Dis* **15:**1183–1189.

28. Fujimoto T, Imaeda H, Takahashi K, Kasumi E, Bamba S, Fujiyama Y, Andoh A. 2013. Decreased abundance of *Faecalibacterium prausnitzii* in the gut microbiota of Crohn's disease. *J Gastroenterol Hepatol* **28:**613–619.

29. Duncan SH, Hold GL, Harmsen HJ, Stewart CS, Flint HJ. 2002. Growth requirements and fermentation products of *Fusobacterium prausnitzii*, and a proposal to reclassify it as Faecalibacterium prausnitzii gen. nov., comb. nov. *Int J Syst Evol Microbiol* **52:**2141–2146.

30. Bibiloni R, Mangold M, Madsen KL, Fedorak RN, Tannock GW. 2006. The bacteriology of biopsies differs between newly diagnosed, untreated, Crohn's disease and ulcerative colitis patients. *J Med Microbiol* **55:**1141–1149.

31. Swidsinski A, Weber J, Loening-Baucke V, Hale LP, Lochs H. 2005. Spatial organization and composition of the mucosal flora in patients with inflammatory bowel disease. *J Clin Microbiol* **43:**3380–3389.

32. Darfeuille-Michaud A, Boudeau J, Bulois P, Neut C, Glasser AL, Barnich N, Bringer MA, Swidsinski A, Beaugerie L, Colombel JF. 2004. High prevalence of adherent-invasive *Escherichia coli* associated with ileal mucosa in Crohn's disease. *Gastroenterology* **127:**412–421.

33. Kotlowski R, Bernstein CN, Sepehri S, Krause DO. 2007. High prevalence of *Escherichia coli* belonging to the B2+D phylogenetic group in inflammatory bowel disease. *Gut* **56:**669–675.

34. Maharshak N, Packey CD, Ellermann M, Manick S, Siddle JP, Huh EY, Plevy S, Sartor RB, Carroll IM. 2013. Altered enteric microbiota ecology in interleukin 10-deficient mice during development and progression of intestinal inflammation. *Gut Microbes* **4:**316–324.

35. Garrett WS, Gallini CA, Yatsunenko T, Michaud M, DuBois A, Delaney ML, Punit S, Karlsson M, Bry L, Glickman JN, Gordon JI, Onderdonk AB, Glimcher LH. 2010. Enterobacteriaceae act in concert with the gut microbiota to induce spontaneous and maternally transmitted colitis. *Cell Host Microbe* **8:**292–300.

36. Lupp C, Robertson ML, Wickham ME, Sekirov I, Champion OL, Gaynor EC, Finlay BB. 2007. Host-mediated inflammation disrupts the intestinal microbiota and promotes the overgrowth of enterobacteriaceae. *Cell Host Microbe* **2:**119–129.

37. Berry D, Schwab C, Milinovich G, Reichert J, Ben Mahfoudh K, Decker T, Engel M, Hai B, Hainzl E, Heider S, Kenner L, Muller M, Rauch I, Strobl B, Wagner M, Schleper C, Urich T, Loy A. 2012. Phylotype-level 16S rRNA analysis reveals new bacterial indicators of health state in acute murine colitis. *ISME J* **6:**2091–2096.

38. Stecher B, Robbiani R, Walker AW, Westendorf AM, Barthel M, Kremer M, Chaffron S, Macpherson AJ, Buer J, Parkhill J, Dougan G, von Mering C, Hardt WD. 2007. *Salmonella enterica* Serovar Typhimurium exploits inflammation to compete with the intestinal microbiota. *PLoS Biol* **5:**e244.

39. Barman M, Unold D, Shifley K, Amir E, Hung K, Bos N, Salzman N. 2008. Enteric salmonellosis disrupts the microbial ecology of the murine gastrointestinal tract. *Infect Immun* **76:**907–915.

40. Molloy MJ, Grainger JR, Bouladoux N, Hand TW, Koo LY, Naik S, Quinones M, Dzutsev AK, Gao JL, Trinchieri G, Murphy PM, Belkaid Y. 2013. Intraluminal containment of commensal outgrowth in the gut during infection-induced dysbiosis. *Cell Host Microbe* **14:**318–328.

41. Ma AT, Mekalanos JJ. In vivo actin crosslinking induced by Vibrio cholerae type VI secretion system is associated with intestinal inflammation. *Proc Natl Acad Sci U S A* **107:** 4365–4370.

42. Lawley TD, Clare S, Walker AW, Goulding D, Stabler RA, Croucher N, Mastroeni P, Scott P, Raisen C, Mottram L, Fairweather NF, Wren BW, Parkhill J, Dougan G. 2009. Antibiotic treatment of *Clostridium difficile* carrier mice triggers a supershedder state, spore-mediated transmission, and severe disease in immunocompromised hosts. *Infect Immun* **77:**3661–3669.

43. Gevers D, Kugathasan S, Denson LA, Vazquez-Baeza Y, Van Treuren W, Ren B, Schwager E, Knights D, Song SJ, Yassour M, Morgan XC, Kostic AD, Luo C, Gonzalez A, McDonald D, Haberman Y, Walters T, Baker S, Rosh J, Stephens M, Heyman M, Markowitz J, Baldassano R, Griffiths A, Sylvester F, Mack D, Kim S, Crandall W, Hyams J, Huttenhower C, Knight R, Xavier RJ. 2014. The treatment-naive microbiome in new-onset Crohn's disease. *Cell Host Microbe* **15:**382–392.

44. Adlerberth I. 2008. Factors influencing the establishment of the intestinal microbiota in infancy. *Nestle Nutr Workshop Ser Pediatr Program* **62:**13–29; discussion 29–33.

45. Palmer C, Bik EM, Digiulio DB, Relman DA, Brown PO. 2007. Development of the human infant intestinal microbiota. *PLoS Biol* **5:**e177.

46. Dominguez-Bello MG, Blaser MJ, Ley RE, Knight R. 2011. Development of the human gastrointestinal microbiota and insights from high-throughput sequencing. *Gastroenterology* **140:**1713–1719.

47. Arumugam M, Raes J, Pelletier E, Le Paslier D, Yamada T, Mende DR, Fernandes GR, Tap J, Bruls T, Batto JM, Bertalan M, Borruel N, Casellas F, Fernandez L, Gautier L, Hansen T, Hattori M, Hayashi T, Kleerebezem M, Kurokawa K, Leclerc M, Levenez F, Manichanh C, Nielsen HB, Nielsen T, Pons N, Poulain J, Qin J, Sicheritz-Ponten T, Tims S, Torrents D, Ugarte E, Zoetendal EG, Wang J, Guarner F, Pedersen O, de Vos WM, Brunak S, Dore J, Consortium M, Weissenbach J, Ehrlich SD, Bork P, Antolin M, Artiguenave F, Blottiere HM, Almeida M, Brechot C, Cara C, Chervaux C, Cultrone A, Delorme C, Denariaz G, Dervyn R, Foerstner KU, Friss C, van de Guchte M, Guedon E, Haimet F, Huber W, van Hylckama-Vlieg J, Jamet A, Juste C, Kaci G, Knol J, Lakhdari O, Layec S, Le Roux K, Maguin E, Merieux A, Melo Minardi R, M'Rini C, Muller J, Oozeer R, Parkhill J, Renault P, Rescigno M, Sanchez N, Sunagawa S, Torrejon A, Turner K, Vandemeulebrouck G, Varela E, Winogradsky Y, Zeller G. 2011. Enterotypes of the human gut microbiome. *Nature* **473:** 174–180.

48. Vollaard EJ, Clasener HA. 1994. Colonization resistance. *Antimicrob Agents Chemother* **38:** 409–414.

49. Buffie CG, Jarchum I, Equinda M, Lipuma L, Gobourne A, Viale A, Ubeda-Morant C, Xavier J, Pamer EG. 2012. Profound alterations of intestinal microbiota following a single dose of Clindamycin results in sustained susceptibility to *C. difficile*-induced colitis. *Infect Immun* **80:**62–73.

50. Taur Y, Xavier JB, Lipuma L, Ubeda C, Goldberg J, Gobourne A, Lee YJ, Dubin KA, Socci ND, Viale A, Perales MA, Jenq RR, van den Brink MR, Pamer EG. 2012. Intestinal domination and the risk of bacteremia in patients undergoing allogeneic hematopoietic stem cell transplantation. *Clin Infect Dis* **55:**905–914.

51. Keseler IM, Bonavides-Martinez C, Collado-Vides J, Gama-Castro S, Gunsalus RP, Johnson DA, Krummenacker M, Nolan LM, Paley S, Paulsen IT, Peralta-Gil M, Santos-Zavaleta A, Shearer AG, Karp PD. 2009. EcoCyc: a comprehensive view of *Escherichia coli* biology. *Nucleic Acids Res* **37:**D464–D470.

52. Rasko DA, Rosovitz MJ, Myers GS, Mongodin EF, Fricke WF, Gajer P, Crabtree J, Sebaihia M, Thomson NR, Chaudhuri R, Henderson IR, Sperandio V, Ravel J. 2008. The pangenome structure of *Escherichia coli*: comparative

genomic analysis of E. coli commensal and pathogenic isolates. *J Bacteriol* **190:**6881–6893.

53. **Vieira G, Sabarly V, Bourguignon PY, Durot M, Le Fevre F, Mornico D, Vallenet D, Bouvet O, Denamur E, Schachter V, Medigue C.** 2011. Core and panmetabolism in *Escherichia coli*. *J Bacteriol* **193:**1461–1472.

54. **Chang DE, Smalley DJ, Tucker DL, Leatham MP, Norris WE, Stevenson SJ, Anderson AB, Grissom JE, Laux DC, Cohen PS, Conway T.** 2004. Carbon nutrition of *Escherichia coli* in the mouse intestine. *Proc Natl Acad Sci U S A* **101:**7427–7432.

55. **Unden G, Bongaerts J.** 1997. Alternative respiratory pathways of *Escherichia coli:* energetics and transcriptional regulation in response to electron acceptors. *Biochim Biophys Acta* **1320:**217–234.

56. **Neidhardt FC CIR, Ingraham JL, Edmund CCL, Brooks Low K, Magasanik B, Reznikoff WS, Riley M, Schaechter M, Umbarger HE.** 1995. *Escherichia coli and Salmonella.* ASM Press, Washington, DC.

57. **Qin X, Galloway-Pena JR, Sillanpaa J, Roh JH, Nallapareddy SR, Chowdhury S, Bourgogne A, Choudhury T, Muzny DM, Buhay CJ, Ding Y, Dugan-Rocha S, Liu W, Kovar C, Sodergren E, Highlander S, Petrosino JF, Worley KC, Gibbs RA, Weinstock GM, Murray BE.** 2012. Complete genome sequence of *Enterococcus faecium* strain TX16 and comparative genomic analysis of *Enterococcus faecium* genomes. *BMC Microbiol* **12:**135.

58. **Ramsey M, Hartke A, Huycke M.** 2014. The physiology and metabolism of Enterococci. *In* Gilmore MS, Clewell DB, Ike Y, Shankar N (ed), Enterococci: From commensals to leading causes of drug resistant infection [Internet]. Boston: Massachusetts Eye and Ear Infirmary; 2014–. 2014 .

59. **Pritchard GG, Wimpenny JW.** 1978. Cytochrome formation, oxygen-induced proton extrusion and respiratory activity in *Streptococcus faecalis* var. zymogenes grown in the presence of haematin. *J Gen Microbiol* **104:**15–22.

60. **Bourgogne A, Garsin DA, Qin X, Singh KV, Sillanpaa J, Yerrapragada S, Ding Y, Dugan-Rocha S, Buhay C, Shen H, Chen G, Williams G, Muzny D, Maadani A, Fox KA, Gioia J, Chen L, Shang Y, Arias CA, Nallapareddy SR, Zhao M, Prakash VP, Chowdhury S, Jiang H, Gibbs RA, Murray BE, Highlander SK, Weinstock GM.** 2008. Large scale variation in *Enterococcus faecalis* illustrated by the genome analysis of strain OG1RF. *Genome Biol* **9:**R110.

61. **Rigottier-Gois L.** 2013. Dysbiosis in inflammatory bowel diseases: the oxygen hypothesis. *ISME J* **7:**1256–1261.

62. **Karhausen J, Furuta GT, Tomaszewski JE, Johnson RS, Colgan SP, Haase VH.** 2004. Epithelial hypoxia-inducible factor-1 is protective in murine experimental colitis. *Journal Clin Invest* **114:**1098–1106.

63. **He G, Shankar RA, Chzhan M, Samouilov A, Kuppusamy P, Zweier JL.** 1999. Noninvasive measurement of anatomic structure and intraluminal oxygenation in the gastrointestinal tract of living mice with spatial and spectral EPR imaging. *Proc Natl Acad Sci U S A* **96:**4586–4591.

64. **Handa K, Ohmura M, Nishime C, Hishiki T, Nagahata Y, Kawai K, Suemizu H, Nakamura M, Wakui M, Kitagawa Y, Suematsu M, Tsukada K.** 2010. Phosphorescence-assisted microvascular O(2) measurements reveal alterations of oxygen demand in human metastatic colon cancer in the liver of superimmunodeficient NOG mice. *Adv Exp Med Biol* **662:**423–429.

65. **Hartman AL, Lough DM, Barupal DK, Fiehn O, Fishbein T, Zasloff M, Eisen JA.** 2009. Human gut microbiome adopts an alternative state following small bowel transplantation. *Proc Natl Acad Sci U S A* **106:**17187–17192.

66. **Hensel M, Hinsley AP, Nikolaus T, Sawers G, Berks BC.** 1999. The genetic basis of tetrathionate respiration in *Salmonella* Typhimurium. *Molecular Microbiol* **32:**275–287.

67. **Fang FC.** 2004. Antimicrobial reactive oxygen and nitrogen species: concepts and controversies. *Nat Rev Microbiol* **2:**820–832.

68. **Kuwano Y, Kawahara T, Yamamoto H, Teshima-Kondo S, Tominaga K, Masuda K, Kishi K, Morita K, Rokutan K.** 2006. Interferon-gamma activates transcription of NADPH oxidase 1 gene and upregulates production of superoxide anion by human large intestinal epithelial cells. *Am J Physiol Cell Physiol* **290:**C433–C443.

69. **Harper RW, Xu C, Eiserich JP, Chen Y, Kao CY, Thai P, Setiadi H, Wu R.** 2005. Differential regulation of dual NADPH oxidases/peroxidases, Duox1 and Duox2, by Th1 and Th2 cytokines in respiratory tract epithelium. *FEBS Lett* **579:**4911–4917.

70. **Nathan C.** 2006. Neutrophils and immunity: challenges and opportunities. *Nat Rev Immunol* **6:**173–182.

71. **Zhu H, Li YR.** 2012. Oxidative stress and redox signaling mechanisms of inflammatory bowel disease: updated experimental and clinical evidence. *Exp Biol Med (Maywood)* **237:**474–480.

72. **Lundberg JO, Hellstrom PM, Lundberg JM, Alving K.** 1994. Greatly increased luminal nitric oxide in ulcerative colitis. *Lancet* **344:** 1673–1674.

73. **Songhet P, Barthel M, Rohn TA, Van Maele L, Cayet D, Sirard JC, Bachmann M, Kopf M, Hardt WD.** 2010. IL-17A/F-signaling does not contribute to the initial phase of mucosal inflammation triggered by *S.* Typhimurium. *PLoS ONE* **5:**e13804.

74. **Winter SE, Thiennimitr P, Winter MG, Butler BP, Huseby DL, Crawford RW, Russell JM, Bevins CL, Adams LG, Tsolis RM, Roth JR, Baumler AJ.** 2010. Gut inflammation provides a respiratory electron acceptor for Salmonella. *Nature* **467:**426–429.

75. **Levitt MD, Furne J, Springfield J, Suarez F, DeMaster E.** 1999. Detoxification of hydrogen sulfide and methanethiol in the cecal mucosa. *J ClinInvest* **104:**1107–1114.

76. **Szabo C, Ischiropoulos H, Radi R.** 2007. Peroxynitrite: biochemistry, pathophysiology and development of therapeutics. *Nat Rev Drug Discov* **6:**662–680.

77. **Rajagopalan KV, Johnson JL.** 1992. The pterin molybdenum cofactors. *J Biol Chem* **267:**10199–10202.

78. **Winter SE, Winter MG, Xavier MN, Thiennimitr P, Poon V, Keestra AM, Laughlin RC, Gomez G, Wu J, Lawhon SD, Popova IE, Parikh SJ, Adams LG, Tsolis RM, Stewart VJ, Baumler AJ.** 2013. Host-derived nitrate boosts growth of *E. coli* in the inflamed gut. *Science* **339:**708–711.

79. **Bertin Y, Girardeau JP, Chaucheyras-Durand F, Lyan B, Pujos-Guillot E, Harel J, Martin C.** 2011. Enterohaemorrhagic *Escherichia coli* gains a competitive advantage by using ethanolamine as a nitrogen source in the bovine intestinal content. *Environ Microbiol* **13:**365–377.

80. **Thiennimitr P, Winter SE, Winter MG, Xavier MN, Tolstikov V, Huseby DL, Sterzenbach T, Tsolis RM, Roth JR, Baumler AJ.** 2011. Intestinal inflammation allows Salmonella to use ethanolamine to compete with the microbiota. *Proc Natl Acad Sci U S A* **108:** 17480–17485.

81. **Kofoid E, Rappleye C, Stojiljkovic I, Roth J.** 1999. The 17-gene ethanolamine (eut) operon of *Salmonella* Typhimurium encodes five homologues of carboxysome shell proteins. *J Bacteriol* **181:**5317–5329.

82. **Price-Carter M, Tingey J, Bobik TA, Roth JR.** 2001. The alternative electron acceptor tetrathionate supports B12-dependent anaerobic growth of *Salmonella enterica* serovar typhimurium on ethanolamine or 1,2-propanediol. *J Bacteriol* **183:**2463–2475.

83. **Roof DM, Roth JR.** 1988. Ethanolamine utilization in *Salmonella* typhimurium. *J Bacteriol* **170:**3855–3863.

84. **Tsoy O, Ravcheev D, Mushegian A.** 2009. Comparative genomics of ethanolamine utilization. *J Bacteriol* **191:**7157–7164.

85. **Garsin DA.** 2010. Ethanolamine utilization in bacterial pathogens: roles and regulation. *Nat Rev Microbiol* **8:**290–295.

86. **Pitts AC, Tuck LR, Faulds-Pain A, Lewis RJ, Marles-Wright J.** 2012. Structural insight into the *Clostridium difficile* ethanolamine utilisation microcompartment. *PLoS ONE* **7:** e48360.

87. **Del Papa MF, Perego M.** 2008. Ethanolamine activates a sensor histidine kinase regulating its utilization in *Enterococcus faecalis. Journal Bacteriol* **190:**7147–7156.

88. **Bradbeer C.** 1965. The clostridial fermentations of choline and ethanolamine. II. Requirement for a cobamide coenzyme by an ethanolamine deaminase. *J Biol Chem* **240:** 4675–4681.

89. **Johansson ME, Ambort D, Pelaseyed T, Schutte A, Gustafsson JK, Ermund A, Subramani DB, Holmen-Larsson JM, Thomsson KA, Bergstrom JH, van der Post S, Rodriguez-Pineiro AM, Sjovall H, Backstrom M, Hansson GC.** 2011. Composition and functional role of the mucus layers in the intestine. *Cell Mol Life Sci* **68:**3635–3641.

90. **McGuckin MA, Linden SK, Sutton P, Florin TH.** 2011. Mucin dynamics and enteric pathogens. *Nat Rev Microbiol* **9:**265–278.

91. **Allen A, Cunliffe WJ, Pearson JP, Sellers LA, Ward R.** 1984. Studies on gastrointestinal mucus. *Scand J Gastroenterol Suppl* **93:**101–113.

92. **Stecher B, Barthel M, Schlumberger MC, Haberli L, Rabsch W, Kremer M, Hardt WD.** 2008. Motility allows *S.* Typhimurium to benefit from the mucosal defence. *Cell Microbiol* **10:**1166–1180.

93. **Zarepour M, Bhullar K, Montero M, Ma C, Huang T, Velcich A, Xia L, Vallance BA.** 2013. The mucin Muc2 limits pathogen burdens and epithelial barrier dysfunction during *Salmonella enterica* serovar Typhimurium colitis. *Infect Immun* **81:**3672–3683.

94. **Bergstrom KS, Kissoon-Singh V, Gibson DL, Ma C, Montero M, Sham HP, Ryz N, Huang T, Velcich A, Finlay BB, Chadee K, Vallance BA.** 2010. Muc2 protects against lethal infectious colitis by disassociating pathogenic and commensal bacteria from the colonic mucosa. *PLoS Pathog* **6:**e1000902.

95. **Songhet P, Barthel M, Stecher B, Muller AJ, Kremer M, Hansson GC, Hardt WD.** 2011. Stromal IFN-gammaR-signaling modulates goblet cell function during *Salmonella* Typhimurium infection. *PLoS ONE* **6:**e22459.

96. **Png CW, Linden SK, Gilshenan KS, Zoetendal EG, McSweeney CS, Sly LI, McGuckin MA, Florin TH.** 2010. Mucolytic bacteria with increased prevalence in IBD mucosa augment in vitro utilization of mucin by other bacteria. *Am J Gastroenterol* **105:**2420–2428.

97. **Schwab C, Berry D, Rauch I, Rennisch I, Ramesmayer J, Hainzl E, Heider S, Decker T, Kenner L, Muller M, Strobl B, Wagner M, Schleper C, Loy A, Urich T.** 2014. Longitudinal study of murine microbiota activity and interactions with the host during acute inflammation and recovery. *ISME J* **8:**1101–1114.

98. **Derrien M, Vaughan EE, Plugge CM, de Vos WM.** 2004. *Akkermansia muciniphila* gen. nov., sp. nov., a human intestinal mucin-degrading bacterium. *Int J Syst Evol Microbiol* **54:**1469–1476.

99. **Robertson BR, O'Rourke JL, Neilan BA, Vandamme P, On SL, Fox JG, Lee A.** 2005. *Mucispirillum schaedleri* gen. nov., sp. nov., a spiral-shaped bacterium colonizing the mucus layer of the gastrointestinal tract of laboratory rodents. *Int J Syst Evol Microbiol* **55:**1199–1204.

100. **Berry D, Stecher B, Schintlmeister A, Reichert J, Brugiroux S, Wild B, Wanek W, Richter A, Rauch I, Decker T, Loy A, Wagner M.** 2013. Host-compound foraging by intestinal microbiota revealed by single-cell stable isotope probing. *Proc Natl Acad Sci U S A* **110:**4720–4725.

101. **Ruseler-van Embden JG, van Lieshout LM.** 1987. Increased faecal glycosidases in patients with Crohn's disease. *Digestion* **37:**43–50.

102. **Deatherage Kaiser BL, Li J, Sanford JA, Kim YM, Kronewitter SR, Jones MB, Peterson CT, Peterson SN, Frank BC, Purvine SO, Brown JN, Metz TO, Smith RD, Heffron F, Adkins JN.** 2013. A multi-omic view of host-pathogen-commensal interplay in mediated intestinal infection. *PLoS ONE* **8:**e67155.

103. **Ng KM, Ferreyra JA, Higginbottom SK, Lynch JB, Kashyap PC, Gopinath S, Naidu N, Choudhury B, Weimer BC, Monack DM, Sonnenburg JL.** 2013. Microbiota-liberated host sugars facilitate post-antibiotic expansion of enteric pathogens. *Nature* **502:**96–99.

104. **Rivera-Chavez F, Winter SE, Lopez CA, Xavier MN, Winter MG, Nuccio SP, Russell JM, Laughlin RC, Lawhon SD, Sterzenbach T, Bevins CL, Tsolis RM, Harshey R, Adams LG, Baumler AJ.** 2013. Salmonella uses energy taxis to benefit from intestinal inflammation. *PLoS Pathog* **9:**e1003267.

105. **Zipfel PF, Hallstrom T, Riesbeck K.** 2013. Human complement control and complement evasion by pathogenic microbes–tipping the balance. *Mol Immunol* **56:**152–160.

106. **Stecher B, Hardt WD.** 2008. The role of microbiota in infectious disease. *Trends Microbiol* **16:**107–114.

107. **Andrews NC, Schmidt PJ.** 2007. Iron homeostasis. *Annu Rev Physiol* **69:**69–85.

108. **Galy B, Ferring-Appel D, Becker C, Gretz N, Grone HJ, Schumann K, Hentze MW.** 2013. Iron regulatory proteins control a mucosal block to intestinal iron absorption. *Cell Rep* **3:**844–857.

109. **De Domenico I, Ward DM, Kaplan J.** 2011. Hepcidin and ferroportin: the new players in iron metabolism. *Semin Liver Dis* **31:**272–279.

110. **Neilands JB.** 1995. Siderophores: structure and function of microbial iron transport compounds. *J Biol Chem* **270:**26723–26726.

111. **Muller SI, Valdebenito M, Hantke K.** 2009. Salmochelin, the long-overlooked catecholate siderophore of Salmonella. *Biometals* **22:**691–695.

112. **Raffatellu M, George MD, Akiyama Y, Hornsby MJ, Nuccio SP, Paixao TA, Butler BP, Chu H, Santos RL, Berger T, Mak TW, Tsolis RM, Bevins CL, Solnick JV, Dandekar S, Baumler AJ.** 2009. Lipocalin-2 resistance confers an advantage to Salmonella enterica serotype Typhimurium for growth and survival in the inflamed intestine. *Cell Host Microbe* **5:**476–486.

113. **Loetscher Y, Wieser A, Lengefeld J, Kaiser P, Schubert S, Heikenwalder M, Hardt WD, Stecher B.** 2012. Salmonella transiently reside in luminal neutrophils in the inflamed gut. *PLoS ONE* **7:**e34812.

114. **Goetz DH, Holmes MA, Borregaard N, Bluhm ME, Raymond KN, Strong RK.** 2002. The neutrophil lipocalin NGAL is a bacteriostatic agent that interferes with siderophore-mediated iron acquisition. *Mol Cell* **10:**1033–1043.

115. **Gill N, Ferreira RB, Antunes LC, Willing BP, Sekirov I, Al-Zahrani F, Hartmann M, Finlay BB.** 2012. Neutrophil elastase alters the murine gut microbiota resulting in enhanced salmonella colonization. *PLoS ONE* **7:**e49646.

116. **Corbin BD, Seeley EH, Raab A, Feldmann J, Miller MR, Torres VJ, Anderson KL, Dattilo BM, Dunman PM, Gerads R, Caprioli RM, Nacken W, Chazin WJ, Skaar EP.** 2008. Metal chelation and inhibition of bacterial

growth in tissue abscesses. *Science* **319**:962–965.

117. **Kehl-Fie TE, Chitayat S, Hood MI, Damo S, Restrepo N, Garcia C, Munro KA, Chazin WJ, Skaar EP.** 2011. Nutrient metal sequestration by calprotectin inhibits bacterial superoxide defense, enhancing neutrophil killing of *Staphylococcus aureus*. *Cell Host Microbe* **10**:158–164.

118. **Liu JZ, Jellbauer S, Poe AJ, Ton V, Pesciaroli M, Kehl-Fie TE, Restrepo NA, Hosking MP, Edwards RA, Battistoni A, Pasquali P, Lane TE, Chazin WJ, Vogl T, Roth J, Skaar EP, Raffatellu M.** 2012. Zinc sequestration by the neutrophil protein calprotectin enhances Salmonella growth in the inflamed gut. *Cell Host Microbe* **11**:227–239.

119. **Wehkamp J, Salzman NH, Porter E, Nuding S, Weichenthal M, Petras RE, Shen B, Schaeffeler E, Schwab M, Linzmeier R, Feathers RW, Chu H, Lima H Jr, Fellermann K, Ganz T, Stange EF, Bevins CL.** 2005. Reduced Paneth cell alpha-defensins in ileal Crohn's disease. *Proc Natl Acad Sci U S A* **102**:18129–18134.

120. **Salzman NH, Hung K, Haribhai D, Chu H, Karlsson-Sjoberg J, Amir E, Teggatz P, Barman M, Hayward M, Eastwood D, Stoel M, Zhou Y, Sodergren E, Weinstock GM, Bevins CL, Williams CB, Bos NA.** 2010. Enteric defensins are essential regulators of intestinal microbial ecology. *Nat Immunol* **11**:76–83.

121. **Rahman A, Fahlgren A, Sundstedt C, Hammarstrom S, Danielsson A, Hammarstrom ML.** 2011. Chronic colitis induces expression of beta-defensins in murine intestinal epithelial cells. *Clin Exp Immunol* **163**:123–130.

122. **Ho S, Pothoulakis C, Koon HW.** 2013. Antimicrobial peptides and colitis. *Curr Pharm Des* **19**:40–47.

123. **Joly S, Maze C, McCray PB Jr, Guthmiller JM.** 2004. Human beta-defensins 2 and 3 demonstrate strain-selective activity against oral microorganisms. *J Clin Microbiol* **42**:1024–1029.

124. **Salzman NH, Ghosh D, Huttner KM, Paterson Y, Bevins CL.** 2003. Protection against enteric salmonellosis in transgenic mice expressing a human intestinal defensin. *Nature* **422**:522–526.

125. **Sperandio B, Regnault B, Guo J, Zhang Z, Stanley SL Jr, Sansonetti PJ, Pedron T.** 2008. Virulent *Shigella flexneri* subverts the host innate immune response through manipulation of antimicrobial peptide gene expression. *Journal Exp Med* **205**:1121–1132.

126. **Pazgier M, Hoover DM, Yang D, Lu W, Lubkowski J.** 2006. Human beta-defensins. *Cell Mol Life Sci* **63**:1294–1313.

127. **Cash HL, Whitham CV, Behrendt CL, Hooper LV.** 2006. Symbiotic bacteria direct expression of an intestinal bactericidal lectin. *Science* **313**:1126–1130.

128. **Stelter C, Kappeli R, Konig C, Krah A, Hardt WD, Stecher B, Bumann D.** 2011. Salmonella-induced mucosal lectin regIIIbeta kills competing gut microbiota. *PLoS ONE* **6**:e20749.

129. **Miki T, Holst O, Hardt WD.** 2012. The bactericidal activity of the C-type lectin RegIIIbeta against Gram-negative bacteria involves binding to lipid A. *J Biol Chem* **287**:34844–34855.

130. **Vaishnava S, Yamamoto M, Severson KM, Ruhn KA, Yu X, Koren O, Ley R, Wakeland EK, Hooper LV.** 2011. The antibacterial lectin RegIIIgamma promotes the spatial segregation of microbiota and host in the intestine. *Science* **334**:255–258.

131. **Craven M, Egan CE, Dowd SE, McDonough SP, Dogan B, Denkers EY, Bowman D, Scherl EJ, Simpson KW.** 2012. Inflammation drives dysbiosis and bacterial invasion in murine models of ileal Crohn's disease. *PLoS ONE* **7**:e41594.

132. **Chow J, Mazmanian SK.** 2010. A pathobiont of the microbiota balances host colonization and intestinal inflammation. *Cell Host Microbe* **7**:265–276.

133. **Ayres JS, Trinidad NJ, Vance RE.** 2012. Lethal inflammasome activation by a multi-drug-resistant pathobiont upon antibiotic disruption of the microbiota. *Nat Med* **18**:799–806.

134. **Barnich N, Darfeuille-Michaud A.** 2007. Adherent-invasive *Escherichia coli* and Crohn's disease. *Curr Opin Gastroenterol* **23**:16–20.

135. **Darfeuille-Michaud A.** 2002. Adherent-invasive *Escherichia coli*: a putative new *E. coli* pathotype associated with Crohn's disease. *Int J Med Microbiol* **292**:185–193.

136. **Tabaqchali S, O'Donoghue DP, Bettelheim KA.** 1978. *Escherichia coli* antibodies in patients with inflammatory bowel disease. *Gut* **19**:108–113.

137. **Wehkamp J, Stange EF.** 2006. A new look at Crohn's disease: breakdown of the mucosal antibacterial defense. *Ann N Y Acad Sci* **1072**:321–331.

138. **Barnich N, Carvalho FA, Glasser AL, Darcha C, Jantscheff P, Allez M, Peeters H, Bommelaer G, Desreumaux P, Colombel JF, Darfeuille-Michaud A.** 2007. CEACAM6 acts as a receptor for adherent-invasive *E. coli*, supporting ileal mucosa colonization in Crohn disease. *Journal Clin Invest* **117**:1566–1574.

139. **Carvalho FA, Barnich N, Sivignon A, Darcha C, Chan CH, Stanners CP, Darfeuille-Michaud A.** 2009. Crohn's disease adherent-invasive *Escherichia coli* colonize and induce strong gut inflammation in transgenic mice expressing human CEACAM. *J Exp Med* **206:**2179–2189.

140. **Vijay-Kumar M, Sanders CJ, Taylor RT, Kumar A, Aitken JD, Sitaraman SV, Neish AS, Uematsu S, Akira S, Williams IR, Gewirtz AT.** 2007. Deletion of TLR5 results in spontaneous colitis in mice. *J Clinical Invest* **117:**3909–3921.

141. **Carvalho FA, Koren O, Goodrich JK, Johansson ME, Nalbantoglu I, Aitken JD, Su Y, Chassaing B, Walters WA, Gonzalez A, Clemente JC, Cullender TC, Barnich N, Darfeuille-Michaud A, Vijay-Kumar M, Knight R, Ley RE, Gewirtz AT.** 2012. Transient inability to manage proteobacteria promotes chronic gut inflammation in TLR5-deficient mice. *Cell Host Microbe* **12:**139–152.

142. **Kuhn R, Lohler J, Rennick D, Rajewsky K, Muller W.** 1993. Interleukin-10-deficient mice develop chronic enterocolitis. *Cell* **75:**263–274.

143. **Kim SC, Tonkonogy SL, Albright CA, Tsang J, Balish EJ, Braun J, Huycke MM, Sartor RB.** 2005. Variable phenotypes of enterocolitis in interleukin 10-deficient mice monoassociated with two different commensal bacteria. *Gastroenterology* **128:**891–906.

144. **Kim SC, Tonkonogy SL, Karrasch T, Jobin C, Sartor RB.** 2007. Dual-association of gnotobiotic IL-10-/- mice with 2 nonpathogenic commensal bacteria induces aggressive pancolitis. *Inflamm Bowel Dis* **13:**1457–1466.

145. **Steck N, Hoffmann M, Sava IG, Kim SC, Hahne H, Tonkonogy SL, Mair K, Krueger D, Pruteanu M, Shanahan F, Vogelmann R, Schemann M, Kuster B, Sartor RB, Haller D.** 2011. *Enterococcus faecalis* metalloprotease compromises epithelial barrier and contributes to intestinal inflammation. *Gastroenterology* **141:**959–971.

146. **Cahill RJ, Foltz CJ, Fox JG, Dangler CA, Powrie F, Schauer DB.** 1997. Inflammatory bowel disease: an immunity-mediated condition triggered by bacterial infection with *Helicobacter hepaticus*. *Infect Immun* **65:**3126–3131.

147. **Dieleman LA, Arends A, Tonkonogy SL, Goerres MS, Craft DW, Grenther W, Sellon RK, Balish E, Sartor RB.** 2000. *Helicobacter hepaticus* does not induce or potentiate colitis in interleukin-10-deficient mice. *Infect Immun* **68:**5107–5113.

148. **Yang I, Eibach D, Kops F, Brenneke B, Woltemate S, Schulze J, Bleich A, Gruber AD, Muthupalani S, Fox JG, Josenhans C, Suerbaum S.** 2013. Intestinal microbiota composition of interleukin-10 deficient C57BL/6J mice and susceptibility to *Helicobacter hepaticus*-induced colitis. *PLoS ONE* **8:**e70783.

149. **Deriu E, Liu JZ, Pezeshki M, Edwards RA, Ochoa RJ, Contreras H, Libby SJ, Fang FC, Raffatellu M.** 2013. Probiotic bacteria reduce *salmonella* typhimurium intestinal colonization by competing for iron. *Cell Host Microbe* **14:**26–37.

150. **Valdebenito M, Crumbliss AL, Winkelmann G, Hantke K.** 2006. Environmental factors influence the production of enterobactin, salmochelin, aerobactin, and yersiniabactin in *Escherichia coli* strain Nissle 1917. *Int J Med Microbiol* **296:**513–520.

151. **Hibbing ME, Fuqua C, Parsek MR, Peterson SB.** 2010. Bacterial competition: surviving and thriving in the microbial jungle. *Nat Rev Microbiol* **8:**15–25.

152. **Cascales E, Buchanan SK, Duche D, Kleanthous C, Lloubes R, Postle K, Riley M, Slatin S, Cavard D.** 2007. Colicin biology. *Microbiol Mol Biol Rev* **71:**158–229.

153. **Riley MA, Gordon DM.** 1999. The ecological role of bacteriocins in bacterial competition. *Trends Microbiol* **7:**129–133.

154. **Riley MA, Gordon DM.** 1992. A survey of Col plasmids in natural isolates of *Escherichia coli* and an investigation into the stability of Col-plasmid lineages. *J Gen Microbiol* **138:**1345–1352.

155. **Nedialkova LP, Denzler R, Koeppel MB, Diehl M, Ring D, Wille T, Gerlach RG, Stecher B.** 2014. Inflammation fuels colicin Ib-dependent competition of *Salmonella* serovar Typhimurium and *E. coli* in Enterobacterial blooms. *PLoS Pathog* **10:**e1003844.

156. **Rakin A, Saken E, Harmsen D, Heesemann J.** 1994. The pesticin receptor of *Yersinia enterocolitica*: a novel virulence factor with dual function. *Mol Microbiol* **13:**253–263.

157. **Braun V, Hantke K, Koster W.** 1998. Bacterial iron transport: mechanisms, genetics, and regulation. *Met Ions Biol Syst* **35:**67–145.

158. **Butala M, Zgur-Bertok D, Busby SJ.** 2009. The bacterial LexA transcriptional repressor. *Cell Mol Life Sci* **66:**82–93.

159. **Kamensek S, Podlesek Z, Gillor O, Zgur-Bertok D.** 2010. Genes regulated by the *Escherichia coli* SOS repressor LexA exhibit heterogeneous expression. *BMC Microbiol* **10:**283.

160. **Mrak P, Podlesek Z, van Putten JP, Zgur-Bertok D.** 2007. Heterogeneity in expression of the *Escherichia coli* colicin K activity gene cka is controlled by the SOS system and

stochastic factors. *Mol Genet Genomics* **277:** 391–401.

161. **Smillie CS, Smith MB, Friedman J, Cordero OX, David LA, Alm EJ.** 2011. Ecology drives a global network of gene exchange connecting the human microbiome. *Nature* **480:**241–244.

162. **Kelly BG, Vespermann A, Bolton DJ.** 2009. Gene transfer events and their occurrence in selected environments. *Food Chem Toxicol* **47:**978–983.

163. **Brown Kav A, Sasson G, Jami E, Doron-Faigenboim A, Benhar I, Mizrahi I.** 2012. Insights into the bovine rumen plasmidome. *Proc Natl Acad Sci U S A* **109:**5452–5457.

164. **Stecher B, Denzler R, Maier L, Bernet F, Sanders MJ, Pickard DJ, Barthel M, Westendorf AM, Krogfelt KA, Walker AW, Ackermann M, Dobrindt U, Thomson NR, Hardt WD.** 2012. Gut inflammation can boost horizontal gene transfer between pathogenic and commensal Enterobacteriaceae. *Proc Natl Acad Sci U S A* **109:**1269–1274.

165. **Hoiseth SK, Stocker BA.** 1981. Aromatic-dependent *Salmonella typhimurium* are non-virulent and effective as live vaccines. *Nature* **291:**238–239.

166. **Stecher B, Chaffron S, Kappeli R, Hapfelmeier S, Freedrich S, Weber TC, Kirundi J, Suar M, McCoy KD, von Mering C, Macpherson AJ, Hardt WD.** 2010. Like will to like: abundances of closely related species can predict susceptibility to intestinal colonization by pathogenic and commensal bacteria. *PLoS Pathog* **6:**e1000711.

167. **Christenson JK, Gordon DM.** 2009. Evolution of colicin BM plasmids: the loss of the colicin B activity gene. *Microbiology* **155:**1645–1655.

168. **Reid SD, Herbelin CJ, Bumbaugh AC, Selander RK, Whittam TS.** 2000. Parallel evolution of virulence in pathogenic *Escherichia coli*. *Nature* **406:**64–67.

169. **Welch RA, Burland V, Plunkett G 3rd, Redford P, Roesch P, Rasko D, Buckles EL, Liou SR, Boutin A, Hackett J, Stroud D, Mayhew GF, Rose DJ, Zhou S, Schwartz DC, Perna NT, Mobley HL, Donnenberg MS, Blattner FR.** 2002. Extensive mosaic structure revealed by the complete genome sequence of uropathogenic *Escherichia coli*. *Proc Natl Acad Sci U S A* **99:**17020–17024.

170. **Brzuszkiewicz E, Bruggemann H, Liesegang H, Emmerth M, Olschlager T, Nagy G, Albermann K, Wagner C, Buchrieser C, Emody L, Gottschalk G, Hacker J, Dobrindt U.** 2006. How to become a uropathogen: comparative genomic analysis of extraintestinal pathogenic *Escherichia coli* strains. *Proc Natl Acad Sci U S A* **103:**12879–12884.

171. **McClelland M, Sanderson KE, Spieth J, Clifton SW, Latreille P, Courtney L, Porwollik S, Ali J, Dante M, Du F, Hou S, Layman D, Leonard S, Nguyen C, Scott K, Holmes A, Grewal N, Mulvaney E, Ryan E, Sun H, Florea L, Miller W, Stoneking T, Nhan M, Waterston R, Wilson RK.** 2001. Complete genome sequence of *Salmonella enterica* serovar Typhimurium LT2. *Nature* **413:**852–856.

172. **McClelland M, Florea L, Sanderson K, Clifton SW, Parkhill J, Churcher C, Dougan G, Wilson RK, Miller W.** 2000. Comparison of the *Escherichia coli* K-12 genome with sampled genomes of a *Klebsiella pneumoniae* and three *salmonella enterica* serovars, Typhimurium, Typhi and Paratyphi. *Nucleic Acids Res* **28:**4974–4986.

173. **Baumler AJ.** 1997. The record of horizontal gene transfer in Salmonella. *Trends Microbiol* **5:**318–322.

174. **Poppe C, Smart N, Khakhria R, Johnson W, Spika J, Prescott J.** 1998. *Salmonella* typhimurium DT104: a virulent and drug-resistant pathogen. *Can Vet J* **39:**559–565.

175. **Threlfall EJ.** 2000. Epidemic salmonella typhimurium DT 104--a truly international multiresistant clone. *J Antimicrob Chemother* **46:**7–10.

176. **Poppe C, Ziebell K, Martin L, Allen K.** 2002. Diversity in antimicrobial resistance and other characteristics among *Salmonella* typhimurium DT104 isolates. *Microb Drug Resist* **8:**107–122.

177. **Mather AE, Reid SW, Maskell DJ, Parkhill J, Fookes MC, Harris SR, Brown DJ, Coia JE, Mulvey MR, Gilmour MW, Petrovska L, de Pinna E, Kuroda M, Akiba M, Izumiya H, Connor TR, Suchard MA, Lemey P, Mellor DJ, Haydon DT, Thomson NR.** 2013. Distinguishable epidemics of multidrug-resistant *Salmonella* Typhimurium DT104 in different hosts. *Science* **341:**1514–1517.

178. **Gordon MA, Graham SM, Walsh AL, Wilson L, Phiri A, Molyneux E, Zijlstra EE, Heyderman RS, Hart CA, Molyneux ME.** 2008. Epidemics of invasive *Salmonella enterica* serovar enteritidis and S. enterica Serovar typhimurium infection associated with multidrug resistance among adults and children in Malawi. *Clin Infect Dis* **46:**963–969.

179. **Kruger T, Szabo D, Keddy KH, Deeley K, Marsh JW, Hujer AM, Bonomo RA, Paterson DL.** 2004. Infections with nontyphoidal Salmonella species producing TEM-63 or a novel TEM enzyme, TEM-131, in South Africa. *Antimicrob Agents Chemother* **48:**4263–4270.

180. **Kim EB, Marco ML.** 2014. Nonclinical and clinical *Enterococcus faecium* strains, but not Enterococcus faecalis strains, have distinct structural and functional genomic features. *Appl Environ Microbiol* **80:**154–165.

181. **Clewell DB, Weaver KE, Dunny GM, Coque TM, Francia MV, Hayes F.** 2014. Extrachromosomal and mobile elements in enterococci: Transmission, maintenance, and epidemiology. *In* Gilmore MS, Clewell DB, Ike Y, Shankar N (ed), Enterococci: From commensals to leading causes of drug resistant infection *[Internet]*. Boston: Massachusetts Eye and Ear Infirmary; 2014–. 2014 Feb 9.

Host Sialic Acids: A Delicacy for the Pathogen with Discerning Taste

15

BRANDY L. HAINES-MENGES,[1] W. BRIAN WHITAKER,[1] J.B. LUBIN,[1] and
E. FIDELMA BOYD[1]

SIALIC ACIDS (NONULOSONIC ACIDS)

Sialic acids (neuraminic acids) are a diverse family of nine carbon (nonulosonic) α-keto acidic carbohydrates. The canonical sialic acid, 2-keto-3-deoxy-5-acetamido-D-*glycero*-D-*galacto*-nonulosonic acid, also known as *N*-acetylneuraminic acid (Neu5Ac) is the backbone on which a large number of known modifications are made (1). The Neu5Ac structure is typified by a 6-carbon carboxylic acid ring structure with a glycerol tail, an acetamido at the C-5 position and hydroxyl groups present on C-4, C-7, C-8, and C-9. Modifications occur primarily on the hydroxyl groups, with *O*-acetylation being the most common alteration, and substitutions have been shown to occur after the completion of the core structure (2). Other modifications such as *O*-methylation, *O*-lactylation, and *O*-sulfation add to the diversity of this molecule *in vivo*. Two structurally similar sialic acids, *N*-glycolylneuraminic acid (Neu5Gc), which differs from Neu5Ac by the presence of a hydroxyl group on the N-5 acetyl moiety, and 2-keto-3-deoxyl-D-*glycero*-D-*galacto*-nonulosonic acid (KDN), a deaminated form of Neu5Ac, also occur in nature and similar modifications are made to their

[1]Department of Biological Sciences, University of Delaware, Newark, DE 19716, United States.
Metabolism and Bacterial Pathogenesis
Edited by Tyrrell Conway and Paul Cohen
© 2015 American Society for Microbiology, Washington, DC
doi:10.1128/microbiolspec.MBP-0005-2014

core structure (1). These three main structures (Neu5Ac, Neu5Gc, and KDN) encompass the family of sialic acids due to their retention of the same stereochemical configuration of the 9-carbon backbone.

Significance of Sialic Acids in Vertebrates

Among metazoans, sialic acid is primarily limited to members of the deuterostome lineage of the phyla Chordata and Echinodermata. Neu5Ac is the most widely synthesized of the family, and the most studied form. Neu5Gc is common among mammals but conspicuously absent in humans, due to loss of the hydroxylase gene required for its formation (3). KDN was once thought to be exclusive to lower order vertebrates, such as fish and amphibians, but recent studies have found it to be present in humans in an unbound form, and its presence in the gametes of fish as well as human ovarian cells and fetal serum may indicate that KDN plays a role in development (4). Sialic acids in both eukaryotes and prokaryotes are typically positioned at the terminal end of glycoconjugates allowing them to interact with the external environment and play a role in cell-to-cell communication as well as self-recognition. In particular, in eukaryotes the self-recognition function of sialic acid is shown in its role as a modulator of immune function. An example of this would include the signaling molecule, Factor H, which preferentially binds to C3b on cell surfaces containing sialic acid glycoconjugates, preventing the binding of Factor B and halting the alternative complement cascade (5). Sialic acids also bind to a family of cell surface proteins known as sialic acid-binding immunoglobulin-like lectins (Siglecs). This interaction between sialic acid glycoconjuates and Siglecs has been reported to dampen immune function in macrophages, natural killer cells, neutrophils, and B-cells (6–9). The highest concentration of sialic acid in humans, however, is found in the brain, and is heavily utilized throughout the central nervous system (CNS), yielding the name neuraminic acid (neurons). Within the brain and CNS polysialic acid chains are associated with the neural cell adhesion molecules of neurons, glia cells, and ganglions (10, 11). Lastly, a major reservoir of sialic acids is located on mucosa surfaces. Sialic acid's common name is derived from the Greek word sialon, meaning saliva, as it was first isolated in bovine submaxillary mucin (12). In addition to salivary mucins, sialic acid has also been shown to be present in intestinal, lung and vaginal mucin glycans (13–15). In intestinal mucin, more than 65% of glycans contain sialic acid residues (16). The addition of sialic acid on mucin glycans is thought to play a role in protecting the underlying peptides from proteolysis, and it has also been implicated in playing a role in mucin-mediated bacterial aggregation and hydroxyl radical scavenging (17–19).

Sialic Acid Biosynthesis

In humans, sialic acid is generated by the function of four primary genes. The precursor molecule in this pathway is the 6-carbon sugar UDP-*N*-acetylglucosamine, which is converted by the bifunctional enzyme, UDP-*N*-acetylglucosamine-2-epimerase (NeuC)/*N*-acetylmannosamine kinase, to *N*-acetylmannosamine-6-phosphate (ManAc-6P). ManAc-6P, with the addition of phosphoenolpyruvate (PEP), is combined by the condensation action of *N*-acetylneuraminic-9-phosphate synthase (NeuB) to form *N*-acetylneuraminic-9-phosphate (Neu5Ac-9-P). Neu5Ac-9-P is converted to Neu5Ac by a phosphatase enzyme and the final 9-carbon product is activated by CMP-*N*-acetylneuraminic synthase (NeuA) generating CMP-Neu5Ac. This activated form is then recognized by sialylotransferases for ensuing glycosylation.

Bacterial sialic acid synthesis follows a similar pathway with homologs of NeuC, NeuB, and NeuA required for *de novo* synthesis (20–22). Two sialic-acid–like molecules, legionaminic and pseudaminic acid, are exclusively synthesized by bacteria (23, 24). Bacteria have been observed to decorate three different surface structures depending on the species and the strain: lipopolysaccharide (LPS) in Gram-negative bacteria; the flagellum; and capsular polysaccharides (Fig. 1). It has been proposed that one of the roles of sialic acid surface decoration in bacteria is to mimic the eukaryotic host cells leading to a dampening of immune responses as described above (25–28). In addition, it has been proposed that sialylation plays a

role in biofilm formation (29–32). The ability of bacteria to biosynthesize sialic acid was once thought to be limited to a few pathogenic and commensal species; however, more recent phylogenetic analysis of sialic acid biosynthesis genes indicates that this ability is highly prevalent across a large number of diverse bacterial lineages (33).

Sialic Acid Catabolism

Conversely, the catabolism of Neu5Ac appears to be mainly the purview of bacterial species that associate with eukaryotic hosts,

Bacterial sialylation of surface components

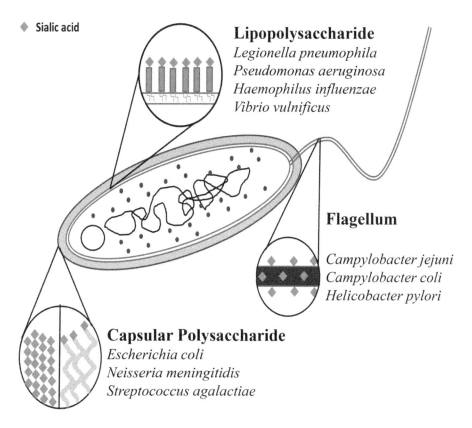

FIGURE 1 Bacterial sialylation of surface components. This diagram depicts the different surface structures in bacteria that are known to be decorated with nonulosonic acids (neuraminic, pseudaminic, or legionaminic). Also indicated are bacterial species demonstrated to have different surfaces sialylated. doi:10.1128/microbiolspec.MBP-0005-2014.f1

either as a commensal or as a pathogen (34, 35). In theory, all mucous-covered epithe-lium surfaces are potential food sources for sialic acid catabolizers since, as stated previ-ously in intestinal mucin, a key component of mucous, more than 65% of glycans con-tain sialic acid residues (16). In bacteria, Neu5Ac catabolism is dependent on the activity of *N*-acetylneuraminic acid lyase (NanA) (36). NanA cleaves pyruvate from Neu5Ac generating ManNAc. ManNAc is then converted to GlcNAc-6-P by the action of *N*-acetylmannosamine kinase (NanK) and *N*-acetylmannosamine 6-phosphate epimer-ase (NanE). GlcNAc-6-P is broken down further; first deacetylated by *N*-acetylgluco-samine-6-phosphate deacetylase (NagA), and subsequently deaminated by glucosmine-6-phosphate deaminase (NagB) yielding fructose-6-phosphate, and ammonia (36) (Fig. 2). Possessing Neu5Ac catabolic capa-bilities becomes significant when dealing with environments within eukaryotic hosts, where competition for space and nutrients is intense. Bacteria acquire sialic acid from their eukaryotic hosts either through the synthesis of a sialidase, a glycoside hydro-lase, which cleaves terminal Neu5Ac resi-dues from host glycoconjugates or simply by scavenging free Neu5Ac released by other bacterial species (37). In Gram-negative bac-teria, uptake of sialic acids requires transport across the outer membrane by either a gen-eral porin, such as OmpC/F, or a sialic-acid–specific porin, such as NanC (38) (Fig. 2). Transport across the inner membrane occurs by a diverse range of transporters. Four main systems have been shown to transport Neu5Ac: a major facilitator superfamily (MFS) permease designated as NanT, a tripartite ATP-independent periplasmic (TRAP) trans-porter SiaPQM, a sodium solute symporter (SSS) SiaT, and an ATP-binding cassette (ABC) transport system SatABCD (34, 39–46) (Fig. 2). The first characterized bacterial Neu5Ac transporter was NanT from *E. coli* by the Vimr group, which contains a hydrophilic domain unique among sugar permeases,

thought to facilitate the interaction with Neu5Ac (39, 40). The TRAP transporter sys-tems, which were originally shown to be carboxylic acid transporters, were demon-strated to transport Neu5Ac in *H. influenzae* by the Apicella and Thomas groups (41, 43, 45). In *H. ducreyi* an ABC-type transporter is proposed to be required for Neu5Ac transport (42). Thomas and colleagues have also dem-onstrated that *Salmonella enterica* encodes a SSS transporter along with NanT that both transport Neu5Ac into the cell (46).

Distribution of sialic acid catabolism gene clusters among sequenced bacteria

An investigation of the phylogenetic dis-tribution of sialic acid catabolism among nearly 2000 finished and unfinished bac-terial genomes was completed in 2009 (47). In that study, sialic acid gene clusters, con-sisting of NanA, NanE, and NanK, were shown to be present in 46 species. These 46 species encompassed six bacterial families of Gamma-Proteobacteria (*Enterobacteriaceae, Pasteurellaceae, Pseudoalteromonadaceae, Psychromonadaceae, Shewanellaceae,* and *Vibrionaceae*), one member of the genus *Fuso-bacterium,* and six Gram-positive families (*Clostridiaceae, Lactobacillaceae, Lachnospira-ceae, Mycoplasmataceae, Staphylococcaceae,* and *Streptococcaceae*), encompassing both low and high GC representatives. A follow up study examined the distribution and phylogeny of NanA only and identified 86 species that contained this protein (35). A search was conducted to determine whether this distribution of sialic acid catabolism genes had expanded by examining the newly sequenced species from all completed and draft genomes published on NCBI (4497 genomes). The search was conducted using pBLAST with NanA of *Vibrio cholerae* N16961 (VC1776) and *Staphylococcus aureus* N315 (SA0304) as seeds. Candidate species were determined by the presence of NanA. To eliminate false positives due to simi-larity to dihydropicolinate synthase (DapA), the presence of homologs of a sialic acid

Sialic Acid Catabolism Pathway

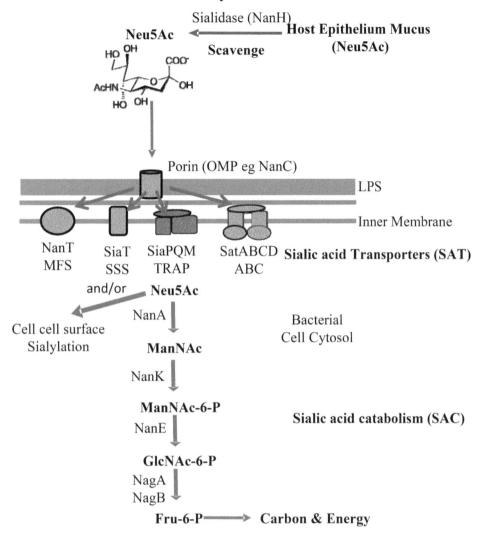

FIGURE 2 Schematic representation of the catabolism of sialic acid in Bacteria. The first step in catabolism is the uptake into the bacterial cell of free sialic acid molecules across the cell wall and cell membrane. Four type of transport systems have been described for transport across the cell membrane, MFS, SSS, TRAP, and ABC systems. NanH, Neuraminidase; Neu5Ac, *N*-acetylneuraminic acid. The sialic acid catabolic pathway involves several steps beginning with NanA. NanA, *N*-acetylneuraminic acid lyase; ManNAc, *N*-acetylmannosamine; NanK, *N*-acetylmannosamine kinase; ManNAc-6-P, *N*-acetylmannosamine-6-phosphate; NanE, *N*-acetylmannosamine-6-P epimerase; GlcNAc-6-P, *N*-acetyl-glucosamine-6-phosphate; NagA, *N*-acetylglucosamine-6-phosphate deacteylase; GlcN-6-P, Glucosamine-6-phosphate; NagB, Glucosamine-6-phosphate deaminase; Fru-6-P, Fructose-6-phosphate; LPS, lipopolysaccharide. doi:10.1128/microbiolspec.MBP-0005-2014.f2

transporter as well as putative *nanE/nanK* genes or a sialidase was required. From this stringent analysis, 265 species were identified, encompassing nine phyla (Actinobacteria, Bacteroidetes, Firmicutes, Fusobacteria, Proteobacteria, Planctomycetes

Spirochaetes, Tenericutes, and Verruco-microbia) with the potential to utilize sialic acid as a carbon and energy source. The total number of species that can catabolize sialic acid is probably much larger if the distribution of NanA, only, was examined. The distribution of sialic acid catabolizers among bacteria is composed primarily, but not exclusively, of commensal and pathogenic species of humans and animals. A major addition to the list of sialic acid catabolizers is 30 species from the phylum Actinobacteria, represented by six families (*Arthrobacter-aceae, Bifidobacteriaceae, Coriobacteriaceae, Corynebacteriaceae, Intrasporangiaceae,* and *Streptomycetaceae*). Many of the species within the families that contain sialic acid catabolism genes are associated with the oral cavity, the respiratory tract, the gastrointestinal tract, or the urogenital tract of humans. Examples include *Bifidobacterium breve* (commensal gut), *Corynebacterium diphtheria* (diphtheria), *C. ulcerans* (mastitis), *C. pseudotuberculosis* (lymphadenitis), *Eggerthella lenta* (commensal gut), *Gardenella vaginalis* (bacterial vaginosis), as well as soil dwellers, including *Streptomycetes* spp and *Kitasatospora setae*. A greatly expanded group of potential sialic acid catabolizers is within the phylum Bacteroidetes, represented by 41 species from 8 families (*Bacteroideteceae, Cytophagaceae, Cyclobacteriaceae, Chitinophagaceae, Flavobacteriaceae, Prevotellaceae, Porphyromonadaceae,* and *Sphingobacteriaceae*). Examples of species within this group that contain sialic acid catabolism genes include *Bacteroides fragilis* (commensal gut), *B. dorei* (commensal gut), *B. vulgatus* (commensal gut), *Capnocytophaga ochracea* (oral cavity), *C. sputigena* (periodontal disease), *Cryptobacterium curtum* (isolated from periodontal lesions), *Tannerella forsythia* (periodontal disease), *Prevotella oralis* (oral cavity), *P. denticola* (oral cavity), and *P. salivae* (oral cavity). *Fusobacterium nucleatum* from the phylum Fusobacteria (mucous membrane dwellers) was previously reported as a putative sialic acid

utilizer, and this phylum now includes species *F. varium, F. gonidiaformans,* and *F. necrophorum*. The phyla Verrucomicrobia and Planctomycetes, are represented by four and two species, respectively, that have the potential to use sialic acid as a carbon source.

The phylum Firmicutes was also well represented with species capable of sialic acid catabolism; 76 species in total from 9 bacterial families, including 21 species from the family *Streptococcaceae*, 10 species from the family *Staphylococcaceae*, 11 species from *Clostridiaceae*, and 7 species from *Lactobacillaceae*. Eighty-eight species were identified from the phylum Proteobacteria that contain sialic acid catabolism genes, which included the well-studied pathogens *E. coli, Salmonella enterica, H. influenzae,* and *Shigella* spp. Members of potential sialic acid catabolizers within this phylum include several *Citrobacter* species, 9 *Yersinia* species, 8 species of *Haemophilus*, and *Raoultella ornithinolytica* (formally *Klebsiella*), *Klebsiella pneumoniae,* and *K. oxytoca*. Within the family Vibrionaceae, 19 species contained sialic acid catabolism genes, including several pathogens of humans such as *V. cholerae, V. mimicus,* and *V. vulnificus*.

The ability to catabolize sialic acid allows these species to utilize a ubiquitous carbon source found in the mucus rich surfaces of their eukaryotic hosts. In regards to these bacteria, the human body is an array of diverse ecosystems that houses and maintains a diverse consortium of bacterial species that compete for these nutritional resources. While the gastrointestinal tract is perhaps the best studied of these ecosystems, there exist a number of other locations that also contain a large amount of diverse species and, that can be invaded and colonized by pathogens that rely on sialic acid as an important carbon source. The role of sialic acid catabolism in pathogenesis in a range of pathogens found in different regions of the human body is discussed in the next sections (Table 1).

Oral pathogens and sialic acid catabolism

The human oral microbiome is comprised of both eukaryotic and prokaryotic organisms, with approximately 1,000 species of bacteria comprising the dominant domain (48). Like other niches within the human body, the oral commensal microbiota plays an important role in preventing the colonization of harmful pathogenic bacteria in large part by outcompeting new comers for nutrients as proposed by Freter more than 30 years ago (49, 50). Unfortunately, these commensals are not always successful in preventing harmful bacteria from taking root in the oral cavity, and a number of oral diseases are associated with bacterial pathogens. One of these diseases is periodontitis, which is the result of inflammation in the tissues that surround the teeth. This inflammation leads to degradation of the gum and bone and culminates with loss of the tooth. It is estimated that periodontitis affects 300 million individuals worldwide (51, 52). One of the culprits of periodontitis is the Gram-negative bacterium *Tannerella forsythia*, a Bacteroidetes, and a member of the family Porphyromonadaceae. This anaerobic bacterium is a dominant member of the subgingival plaque (biofilm), the presence of which is typically associated with periodontitis. Under *in vitro* culture conditions (liquid broth and plates), *T. forsythia* is fastidious and has traditionally been thought to only utilize the bacterial cell wall component *N*-acetylmuramic acid (NAM) as its sole carbon source (51). Additional analysis of the *T. forsythia* genome revealed the presence of a putative sialic acid utilization cluster (51, 52). Contained in this cluster are genes for sialic acid scavenging (including the sialidase, NanH), inner and outer membrane transport, and genes for the conversion of Neu5Ac to the preferentially used carbon source NAM (51, 52). Furthermore, encoded elsewhere on the *T. forsythia* genome is the gene for a second sialadase, SiaHI. Similar to *E. coli*, *T. forsythia* contains the gene that encodes the enzyme (NanA) necessary for the conversion of sialic acid to ManNAc.

While *T. forsythia* contains a copy of the *nanE* gene, it does not contain *nanK*. The NanE of *T. forsythia* is highly homologous (83% amino acid identity) to the NanE of *Bacteroides fragilis*, which also lacks the *nanK* gene. In *B. fragilis*, NanE has the ability to convert ManNAc to GlcNAc, after which it is then phosphorylated by the hexokinase RokA (51, 53, 54). The genome of *T. forsythia* contains a gene that is 82% homologous, on the amino acid level, to RokA of *B. fragilis*, indicating that *T. forsythia* most likely utilizes sialic acid via the *B. fragilis* utilization pathway versus the canonical pathway observed in *E. coli* (51). Bioinformatics analysis identified this pathway in eight species of the genus *Prevotella* from the phylum Bacteroidetes that were all isolated from the oral cavity suggesting that this is an important phenotypic trait in this environment.

Perplexingly, *T. forsythia* was shown to not utilize sialic acid as a nutritional source when grown on plates or in planktonic culture. In contrast, sialic acid does appear to stimulate growth of *T. forsythia* under static growth conditions that is in biofilm (51). When statically grown cultures were supplemented with increasing concentrations of sialic acid, these cultures began to exceed the biomass of control cultures supplemented with NAM. It should be noted that the concentration of sialic acid required to exceed growth observed in cells utilizing NAM was much higher (6 mM [milliMolar] vs 0.17 mM) (51).

Roy and colleagues also investigated the range of sialic acids that could promote the growth of *T. forsythia* in biofilm and found that growth was also supported by sialyllactose, which was a mixture of 2,3- and 2,6-sialyllactose, as well as glycolyl sialic acid (Neu5Gc). For both sialyllactose and Neu5Gc, growth of *T. forsythia* was comparable to levels reached using Neu5Ac as the sole carbon source. The addition of a sialidase inhibitor could abolish growth of *T. forsythia* in sialyllactose, indicating that the use of a sialidase was necessary for sialyllactose

TABLE 1 Bacterial species in which sialic acid catabolism was experimentally examined

Species	Transporter	Sialidase	Location	Reference
Bacteroides fragilis	NanT(MSF)	NanH		(53)
Bifidobacterium longum	SAT (ABC)	NanH	Intestine	(122), (123)
Clostridium difficile	SSS	NanH	Intestine	(99)
Clostridium perfringens	SSS	NanH	Intestine	(117)
Corynebacterium glutamicum	SAT (ABC)	NanP		(124)
Cronobacter sakazakii	NanT(MSF)	–	Brain	(125, 126)
Escherichia coli	NanT(–MSF)	–	Intestine	(39, 86)
Gardnerella vaginalis	NK	NanH	Urogenital tract	(37)
Haemophilus influenzae	SiaPQM (TRAP)		Respiratory tract	(43)
Lactobacillus plantarum	SSS			(127)
Mycoplasma sinoviae		NanH		(128)
Pasteurella multocida	SiaPQM (TRAP)		Intestine	(129)
Ruminococcus gnavus	SAT2	NanH	Intestine	(130)
Salmonella enterica	NanT(MSF) SiaT (SSS)	–	Intestine	(47), (46, 99)
Staphylococcus aureus	Sym		Nose	(131)
Streptococcus pneumoniae	SAT 2/3 (ABC)	NanA,B,C	Respiratory tract	(132), (133)
Oralis group streptococci	NK	NK	Oral Cavity Endocarditis	(60, 61)
Milleri group streptococci	NK	NK	Oral Cavity Endocarditis	(60)
Group B streptococci	SAT (ABC)	–	Intestine Genital tract	(82)
Tannerella forsythia	NanT(MSF)	NanH	Oral Cavity	(51)
Treponema denticola	NanT(MSF)	NanH	Oral Cavity	(59)
Vibrio cholerae N16961	SiaPQM (TRAP)	NanH	Intestine	(34)
Vibrio mimicus	SiaPQM (TRAP)	NanH	Intestine	Boyd, unpublished
Vibrio fischeri ES114	SSS	–	Light organ of squid	(34)
Vibrio vulnificus YJ016	SiaPQM (TRAP)	–	Intestine/Blood	(113)
Yersinia enterocolitica	NanT(MSF)	–	Intestine	(47)
Yersinia pestis KIM	NanT(MSF)	–	Intestine Blood	(47)

utilization (51). Given that sialyllactose is commonly found on the surface of host glycoproteins, this could indicate that the bacterial sialidase would also be important during colonization of the subgingival cavity.

In addition to playing roles in catabolism, the two known sialidases (NanH and SiaHI) of *T. forsythia* were also investigated for their role in the pathogenesis of this bacterium (55–57). Single mutants in both the *nanH* and *siaHI* genes were constructed and both strains were found to be defective in sialidase activity, though the defect was more severe for the *nanH* mutant than the *siaHI* mutant (55). The two mutants and the wild-type strain were assayed for their ability to adhere to epithelial cells, and it was found that all three stains possessed the ability to adhere to and invade the cells. However, the *nanH* deletion mutant exhibited a significant defect in both adhesion and subsequent invasion of epithelial cells (55). Conversely, the *siaHI* deletion mutant demonstrated levels of adherence and invasion on par with the wild-type strain. Thus, this group demonstrates that NanH is the principal sialidase while SiaHI likely does not play a role in the cleavage of extracellular sialic acids (55). Another species associated with subgingival plaque is *Treponema denticola*, a Spirochaete, which has been shown to possess sialidase activity (58, 59). The sialidase of *T. denticola* cleaves both α2,3- and α2,6-linked sialic acid from glycoproteins. Additionally, a sialidase deletion mutant failed to grow in serum growth media, presumably due to its inability to cleave sialic acid from the serum proteins (59). This mutant was more susceptible to complement deposition and exhibited a defect in the mouse skin-infection model (59).

Many species of *Streptococcus* are associated with the oral cavity. Byers and colleagues demonstrated that *S. oralis*, *S. sanguis*, *S. gordonii*, *S. mitis*, *S. intermedius*, *S. anginosus*, *S. constellatus*, and *S. defectivus* use Neu5Ac as a sole carbon source (60, 61). The genome sequence database was used to identify a total of 10 species that contain sialic acid catabolism genes within the genus; *S. gordonii*, *S. infantis*, *S. iniae*, *S. intermedius*, *S. mitis*, *S. oralis*, *S. parasanguinis*, *S. pneumoniae*, *S. pyogenes*, and *S. sanguinis*. Many of these species were isolated from the oral cavity and can cause diseases such as endocarditis.

Respiratory Tract Pathogens and Sialic Acid Catabolism

The human lung has traditionally been considered a sterile environment in healthy individuals. However, recent metagenomic research indicates that the lungs do support their own consortium of microbiota, adding yet another microbial niche within the human body (62–64). Mucus production is important for lung tissue as it helps maintain optimum function by preventing the epithelium from becoming too dry. The lungs produce approximately 2 liters of mucus per day, which can serve as a rich nutrient source for pathogenic bacteria with the ability to cleave sugars from the glycoproteins found in host mucus. In cystic fibrosis patients, lung disease caused by polymicrobial infections is a serious concern. Traditionally the main players in CF lung infections were Proteobacteria, such as *Pseudomonas aeruginosa* (NanH), *Burkholderia cepacia* complex, *Haemophilus influenzae* (NanA), *Alcaligenes xylosoxidans*, *Stenotrophomonas maltophilia*, and the Firmicute *Staphylococcus aureus* (NanA). More recent metagenomic studies have shown that members of the phyla Actinobacteria, Bacteroidetes, Spirochaetes, and Fusobacteria, are also present in the CF lung (62).

H. influenzae is a Gamma-Proteobacteria, a member of the family Pasteurellaceae and is an inhabitant of the human respiratory tract, associated with both upper and lower respiratory infections. *H. influenzae* strains can be subdivided into two groupings: encapsulated and non-encapsulated strains. Encapsulated strains are often associated with causing meningitis, while the non-encapsulated strains are commonly associated with mucosal diseases, such as otitis media and bronchitis. *H. influenzae* strains have the ability to scavenge host sialic acid for both catabolism and sialylation of its LPS (28, 41, 43, 65–68). Genes for the catabolism of sialic acid (*nanEK*, *nanA*, *siaR*, *nagBA*) are present in the genome of *H. influenzae* within a single gene cluster, which has a different orientation than the cluster found in *E. coli* (41, 43, 67, 68). Located downstream of the catabolism gene cluster are the genes encoding for a sialic acid TRAP transporter (*siaPQM*) (41, 43, 68). Even though the catabolic pathway for sialic acid degradation is functional in *H. influenzae*, no defect was seen *in vivo* when comparing the infectious dose of the wild-type and the *nanA* deletion mutant strains using the intraperitoneal infant rat model (67). Additionally, when the wild-type and the *nanA* mutant were co-infected in the same mouse, the *nanA* mutant out-competed the wild-type strain, indicating that loss of sialic acid catabolism led to an increased fitness of the *nanA* deletion strain (67). Previous work demonstrated the importance of *H. influenzae* lipooligosaccharides (LOS) *in vivo* and that host-derived sialic acid is transported into the bacterial cell and incorporated into LPS, which is essential for resistance to host serum (28, 41, 43, 66). Interestingly, deletion of *nanA* appears to increase the amount of sialylation of *H. influenzae* LOS. Hyper-sialylation could confer more resistance to host serum and lead to a competitive advantage *in vivo*. It is likely, based on the results observed above, that sialic acid catabolism does not play a role in pathogenesis. However, all seven additional sequenced *Haemophilus* species: *H. haemolyticus*, *H. parahaemolyticus*,

H. parainfluenzae, H. parasuis, H. pittmaniae, H. ducreyi, and *H. sputorum,* contain the Neu5Ac catabolism gene cluster similar to *H. influenzae,* which suggests that sialic acid is an important carbon source but not under the conditions examined in the above studies.

Streptococcus pneumoniae is a Gram-positive inhabitant of the human naso-oropharanx in healthy individuals. However, in some individuals the bacterium can progress from an asymptomatic resident to a pathogen, causing primarily pneumococcal pneumonia and an assortment of other diseases such as otitis media, bacteremia, and meningitis. Strains of *S. pneumoniae* are especially adapted for life in an environment where sugar sources are bound to host proteins as they express up to 9 different glycosylases, 3 of which have been shown to have sialidase activity (NanA, NanB, and NanC) (69, 70). Bioinformatics analysis predicted that *S. pneumoniae* encodes the genes required for sialic acid catabolism and transport into the cell (47). Similar to *H. influenzae,* *S. pneumoniae* can scavenge Neu5Ac for either catabolism and/or sialylation of their cell surface, which serves as a key determinant of pathogenesis. And a recent study demonstrated that *S. pneumoniae* can utilize sialic acid as a carbon source and that an ABC type sialic acid transporter contributes to growth on a human glycoprotein and colonization *in vivo* (71).

Of the three sialidases identified in *S. pneumoniae, NanA* is present in all strains and is a surface associated sialidase. Studies have determined that NanA has activity common to other exosialidases in that it can cleave both α2,3- or α2,6-sialyllactose to release free sialic acid (70, 72). NanB is present in approximately 96% of *S. pneumoniae* strains and has been shown to act as a secreted sialidase (70, 72). And this sialidase has the ability to cleave α2,3-sialic acids to release 2,7-anhydro-sialic acid (70, 72). Lastly, NanC is present only in approximately 50% of *S. pneumoniae* isolates and is a *trans*-sialidase (72). NanC has been demonstrated to cleave α2,3-sialic acids to release 2,7-anhydro-sialic acid, and has been demonstrated to act as a sialidase inhibitor (72).

The roles of the *S. pneumoniae* sialidases in pathogenesis have been extensively investigated *in vitro* and *in vivo,* with a particular emphasis on the more widely distributed NanA and NanB enzymes that contribute to the bacterium's ability to adhere to various epithelial cell lines (73). Overexpression of NanA also contributed to biofilm formation and growth on saliva-coated glass cover slips in *S. pneumoniae* (73). Lastly, *nanA* and *nanB* deletion mutants are defective for host colonization and the ability to cause sepsis following intranasal dosage (74–76).

Urogenital pathogens and sialic acid catabolism

The human vagina is another mucus-covered surface that serves as a distinct ecosystem for normal host microbiota. Vaginal mucus secretions are rich with sialic acids, which serve as a nutrient source for the various bacterial species that inhabit this niche. Again, as in other locations on and in the human body, perturbations of the vaginal microbiota have been associated with the onset of disease. Specifically, bacterial vaginosis is a condition affecting the vaginal tract characterized by increased pH, thinning of vaginal secretions, and a fishy odor on hydrogen peroxide treatment of vaginal samples. Women who have bacterial vaginosis are more at risk for pelvic inflammatory disease, sexually transmitted diseases, postsurgical complications, and pregnancy complications (77). The vaginal microbiota in healthy individuals is comprised mostly of lactobacilli and it is believed that bacterial vaginosis is caused by the loss of the normal microbiota (lactobacillus) and due to the overgrowth of a number of different anaerobic bacterial species (37, 78). Another hallmark of vaginosis is sialidase activity in vaginal secretions, which is not present in healthy individuals. Presumably, the presence of sialidases in vaginal secretion from individuals with vaginosis is due to the fluctuations in vaginal microbiota (37, 78).

A common bacterial species present in the microbiota of patients with bacterial vaginosis is *Gardnerella vaginalis* (77). *Gardnerella vaginalis* is an obligate anaerobe that belongs to the family Bifidobacteriaceae that produces sialidase and utilizes both Neu5**A**c and Neu5**G**c as carbon sources (37, 79). Recently, Lewis and colleagues characterized the role of sialidase in *G. vaginalis* in host vaginal colonization. They demonstrated that *G. vaginalis* strains that were sialidase positive could deplete sialic acid from bound sialoglycans in specialized culture media. However, sialic acid depletion and utilization by *G. vaginalis* was shown to be dependent on the host source of sialic acid: *G. vaginalis* was able to utilize human IgA as a source of sialic acid but was deficient in sialic acid utilization when grown using bovine submaxillary mucin (37). While free sialic acid is liberated from bovine mucin, the amount of sialic acid utilized by *G. vaginalis* is much less than when grown using human mucin. Nonhuman mucin contains, in addition to Neu5**A**c, a derivative Neu5**G**c. The addition of Neu5**G**c to media containing human IgA did not inhibit the increase of free Neu5**A**c after bacterial inoculation (due to the presence of sialidase), but Neu5**G**c did appear to inhibit the bacteria from subsequently up-taking and utilizing the free Neu5**A**c (37). Furthermore, inhibition of sialic acid catabolism by Neu5**G**c was demonstrated to be at the level of transport and not by inhibiting sialidase or lyase enzymatic activity. It will be of interest to determine what type of sialic acid transporter is present in this species since NanT, TRAP and SSS systems were previously shown to transport Neu5**A**c and Neu5**G**c (80). Lastly, Lewis and colleagues demonstrated that *G. vaginalis* could free bound sialic acid *in vivo* using a vaginally-infected mouse model, mimicking what is observed in human patients presenting with bacterial vaginosis (37, 81).

Group B streptococci (GBS) are a devastating pathogen of newborns and infants and transmission from mother to child is an enormous concern. A recent study by Pezzicoli and colleagues demonstrated that GBS utilize sialic acid as a carbon source and this ability was dependent upon an ABC sialic acid transporter (82). This group also demonstrated that the ability to catabolize sialic acid was important during *in vivo* mucosal colonization using an *in vivo* mouse model of intranasal and intravaginal GBS infection in which sialidase release of sialic acid was simulated by adding free Neu5Ac. The *in vivo* data demonstrated that exogenous sialic acid significantly increased the capacity of GBS to infect mice at the mucosal level.

Utilization of sialic acid as a carbon source by intestinal pathogens

The gastrointestinal tract is home to many commensals but pathogens also find their way into this environment, with the potential of causing disease. As with all epithelial cells, the intestinal tract is cover in a protective mucosal layer, which aids in preventing infection by the vast number of microorganisms that reside within a healthy human gut (83). As discussed previously, there is evidence which shows that both the secreted and cell surface intestinal mucus, the main component of which is mucin glycoproteins, act as mucosal barriers against potential pathogens (83). However, there are many enteric pathogens that have found ways to circumvent this mucosal barrier (83). Mucin glycoproteins, can serve as ligands for microbial adhesions and can also be utilized as an energy source by both commensal organisms as well as enteric pathogens (83).

There is a complex array of oligosaccharides present on the glycosylated domains of mucin and mucolytic bacteria release these mucin glycoproteins that can then be used as nutrient sources by commensals and pathogens alike. Some of the sugars available to enteric pathogens as food sources include fucose, galactose, galactosamine, glucosamine, Neu5**A**c, mannose, glucose, glucuronate, gluconate, and galacturonate (84–86). Neu5**A**c is an excellent source of carbon and energy for intestinal commensals and pathogens, thus it is not surprising that there are a number of commensal species that encode the genes required to utilize Neu5Ac as a carbon source, such as, *Bacteroides caccae*, *B. fragilis*, *B. ovatus*, *B. stercoris*, *B. uniformis*, *B. vulgatus*, *Parabacteroides distasonis*, *B. breve*, *E. coli*, *E. blattae*, *Edwardsiella tarda*, *Dorea formicigenerans*, *D. longicatena*, *Faecalibacterium prausnitzii*, *Fusobacterium nucleatum*, *Proteus mirabilis*, *Providencia rettgeri*, *Ruminococcus gnavus*, *Lactobacillus plantarum*, *L. sakei*, *L. salivarius*, *L. vadensis*, and *Yokenella regensburgei*. Among these species, only a handful have been experimentally shown to catabolize sialic acid. Examples of enteric pathogens that can utilize sialic acid as a carbon source include *Citrobacter spp.*, *Clostridium spp.*, *E. coli*, *E. hermannii*, *Enterobacter cloacae*, *E. cancerogenus*, *E. asburiae*, *E. aerogenes*, *Klebsiella pneumoniae*, *K. oxytoca*, *Salmonella enterica*, *Shigella spp.*, *Vibrio cholerae*, *V. vulnificus*, and 9 of 13 sequenced *Yersinia* spp., (*Y. pestis*, *Y. enterocolitica*, *Y. pseudotuberculosis*, *Y. bercovieri*, *Y. kristensenii*, *Y. mollaretii*, *Y. rohdei*, *Y. frederiksenii*, and *Y. ruckeri*). Similar to commensal species, sialic acid metabolism has only been investigated in a handful of species (44, 47, 87).

E. coli, the most abundant Gram-negative facultative commensal in the intestinal tract, can also be a pathogen, and utilizes many of the sugars present in the intestinal tract including sialic acid (86). Vimr and colleagues demonstrated the ability of *E. coli* K-12 to utilize sialic acid as a sole carbon source and identified an inducible catabolic system for sialic acids termed Nan (39). Since this initial work, it has been shown that sialic acid utilization is important for the *in vivo* survival of *E. coli* isolates (86, 88, 89). It was demonstrated that sialic acid catabolism was important for colonization initiation by *E. coli* MG1655 but not maintenance in the streptomycin treated mouse model of colonization and additionally, sialic acid was shown

in vitro to be third in the order of preference of nutrients available within the intestinal tract (86).

In addition to looking at the ability of *E. coli* K-12 strains to utilize sialic acid, there have been studies comparing carbon metabolism between K-12 and pathogenic O157:H7 EDL933 strains (88–90). A carbon metabolism comparison between these two strains demonstrated a similar but not identical preference of nutrients *in vitro*, with both strains having Neu5Ac as their fourth most preferred nutrient (88). Catabolic genes for utilizing Neu5Ac were up-regulated in the presence of sialic acid in *E. coli* EDL933, however, mutation of the pathway for Neu5Ac catabolism caused colonization defects for *E. coli* MG1655 but not for *E. coli* EDL933, suggesting that sialic acid is not as important of a nutrient *in vivo* for *E. coli* EDL933 (88). It may be that the two strains utilize different carbon sources in order to occupy different intestinal niches and not compete for nutrients when both strains are present. A recent study using bovine small intestine contents as a growth media demonstrated that in *E. coli* O157:H7 the genes required for Neu5Ac catabolism were more highly expressed than in *E. coli* K12 and that Neu5Ac catabolism conferred a competitive growth advantage to the O157:H7 strain (89). Together these studies show the importance of sialic acid catabolism for *E. coli* colonization and demonstrate that there may be an interesting phenomenon of carbon preferences between commensal and pathogenic strains of the same species and in the same strain in different hosts.

In addition to catabolizing Neu5Ac, the most prevalent sialic acid in nature, *E. coli* can also grow on alternative sialic acids such as 9-O-acetyl N-acetylneuraminic acid (91). The ability to utilize 9-O-acetyl N-acetylneuraminic requires YjhS (NanS), a 9-O-acetyl N-acetylneuraminic esterase, and has relevance to pathogenicity as this alternative sialic acid is commonly found in mammalian host mucosal sites (91). Additional alternative sialic acids that can be utilized and transported by *E. coli* K-12 include Neu5Gc and KDN, which are transported via NanT and catabolized using the – aldolase, NanA (80). Hopkins and colleagues demonstrated that an *E. coli nanT* deletion strain could utilize Neu5Gc and KDN when expressing sialic acid transporters from two other human pathogens: the TRAP SiaPQM from *H. influenzae* and the SSS transporter from *Salmonella enterica*, demonstrating that potentially many human pathogens may be able to utilize KDN and Neu5Gc as carbon sources (80).

Another human enteric pathogen that utilizes sialic acid is *S. enterica*, which causes enterocolitis/diarrhea and infections have an incidence rate in the United States of 16.42 cases per 100,000 individuals (92, 93). It was shown in *S. enterica* serovar Typhimurium strain LT2 that this strain possesses a sialidase (NanH) that is absent from most other isolates (94, 95). NanH was shown to be homologous to clostridial sialidases (96). The function of NanH *in vivo* has not been studied. However, sialic acid was shown to be important for adherence of *Salmonella* to colonic cells (97). Sakarya and colleagues demonstrated that sialic acid is important for *S. enterica* serovar Typhi to adhere to Caco2 cells; when sialic acid was removed via sialidase treatment, adherence was reduced by 41% (97).

The Kingsley group identified genes important for host colonization via ChIP-seq and transcriptome analysis of the OmpR regulon and two of the operons subsequently identified (SL1068-71 and SL1066-67) were shown to be required for growth on sialic acid (98). They demonstrated that in a mixed inoculum experiment of the streptomycin treated mouse model of colitis that a deletion of operon SL1068-71 exhibited a significant reduction in the ability to colonize the cecum and ileum, and this operon exhibits sequence similarity to sialic acid uptake systems (98). In another recent paper examining expansion of enteric pathogens following antibiotic treatment and their ability to utilize microbiota-liberated host sugars, it was shown that *S. enterica* serovar Typhimurium utilized sialic acid and if this pathway was abolished through gene deletion, the competitiveness of the organism *in vivo* was reduced (99). The described studies demonstrate that *Salmonella* can utilize sialic acid and that it is an important nutrient in host colonization.

Vibrio cholerae, is the causative agent of the profuse secretory diarrheal disease cholera and it is estimated, conservatively, that there are more than a million cases of cholera worldwide annually. Pathogenic isolates of this enteric extracellular pathogen are capable of utilizing sialic acid as a sole carbon source (34) (Fig. 3). *V. cholerae* contains a TRAP transporter SiaPQM (VC1777-VC1779), which is required for uptake of sialic acid (100, 101). A recent paper had suggested that an entirely different TRAP transporter (VC1927-VC1929) was the sole Neu5Ac transporter in *V. cholerae* (102). However, bioinformatics, genomic and genetic analyses clearly demonstrated that this is not the case (100, 103). This data demonstrates that VC1927-VC1929 encode a C_4-dicarboxylate-specific TRAP transporter. A deletion of VC1929 resulted in a defect in growth on C_4-dicarboxylates but not Neu5Ac as the sole carbon source whereas deletion of *siaP* (VC1777) resulted in a mutant strain that was unable to support growth on Neu5Ac as the sole carbon source. These data unequivocally show that *siaPQM* (VC1777-1779) encoded a TRAP transporter and is the sole sialic acid transporter in *V. cholerae* (100).

In addition to the catabolism and transporter genes, *V. cholerae*, also possesses a sialidase encoded by the *nanH* gene (104) (Fig. 4). It has been established that NanH removes two molecules of sialic acid from sialylated gangliosides found in intestinal epithelium, unmasking the GM1 ganglioside, the receptor for cholera toxin (105–108). The sialidase along with the catabolism genes *nanA*, *nanEK*, and the TRAP transporter genes *siaPQM* are contained within a 57 kb

Vibrio cholerae – in vivo sialic acid catabolism

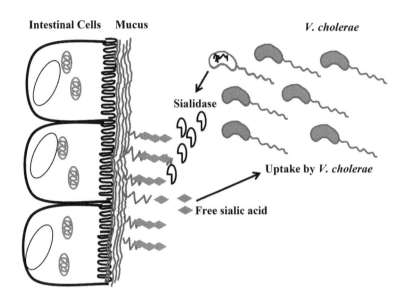

FIGURE 3 *Vibrio cholerae* sialic acid metabolism. *V. cholerae* is an extracellular intestinal pathogen that colonizes the mucus layer of the small intestine and elaborates cholera toxin and sialidase. Sialidase cleaves sialic acid from high order gangliosides to release sialic acid and expose the GM1 ganglioside, the receptor for cholera toxin. Free sialic acid can be transported into the *V. cholerae* cell via the TRAP transporter SiaPQM contained on the pathogenicity island VPI-2, which also contains the genes for sialic acid catabolism. doi:10.1128/microbiolspec.MBP-0005-2014.f3

pathogenicity island, named Vibrio Pathogenicity Island-2 (VPI–2) (Fig. 4) (104). This pathogenicity island is only present in pathogenic isolates and consists of 52 open reading frames (ORFs) in choleragenic *V. cholerae* isolates. Previously, studies showed that VPI-2 displays all of the characteristics of a horizontally acquired region; a lower G+C content compared to the entire genome; the presence of a tyrosine recombinase integrase; and a tRNA chromosomal insertion site. VPI-2 in choleragenic isolates encodes a type 1 restriction modification system, the sialic acid scavenging, transport and catabolism genes and a region containing several hypothetical proteins (104). In pathogenic *V. cholerae* isolates that cause inflammatory diarrhea, VPI-2 is also present but in these isolates the restriction modification system is replaced by a type three secretion

system (109). Additionally, in *V. mimicus* isolates that cause inflammatory diarrhea, VPI-2 is present and also contains a type three secretion system. The key point is that the ability to catabolize sialic acid is retained and present in these enteric pathogens.

A role for sialic acid catabolism in *V. cholerae* pathogenesis has been demonstrated in choleragenic isolates (34). This bacterium colonizes the heavily sialylated mucus of the human gut and the ability to catabolize sialic acid as a carbon and energy source should confer the organism with a growth advantage compared to strains unable to utilize sialic acid. In single infection assays in the infant mouse model of cholera, which compared a wild-type strain with a *nanA* mutant, it was found that in the early stages of infection (3 hours and 6 hours post infection) the mutant was unable to maintain the same high cell

Vibrio Pathogenicity island-2 (VPI-2) 57 Kb

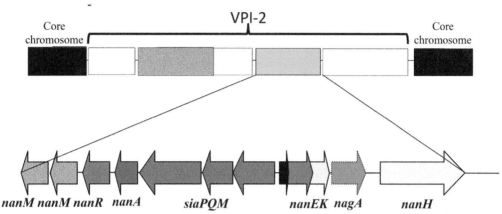

Sialic acid scavenging, transport and catabolism genes 12 kb

FIGURE 4 **Schematic diagram of the pathogenicity island VPI-2 in** *V. cholerae* **choleragenic isolates. ORFs of interest are depicted as arrows indicting the direction of transcription. The entire sialic acid catabolism cluster encompasses 12 kb on the 57 kb VPI-2 region. doi:10.1128/microbiolspec.MBP-0005-2014.f4**

density compared to the wild-type strain (34). This suggests that the ability to catabolize sialic acid as a carbon source in the early stages of infection increases the fitness and likelihood of colonization in the highly populated environment of the human gut (34). Similarly, during *in vivo* co-infection competition assays, it was shown that the competitive index of the mutant versus wild-type was 0.06 indicating a decrease in fitness of the mutant. A recent study demonstrated that sialic acid transport is important in *in vivo* fitness in a streptomycin treated adult mouse model of colonization. This study found that a sialic acid transporter deficient strain was out-competed by wild-type in colonization persistence assays (110). These data show that the ability to utilize sialic acid as a carbon source by *V. cholerae* confers them with a competitive advantage in the sialic acid rich environment of the gut.

Another pathogenic *Vibrio* species which can utilize sialic acid as a sole carbon source

is *V. vulnificus* (47). Previous phylogenetic analysis revealed that sialic acid transport and catabolism genes are present predominantly in clinical isolates and much less frequently in environmental isolates (111). This analysis demonstrated that the TRAP transporter, SiaPQM that clusters with the catabolism genes, is essential for sialic acid uptake in this species (111). Interestingly, unlike *V. cholerae, V. vulnificus* is also capable of biosynthesis of nonulosonic acid a phenotype that is present in all *V. vulnificus* strains (112). However, this data suggest that *V. vulnificus* does not synthesize Neu5Ac but rather either of two sialic acid-like molecules, legionaminic or pseudaminic acid (112). This would indicate that transport of Neu5Ac into the cell is for the sole purpose of catabolism and not sialylation. More recently, an unpublished study showed that *V. vulnificus* decorates its LPS with nonulosonic acids and expression of these proteins is required for biofilm formation, motility, and flagellar expression. An *in vivo* mouse model of septicemia demonstrated that a sialylation defective mutant had a significant survival disadvantage compared wild-type strain (J.B. Lubin, E.F. Boyd, and A.L. Lewis, unpublished data).

It has also been proposed for *V. vulnificus*, that catabolism of Neu5Ac is important for enteropathogenesis (113). A *nanA* mutant was shown to be incapable of utilizing Neu5Ac as a carbon source and was defective for intestinal colonization (113). Additionally, it was proposed that sialic acid catabolism is important in virulence as the *nanA* mutant exhibited decreased cytotoxicity towards INT-407 epithelial cells, decreased adherence *in vitro*, and also displayed a decrease in histopathological damage in jejunum and colon tissues from the mouse intestine (113). The Nan cluster of *V. vulnificus* has been shown to be transcriptionally repressed by NanR, and is induced by the presence of N-acetylmannosamine-6-phosphate (ManNAc-6P) which specifically binds to NanR (114, 115). The interaction between ManNAc-6P and NanR has been shown to be important in *V. vulnificus* pathogenesis. A mutant strain containing a NanR mutation that prevented binding of the ligand, ManNAc-6P, was shown to have growth impairment when sialic acid was the sole carbon source and the mutant strain was also less fit than the wild-type strain *in vivo*, displaying decreased virulence (114). Together, these results demonstrate that catabolism of sialic acid is a feature of clinical isolates and is an important component of survival and virulence *in vivo* in *V. vulnificus*, a food-borne pathogen.

One of the first bacteria in which the ability to utilize sialic acid as a carbon source was demonstrated was *Clostridium perfringens*, a bacterium associated with food poisoning (116, 117). In *C. perfringens*, NanA, the sialic acid lyase, is necessary for growth in minimal medium supplemented with sialic acid as the sole carbon source and transcription of *nanA* is induced by sialic acid (118). Bioinformatics analysis identified putative sialic acid catabolism genes among 10 *Clostridium* species, including *C. difficile*, a leading cause of antibiotic-associated diarrhea and colitis (119). A recent study of *C. difficile* demonstrated that post-antibiotic expansion of this organism is aided by the elevation of sialic acid levels *in vivo* (99). This study showed that not only do sialic acid levels have an impact on expansion of *C. difficile* but strains of *C. difficile* unable to utilize sialic acid exhibited reduced colonization following antibiotic treatment of mice, despite a spike in sialic acid levels (99). Taken together these results demonstrate that intestinal pathogenic species of Clostridia catabolize sialic acid and this is an important component of pathogenesis and host interaction of these bacteria.

Clostridia also produce a number of sialidases, the function of which was first examined in *C. perfringens* and was found to be variably present among strains (116, 117). A recent study examined *C. perfringens* Type D Strain CN3718, which encodes three sialidases, NanI, NanJ, and NanH, and demonstrated that secreted sialidiase, NanI, was important for enhanced binding to MDCK cells and cytotoxic effects of the episilon toxin, ETX (120) suggesting NanI may contribute to intestinal colonization and increased ETX action.

In summary, this chapter highlights some of the research demonstrating the importance of sialic acid as a carbon source for pathogens and that this ability can play a significant role in pathogenesis. One of the main challenges invading pathogens face is the limited availability of nutrient sources and in order to survive, they must be able to compete for these resources. This can be achieved through the use of alternative carbon sources such as amino sugars like sialic acids (49, 50, 86, 88, 99, 121). In conclusion, these studies on sialic acid further understanding of the biology of sialic acid and how it is exploited in a variety of ways, including as an alternative carbon nutrient source, by pathogens *in vivo*.

ACKNOWLEDGMENTS

Research on *Vibrio* species in the Boyd group is supported by a National Science Foundation CAREER award DEB-0844409. We thank members of our group for their enthusiasm and hard work, Megan R. Carpenter, Sai S. Kalburge, Nathan McDonald, Serge Y. Ongagna-Yhombi, Abish Regmi, Molly Peters, and Laura Powell.

CITATION

Haines-Menges BL, Whitaker WB, Lubin JB, Boyd EF. 2015. Host sialic acids: a delicacy for the pathogen with discerning taste. Microbiol Spectrum 3(4):MBP-0005-2014.

REFERENCES

1. **Angata T, Varki A.** 2002. Chemical diversity in the sialic acids and related alpha-keto acids: an evolutionary perspective. *Chem Rev* **102:** 439–469.

2. **Butor C, Diaz S, Varki A.** 1993. High level O-acetylation of sialic acids on N-linked oligosaccharides of rat liver membranes. Differential subcellular distribution of 7- and 9-O-acetyl groups and of enzymes involved in their regulation. *J Biol Chem* **268:**10197–10206.

3. **Varki A.** 2001. Loss of N-glycolylneuraminic acid in humans: Mechanisms, consequences, and implications for hominid evolution. *Am J Phys Anthropol* Suppl 33:54–69.

4. **Inoue S, Kitajima K.** 2006. KDN (deaminated neuraminic acid): dreamful past and exciting future of the newest member of the sialic acid family. *Glycoconj J* **23:**277–290.

5. **Kazatchkine MD, Fearon DT, Austen KF.** 1979. Human alternative complement pathway: membrane-associated sialic acid regulates the competition between B and beta1 H for cell-bound C3b. *J Immunol* **122:**75–81.

6. **Bennett M, Schmid K.** 1980. Immunosuppression by human plasma alpha 1-acid glycoprotein: importance of the carbohydrate moiety. *Proc Natl Acad Sci USA* **77:**6109.

7. **Ozkan AN, Ninnemann JL.** 1985. Suppression of in vitro lymphocyte and neutrophil responses by a low molecular weight suppressor active peptide from burn-patient sera. *J Clin Immunol* **5:**172–179.

8. **Cameron DJ, Churchill WH.** 1982. Specificity of macrophage mediated cytotoxicity: role of target cell sialic acid. *Jpn J Exp Med* **52:** 9–16.

9. **Lanoue A, Batista FD, Stewart M, Neuberger MS.** 2002. Interaction of CD22 with alpha2,6-linked sialoglycoconjugates: innate recognition of self to dampen B cell autoreactivity? *Eur J Immunol* **32:**348–355.

10. **Rutishauser U.** 2008. Polysialic acid in the plasticity of the developing and adult vertebrate nervous system. *Nat Rev Neurosci* **9:**26–35.

11. **Wang B.** 2012. Molecular mechanism underlying sialic acid as an essential nutrient for brain development and cognition. *Adv Nutr* **3:**465S–472S.

12. **Blix G.** 1936. Concerning the carbohydrate groups of submaxillary mucin. *Hoppe-Seylers Zeitschrift Fur Physiologische Chemie* **240:**43–54.

13. **Culling CFA, Reid PE, Clay MG, Dunn WL.** 1974. The histochemical demonstration of O-acylated sialic acid in gastrointestinal mucins. Their association with the potassium hydroxide-periodic acid-schiff effect. *J Histochem Cytochem* **22:**826–831.

14. **Thornton DJ, Carlstedt I, Howard M, Devine PL, Price MR, Sheehan JK.** 1996. Respiratory mucins: identification of core proteins and glycoforms. *Biochem J* **316**(Pt 3):967–975.

15. **Scudder PR, Chantler EN.** 1982. Control of human cervical mucin glycosylation by endogenous fucosyl and sialyltransferases. *Adv Exp Med Biol* **144:**265–267.

16. **Robbe C, Capon C, Coddeville B, Michalski JC.** 2004. Structural diversity and specific distribution of O-glycans in normal human mucins along the intestinal tract. *Biochem J* **384:**307–316.

17. **Slomiany BL, Murty VLN, Piotrowski J, Slomiany A.** 1996. Salivary mucins in oral mucosal defense. *Gen Pharmacol—Vascular Syst* **27:**761–771.

18. **Ho JJ, Cheng S, Kim YS.** 1995. Access to peptide regions of a surface mucin (MUC1) is reduced by sialic acids. *Biochem Biophys Res Commun* **210:**866–873.

19. **Ogasawara Y, Namai T, Yoshino F, Lee MC, Ishii K.** 2007. Sialic acid is an essential moiety of mucin as a hydroxyl radical scavenger. *FEBS Letters* **581:**2473–2477.

20. **Vann WF, Daines DA, Murkin AS, Tanner ME, Chaffin DO, Rubens CE, Vionnet J, Silver RP.** 2004. The NeuC protein of *Escherichia coli* K1 is a UDP N-acetylglucosamine 2-epimerase. *J Bacteriol* **186:**706–712.

21. **Vann WF, Silver RP, Abeijon C, Chang K, Aaronson W, Sutton A, Finn CW, Lindner W,**

Kotsatos M. 1987. Purification, properties, and genetic location of *Escherichia coli* cytidine 5'-monophosphate N-acetylneuraminic acid synthetase. *J Biol Chem* **262**:17556–17562.

22. Vann WF, Tavarez JJ, Crowley J, Vimr E, Silver RP. 1997. Purification and characterization of the *Escherichia coli* K1 neuB gene product N-acetylneuraminic acid synthetase. *Glycobiology* **7**:697–701.

23. Knirel YA, Rietschel ET, Marre R, ZÄHringer U. 1994. The structure of the O-specific chain of *Legionella pneumophila* serogroup 1 lipopolysaccharide. *Eur J Biochem* **221**:239–245.

24. Knirel YA, Vinogradov EV, L'Vov VL, Kocharova NA, Shashkov AS, Dmitriev BA, Kochetkov NK. 1984. Sialic acids of a new type from the lipopolysaccharides of *Pseudomonas aeruginosa* and *Shigella boydii*. *Carbohydr Res* **133**:C5–C8.

25. Parsons NJ, Patel PV, Tan EL, Andrade JR, Nairn CA, Goldner M, Cole JA, Smith H. 1988. Cytidine 5'-monophospho-N-acetyl neuraminic acid and a low molecular weight factor from human blood cells induce lipopolysaccharide alteration in gonococci when conferring resistance to killing by human serum. *Microb Pathog* **5**:303–309.

26. Mushtaq N, Redpath MB, Luzio JP, Taylor PW. 2004. Prevention and Cure of Systemic *Escherichia coli* K1 Infection by Modification of the Bacterial Phenotype. *Antimicrob Agents Chemother* **48**:1503–1508.

27. Kline KA, Schwartz DJ, Lewis WG, Hultgren SJ, Lewis AL. 2011. Immune activation and suppression by group B streptococcus in a murine model of urinary tract infection. *Infect Immun* **79**:3588–3595.

28. Bouchet V, Hood DW, Li J, Brisson JR, Randle GA, Martin A, Li Z, Goldstein R, Schweda EKH, Pelton SI, Richards JC, Moxon ER. 2003. Host-derived sialic acid is incorporated into *Haemophilus influenzae* lipopolysaccharide and is a major virulence factor in experimental otitis media. *Proc Natl Acad Sci USA* **100**:8898–8903.

29. Soong G, Muir A, Gomez MI, Waks J, Reddy B, Planet P, Singh PK, Kaneko Y, Wolfgang MC, Hsiao YS, Tong L, Prince A. 2006. Bacterial neuraminidase facilitates mucosal infection by participating in biofilm production. *J Clin Invest* **116**:2297–2305.

30. Jurcisek J, Greiner L, Watanabe H, Zaleski A, Apicella MA, Bakaletz LO. 2005. Role of sialic acid and complex carbohydrate biosynthesis in biofilm formation by nontypeable *Haemophilus influenzae* in the chinchilla middle ear. *Infect Immun* **73**:3210–3218.

31. Trappetti C, Kadioglu A, Carter M, Hayre J, Iannelli F, Pozzi G, Andrew PW, Oggioni MR. 2009. Sialic acid: a preventable signal for pneumococcal biofilm formation, colonization, and invasion of the host. *J Infect Dis* **199**:1497–1505.

32. Parker D, Soong G, Planet P, Brower J, Ratner AJ, Prince A. 2009. The NanA neuraminidase of *Streptococcus pneumoniae* is involved in biofilm formation. *Infect Immun* **77**:3722–3730.

33. Lewis AL, Desa N, Hansen EE, Knirel YA, Gordon JI, Gagneux P, Nizet V, Varki A. 2009. Innovations in host and microbial sialic acid biosynthesis revealed by phylogenomic prediction of nonulosonic acid structure. *Proc Natl Acad Sci USA* **106**:13552–13557.

34. Almagro-Moreno S, Boyd EF. 2009. Sialic acid catabolism confers a competitive advantage to pathogenic *Vibrio cholerae* in the mouse intestine. *Infect Immun* **77**:3807–3816.

35. Almagro-Moreno S, Boyd EF. 2010. Bacterial catabolism of nonulosonic (sialic) acid and fitness in the gut. *Gut Microbes* **1**:45–50.

36. Vimr ER, Troy FA. 1985. Regulation of sialic acid metabolism in *Escherichia coli*: role of N-acylneuraminate pyruvate-lyase. *J Bacteriol* **164**:854–860.

37. Lewis WG, Robinson LS, Gilbert NM, Perry JC, Lewis AL. 2013. Degradation, foraging, and depletion of mucus sialoglycans by the vagina-adapted Actinobacterium *Gardnerella vaginalis*. *J Biol Chem* **288**:12067–12079.

38. Condemine G, Berrier C, Plumbridge J, Ghazi A. 2005. Function and Expression of an N-Acetylneuraminic Acid-Inducible Outer Membrane Channel in *Escherichia coli*. *J Bacteriol* **187**:1959–1965.

39. Vimr ER, Troy FA. 1985. Identification of an inducible catabolic system for sialic acids (nan) in *Escherichia coli*. *J Bacteriol* **164**:845–853.

40. Martinez J, Steenbergen S, Vimr E. 1995. Derived structure of the putative sialic acid transporter from *Escherichia coli* predicts a novel sugar permease domain. *J Bacteriol* **177**:6005–6010.

41. Allen S, Zaleski A, Johnston JW, Gibson BW, Apicella MA. 2005. Novel sialic acid transporter of *Haemophilus influenzae*. *Infect Immun* **73**:5291–5300.

42. Post DM, Mungur R, Gibson BW, Munson RS Jr. 2005. Identification of a novel sialic acid transporter in *Haemophilus ducreyi*. *Infect Immun* **73**:6727–6735.

43. Severi E, Randle G, Kivlin P, Whitfield K, Young R, Moxon R, Kelly D, Hood D, Thomas

GH. 2005. Sialic acid transport in *Haemophilus influenzae* is essential for lipopolysaccharide sialylation and serum resistance and is dependent on a novel tripartite ATP-independent periplasmic transporter. *Mol Microbiol* **58**:1173–1185.

44. **Severi E, Hood DW, Thomas GH.** 2007. Sialic acid utilization by bacterial pathogens. *Microbiology* **153**:2817–2822.

45. **Mulligan C, Geertsma ER, Severi E, Kelly DJ, Poolman B, Thomas GH.** 2009. The substrate-binding protein imposes directionality on an electrochemical sodium gradient-driven TRAP transporter. *Proc Natl Acad Sci USA* **106**:1778–1783.

46. **Severi E, Hosie AH, Hawkhead JA, Thomas GH.** 2010. Characterization of a novel sialic acid transporter of the sodium solute symporter (SSS) family and in vivo comparison with known bacterial sialic acid transporters. *FEMS Microbiol Lett* **304**:47–54.

47. **Almagro-Moreno S, Boyd EF.** 2009. Insights into the evolution of sialic acid catabolism among bacteria. *BMC Evol Biol* **9**:118.

48. **Wade WG.** 2013. The oral microbiome in health and disease. *Pharmacol Res* **69**:137–143.

49. **Freter R.** 1983. Mechanisms that control the microflora in the large intestine, p 33–54. *In* Hentges DJ (ed), *Human intestinal microflora in health and disease*. Academic Press, Inc., New York.

50. **Freter R.** 1988. Mechanisms of bacterial colonization of the mucosal surfaces of the gut, p 45–60. *Virulence mechanisms of bacterial pathogens*. American Society for Microbiology, Washington, DC.

51. **Roy S, Douglas CW, Stafford GP.** 2010. A novel sialic acid utilization and uptake system in the periodontal pathogen *Tannerella forsythia. J Bacteriol* **192**:2285–2293.

52. **Stafford G, Roy S, Honma K, Sharma A.** 2012. Sialic acid, periodontal pathogens and *Tannerella forsythia*: stick around and enjoy the feast! *Mol Oral Microbiol* **27**:11–22.

53. **Brigham C, Caughlan R, Gallegos R, Dallas MB, Godoy VG, Malamy MH.** 2009. Sialic acid (N-acetyl neuraminic acid) utilization by *Bacteroides fragilis* requires a novel N-acetyl mannosamine epimerase. *J Bacteriol* **191**:3629–3638.

54. **Brigham CJ, Malamy MH.** 2005. Characterization of the RokA and HexA broad-substrate-specificity hexokinases from *Bacteroides fragilis* and their role in hexose and N-acetylglucosamine utilization. *J Bacteriol* **187**:890–901.

55. **Honma K, Mishima E, Sharma A.** 2011. Role of *Tannerella forsythia* NanH sialidase in epithelial cell attachment. *Infect Immun* **79**: 393–401.

56. **Thompson H, Homer KA, Rao S, Booth V, Hosie AH.** 2009. An orthologue of *Bacteroides fragilis* NanH is the principal sialidase in *Tannerella forsythia. J Bacteriol* **191**:3623–3628.

57. **Ishikura H, Arakawa S, Nakajima T, Tsuchida N, Ishikawa I.** 2003. Cloning of the *Tannerella forsythensis* (*Bacteroides forsythus*) siaHI gene and purification of the sialidase enzyme. *J Med Microbiol* **52**:1101–1107.

58. **Wyss C, Moter A, Choi BK, Dewhirst FE, Xue Y, Schupbach P, Gobel UB, Paster BJ, Guggenheim B.** 2004. *Treponema putidum* sp. nov., a medium-sized proteolytic spirochaete isolated from lesions of human periodontitis and acute necrotizing ulcerative gingivitis. *Int J Syst Evol Microbiol* **54**:1117–1122.

59. **Kurniyati K, Zhang W, Zhang K, Li C.** 2013. A surface-exposed neuraminidase affects complement resistance and virulence of the oral spirochaete *Treponema denticola. Mol Microbiol* **89**:842–856.

60. **Byers HL, Homer KA, Beighton D.** 1996. Utilization of sialic acid by viridans streptococci. *J Dent Res* **75**:1564–1571.

61. **Byers HL, Tarelli E, Homer KA, Hambley H, Beighton D.** 1999. Growth of Viridans streptococci on human serum alpha1-acid glycoprotein. *J Dent Res* **78**:1370–1380.

62. **Beck JM, Young VB, Huffnagle GB.** 2012. The microbiome of the lung. *Transl Res* **160**: 258–266.

63. **Cui L, Morris A, Ghedin E.** 2013. The human mycobiome in health and disease. *Genome Med* **5**:63.

64. **Huang YJ, Lynch SV.** 2011. The emerging relationship between the airway microbiota and chronic respiratory disease: clinical implications. *Expert Rev Respir Med* **5**:809–821.

65. **Lilley GG, Barbosa JA, Pearce LA.** 1998. Expression in *Escherichia coli* of the putative N-acetylneuraminate lyase gene (nanA) from *Haemophilus influenzae*: overproduction, purification, and crystallization. *Protein Expr Purif* **12**:295–304.

66. **Hood DW, Makepeace K, Deadman ME, Rest RF, Thibault P, Martin A, Richards JC, Moxon ER.** 1999. Sialic acid in the lipopolysaccharide of *Haemophilus influenzae*: strain distribution, influence on serum resistance and structural characterization. *Mol Microbiol* **33**:679–692.

67. **Vimr E, Lichtensteiger C, Steenbergen S.** 2000. Sialic acid metabolism's dual function in *Haemophilus influenzae. Mol Microbiol* **36**:1113–1123.

68. Johnston JW, Zaleski A, Allen S, Mootz JM, Armbruster D, Gibson BW, Apicella MA, Munson RS Jr. 2007. Regulation of sialic acid transport and catabolism in *Haemophilus influenzae*. *Mol Microbiol* **66:**26–39.

69. Pettigrew MM, Fennie KP, York MP, Daniels J, Ghaffar F. 2006. Variation in the presence of neuraminidase genes among *Streptococcus pneumoniae* isolates with identical sequence types. *Infect Immun* **74:**3360–3365.

70. King SJ. 2010. Pneumococcal modification of host sugars: a major contributor to colonization of the human airway? *Mol Oral Microbiol* **25:**15–24.

71. Marion C, Aten AE, Woodiga SA, King SJ. 2011. Identification of an ATPase, MsmK, which energizes multiple carbohydrate ABC transporters in *Streptococcus pneumoniae*. *Infect Immun* **79:**4193–4200.

72. Xu G, Kiefel MJ, Wilson JC, Andrew PW, Oggioni MR, Taylor GL. 2011. Three *Streptococcus pneumoniae* sialidases: three different products. *J Am Chem Soc* **133:**1718–1721.

73. Brittan JL, Buckeridge TJ, Finn A, Kadioglu A, Jenkinson HF. 2012. Pneumococcal neuraminidase A: an essential upper airway colonization factor for *Streptococcus pneumoniae*. *Mol Oral Microbiol* **27:**270–283.

74. Manco S, Hernon F, Yesilkaya H, Paton JC, Andrew PW, Kadioglu A. 2006. Pneumococcal neuraminidases A and B both have essential roles during infection of the respiratory tract and sepsis. *Infect Immun* **74:**4014–4020.

75. Orihuela CJ, Gao G, Francis KP, Yu J, Tuomanen EI. 2004. Tissue-specific contributions of pneumococcal virulence factors to pathogenesis. *J Infect Dis* **190:**1661–1669.

76. Tong HH, Blue LE, James MA, DeMaria TF. 2000. Evaluation of the virulence of a *Streptococcus pneumoniae* neuraminidase-deficient mutant in nasopharyngeal colonization and development of otitis media in the chinchilla model. *Infect Immun* **68:**921–924.

77. Srinivasan S, Hoffman NG, Morgan MT, Matsen FA, Fiedler TL, Hall RW, Ross FJ, McCoy CO, Bumgarner R, Marrazzo JM, Fredricks DN. 2012. Bacterial communities in women with bacterial vaginosis: high resolution phylogenetic analyses reveal relationships of microbiota to clinical criteria. *PLoS One* **7:**e37818.

78. Lewis AL, Lewis WG. 2012. Host sialoglycans and bacterial sialidases: a mucosal perspective. *Cell Microbiol* **14:**1174–1182.

79. von Nicolai H, Hammann R, Salehnia S, Zilliken F. 1984. A newly discovered sialidase from *Gardnerella vaginalis*. *Zentralbl Bakteriol Mikrobiol Hyg A* **258:**20–26.

80. Hopkins AP, Hawkhead JA, Thomas GH. 2013. Transport and catabolism of the sialic acids N-glycolylneuraminic acid and 3-keto-3-deoxy-D-glycero-D-galactonon(onic) acid by *Escherichia coli* K-12. *FEMS Microbiol Lett* **347:**14–22.

81. Gilbert NM, Lewis WG, Lewis AL. 2013. Clinical features of bacterial vaginosis in a murine model of vaginal infection with *Gardnerella vaginalis*. *PLoS One* **8:**e59539.

82. Pezzicoli A, Ruggiero P, Amerighi F, Telford JL, Soriani M. 2012. Exogenous sialic acid transport contributes to group B streptococcus infection of mucosal surfaces. *J Infect Dis* **206:**924–931.

83. McGuckin MA, Linden SK, Sutton P, Florin TH. 2011. Mucin dynamics and enteric pathogens. *Nat Rev Microbiol* **9:**265–278.

84. LaMont JT, Ventola AS. 1980. Purification and composition of colonic epithelial mucin. *Biochim Biophys Acta* **626:**234–243.

85. Peekhaus N, Conway T. 1998. What's for dinner?: Entner-Doudoroff metabolism in *Escherichia coli*. *J Bacteriol* **180:**3495–3502.

86. Chang DE, Smalley DJ, Tucker DL, Leatham MP, Norris WE, Stevenson SJ, Anderson AB, Grissom JE, Laux DC, Cohen PS, Conway T. 2004. Carbon nutrition of *Escherichia coli* in the mouse intestine. *Proc Natl Acad Sci USA* **101:**7427–7432.

87. Vimr ER, Kalivoda KA, Deszo EL, Steenbergen SM. 2004. Diversity of microbial sialic acid metabolism. *Microbiol Mol Biol Rev* **68:**132–153.

88. Fabich AJ, Jones SA, Chowdhury FZ, Cernosek A, Anderson A, Smalley D, McHargue JW, Hightower GA, Smith JT, Autieri SM, Leatham MP, Lins JJ, Allen RL, Laux DC, Cohen PS, Conway T. 2008. Comparison of carbon nutrition for pathogenic and commensal *Escherichia coli* strains in the mouse intestine. *Infect Immun* **76:**1143–1152.

89. Bertin Y, Chaucheyras-Durand F, Robbe-Masselot C, Durand A, de la Foye A, Harel J, Cohen PS, Conway T, Forano E, Martin C. 2013. Carbohydrate utilization by enterohaemorrhagic *Escherichia coli* O157:H7 in bovine intestinal content. *Environ Microbiol* **15:**610–622.

90. Polzin S, Huber C, Eylert E, Elsenhans I, Eisenreich W, Schmidt H. 2013. Growth media simulating ileal and colonic environments affect the intracellular proteome and carbon fluxes of enterohemorrhagic *Escherichia coli* O157:H7 strain EDL933. *Appl Environ Microbiol* **79:**3703–3715.

91. Steenbergen SM, Jirik JL, Vimr ER. 2009. YjhS (NanS) is required for *Escherichia coli* to

grow on 9-O-acetylated N-acetylneuraminic acid. *J Bacteriol* **191**:7134–7139.

92. 2013. Incidence and trends of infection with pathogens transmitted commonly through food — foodborne diseases active surveillance network, 10 U.S. sites, 1996–2012. *Morb Mortal Wkly Rep* **62**:283–287.

93. **Coburn B, Grassl GA, Finlay BB.** 2007. Salmonella, the host and disease: a brief review. *Immunol Cell Biol* **85**:112–118.

94. **Crennell SJ, Garman EF, Philippon C, Vasella A, Laver WG, Vimr ER, Taylor GL.** 1996. The structures of *Salmonella typhimurium* LT2 neuraminidase and its complexes with three inhibitors at high resolution. *J Mol Biol* **259**:264–280.

95. **Hoyer LL, Roggentin P, Schauer R, Vimr ER.** 1991. Purification and properties of cloned *Salmonella typhimurium* LT2 sialidase with virus-typical kinetic preference for sialyl alpha 2—3 linkages. *J Biochem* **110**:462–467.

96. **Hoyer LL, Hamilton AC, Steenbergen SM, Vimr ER.** 1992. Cloning, sequencing and distribution of the *Salmonella typhimurium* LT2 sialidase gene, nanH, provides evidence for interspecies gene transfer. *Mol Microbiol* **6**:873–884.

97. **Sakarya S, Gokturk C, Ozturk T, Ertugrul MB.** 2010. Sialic acid is required for nonspecific adherence of *Salmonella enterica* ssp. *enterica* serovar Typhi on Caco-2 cells. *FEMS Immunol Med Microbiol* **58**:330–335.

98. **Perkins TT, Davies MR, Klemm EJ, Rowley G, Wileman T, James K, Keane T, Maskell D, Hinton JC, Dougan G, Kingsley RA.** 2013. ChIP-seq and transcriptome analysis of the OmpR regulon of *Salmonella enterica* serovars Typhi and Typhimurium reveals accessory genes implicated in host colonization. *Mol Microbiol* **87**:526–538.

99. **Ng KM, Ferreyra JA, Higginbottom SK, Lynch JB, Kashyap PC, Gopinath S, Naidu N, Choudhury B, Weimer BC, Monack DM, Sonnenburg JL.** 2013. Microbiota-liberated host sugars facilitate post-antibiotic expansion of enteric pathogens. *Nature* **502**:96–99.

100. **Chowdhury N, Norris J, McAlister E, Lau SY, Thomas GH, Boyd EF.** 2012. The VC1777–VC1779 proteins are members of a sialic acid-specific subfamily of TRAP transporters (SiaPQM) and constitute the sole route of sialic acid uptake in the human pathogen *Vibrio cholerae*. *Microbiology* **158**:2158–2167.

101. **Mulligan C, Leech AP, Kelly DJ, Thomas GH.** 2012. The membrane proteins SiaQ and SiaM form an essential stoichiometric complex in the sialic acid tripartite ATP-independent periplasmic (TRAP) transporter SiaPQM (VC1777–1779) from *Vibrio cholerae*. *J Biol Chem* **287**:3598–3608.

102. **Sharma SK, Moe TS, Srivastava R, Chandra D, Srivastava BS.** 2011. Functional characterization of VC1929 of *Vibrio cholerae* El Tor: role in mannose-sensitive haemagglutination, virulence and utilization of sialic acid. *Microbiology* **157**:3180–3186.

103. **Thomas GH, Boyd EF.** 2011. On sialic acid transport and utilization by *Vibrio cholerae*. *Microbiology* **157**:3253–3254; discussion 3254–3255.

104. **Jermyn WS, Boyd EF.** 2002. Characterization of a novel Vibrio pathogenicity island (VPI-2) encoding neuraminidase (nanH) among toxigenic *Vibrio cholerae* isolates. *Microbiology* **148**:3681–3693.

105. **Holmgren J, Lonnroth I, Mansson J, Svennerholm L.** 1975. Interaction of cholera toxin and membrane GM1 ganglioside of small intestine. *Proc Natl Acad Sci USA* **72**:2520–2524.

106. **Moustafa I, Connaris H, Taylor M, Zaitsev V, Wilson JC, Kiefel MJ, Von Itzstein M, Taylor G.** 2004. Sialic acid recognition by *Vibrio cholerae* neuraminidase. *J Biol Chem* **279**:40819–40826.

107. **Galen JE, Ketley JM, Fasano A, Richardson SH, Wasserman SS, Kaper JB.** 1992. Role of *Vibrio cholerae* neuraminidase in the function of cholera toxin. *Infect Immun* **60**:406–415.

108. **Holmgren J, Lonnroth I, Svennerholm L.** 1973. Fixation and inactivation of cholera toxin by GM1 ganglioside. *Scand J Infect Dis* **5**:77–78.

109. **Murphy RA, Boyd EF.** 2008. Three pathogenicity islands of *Vibrio cholerae* can excise from the chromosome and form circular intermediates. *J Bacteriol* **190**:636–647.

110. **Boyd EF, Chowdhury N, McDonald ND, Lubin JB.** 2014. Host sialic acids are an important bacterial nutrient source that increase fitness of intestinal pathogens *in vivo*, 114th General Meeting of the American Society for Microbiology. ASM Abstracts.

111. **Lubin JB, Kingston JJ, Chowdhury N, Boyd EF.** 2012. Sialic acid catabolism and transport gene clusters are lineage specific in *Vibrio vulnificus*. *Appl Environ Microbiol* **78**:3407–3415.

112. **Lewis AL, Lubin JB, Argade S, Naidu N, Choudhury B, Boyd EF.** 2011. Genomic and metabolic profiling of nonulosonic acids in Vibrionaceae reveal biochemical phenotypes of allelic divergence in *Vibrio vulnificus*. *Appl Environ Microbiol* **77**:5782–5793.

113. **Jeong HG, Oh MH, Kim BS, Lee MY, Han HJ, Choi SH.** 2009. The capability of catabolic

utilization of N-acetylneuraminic acid, a sialic acid, is essential for *Vibrio vulnificus* pathogenesis. *Infect Immun* **77**:3209–3217.

114. **Hwang J, Kim BS, Jang SY, Lim JG, You DJ, Jung HS, Oh TK, Lee JO, Choi SH, Kim MH.** 2013. Structural insights into the regulation of sialic acid catabolism by the *Vibrio vulnificus* transcriptional repressor NanR. *Proc Natl Acad Sci USA* **110**:E2829–E2837.

115. **Kim BS, Hwang J, Kim MH, Choi SH.** 2011. Cooperative regulation of the *Vibrio vulnificus* nan gene cluster by NanR protein, cAMP receptor protein, and N-acetylmannosamine 6-phosphate. *J Biol Chem* **286**:40889–40899.

116. **Fraser AG, Collee JG.** 1975. The production of neuraminidase by food poisoning strains of *Clostridium welchii* (*C. perfringens*). *J Med Microbiol* **8**:251–263.

117. **Nees S, Schauer R, Mayer F.** 1976. Purification and characterization of N-acetylneuraminate lyase from *Clostridium perfringens*. *Hoppe Seylers Z Physiol Chem* **357**:839–853.

118. **Walters DM, Stirewalt VL, Melville SB.** 1999. Cloning, sequence, and transcriptional regulation of the operon encoding a putative N-acetylmannosamine-6-phosphate epimerase (nanE) and sialic acid lyase (nanA) in *Clostridium perfringens*. *J Bacteriol* **181**:4526–4532.

119. **Borriello SP.** 1995. Clostridial disease of the gut. *Clin Infect Dis* **20**(Suppl 2):S242–S250.

120. **Li J, Sayeed S, Robertson S, Chen J, McClane BA.** 2011. Sialidases affect the host cell adherence and epsilon toxin-induced cytotoxicity of *Clostridium perfringens* type D strain CN3718. *PLoS Pathog* **7**:e1002429.

121. **Hooper LV, Midtvedt T, Gordon JI.** 2002. How host-microbial interactions shape the nutrient environment of the mammalian intestine. *Annu Rev Nutr* **22**:283–307.

122. **Ward RE, Ninonuevo M, Mills DA, Lebrilla CB, German JB.** 2007. In vitro fermentability of human milk oligosaccharides by several strains of bifidobacteria. *Mol Nutr Food Res* **51**:1398–1405.

123. **Sela DA, Mills DA.** 2010. Nursing our microbiota: molecular linkages between bifidobacteria and milk oligosaccharides. *Trends Microbiol* **18**:298–307.

124. **Gruteser N, Marin K, Krämer R, Thomas GH.** 2012. Sialic acid utilization by the soil bacterium *Corynebacterium glutamicum*. *FEMS Microbiol Lett* **336**:131–138.

125. **Joseph S, Desai P, Ji Y, Cummings CA, Shih R, Degoricija L, Rico A, Brzoska P, Hamby SE, Masood N, Hariri S, Sonbol H, Chuzhanova N, McClelland M, Furtado MR, Forsythe SJ.** 2012. Comparative analysis of genome sequences covering the seven cronobacter species. *PLoS One* **7**:e49455.

126. **Joseph S, Hariri S, Masood N, Forsythe S.** 2013. Sialic acid utilization by *Cronobacter sakazakii*. *Microb Inform Exp* **3**:3.

127. **Sanchez-Carron G, Garcia-Garcia MI, Lopez-Rodriguez AB, Jimenez-Garcia S, Sola-Carvajal A, Garcia-Carmona F, Sanchez-Ferrer A.** 2011. Molecular characterization of a novel N-acetylneuraminate lyase from *Lactobacillus plantarum* WCFS1. *Appl Environ Microbiol* **77**:2471–2478.

128. **May M, Brown DR.** 2008. Genetic variation in sialidase and linkage to N-acetylneuraminate catabolism in *Mycoplasma synoviae*. *Microb Pathog* **45**:38–44.

129. **Steenbergen SM, Lichtensteiger CA, Caughlan R, Garfinkle J, Fuller TE, Vimr ER.** 2005. Sialic Acid metabolism and systemic pasteurellosis. *Infect Immun* **73**:1284–1294.

130. **Crost EH, Tailford LE, Le Gall G, Fons M, Henrissat B, Juge N.** 2013. Utilisation of Mucin Glycans by the Human Gut Symbiont *Ruminococcus gnavus* Is Strain-Dependent. *PLoS One* **8**:e76341.

131. **Olson ME, King JM, Yahr TL, Horswill AR.** 2013. Sialic acid catabolism in *Staphylococcus aureus*. *J Bacteriol* **195**:1779–1788.

132. **Marion C, Burnaugh AM, Woodiga SA, King SJ.** 2011. Sialic acid transport contributes to pneumococcal colonization. *Infect Immun* **79**:1262–1269.

133. **Burnaugh AM, Frantz LJ, King SJ.** 2008. Growth of *Streptococcus pneumoniae* on human glycoconjugates is dependent upon the sequential activity of bacterial exoglycosidases. *J Bacteriol* **190**:221–230.

Commensal and Pathogenic *Escherichia coli* Metabolism in the Gut

16

TYRRELL CONWAY[1] and PAUL S. COHEN[2]

INTRODUCTION

Every mammal on the planet is colonized with *E. coli* (1), as well as cold-blooded animals (e.g., fish) at an appropriately warm temperature (2). We estimate there are 10^{21} *E. coli* cells in the human population alone. *E. coli* is frequently the first bacterium to colonize human infants and is a lifelong colonizer of adults (3). *E. coli* is arguably the best understood of all model organisms (4). Yet the essence of how *E. coli* colonizes and/or causes disease is still not completely understood. Certainly, innate immunity, adaptive immunity, and bacterial cell-to-cell communication play important roles in modulating the populations of the 500–1000 different commensal species in the intestine (5–11); however, these topics will not be a focus of this chapter. We have reviewed the mucus layer as habitat for *E. coli* to colonize the intestine, aspects of *E. coli* physiology that enable its success, and the model systems employed for colonization research (12–14). Here, we focus on *E. coli* metabolism in the intestinal mucus layer. We discuss evidence that *E. coli* must obtain nutrients in the mucus layer to colonize, that it resides in the mucus layer as a member of mixed biofilms, and that each *E. coli* strain displays a unique nutritional program in the intestine. We also discuss evidence

[1]Department of Microbiology and Molecular Genetics, Oklahoma State University, Stillwater, OK 74078;
[2]Department of Cell and Molecular Biology, University of Rhode Island, Kingston, RI.
Metabolism and Bacterial Pathogenesis
Edited by Tyrrell Conway and Paul Cohen
© 2015 American Society for Microbiology, Washington, DC
doi:10.1128/microbiolspec.MBP-0006-2014

supporting the "Restaurant" hypothesis for commensal *E. coli* strains, i.e., that they colonize the intestine as sessile members of mixed biofilms obtaining the nutrients they need for growth locally, but compete for nutrients with invading *E. coli* pathogens planktonically.

FROM INGESTION TO COLONIZATION

When *E. coli* is eliminated by a host animal, it is not growing because it cannot grow in the luminal contents of the intestine (15). *E. coli* persists in the environment until its next host consumes viable bacteria in contaminated water or adulterated food. Following ingestion, a stressor faced by *E. coli* is acidity in the stomach, which it survives because stationary phase bacteria induce protective acid-resistance systems (16). Extreme acid tolerance makes *E. coli* transmissible by as few as ten bacterial cells (17). Upon reaching the colon, *E. coli* must find the nutrients it needs to exit lag phase and grow from low to high numbers. Failure to transition from lag phase to logarithmic phase will lead to elimination of the invading *E. coli* bacteria (18). Successful colonization of the colon by *E. coli* depends upon competition for nutrients with a dense and diverse microbiota (18), penetration of the mucus layer (19) (but not motility [20]), avoid host defenses (21, 22), and grow rapidly, exceeding the turnover rate of the mucus layer (23). *E. coli* resides in mucus until being sloughed into the lumen of the intestine (24, 25), from whence some cells are eliminated in the host feces and the cycle begins again. This circle of colonization and extra-intestinal survival is the reality for commensal and pathogenic *E. coli* alike.

BASIC PRINCIPLES OF COLONIZATION

Colonization is defined as the indefinite persistence of a particular bacterial population without reintroduction of that bacterium. We agree with Rolf Freter, a true pioneer in the field of intestinal colonization, who concluded that although several factors could theoretically contribute to an organism's ability to colonize, competition for nutrients is paramount for success in the intestinal ecosystem (26). According to Freter's nutrient-niche hypothesis, the mammalian intestine is analogous to a chemostat in which several hundreds of species of bacteria are in equilibrium. To co-colonize, each species must use at least one limiting nutrient better than all the other species (18, 27, 28). The nutrient-niche hypothesis further predicts that invading species will have difficulty colonizing a stable ecosystem, such as the healthy intestine. The ability of the microbiota to resist invasion is termed colonization resistance (29), an example of which being that when human volunteers were fed *E. coli* strains isolated from their own feces, those *E. coli* failed to colonize (30). Yet, despite colonization resistance, humans are colonized on average with five different *E. coli* strains and there is a continuous succession of strains in individuals (30). This suggests that diversity exists among commensal *E. coli* strains and that different strains may possess different strategies for utilizing growth-limiting nutrients.

If diversity amongst *E. coli* commensal strains plays a role in colonization resistance, then mice pre-colonized with a human *E. coli* commensal strain would resist colonization by the same strain (isogenic challenge strain) because bacteria that consume the nutrients it needs to colonize already occupy its preferred niche. However, if mice pre-colonized with one human *E. coli* commensal strain were subsequently fed a different *E. coli* strain (non-isogenic challenge strain) then, if the second strain could occupy a distinct niche in the intestine, it would co-colonize with the first strain. The results of such experiments showed that each of several precolonized *E. coli* strains nearly eliminated its isogenic challenge strain from the intestine, confirming that colonization resistance can be modeled in mice, but non-isogenic challenge *E. coli* strains grew to higher numbers in the presence of different pre-colonized strains, suggesting that the newly introduced non-isogenic challenge strain either grows faster than the pre-colonized strain on one or more nutrients or uses nutrient(s) not being used by the pre-colonized strain (31).

How might an invading enteric pathogen subvert colonization resistance? According to the nutrient-niche hypothesis, upon reaching the intestine the pathogen would first have to outcompete the resident microbiota for at least one nutrient, allowing it initially to colonize the intestine. However, colonization would not in itself result in pathogenesis if the pathogen must reach the epithelium and either bind to epithelial cells or invade the epithelium. In such instances, the pathogen must presumably penetrate the mucus layer. In a series of groundbreaking studies (32–34), Stecher, Hardt, and colleagues showed that when *Salmonella enterica* serovar Typhimurium induces inflammation in a mouse colitis model, the composition of the microbiota is changed and its growth is suppressed while serovar Typhimurium growth is enhanced. The authors also showed that serovar Typhimurium is attracted by chemotaxis to galactose-containing nutrients on the mucosal surface (e.g., galactose-containing glycoconjugates and mucin) and, as expected, flagella and motility were required (32). Thus, to quote the authors (34), "Triggering the host's immune defense can shift the balance between the protective microbiota and the pathogen in favor of the pathogen."

In streptomycin-treated mice, nutrient consumption by colonized *E. coli* strains can prevent invading *E. coli* strains from colonizing (35). By examining the sugars used by various human commensal *E. coli* strains to colonize, we identified a pair of strains (*E. coli* HS and *E. coli* Nissle 1917) that together use the five sugars previously found to be most important for colonization by the enterohemorrhagic *E. coli* (EHEC) strain EDL933 (O157:H7) (36). When mice were pre-colonized with *E. coli* HS

and *E. coli* Nissle 1917, invading *E. coli* EDL933 was eliminated from the intestine (35). Clearly, one therapeutic strategy to prevent pathogenesis would be to outcompete the pathogen for nutrients normally present in the intestine and eliminate it before it can colonize and subsequently cause inflammation (5, 6, 37).

Implicit in the nutrient-niche hypothesis is the idea that different species compete for preferred nutrients from a mixture that is equally available to all species. However, there is growing evidence that, at least under some circumstances, *E. coli* receives the nutrients it needs through direct interactions with neighboring microbes in the intestinal community. Thus, we take a renewed look at the metabolism of and nutrient flow between members of the intestinal microbiota.

CENTRAL METABOLISM AND INTESTINAL COLONIZATION

E. coli is a Gram-negative, prototrophic, facultative anaerobe with the ability to respire oxygen, use alternative anaerobic electron acceptors, or ferment, depending on electron-acceptor availability. Central metabolism in *E. coli* consists of the Embden-Meyerhof-Parnas glycolytic pathway (EMP), the pentose phosphate pathway (PP), the Entner-Doudoroff pathway (ED), the TCA cycle, and diverse fermentation pathways. *E. coli* grows best on sugars, including a wide range of mono- and disaccharides, but it cannot grow on complex polysaccharides because it lacks the necessary hydrolase enzymes (36). *E. coli* also can grow on amino acids and dicarboxylates that feed into the TCA cycle; the metabolism of these nutrients requires gluconeogenesis, the biosynthesis of glucose phosphate to be used as precursors of macromolecules such as LPS and peptidoglycan. Central metabolic pathways in *E. coli* are highly conserved, constituting a significant part of the core *E. coli* genome (38). The role of central metabolism during intestinal colonization has been studied in *E. coli*. The results of these experiments are summarized below (Table 1).

Mutants blocked in glycolysis or the ED pathway, but not the PP pathway, have major colonization defects in competition with their wild type parents (39). Given its role in hexose metabolism, it is expected that glycolysis is important for colonization. Indeed, a *pgi* mutant lacking the key enzyme, phosphoglucose isomerase, of the EMP glycolytic pathway has a substantial colonization defect when competed against its wild-type *E. coli* K-12 parent (Table 1). The role of the EMP pathway goes beyond colonization by *E. coli*. For example, glucose catabolism and glycolysis are known to play a role in intracellular growth of serovar Typhimurium within macrophage vacuoles (40), and proper regulation of glucose catabolism and glycolysis are coupled to virulence-factor expression in EHEC (41). A recent study of *Shigella flexneri* revealed similar usage of these central metabolic pathways to support replication within host cells (42). We conclude that glycolysis is important for *E. coli* colonization and other aspects of enteric pathogenesis.

Gluconate was the first nutrient that was shown to be used by *E. coli* to colonize the streptomycin-treated mouse intestine (43). Since gluconate and other sugar acids are primarily catabolized via the ED pathway, it is reasonable to expect that mutants lacking the pathway will be defective in colonization

TABLE 1 Central metabolism mutants tested for colonization defects in the mouse intestine

Pathway	Gene Defect	MG1655	EDL933
Glyoxylate bypass	*aceA*	No	Yes
ED	*edd*	Yes	Yes
glycolysis	*pgi*	Yes	ND
PPP	*gnd*	No	ND
gluconeogenesis	*ppsA pckA*	No	No
TCA cycle	*frdA*	Yes	Yes
TCA cycle	*sdhAB*	No	Yes
TCA cycle	*frdA sdhAB*	Yes	Yes

Results show the difference in population sizes of wild-type verses mutant strains at Day 9.
Yes indicates the difference exceeds a 0.8 $\log_{10}$ colonization advantage and students *t* test value $P < 0.05$.
ND indicates not determined.

(44). The ED pathway is encoded by the *edd-eda* operon (45). The promoter-proximal *edd* gene encodes 6-phosphogluconate dehydratase, which converts 6-phosphogluconate to 2-keto-3-deoxy-6-phosphogluconate. The *eda* gene encodes 2-keto-3-deoxy-6-phosphogluconate aldolase, which converts 2-keto-3-deoxy-6-phosphogluconate to glyceraldehye-3-phosphate and pyruvate. *E. coli edd* mutants lacking the ED pathway, but retaining the pentose phosphate (PP) pathway, are poor colonizers of the mouse intestine, suggesting that *E. coli* utilizes the ED pathway for growth in the intestine (43). Other enteric bacteria require the ED pathway. For example, intracellular serovar Typhimurium induces genes of the ED pathway and gluconate catabolism during growth in macrophages (46). Moreover, the ED pathway is induced by *Vibrio cholerae in vivo* and an *edd* mutant failed to colonize the mouse intestine (47).

In contrast to the importance of the ED pathway, an *E. coli gnd* mutant, missing 6-phosphogluconate dehydrogenase and therefore deficient in the oxidative branch of the PP pathway, was as good a mouse-intestine colonizer as the wild-type (39). It should be noted that *gnd* mutants retain the non-oxidative PP pathway; therefore, they retain the ability to make essential precursor metabolites (e.g., ribose-5-phosphate) (48). We conclude that *E. coli* has alternative mechanisms for generating reducing power (nicotinamide adenine dinucleotide phosphate; NADPH) other than the oxidative PP pathway, but that the ED pathway for sugar acid catabolism is required to colonize efficiently (Table 1).

The role of the TCA cycle in commensal *E. coli* colonization of the intestine and in *E. coli* pathogenesis is poorly studied. It has been reported that an *sdhB* mutant lacking succinate dehydrogenase colonized as well as its wild-type parent (39). However, *E. coli* has a second isoform of succinate dehydrogenase: fumarate reductase, which provides redundant enzyme function under some circumstances (49). Indeed, an *E. coli sdhAB frdA* double mutant has a significant colonization defect (Table 1). The role of the TCA cycle in colonization and pathogenesis by other Enterobacteriaceae is better understood, as described immediately below.

A fully functional TCA cycle is required for virulence of *Salmonella enterica* serovar Typhimurium via oral infection of BALB/c mice, i.e., a *sucCD* mutant, which prevents the conversion of succinyl coenzyme A to succinate, was attenuated. Also, an *sdhCDA* mutant, which blocks the conversion of succinate to fumarate, was attenuated, whereas both an *aspA* mutant and an *frdABC* mutant, deficient in the ability to run the reductive branch of the TCA cycle, were fully virulent (50). Moreover, although it appears that serovar Typhimurium replenishes TCA cycle intermediates from substrates present in mouse tissues, fatty acid degradation and the glyoxylate bypass are not required, since a *fadD*, *fadF*, and *aceA* mutants were all fully virulent during acute infection (50–52). Interestingly, it appears that the TCA cycle is required for virulence of *Edwardsiella ictaluri* in catfish fingerlings (53) and that the glyoxylate bypass is required for serovar Typhimurium persistent infection of mice (51).

The fact that *E. coli* depends on the TCA cycle for colonization implies that gluconeogenesis also is important. Using mutants that are unable to synthesize glucose from fatty acids, acetate, and TCA cycle intermediates because they are blocked in converting pyruvate to phosphoenolpyruvate (*ppsA pckA*), a critical step in gluconeogenesis, it was shown that neither the commensal *E. coli* K-12 strain MG1655 nor EHEC use gluconeogenesis for growth in the streptomycin-treated mouse intestine when each is the only *E. coli* strain fed to mice (54). However, *E. coli* Nissle 1917, the probiotic strain, does use gluconeogenesis to colonize (55). In addition, while *E. coli* EDL933 did not use gluconeogenic nutrients when it was the only *E. coli* strain in the mouse intestine, it used metabolic flexibility to switch to gluconeogenic nutrients when in competition in the intestine with either

E. coli MG1655 (54) or *E. coli* Nissle 1917 (55). These findings are of extreme interest in view of a recent report showing that *E. coli* EDL933 activates expression of virulence factor genes only under gluconeogenic conditions (41).

CATABOLIC PATHWAY DIVERSITY IN *E. COLI*

The substrate range of *E. coli* is limited to mono-saccharides, disaccharides, a small number of larger sugars, some polyols, and sugar acids (56). Amino acids and carboxylates also are consumed (56, 57). The corresponding catabolic pathways feed these substrates into central metabolism. While the genes encoding central metabolism in *E. coli* fall within the highly conserved core genome (38), there is predicted to be some variation between strains with respect to the catabolic pathways that feed various substrates into central metabolism, as indicated by genome-based metabolic modeling (58). For example, pathogenic *E. coli* strains are predicted to grow on sucrose while commensals are not. In contrast, commensals are predicted to grow on galactonate while pathogens are not. However, most of the substrates predicted by modeling to be used differentially by different *E. coli* strains are not known to be present in the intestine (58). *E. coli* EDL933, the prototypical EHEC strain, is able to grow on sucrose, whereas most commensal strains do not because they lack the *sac* genes, and some strains are missing genes within the N-acetylgalactosamine operon and are thus unable to grow on this substrate (36). Despite the modest differences between strains regarding their substrate range, in laboratory cultures containing a mixture of 13 different sugars known to be present in mucus polysaccharides, *E. coli* EDL933 and *E. coli* MG1655 each use the sugars in the same order (36). However, although *E. coli* strains have nearly identical catabolic potential, they vary significantly in the sugars that support their colonization (35, 36, 39, 43, 54, 59, 60).

NUTRIENT AVAILABILITY
IN THE INTESTINE

Fluorescent *in situ* hybridization (FISH) microscopy of thin sections of the cecum of streptomycin-treated mice shows that colonized *E. coli* are surrounded by other members of the microbiota within the mucus layer (19, 54, 61). Indeed, *E. coli* grows well *in vitro* on cecal mucus, but fails to grow in intestinal contents (15, 19). When a transposon insertion-mutant library was screened for poor growth on mucus agar plates, a *waaQ* mutant of *E. coli* K-12 was isolated that also was defective in lipopolysaccharide biosynthesis, sensitive to detergents, clumped *in vitro* in broth culture, and failed to colonize streptomycin-treated mice (19). While the *waaQ* strain initially (during the first 24 h) grew from low to high numbers in the intestine, it rapidly declined in fecal plate counts and was undetectable by day 7 of the experiment. FISH showed that that the *waaQ* mutant formed clumps in the cecal mucus layer of streptomycin-treated mice at 24 h post-feeding, leading to the conclusion that failure to penetrate mucus and grow as dispersed cells within the mucus layer prevented it from colonizing (19).

The sources of nutrients that support intestinal colonization by *E. coli* are shed epithelial cells, dietary fiber, and mucosal polysaccharides (12–14). Most of the amino acids are available in the cecum, as growth of *E. coli* in mucus results in repression of the majority of genes involved in amino acid biosynthesis (39). In rat, mouse, and human, colonic mucus is organized by Muc2, the major glycoprotein, which is a high molecular weight gel-forming glycoprotein containing L-fucose, D-galactose, D-mannose, N-acetyl-D-glucosamine, N-acetyl-D-galactosamine, and N-acetylneuraminic acid (62). Mucin is 80% polysaccharide and 20% protein and is highly viscous (63). In addition to mucin, the mucus layer contains a number of smaller glycoproteins, proteins, glycolipids, and lipids (62–65). There are two mucus layers, a loosely adherent suction-removable layer closest to the lumen of the intestine and an adherent layer firmly attached to the mucosa (62, 66, 67). In the rat colon, the thickness of the adherent layer is about 100 μm and that of the loose layer about 700 μm (62). In the mouse colon, the thickness of the adherent layer is about 50 μm and that of the loose layer about 100 μm (67). The mucus layer itself is in a dynamic state, constantly being synthesized and secreted by the mucin-secreting, specialized goblet cells and degraded to a large extent by the indigenous intestinal microbes (68, 69). Degraded mucus components are shed into the intestinal lumen forming a part of the luminal contents that is excreted in the feces (68).

The loosely adherent mucus layer contains large numbers of bacteria in the mouse, but the inner adherent mucus layer is largely devoid of bacteria (70) and is not penetrated by beads the size of bacteria (71), suggesting that the inner mucus layer protects the colonic epithelium from the commensal microbiota. Commensal *E. coli* strains do not attach to intestinal epithelial cells and growth takes place predominantly in the mucus layer (19, 24). The mucus layer of the conventional mouse large intestine turns over about every 2 hours (23). Hence, to maintain a stable population, the bacterial growth rate in mucus must keep pace with the turnover rate of the mucus layer. For example, *E. coli* BJ4 has a generation time of 40–80 minutes in the streptomycin-treated mouse cecum, which is more than fast enough to maintain its population (23).

E. coli cannot degrade oligosaccharides or polysaccharides, except dextrin (36, 72). In the intestine, this is the job of anaerobes. To obtain the mono- and di-saccharides it needs for growth, *E. coli* relies on hydrolysis of complex polysaccharides by members of the intestinal community such as *Bacteroides thetaiotaomicron*, a Gram-negative obligate anaerobe and a major member of the human intestinal microbiota (73–78). This symbiotic relationship is illustrated in Fig 1. Complex

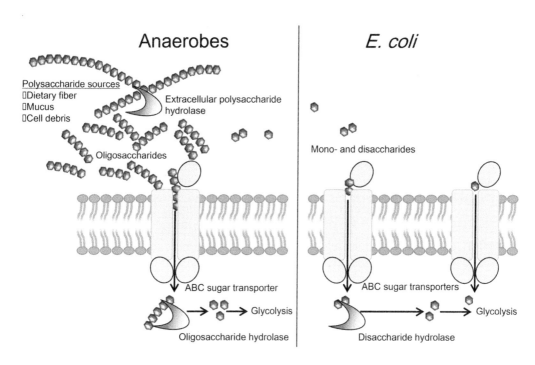

FIGURE 1 Nutrient flow in the intestine. The primary sources of carbohydrates in the large intestine are mucus, dietary fiber, and epithelial cell debris. Mucus and dietary fiber consist of complex polysaccharides. *E. coli* typically cannot degrade complex polysaccharides; that is the job of anaerobes. Hence, degradation of polysaccharides by anaerobes releases oligosaccharides, which are preferred by anaerobes, as well as mono- and disaccharides, which are preferred by *E. coli*. doi:10.1128/microbiolspec. MBP-0006-2014.f1

polysaccharides derived from epithelial cell debris, dietary fiber, or mucus are degraded by extracellular polysaccharide hydrolases secreted by anaerobes. The anaerobes preferentially take up the resultant oligosaccharides, which are further degraded intracellularly to monosaccharides that enter central metabolism. The mono- and di-saccharides that are released by polysaccharide hydrolysis are discarded by the anaerobes and thereby made available to *E. coli* and organisms with similar metabolism.

Evidence for the model shown in Fig 1 is mounting. In response to competition from *Eubacterium rectale*, *B. thetaiotaomicron* upregulates a number of polysaccharide-utilization loci that encode a variety of glycosyl hydrolases (79). In contrast, *E. rectale* responds to *B. thetaiotaomicron* by down-regulating expression of loci encoding glycan-degrading enzymes and

up-regulating expression of a number of sugar transporters, suggesting that *E. rectale* uses metabolic flexibility to take advantage of the superior ability of *B. thetaiotaomicron* to degrade polysaccharides (79). In a *tour de force* series of experiments from the Sonnenberg laboratory, it was recently shown in gnotobiotic mice associated with *B. thetaiotaomicron* and *S.* Typhimurium that fucose and sialic acid catabolic-gene systems are up-regulated in the latter organism and its growth is stimulated (77). Importantly, *Salmonella* mutants lacking the capacity to catabolize either fucose or sialic acid showed substantial fitness defects in streptomycin-treated mice (77). A *B. thetaiotaomicron* mutant lacking a predicted surface-associated sialidase failed to release free sialic acid in mice, whereas the wild-type *B. thetaiotaomicron* released free sialic acid and stimulated intestinal growth of

Clostridium difficile (77). Free sialic acid was very low in conventional mice and greatly elevated in mice that were treated with streptomycin, which is consistent with the idea that the microbiota consumed sialic acid in the conventional animals, but streptomycin treatment removed those members of the microbiota (77). These experiments prove that polysaccharide hydrolysis by anaerobes in the intestine can provide monosaccharides to members of the microbiota that can use them.

NUTRIENT LIMITATION IN THE INTESTINE

Competition for limiting resources drives ecosystems (80). In microbial ecosystems, such as the intestine, the microbial community competes for carbon and energy sources and terminal electron acceptors (27). It appears that *E. coli* uses strategies for maximizing its population in a stiffly competitive environment. *E. coli* can simultaneously utilize a mixture of six sugars under nutrient-limiting conditions in chemostats (81). When presented with a mixture of 13 sugars *in vitro*, *E. coli* uses up to nine of them at a time (36). Bacteria expand their transcriptome to induce a number of gene systems for carbon-source transport and catabolism when growing slowly (82) or when they are nutrient-deprived (83). While there is a fine line between "hunger" and "starvation", in fact, the two are distinct. Hunger is a state that is defined by physiological and genetic changes that expand the metabolic capacity of the cell (84). To better understand the role of the hunger lifestyle *in vivo*, the importance of carbon stores for colonization fitness was examined (60). It is well known that stored glycogen promotes survival during times of carbon limitation. It was found that mutants of commensal and pathogenic *E. coli* that cannot synthesize or degrade glycogen have significant intestinal colonization defects in mice (60). Furthermore, in support of the hypothesis that *E. coli* uses glycogen to withstand

hunger, when a constant supply of a readily metabolized carbon source was supplied in the drinking water of the mice, these colonization defects were rescued (60).

E. coli Nissle 1917 is a non-pathogenic strain that has been used since the early 1920's as a probiotic agent to treat gastrointestinal infections in humans, sold under the name "Mutaflor" (85). Despite having the *E. coli* Nissle 1917 genome sequence to search for clues as to its success as a probiotic agent (86, 87), little attention has been paid to the possibility that it may out-compete pathogens for essential nutrients. When *E. coli* Nissle 1917 is the only *E. coli* strain in the streptomycin-treated mouse intestine, it appears to use arabinose, fucose, galactose, gluconate, mannose, N-acetylgalactosamine, and sialic acid to colonize (Table 2 and [59]). In contrast, *E. coli* Nissle 1917 does not appear to use ribose to colonize (Table 2). Identifying the nutrients used by *E. coli* Nissle 1917 when competing with *E. coli* pathotypes might lead to new approaches to prevent *E. coli* infections.

Metabolic flexibility is also exhibited by uropathogenic *E. coli* (UPEC) strain CFT073, a human urinary-tract pathogen, when its nutritional program in the mouse intestine is compared to that in the mouse urinary tract. *E. coli* CFT073 utilizes several sugars simultaneously to colonize the intestine (Table 2 and [35]), but not to infect the urinary tract. Instead, in the urinary tract, transport of peptides and gluconeogenesis are required for maximum *E. coli* CFT073 growth (88). When *E. coli* EDL933 is the only *E. coli* strain in the mouse intestine, it does not use gluconeogenic substrates to colonize, but it switches to gluconeogenic nutrients when competing in the intestine with *E. coli* MG1655 (55). As an additional example of metabolic flexibility *in vivo*, neither *E. coli* MG1655 nor *E. coli* Nissle 1917 use ribose for growth in the intestine, unless fuculose-1-phosphate accumulates in mutants unable to catabolize it further (59). Moreover, it was shown that fucose at a concentration too low to support growth, stimulated the utilization of ribose

TABLE 2 Sugar utilization in the intestine by *E. coli* strains

Sugar defect	Mutation	MG1655	Nissle	HS	EDL933	UPEC	EPEC
Arabinose	*araBAD*	Yes	Yes	Yes	Yes	Yes	Yes
Fucose	*fucK*	No	Yes	No	No	No	No
Galactose	*galK*	No	Yes	Yes	Yes	Yes	Yes
Gluconate	*gntK/ΔidnK*	Yes	Yes	Yes	No	ND	ND
Hexuronates	*uxaC*	No	No	No	Yes	ND	ND
Lactose	*lacZ*	No	No	Yes	No	ND	ND
Mannose	*manA*	No	Yes	No	Yes	Yes	Yes
N-acetylglucosamine	*nagE*	Yes	No	Yes	Yes	Yes	Yes
N-acetylgalactosamine	*agaWEFA*	NA	Yes	ND	No	ND	ND
N-acetylneuraminate	*nanAT*	Yes	Yes	Yes	No	ND	ND
Ribose	*rbsK*	No	No	Yes	Yes	Yes	Yes
Sucrose	*sacH*	NA	NA	NA	Yes	NA	ND

Results show the difference in population sizes of wild-type verses mutant strains at Day 9.
Yes indicates the difference exceeds a 0.8 $\log_{10}$ colonization advantage and students *t* test value $P < 0.05$.
NA indicates that the pathway is not intact in this genetic background.
ND indicates not determined.

by the wild-type *E. coli* strains *in vitro*, suggesting that fuculose-1-phosphate plays a role in regulating the use of ribose as a carbon source by *E. coli* MG1655 and *E. coli* Nissle 1917 in the mouse intestine (59). In summary, to colonize successfully, *E. coli* must compete for limiting nutrients, so it uses several sugars at a time, is flexible in its nutrient preference, and relies on glycogen-carbon stores in the intestine.

COMPETITION FOR NUTRIENTS IN THE INTESTINE

E. coli competes for nutrients in the intestine in (at least) three ways. First, it can use nutrients that are available because no other community member has used it (Table 2). Second, it can outcompete other strains for the nutrients it prefers by growing faster on them (89). Third, it can enter into a symbiotic association with the anaerobe(s) that releases its preferred sugar(s) (61, 90). Evidence for each of these mechanisms comes from competitive-fitness studies in streptomycin-treated mice. Streptomycin treatment (5 g/l in drinking water) perturbs the microbiota by selectively removing the facultative anaerobes with which *E. coli* competes, making nutrients available and allowing

experimentally introduced strains to overcome colonization resistance (27, 91, 92). Colonization resistance in animal models can also be overcome in other ways, e.g., by gnotobiotic mice in which there is no native microbiota (93), by mice that have a contrived microbiota (94), or by mice with intestinal inflammation, which generates nitrate that is respired by *E. coli* (95).

The three ways that *E. coli* can compete with the microbiota for nutrients are described in detail here. First, different *E. coli* strains use different nutrients *in vivo*, despite using the same nutrients in the same order *in vitro* (Table 2). Mutants with deletions in genes corresponding to metabolic pathways induced in mucus were constructed by allelic replacement (29) and were tested for their ability to compete with their wild-type parent strain when simultaneously fed to mice in low numbers (10^5 CFU/mouse). The data obtained from these studies showed that the human commensals *E. coli* HS, *E. coli* Nissle 1917, *E. coli* MG1655, and the pathogen *E. coli* EDL933 each occupy unique nutritional niches in the mouse intestine (Table 2 and [35, 36]). Of the 12 sugars available in the mucus layer, *E. coli* HS utilizes six for colonization: arabinose, galactose, gluconate, N-acetylglucosamine, lactose, and ribose (Table 2). *E. coli* Nissle 1917 uses a different list of seven carbon sources to support colonization, including arabinose, fucose, galactose, gluconate, N-acetylglucosamine, and N-acetylneuraminate, and mannose (Table 2) and *E. coli* MG1655 utilizes five sugars for colonization: arabinose, fucose, gluconate, N-acetylglucosamine, and N-acetylneuraminate (Table 2). Each of these commensals is capable of colonizing mice that are pre-colonized with any one of the others (31), and each strain is capable of utilizing at least one sugar not used by the others *in vivo*, which suggests that differences in their *in vivo* sugar preferences allows them to occupy distinct nutrient-defined niches in the intestine. Furthermore, the intestinal niche occupied by pathogenic *E. coli* EDL933 is also unique and is defined by utilization of seven sugars: arabinose, galactose, hexuronates, mannose, N-acetylglucosamine, ribose, and sucrose (Table 2).

That *E. coli* MG1655 and *E. coli* EDL933 display different nutritional programs in the mouse intestine, e.g., *E. coli* MG1655 uses N-acetylneuraminate but not mannose whereas *E. coli* EDL933 uses mannose but not N-acetylneuraminate (Table 2), was surprising in view of the fact that *E. coli* MG1655 and *E. coli* EDL933 utilize them equally well *in vitro* and display identical nutritional preferences *in vitro*, i.e., they use sugars in the same order *in vitro* as follows: N-acetylglucosamine, gluconate, ribose, sialic acid, mannose, arabinose, maltose, and fucose (36). These findings will be discussed in the context of the "Restaurant" hypothesis below.

Since different commensal *E. coli* strains use different sugars to colonize the intestine, it seemed reasonable that a potential strategy for preventing colonization by the enterohemorrhagic *E. coli* EDL933 would be to pre-colonize mice with a combination of commensal strains that would fill the sugar-defined nutritional niches normally available to the invading pathogen. When mice were pre-colonized for 10 days with either the commensal *E. coli* MG1655, *E. coli* HS, or *E. coli* Nissle 1917 and then fed 10^5 CFU of

the pathogenic *E. coli* EDL933, *E. coli* Nissle 1917 limited growth of *E. coli* EDL933 in the intestine (10^3 to 10^4 CFU/gram of feces), whereas *E. coli* MG1655 and *E. coli* HS allowed growth to higher numbers (10^6 to 10^7 CFU/gram of feces). However, when *E. coli* EDL933 was fed to mice previously pre-colonized with the three commensal *E. coli* strains (MG1655, HS, and Nissle 1917), each of which displays a different nutritional program *in vivo* (Table 2), *E. coli* EDL933 was eliminated from the intestine (31). Therefore, a combination of as few as three commensal *E. coli* strains provided a barrier to *E. coli* EDL933 infection.

If the basis for exclusion of *E. coli* EDL933 was because the three commensal *E. coli* strains utilize the nutrients needed by *E. coli* EDL933 to compete and colonize, then any *E. coli* commensal strain or combination of strains that effectively catabolizes the sugars used by *E. coli* EDL933 would prevent its colonization. Indeed, when the ability of *E. coli* EDL933 to colonize mice that were pre-colonized with *E. coli* HS and *E. coli* Nissle 1917 was tested, which the data in Table 2 indicate should be equally effective without *E. coli* MG1655 present, *E. coli* EDL933, was indeed eliminated 5 days following association (35). However, the same two commensal *E. coli* strains could not prevent colonization of *E. coli* CFT073, an uropathogenic strain, and *E. coli* E2348/69, an enteropathogenic strain (96). Therefore, it is unlikely that any particular commensal strain(s) of *E. coli* will be generally effective as a probiotic to prevent colonization by enteric pathogens. Nevertheless, the data support the hypothesis that nutrient consumption by commensal *E. coli* can limit nutrient availability to pathogens, which in turn points to the potential of probiotics for preventing disease.

A second way of competing for nutrients in the intestine is illustrated by what happens when *E. coli* MG1655 adapts to the mouse intestine. When mice were fed *E. coli* MG1655, non-motile *flhDC* deletion mutants appeared in the feces 3 days post-feeding and reached approximately 90% of the total population by day 15 of the experiment (97). These mutants had a striking colonization advantage over the wild-type *E. coli* MG1655 parent strain. The deletions were of varying length and began immediately downstream of the IS*1* element in the *flhDC* promoter region. One such mutant, designated MG1655*, was a better colonizer than its parent, grew in cecal mucus faster than its parent *in vitro* (90 ± 2 min generation time vs 105 ± 2 min, *P* < 0.001), and grew 15–30% faster than its parent on a number of sugars present in the mouse intestine (98). The *E. coli flhDC* operon encodes the $FlhD_4C_2$ regulatory complex, which is the master positive regulator of the more than 40 gene flagella regulon (99). The $FlhD_4C_2$ complex has also been reported to negatively regulate *E. coli* K-12 genes involved in galactose transport, the ED pathway, and the TCA cycle and positively regulate genes involved in ribose transport (89, 100). *E. coli* MG1655 *flhDC* deletion mutants have also been reported to be selected in the intestines of ex-germfree mono-associated mice (101).

Several high-throughput genomic approaches were taken to further characterize *E. coli* MG1655*. Whole-genome pyrosequencing did not reveal any changes on its genome, aside from the deletion at the *flhDC* locus, that could explain the colonization advantage of *E. coli* MG1655* (89). Microarray analysis revealed modest yet significant induction of catabolic gene systems across the genome in both *E. coli* MG1655* and an isogenic *flhD* mutant constructed in the laboratory (89). Catabolome analysis with Biolog GN2 microplates revealed an enhanced ability of both *E. coli* MG1655* and the isogenic *flhD* mutant to oxidize a variety of carbon sources (89). Collectively, the results showed that intestine-adapted *E. coli* MG1655* is more fit than the wild-type for intestinal colonization, because loss of FlhD results in elevated expression of genes involved in carbon and energy metabolism, allowing the mutants to outcompete their wild-type parent for the

same nutrients. Hence, a second strategy for gaining a colonization advantage is to outcompete other members of the microbiota for their preferred nutrients.

There is a third way that *E. coli* competes for nutrients in the intestine. The intestine selects for mutants that gain a colonization advantage by promoting occupation of a distinct niche. The selection of non-motile *E. coli* MG1655 *flhDC* mutants by the streptomycin-treated mouse intestine is easily explained by the nutrient-niche hypothesis, i.e., the mutants grow 15% faster *in vitro* in mouse cecal mucus and 15%–30% faster on several sugars present in cecal mucus than *E. coli* MG1655 (97). In addition to *flhDC* mutants, *E. coli* MG1655 mutants with reduced motility also were selected by adaptation in the streptomycin-treated mouse intestine and these turned out to be *E. coli* MG1655 *envZ* missense mutants (61). *E. coli* MG1655 *envZ* missense mutants have also been reported to be selected in the intestines of ex-germfree mono-associated mice (102). The *envZ* gene encodes a histidine kinase that is a member of the *envZ/ompR* two-component signal-transduction system that modulates gene expression in response to osmolarity. The genes regulated include *flhDC*, the porin genes *ompC* and *ompF*, and several other genes encoding outer-membrane proteins (103, 104). These *E. coli* MG1655 *envZ* missense mutants produced more phosphorylated OmpR than both *E. coli* MG1655 and the *E. coli* MG1655 *flhDC* deletion mutants and produced more of the outer-membrane porin OmpC and less of the outer-membrane porin OmpF (90). As a result, the *E. coli* MG1655 *envZ* missense mutants were more resistant to bile salts and colicin V than *E. coli* MG1655 and the *E. coli* MG1655 *flhDC* deletion mutants (61). One of the *E. coli* MG1655 *envZ* missense mutants, which was studied further, contained the $envZ_{P41L}$ missense mutation and grew about 15% slower *in vitro* in mouse cecal mucus and on several sugars present in mucus compared to the *flhDC* deletion mutants, yet was as good an intestinal colonizer as the *flhDC* deletion mutants and far better than *E. coli* MG1655 (61). Moreover, *E. coli* MG1655 envZ$_{P41L}$ and the *E. coli* MG1655 *flhDC* deletion mutants appeared to colonize equally well in one major intestinal niche, but *E. coli* MG1655 envZ$_{P41L}$ appeared to use galactose to colonize a second intestinal niche either not colonized or colonized poorly by the *E. coli* MG1655 *flhDC* deletion mutants. These data are not consistent with the nutrient-niche hypothesis, but they are consistent with what we call the "Restaurant" hypothesis, which will be discussed below.

Since *E. coli* MG1655 envZ$_{P41L}$ was a far better intestinal colonizer than wild-type *E. coli* MG1655, the envZ$_{P41L}$ gene was transferred for further study into *E. coli* Nissle 1917, the human probiotic strain used to treat gastrointestinal infections. Like *E. coli* MG1655 envZ$_{P41L}$, *E. coli* Nissle 1917 envZ$_{P41L}$ produced more phosphorylated OmpR than its parent and produced more of the outer membrane porin OmpC and less of the outer membrane porin OmpF (90). It also became more resistant to bile salts and colicin V, grew 50% slower *in vitro* in mucus and 15%-30% slower on several sugars present in mucus, yet was a 10-fold better colonizer than *E. coli* Nissle 1917 (90). Furthermore, like *E. coli* MG1655 envZ$_{P41L}$, *E. coli* Nissle 1917 envZ$_{P41L}$ appeared to use galactose to colonize a second intestinal niche either not colonized or colonized poorly by wild-type *E. coli* Nissle 1917, despite not growing as well as *E. coli* Nissle 1917 on galactose as a sole carbon source (90). Moreover, despite being a better colonizer, *E. coli* Nissle 1917 envZ$_{P41L}$ was not better than its parent at preventing colonization by enterohemorrhagic *E. coli* EDL933 and, in fact, appeared to be worse (90). The data can be explained according to our "Restaurant" hypothesis for commensal *E. coli* strains, i.e., that they colonize the intestine as sessile members of mixed biofilms obtaining the sugars they need for growth locally, but compete for sugars with invading *E. coli* pathogens planktonically as described below.

BIOFILMS IN THE INTESTINE

Much attention has been given to the role played by biofilms in bacterial colonization of many environments, but until recently surprisingly little information was available regarding biofilms in the intestine. On the one hand, the transit time of intestinal contents is short compared to the timescale of biofilm development (105), so it was hard to imagine how a stable biofilm might be maintained in the intestine, yet the kinetics of plasmid transfer between *E. coli* strains in the streptomycin-treated mouse intestine suggested that *E. coli* resides in biofilms *in vivo* (106). Moreover, the mucus layer itself has many of the characteristics of a secreted biofilm matrix. Add to this the concept of bacterial binding to mucus components and it became reasonable to consider the possibility of bacterial biofilms in the gastrointestinal tract. Indeed, it's been shown that biofilms form in the mucus layers of the large intestines of healthy humans, rats, baboons, and mice (107–110) and that mixed biofilms consisting of *Bacteroides*, *Enterobacter*, and *Clostridia* species form rapidly on strands of mucin in mucus introduced into a growing human microbiota contained in a continuous-flow culture system constructed to mimic the human intestine (111). Furthermore, it appears that human colonic-mucosal biofilms and bacterial communities in feces differ greatly in composition (112) and dysbiosis in the community structure of mucosal biofilms may play an important role in contributing to chronic inflammatory-bowel diseases such as ulcerative colitis and Crohn's disease (107, 108). In *Vibrio cholerae*, biofilm formation is important for pathogenesis (113). Intestinal biofilms would provide the habitat for microbe-microbe interactions such as those that are thought to occur between *E. coli* and the polysaccharide-degrading anaerobes.

THE "RESTAURANT" HYPOTHESIS

As discussed above, commensal strains of *E. coli* appear to reside in mixed biofilms in the large intestines of mice (61, 90). Moreover, commensal and pathogenic strains of *E. coli* use mono- and disaccharides for growth in the intestine (36). However, these sugars are absorbed in the small intestine, whereas dietary fiber reaches the large intestine intact. In contrast to the anaerobes, most *E. coli* strains do not secrete extracellular polysaccharide hydrolases (72, 114) and therefore cannot degrade dietary fiber-derived and mucin-derived oligo- and polysaccharides. Since commensal and pathogenic *E. coli* strains colonize the mouse large intestine by growing in intestinal mucus (43, 115–119), it appears likely that *E. coli* depends on the anaerobes present in mucus that can degrade oligo- and polysaccharides to provide them with the mono-and disaccharides and maltodextrins they need for growth. Indeed, *Salmonella enterica* serovar Typhimurium, which is in the same family as *E. coli* and has very similar metabolism, catabolizes fucose and sialic acid liberated from mucosal polysaccharides by *Bacteroides thetaiotaomicron* (77). It is therefore possible that anaerobes in the mixed biofilms provide *E. coli* with the sugars it needs for growth locally, rather than from a perfectly mixed pool available to all species, which is an assumption of the nutrient-niche hypothesis. We call the mixed biofilms that feed the *E. coli* strains "Restaurants" and we hypothesize that different commensal *E. coli* strains reside in different "Restaurants" interacting physically and metabolically with different anaerobes. Each restaurant might serve different nutrients, i.e., each commensal *E. coli* strain could be exposed to a different menu, which explains why different *E. coli* strains display different nutritional programs in the mouse intestine despite displaying identical nutritional programs *in vitro* (35, 36). The restaurant hypothesis can also explain how *E. coli* strains that grow more slowly in

mucus and on several sugars found in mucus are better colonizers than their parents as long as they have a higher affinity for biofilm-binding sites than their parents. Indeed, the outer membranes of both *E. coli* MG1655 $envZ_{P41L}$ and *E. coli* Nissle 1917 $envZ_{P41L}$ are very different from those of their parents (61, 90), which could result in increased affinities for mixed biofilms.

The "Restaurant" hypothesis can also explain why, despite being a better colonizer than *E. coli* Nissle 1917, *E. coli* Nissle 1917 $envZ_{P41L}$ is not better at limiting enterohemorrhagic *E. coli* EDL933 colonization than *E. coli* Nissle 1917 and, in fact, may be worse (90). It is possible when *E. coli* EDL933 invades the mouse intestine it initially grows planktonically in mucus and not in mixed biofilms. If we are correct that *E. coli* Nissle1917 and *E. coli* Nissle1917 $envZ_{P41L}$ colonize the mouse intestine by being served specific sugars by the anaerobes in the mixed biofilms they inhabit, then small amounts of these sugars that escape the mixed biofilms might be available to invading *E. coli* EDL933 as well as to the small numbers of planktonic *E. coli* Nissle1917 or *E. coli* Nissle1917 $envZ_{P41L}$ that leave the mixed biofilms. Therefore, it may be that both planktonic *E. coli* Nissle 1917 $envZ_{P41L}$ and *E. coli* Nissle 1917 compete directly with planktonic *E. coli* EDL933 for the sugars that escape the biofilms or that are produced by small numbers of planktonic members of the microbiota that leave the biofilms. This scenario would allow planktonic *E. coli* EDL933 to grow to the extent allowed by the available concentrations of those sugars in competition with planktonic *E. coli* Nissle 1917 $envZ_{P41L}$ or *E. coli* Nissle 1917, which could explain why *E. coli* Nissle 1917, the faster grower in perfectly mixed bacteria-free mucus *in vitro*, appears to limit *E. coli* EDL933 growth in the intestine to a greater extent than does *E. coli* Nissle 1917 $envZ_{P41L}$ (90).

We stress that the granularity of mixed intestinal biofilms and nutrient flow between the microbes that reside within them is not known. These interactions could be so finite as to allow two different *E. coli* strains to interact with the same anaerobe cell and each grow on a different preferred nutrient. If so, it will not be possible to find zones within the biofilm (i.e., restaurants) that contain only a single population of *E. coli* and interacting anaerobe partner.

CONCLUSIONS

It is becoming increasingly clear that once *E. coli* strains reach the large intestine, in order to colonize, they must enter the mucus layer and utilize nutrients there for growth. It is also clear that different strains of *E. coli* display different nutritional programs in the intestine. However, it is not known whether a specific *E. coli* commensal strain utilizes the same nutrients when it is the only *E. coli* strain in the intestine compared to a situation in which it colonizes along with several different commensal *E. coli* strains with which it must compete for nutrients. Metabolic flexibility could be a key requirement for successful colonization of the intestine by several *E. coli* strains simultaneously. However, the "Restaurant" hypothesis explains long-term colonization by several established commensal *E. coli* strains without invoking metabolic flexibility, i.e., each commensal *E. coli* strain resides as a sessile member of a mixed biofilm in the intestine and obtains nutrients locally rather than from a perfectly mixed pot of nutrients. However, as described above, it seems likely that when mice colonized long-term with one *E. coli* strain are fed a pathogenic *E. coli* strain, planktonic members of the pre-colonized strain that escape the mixed biofilm compete directly with the invading pathogen for nutrients from the same perfectly mixed pot, according to the Freter nutrient-niche hypothesis. Since it appears likely that a pathogen must be able to grow in the intestine in order to initiate the pathogenic process, we hope that future research will

provide a nutritional framework for colonizing humans with a combination of commensal *E. coli* strains or with one commensal *E. coli* strain that has been engineered to be as effective as several strains and can serve as an effective first line of defense against pathogenic *E. coli* intestinal infections.

ACKNOWLEDGMENTS

The work carried out in the authors' laboratories was supported by U. S. Public Health Service grants AI48945 and GM095370. The authors wish to acknowledge the contributions of the following people who work in the authors' laboratories. In Rhode Island: Jimmy Adediran, James Allen, Steven Autieri, Swati Banergee, Megan Banner, Eric Gauger, Jakob Frimodt-Møller, Mathias Jorgensen, Mary Leatham-Jensen, Regino Mercado-Lubo, Regina Miranda, Matthew Mokszycki, Annette Møller, and Silvia Schinner. In Oklahoma: April Anderson, Amanda Ashby, Matthew Caldwell, Amanda Cernosek, Dong-Eun Chang, Fatema Chowdhury, Andrew Fabich, Terri Gibson, Shari Jones, Amanda Laughlin, Rosalie Maltby, Jessica Meador, Darren Smalley, Stephanie Tison, and Don Tucker.

CITATION

Conway T, Cohen PS. 2015. Commensal and pathogenic *escherichia coli* metabolism in the gut. Microbiol Spectrum 3(2):MBP-0006-2014.

REFERENCES

1. **Finegold SM, Sutter VL, Mathisen GE.** 1983. Normal indigenous intestinal microflora, p 3–31. *In* Hentges DJ (ed), *Human intestinal microflora in health and disease.* Academic Press, Inc, New York, NY.
2. **Huggins C, Rast HV Jr.** 1963. Incidence of coliform bacteria in the intestinal tract of *Gambusia affinis holbrooki* (Girard) and in their habitat water. *J Bacteriol* **85:**489–490.
3. **Palmer C, Bik EM, Digiulio DB, Relman DA, Brown PO.** 2007. Development of the human

infant intestinal microbiota. *PLoS Biol* **5:**e177. doi:10.1371/journal.pbio.0050177

4. **Riley M, Abe T, Arnaud MB, Berlyn MK, Blattner FR, Chaudhuri RR, Glasner JD, Horiuchi T, Keseler IM, Kosuge T, Mori H, Perna NT, Plunkett G III, Rudd KE, Serres MH, Thomas GH, Thomson NR, Wishart D, Wanner BL.** 2006. *Escherichia coli* K-12: a cooperatively developed annotation snapshot--2005. *Nucleic Acids Res* **34:**1–9.

5. **Stecher B, Berry D, Loy A.** 2013. Colonization resistance and microbial ecophysiology: using gnotobiotic mouse models and single-cell technology to explore the intestinal jungle. *FEMS Microbiol Rev* **37:**793–829.

6. **Stecher B, Hardt WD.** 2011. Mechanisms controlling pathogen colonization of the gut. *Curr Opin Microbiol* **14:**82–91.

7. **Clarke MB, Sperandio V.** 2005. Events at the host-microbial interface of the gastrointestinal tract III. Cell-to-cell signaling among microbial flora, host, and pathogens: there is a whole lot of talking going on. *Am J Physiol Gastrointest Liver Physiol* **288:**G1105–G1109.

8. **Cole AM, Ganz T.** 2005. Defensins and other antimicrobial peptides: innate defense of mucosal surfaces, p 17–34. *In* Nataro JP, Cohen PS, Mobley HLT, Weiser JN (ed), *Colonization of mucosal surfaces.* ASM Press, Washington, DC.

9. **Kaper JB, Sperandio V.** 2005. Bacterial cell-to-cell signaling in the gastrointestinal tract. *Infect Immun* **73:**3197–3209.

10. **Pasetti MF, Salerno-Gonçalves R, Sztein MB.** 2005. Mechanisms of adaptive immunity that prevent colonization of mucosal surfaces, p 35–47. *In* Nataro JP, Cohen PS, Mobley HLT, Weiser JN (ed), *Colonization of mucosal surfaces.* ASM Press, Washington, DC.

11. **Sansonetti PJ.** 2004. War and peace at mucosal surfaces. *Nat Rev Immunol* **4:**953–964.

12. **Conway T, Krogfelt KA, Cohen PS.** 2004. Chapter 8.3.1.2, The life of commensal *Escherichia coli* in the mammalian intestine. *In* Kaper JB (ed), *EcoSalPlus Cellular and molecular biology of E. coli Salmonella, and the Enterobacteriaceae:,* 3rd ed, (online). ASM Press, Washington, DC.

13. **Conway T, Krogfelt KA, Cohen PS.** 2007. *Escherichia coli* at the intestinal mucosal surface, p 175–196. *In* Brogden KA, Minion FC, Cornick N, Stanton TB, Zhang Q, Nolan LK, Wannemuehler MJ (ed), *Virulence mechanisms of bacterial pathogens,* 4th ed. ASM Press, Washington, DC.

14. **Laux DC, Cohen PS, Conway T.** 2005. Role of the mucus layer in bacterial colonization of the intestine, p 199–212. *In* Nataro JP, Mobley HLT, Cohen PS (ed), *Colonization of mucosal surfaces.* ASM Press, Washington, DC.

15. **Wadolkowski EA, Laux DC, Cohen PS.** 1988. Colonization of the streptomycin-treated mouse large intestine by a human fecal *Escherichia coli* strain: role of growth in mucus. *Infect Immun* **56:**1030–1035.

16. **Foster JW.** 2004. *Escherichia coli* acid resistance: tales of an amateur acidophile. *Nat Rev Microbiol* **2:**898–907.

17. **Lin J, Smith MP, Chapin KC, Baik HS, Bennett GN, Foster JW.** 1996. Mechanisms of acid resistance in enterohemorrhagic *Escherichia coli. Appl Environ Microbiol* **62:**3094–3100.

18. **Freter R, Brickner H, Fekete J, Vickerman MM, Carey KE.** 1983. Survival and implantation of *Escherichia coli* in the intestinal tract. *Infect Immun* **39:**686–703.

19. **Moller AK, Leatham MP, Conway T, Nuijten PJ, de Haan LA, Krogfelt KA, Cohen PS.** 2003. An *Escherichia coli* MG1655 lipopolysaccharide deep-rough core mutant grows and survives in mouse cecal mucus but fails to colonize the mouse large intestine. *Infect Immun* **71:**2142–2152.

20. **McCormick BA, Laux DC, Cohen PS.** 1990. Neither motility nor chemotaxis plays a role in the ability of *Escherichia coli* F-18 to colonize the streptomycin-treated mouse large intestine. *Infect Immun* **58:**2957–2961.

21. **McGuckin MA, Lindén SK, Sutton P, Florin TH.** 2011. Mucin dynamics and enteric pathogens. *Nat Rev Microbiol* **9:**265–278.

22. **Bergstrom KS, Sham HP, Zarepour M, Vallance BA.** 2012. Innate host responses to enteric bacterial pathogens: a balancing act between resistance and tolerance. *Cell Microbiol* **14:**475–484.

23. **Rang CU, Licht TR, Midtvedt T, Conway PL, Chao L, Krogfelt KA, Cohen PS, Molin S.** 1999. Estimation of growth rates of *Escherichia coli* BJ4 in streptomycin-treated and previously germfree mice by *in situ* rRNA hybridization. *Clin Diagn Lab Immunol* **6:**434–436.

24. **Poulsen LK, Lan F, Kristensen CS, Hobolth P, Molin S, Krogfelt KA.** 1994. Spatial distribution of *Escherichia coli* in the mouse large intestine inferred from rRNA *in situ* hybridization. *Infect Immun* **62:**5191–5194.

25. **Poulsen LK, Licht TR, Rang C, Krogfelt KA, Molin S.** 1995. Physiological state of *Escherichia coli* BJ4 growing in the large intestines of streptomycin-treated mice. *J Bacteriol* **177:**5840–5845.

26. **Freter R.** 1992. Factors affecting the microecology of the gut, p 355–376. *In* Fuller R (ed), *Probiotics. The scientific basis.* Chapman and Hall, London.

27. **Freter R.** 1983. Mechanisms that control the microflora in the large intestine, p 33–54. *In* Hentges DJ (ed), *Human intestinal microflora in health and disease.* Academic Press, Inc., New York, NY.

28. **Freter R.** 1988. Mechanisms of bacterial colonization of the mucosal surfaces of the gut, p 45–60. *In* Roth JA (ed), *Virulence mechanisms of bacterial pathogens.* American Society for Microbiology, Washington, DC.

29. **van der Waaij D, Berghuis-de Vries JM, Lekkerkerk-k-v.** 1971. Colonization resistance of the digestive tract in conventional and antibiotic-treated mice. *J Hyg (Lond)* **69:**405–411.

30. **Apperloo-Renkema HZ, Van der Waaij BD, Van der Waaij D.** 1990. Determination of colonization resistance of the digestive tract by biotyping of Enterobacteriaceae. *Epidemiol Infect* **105:**355–361.

31. **Leatham MP, Banerjee S, Autieri SM, Mercado-Lubo R, Conway T, Cohen PS.** 2009. Precolonized human commensal *Escherichia coli* strains serve as a barrier to *E. coli* O157:H7 growth in the streptomycin-treated mouse intestine. *Infect Immun* **77:**2876–2886.

32. **Stecher B, Barthel M, Schlumberger MC, Haberli L, Rabsch W, Kremer M, Hardt WD.** 2008. Motility allows S. Typhimurium to benefit from the mucosal defence. *Cell Microbiol* **10:**1166–1180.

33. **Stecher B, Hapfelmeier S, Müller C, Kremer M, Stallmach T, Hardt WD.** 2004. Flagella and chemotaxis are required for efficient induction of *Salmonella enterica* serovar Typhimurium colitis in streptomycin-pretreated mice. *Infect Immun* **72:**4138–4150.

34. **Stecher B, Robbiani R, Walker AW, Westendorf AM, Barthel M, Kremer M, Chaffron S, Macpherson AJ, Buer J, Parkhill J, Dougan G, von Mering C, Hardt WD.** 2007. *Salmonella enterica* serovar typhimurium exploits inflammation to compete with the intestinal microbiota. *PLoS Biol* **5:**2177–2189. doi:10.1371/journal.pbio.0050244

35. **Maltby R, Leatham-Jensen MP, Gibson T, Cohen PS, Conway T.** 2013. Nutritional basis for colonization resistance by human commensal *Escherichia coli* strains HS and Nissle 1917 against *E. coli* O157:H7 in the mouse intestine. *PLoS One* **8:**e53957. doi:10.1371/journal.pone.0053957

36. **Fabich AJ, Jones SA, Chowdhury FZ, Cernosek A, Anderson A, Smalley D, McHargue JW, Hightower GA, Smith JT, Autieri SM, Leatham MP, Lins JJ, Allen RL, Laux DC, Cohen PS, Conway T.** 2008. Comparison of carbon nutrition for pathogenic and commensal *Escherichia coli*

strains in the mouse intestine. *Infect Immun* **76:**1143–1152.

37. **Stecher B, Hardt WD.** 2008. The role of microbiota in infectious disease. *Trends Microbiol* **16:**107–114.

38. **Cook H, Ussery DW.** 2013. Sigma factors in a thousand *E. coli* genomes. *Environ Microbiol* **15:**3121–3129.

39. **Chang DE, Smalley DJ, Tucker DL, Leatham MP, Norris WE, Stevenson SJ, Anderson AB, Grissom JE, Laux DC, Cohen PS, Conway T.** 2004. Carbon nutrition of *Escherichia coli* in the mouse intestine. *Proc Natl Acad Sci U S A* **101:**7427–7432.

40. **Bowden SD, Rowley G, Hinton JC, Thompson A.** 2009. Glucose and glycolysis are required for the successful infection of macrophages and mice by *Salmonella enterica* serovar typhimurium. *Infect Immun* **77:**3117–3126.

41. **Njoroge JW, Nguyen Y, Curtis MM, Moreira CG, Sperandio V.** 2012. Virulence meets metabolism: Cra and KdpE gene regulation in enterohemorrhagic *Escherichia coli*. *MBio* **3:**e00280-00212. doi:10.1128/mBio.00280-12

42. **Waligora EA, Fisher CR, Hanovice NJ, Rodou A, Wyckoff EE, Payne SM.** 2014. Role of intracellular carbon metabolism pathways in *Shigella flexneri* virulence. *Infect Immun* **82:**2746–2755.

43. **Sweeney NJ, Laux DC, Cohen PS.** 1996. *Escherichia coli* F-18 and *E. coli* K-12 *eda* mutants do not colonize the streptomycin-treated mouse large intestine. *Infect Immun* **64:**3504–3511.

44. **Peekhaus N, Conway T.** 1998. What's for dinner?: Entner-Doudoroff metabolism in *Escherichia coli*. *J Bacteriol* **180:**3495–3502.

45. **Conway T.** 1992. The Entner-Doudoroff pathway: history, physiology and molecular biology. *FEMS Microbiol Rev* **9:**1–27.

46. **Eriksson S, Lucchini S, Thompson A, Rhen M, Hinton JC.** 2003. Unravelling the biology of macrophage infection by gene expression profiling of intracellular *Salmonella enterica*. *Mol Microbiol* **47:**103–118.

47. **Patra T, Koley H, Ramamurthy T, Ghose AC, Nandy RK.** 2012. The Entner-Doudoroff pathway is obligatory for gluconate utilization and contributes to the pathogenicity of *Vibrio cholerae*. *J Bacteriol* **194:**3377–3385.

48. **Zhao J, Baba T, Mori H, Shimizu K.** 2004. Global metabolic response of *Escherichia coli* to *gnd* or *zwf* gene-knockout, based on 13C-labeling experiments and the measurement of enzyme activities. *Appl Microbiol Biotechnol* **64:**91–98.

49. **Steinsiek S, Frixel S, Stagge S, SUMO, Bettenbrock K.** 2011. Characterization of *E. coli* MG1655 and *frdA* and *sdhC* mutants at various aerobiosis levels. *J Biotechnol* **154:**35–45.

50. **Tchawa Yimga M, Leatham MP, Allen JH, Laux DC, Conway T, Cohen PS.** 2006. Role of gluconeogenesis and the tricarboxylic acid cycle in the virulence of *Salmonella enterica* serovar Typhimurium in BALB/c mice. *Infect Immun* **74:**1130–1140.

51. **Fang FC, Libby SJ, Castor ME, Fung AM.** 2005. Isocitrate lyase (AceA) is required for *Salmonella* persistence but not for acute lethal infection in mice. *Infect Immun* **73:**2547–2549.

52. **Spector MP, DiRusso CC, Pallen MJ, Garcia del Portillo F, Dougan G, Finlay BB.** 1999. The medium-/long-chain fatty acyl-CoA dehydrogenase (fadF) gene of *Salmonella typhimurium* is a phase 1 starvation-stress response (SSR) locus. *Microbiology* **145**(Pt 1):15–31.

53. **Dahal N, Abdelhamed H, Lu J, Karsi A, Lawrence ML.** 2013. Tricarboxylic acid cycle and one-carbon metabolism pathways are important in *Edwardsiella ictaluri* virulence. *PLoS One* **8:**e65973. doi:10.1371/journal.pone.0065973

54. **Miranda RL, Conway T, Leatham MP, Chang DE, Norris WE, Allen JH, Stevenson SJ, Laux DC, Cohen PS.** 2004. Glycolytic and gluconeogenic growth of *Escherichia coli* O157: H7 (EDL933) and *E. coli* K-12 (MG1655) in the mouse intestine. *Infect Immun* **72:**1666–1676.

55. **Schinner SA, Mokszycki ME, Adediran J, Leatham-Jensen M, Conway T, Cohen PS.** 2015. *Escherichia coli* EDL933 Requires Gluconeogenic Nutrients To Successfully Colonize the Intestines of Streptomycin-Treated Mice Precolonized with E. coli Nissle 1917. *Infect Immun* **83:**1983–1991.

56. **Lin ECC.** 1996. Sugars, polyols, and carboxylates, p 307–342. *In* Neidhardt FC, Curtiss R III, Ingraham JL, Lin ECC, Low KB, Magasanik B, Reznikoff WS, Riley M, Schaechter M, Umbarger HE (ed), *Escherichia coli and Salmonella: cellular and molecular biology*, 2nd ed. ASM Press, Washington, DC.

57. **Reitzer L.** 2005. Chapter 3.4.7 Catabolism of Amino Acids and Related Compounds. *In* Böck A, Curtiss R III, Kaper JB, Karp PD, Neidhardt FC, Nyström T, Slauch JM, Squires CL, Usery D (ed), *EcoSal–Escherichia coli and Salmonella: cellular and molecular biology*, 3rd ed (online). ASM Press, Washington, DC.

58. **Monk JM, Charusanti P, Aziz RK, Lerman JA, Premyodhin N, Orth JD, Feist AM, Palsson BØ.** 2013. Genome-scale metabolic reconstructions of multiple *Escherichia coli* strains highlight strain-specific adaptations to nutritional environments. *Proc Natl Acad Sci U S A* **110:**20338–20343.

59. **Autieri SM, Lins JJ, Leatham MP, Laux DC, Conway T, Cohen PS.** 2007. L-fucose stimulates utilization of D-ribose by *Escherichia coli* MG1655 Δ*fucAO* and *E. coli* Nissle 1917 Δ*fucAO* mutants in the mouse intestine and in M9 minimal medium. *Infect Immun* **75:**5465–5475.

60. **Jones SA, Jorgensen M, Chowdhury FZ, Rodgers R, Hartline J, Leatham MP, Struve C, Krogfelt KA, Cohen PS, Conway T.** 2008. Glycogen and maltose utilization by *Escherichia coli* O157:H7 in the mouse intestine. *Infect Immun* **76:**2531–2540.

61. **Leatham-Jensen MP, Frimodt-Møller J, Adediran J, Mokszycki ME, Banner ME, Caughron JE, Krogfelt KA, Conway T, Cohen PS.** 2012. The streptomycin-treated mouse intestine selects *Escherichia coli* envZ missense mutants that interact with dense and diverse intestinal microbiota. *Infect Immun* **80:** 1716–1727.

62. **Atuma C, Strugala V, Allen A, Holm L.** 2001. The adherent gastrointestinal mucus gel layer: thickness and physical state *in vivo*. *Am J Physiol Gastrointest Liver Physiol* **280:**G922–G929.

63. **Allen A.** 1984. The structure and function of gastrointestinal mucus, p 3–11. *In* Boedeker EC (ed), *Attachment of organisms to the gut mucosa*, **vol II**. CRC Press, Boca Raton, FL.

64. **Kim YS, Morita A, Miura S, Siddiqui B.** 1984. Structure of glycoconjugates of intestinal mucosal membranes, p 99–109. *In* Boedeker EC (ed), *Attachment of organisms to the gut mucosa*, **vol II**. CRC Press, Boca Raton, FL.

65. **Slomiany BL, Slomiany A.** 1984. Lipid and mucus secretions of the alimentary tract, p 24–31. *In* Boedeker EC (ed), *Attachment of organisms to the gut mucosa*, **vol II**. CRC Press, Boca Raton, FL.

66. **Johansson ME, Gustafsson JK, Holmén-Larsson J, Jabbar KS, Xia L, Xu H, Ghishan FK, Carvalho FA, Gewirtz AT, Sjövall H, Hansson GC.** 2014. Bacteria penetrate the normally impenetrable inner colon mucus layer in both murine colitis models and patients with ulcerative colitis. *Gut* **63:**281–291.

67. **Johansson ME, Larsson JM, Hansson GC.** 2011. The two mucus layers of colon are organized by the MUC2 mucin, whereas the outer layer is a legislator of host-microbial interactions. *Proc Natl Acad Sci U S A* **108**(Suppl 1):4659–4665.

68. **Hoskins LC.** 1984. Mucin degradation by enteric bacteria: ecological aspects and implications for bacterial attachment to gut mucosa, p 51–65. *In* Boedeker EC (ed), *Attachment of organisms to the gut mucosa*, **vol II**. CRC Press, Boca Raton, FL.

69. **Neutra MR.** 1984. The mechanism of intestinal mucous secretion, p 33–41. *In* Boedeker EC (ed), *Attachment of organisms to the gut mucosa*, vol II. CRC Press, Boca Raton, FL.

70. **Johansson ME, Phillipson M, Petersson J, Velcich A, Holm L, Hansson GC.** 2008. The inner of the two Muc2 mucin-dependent mucus layers in colon is devoid of bacteria. *Proc Natl Acad Sci U S A* **105:**15064–15069.

71. **Holmén Larsson JM, Thomsson KA, Rodriguez-Piñeiro AM, Karlsson H, Hansson GC.** 2013. Studies of mucus in mouse stomach, small intestine, and colon. III. Gastrointestinal Muc5ac and Muc2 mucin O-glycan patterns reveal a regiospecific distribution. *Am J Physiol Gastrointest Liver Physiol* **305:**G357–G363.

72. **Hoskins LC, Agustines M, McKee WB, Boulding ET, Kriaris M, Niedermeyer G.** 1985. Mucin degradation in human colon ecosystems. Isolation and properties of fecal strains that degrade ABH blood group antigens and oligosaccharides from mucin glycoproteins. *J Clin Invest* **75:**944–953.

73 **Moore WE, Holdeman LV.** 1974. Human fecal flora: the normal flora of 20 Japanese-Hawaiians. *Appl Microbiol* **27:**961–979.

74. **Comstock LE, Coyne MJ.** 2003. *Bacteroides thetaiotaomicron*: a dynamic, niche-adapted human symbiont. *Bioessays* **25:**926–929.

75. **Goodman AL, McNulty NP, Zhao Y, Leip D, Mitra RD, Lozupone CA, Knight R, Gordon JI.** 2009. Identifying genetic determinants needed to establish a human gut symbiont in its habitat. *Cell Host Microbe* **6:**279–289.

76. **Martens EC, Chiang HC, Gordon JI.** 2008. Mucosal glycan foraging enhances fitness and transmission of a saccharolytic human gut bacterial symbiont. *Cell Host Microbe* **4:**447–457.

77. **Ng KM, Ferreyra JA, Higginbottom SK, Lynch JB, Kashyap PC, Gopinath S, Naidu N, Choudhury B, Weimer BC, Monack DM, Sonnenburg JL.** 2013. Microbiota-liberated host sugars facilitate post-antibiotic expansion of enteric pathogens. *Nature* **502:**96–99.

78. **Salyers AA, Pajeau M.** 1989. Competitiveness of different polysaccharide utilization mutants of *Bacteroides thetaiotaomicron* in the intestinal tracts of germfree mice. *Appl Environ Microbiol* **55:**2572–2578.

79. **Mahowald MA, Rey FE, Seedorf H, Turnbaugh PJ, Fulton RS, Wollam A, Shah N, Wang C, Magrini V, Wilson RK, Cantarel BL, Coutinho PM, Henrissat B, Crock LW, Russell A, Verberkmoes NC, Hettich RL, Gordon JI.** 2009. Characterizing a model human gut microbiota composed of members of its two dominant bacterial phyla. *Proc Natl Acad Sci U S A* **106:**5859–5864.

80. **Tilman D.** 1982. Resource competition and community structure. *Monogr Popul Biol* **17:**1–296.

81. **Lendenmann U, Snozzi M, Egli T.** 1996. Kinetics of the simultaneous utilization of sugar mixtures by *Escherichia coli* in continuous culture. *Appl Environ Microbiol* **62:**1493–1499.

82. **Ihssen J, Egli T.** 2005. Global physiological analysis of carbon- and energy-limited growing *Escherichia coli* confirms a high degree of catabolic flexibility and preparedness for mixed substrate utilization. *Environ Microbiol* **7:**1568–1581.

83. **Liu M, Durfee T, Cabrera JE, Zhao K, Jin DJ, Blattner FR.** 2005. Global transcriptional programs reveal a carbon source foraging strategy by *Escherichia coli. J Biol Chem* **280:**15921–15927.

84. **Ferenci T.** 2001. Hungry bacteria–definition and properties of a nutritional state. *Environ Microbiol* **3:**605–611.

85. **Sartor RB.** 2005. Probiotic therapy of intestinal inflammation and infections. *Curr Opin Gastroenterol* **21:**44–50.

86. **Grozdanov L, Raasch C, Schulze J, Sonnenborn U, Gottschalk G, Hacker J, Dobrindt U.** 2004. Analysis of the genome structure of the non-pathogenic probiotic *Escherichia coli* strain Nissle 1917. *J Bacteriol* **186:**5432–5441.

87. **Sun J, Gunzer F, Westendorf AM, Buer J, Scharfe M, Jarek M, Gössling F, Blöcker H, Zeng AP.** 2005. Genomic peculiarity of coding sequences and metabolic potential of probiotic *Escherichia coli* strain Nissle 1917 inferred from raw genome data. *J Biotechnol* **117:**147–161.

88. **Alteri CJ, Smith SN, Mobley HL.** 2009. Fitness of *Escherichia coli* during urinary tract infection requires gluconeogenesis and the TCA cycle. *PLoS Pathog* **5:**e1000448. doi:10.1371/journal.ppat.1000448

89. **Fabich AJ, Leatham MP, Grissom JE, Wiley G, Lai H, Najar F, Roe BA, Cohen PS, Conway T.** 2011. Genotype and phenotypes of an intestine-adapted *Escherichia coli* K-12 mutant selected by animal passage for superior colonization. *Infect Immun* **79:**2430–2439.

90. **Adediran J, Leatham-Jensen MP, Mokszycki ME, Frimodt-Møller J, Krogfelt KA, Kazmierczak K, Kenney LJ, Conway T, Cohen PS.** 2014. An *Escherichia coli* Nissle 1917 missense mutant colonizes the streptomycin-treated mouse intestine better than the wild type but is not a better probiotic. *Infect Immun* **82:**670–682.

91. **Bohnhoff M, Drake BL, Miller CP.** 1954. Effect of streptomycin on susceptibility of intestinal tract to experimental *Salmonella* infection. *Proc Soc Exp Biol Med* **86:**132–137.

92. **Hentges DJ, Que JU, Casey SW, Stein AJ.** 1984. The influence of streptomycin on colonization resistance in mice. *Microecol Ther* **14:**53–62.

93. **Sonnenburg JL, Chen CT, Gordon JI.** 2006. Genomic and metabolic studies of the impact of probiotics on a model gut symbiont and host. *PLoS Biol* **4:**e413. doi:10.1371/journal.pbio.0040413

94. **Samuel BS, Gordon JI.** 2006. A humanized gnotobiotic mouse model of host-archaeal-bacterial mutualism. *Proc Natl Acad Sci U S A* **103:**10011–10016.

95. **Winter SE, Winter MG, Xavier MN, Thiennimitr P, Poon V, Keestra AM, Laughlin RC, Gomez G, Wu J, Lawhon SD, Popova IE, Parikh SJ, Adams LG, Tsolis RM, Stewart VJ, Bäumler AJ.** 2013. Host-derived nitrate boosts growth of *E. coli* in the inflamed gut. *Science* **339:**708–711.

96. **Meador JP, Caldwell ME, Cohen PS, Conway T.** 2014. *Escherichia coli* pathotypes occupy distinct niches in the mouse intestine. *Infect Immun* **82:**1931–1938.

97. **Gauger EJ, Leatham MP, Mercado-Lubo R, Laux DC, Conway T, Cohen PS.** 2007. Role of motility and the *flhDC* Operon in *Escherichia coli* MG1655 colonization of the mouse intestine. *Infect Immun* **75:**3315–3324.

98. **Leatham MP, Stevenson SJ, Gauger EJ, Krogfelt KA, Lins JJ, Haddock TL, Autieri SM, Conway T, Cohen PS.** 2005. Mouse intestine selects nonmotile *flhDC* mutants of *Escherichia coli* MG1655 with increased colonizing ability and better utilization of carbon sources. *Infect Immun* **73:**8039–8049.

99. **Bartlett DH, Frantz BB, Matsumura P.** 1988. Flagellar transcriptional activators FlbB and FlaI: gene sequences and 5′ consensus sequences of operons under FlbB and FlaI control. *J Bacteriol* **170:**1575–1581.

100. **Prüss BM, Campbell JW, Van Dyk TK, Zhu C, Kogan Y, Matsumura P.** 2003. FlhD/FlhC is a regulator of anaerobic respiration and the Entner-Doudoroff pathway through induction of the methyl-accepting chemotaxis protein Aer. *J Bacteriol* **185:**534–543.

101. **De Paepe M, Gaboriau-Routhiau V, Rainteau D, Rakotobe S, Taddei F, Cerf-Bensussan N.** 2011. Trade-off between bile resistance and nutritional competence drives *Escherichia coli* diversification in the mouse gut. *PLoS Genet* **7:**e1002107. doi:10.1371/journal.pgen.1002107

102. **Giraud A, Arous S, Paepe MD, Gaboriau-Routhiau V, Bambou JC, Rakotobe S, Lindner AB, Taddei F, Cerf-Bensussan N.** 2008. Dissecting the genetic components of adaptation of *Escherichia coli* to the mouse gut. *PLoS Genet* **4:**e2. doi:10.1371/journal.pgen.0040002

103. **Egger LA, Park H, Inouye M.** 1997. Signal transduction via the histidyl-aspartyl phosphorelay. *Genes Cells* **2:**167–184.

104. **Walthers D, Go A, Kenney LJ.** 2004. Regulation of porin gene expression by the two-component regulatory system EnvZ/OmpR. *In* Benz R (ed), *Bacterial and eukaryotic porins. Structure, function, mechanism.* Wiley-VCH, Germany.

105. **Pratt LA, Kolter R.** 1998. Genetic analysis of *Escherichia coli* biofilm formation: roles of flagella, motility, chemotaxis and type I pili. *Mol Microbiol* **30:**285–293.

106. **Licht TR, Christensen BB, Krogfelt KA, Molin S.** 1999. Plasmid transfer in the animal intestine and other dynamic bacterial populations: the role of community structure and environment. *Microbiology* **145**(Pt 9):2615–2622.

107. **Macfarlane S.** 2008. Microbial biofilm communities in the gastrointestinal tract. *J Clin Gastroenterol* **42**(Suppl 3 Pt 1):S142–S143.

108. **Macfarlane S, Bahrami B, Macfarlane GT.** 2011. Mucosal biofilm communities in the human intestinal tract. *Adv Appl Microbiol* **75:** 111–143.

109. **Palestrant D, Holzknecht ZE, Collins BH, Parker W, Miller SE, Bollinger RR.** 2004. Microbial biofilms in the gut: visualization by electron microscopy and by acridine orange staining. *Ultrastruct Pathol* **28:**23–27.

110. **Swidsinski A, Weber J, Loening-Baucke V, Hale LP, Lochs H.** 2005. Spatial organization and composition of the mucosal flora in patients with inflammatory bowel disease. *J Clin Microbiol* **43:**3380–3389.

111. **Macfarlane S, Woodmansey EJ, Macfarlane GT.** 2005. Colonization of mucin by human intestinal bacteria and establishment of biofilm communities in a two-stage continuous culture system. *Appl Environ Microbiol* **71:**7483–7492.

112. **Zoetendal EG, von Wright A, Vilpponen-Salmela T, Ben-Amor K, Akkermans AD, de Vos WM.** 2002. Mucosa-associated bacteria in the human gastrointestinal tract are uniformly distributed along the colon and differ from the community recovered from feces. *Appl Environ Microbiol* **68:**3401–3407.

113. **Fong JC, Syed KA, Klose KE, Yildiz FH.** 2010. Role of *Vibrio* polysaccharide (vps) genes in VPS production, biofilm formation

and *Vibrio cholerae* pathogenesis. *Microbiology* **156:**2757–2769.

114. Henrissat B, Davies G. 1997. Structural and sequence-based classification of glycoside hydrolases. *Curr Opin Struct Biol* **7:**637–644.

115. Franklin DP, Laux DC, Williams TJ, Falk MC, Cohen PS. 1990. Growth of *Salmonella typhimurium* SL5319 and *Escherichia coli* F-18 in mouse cecal mucus: role of peptides and iron. *FEMS Microbiol Lett* **74:**229–240.

116. Licht TR, Tolker-Nielsen T, Holmstrøm K, Krogfelt KA, Molin S. 1999. Inhibition of *Escherichia coli* precursor-16S rRNA processing by mouse intestinal contents. *Environ Microbiol* **1:**23–32.

117. Newman JV, Kolter R, Laux DC, Cohen PS. 1994. Role of *leuX* in *Escherichia coli* coloni-

zation of the streptomycin-treated mouse large intestine. *Microb Pathog* **17:**301–311.

118. Sweeney NJ, Klemm P, McCormick BA, Moller-Nielsen E, Utley M, Schembri MA, Laux DC, Cohen PS. 1996. The *Escherichia coli* K-12 *gntP* gene allows *E. coli* F-18 to occupy a distinct nutritional niche in the streptomycin-treated mouse large intestine. *Infect Immun* **64:**3497–3503.

119. McCormick BA, Stocker BA, Laux DC, Cohen PS. 1988. Roles of motility, chemotaxis, and penetration through and growth in intestinal mucus in the ability of an avirulent strain of *Salmonella typhimurium* to colonize the large intestine of streptomycin-treated mice. *Infect Immun* **56:**2209–2217.

Index